Advanced Integrated Communication Microsystems

STRIPLINE CIRCULATORS • *Joseph Helszajn*

THE STRIPLINE CIRCULATOR: THEORY AND PRACTICE • *Joseph Helszajn*

LOCALIZED WAVES • *Hugo E. Hernández-Figueroa, Michel Zamboni-Rached, and Erasmo Recami (eds.)*

MICROSTRIP FILTERS FOR RF/MICROWAVE APPLICATIONS • *Jia-Sheng Hong and M. J. Lancaster*

MICROWAVE APPROACH TO HIGHLY IRREGULAR FIBER OPTICS • *Huang Hung-Chia*

NONLINEAR OPTICAL COMMUNICATION NETWORKS • *Eugenio Iannone, Francesco Matera, Antonio Mecozzi, and Marina Settembre*

FINITE ELEMENT SOFTWARE FOR MICROWAVE ENGINEERING • *Tatsuo Itoh, Giuseppe Pelosi, and Peter P. Silvester (eds.)*

INFRARED TECHNOLOGY: APPLICATIONS TO ELECTROOPTICS, PHOTONIC DEVICES, AND SENSORS • *A. R. Jha*

SUPERCONDUCTOR TECHNOLOGY: APPLICATIONS TO MICROWAVE, ELECTRO-OPTICS, ELECTRICAL MACHINES, AND PROPULSION SYSTEMS • *A. R. Jha*

OPTICAL COMPUTING: AN INTRODUCTION • *M. A. Karim and A. S. S. Awwal*

INTRODUCTION TO ELECTROMAGNETIC AND MICROWAVE ENGINEERING • *Paul R. Karmel, Gabriel D. Colef, and Raymond L. Camisa*

MILLIMETER WAVE OPTICAL DIELECTRIC INTEGRATED GUIDES AND CIRCUITS • *Shiban K. Koul*

ADVANCED INTEGRATED COMMUNICATION MICROSYSTEMS • *Joy Laskar, Sudipto Chakraborty, Manos Tentzeris, Franklin Bien, and Anh-Vu Pham*

MICROWAVE DEVICES, CIRCUITS AND THEIR INTERACTION • *Charles A. Lee and G. Conrad Dalman*

ADVANCES IN MICROSTRIP AND PRINTED ANTENNAS • *Kai-Fong Lee and Wei Chen (eds.)*

SPHEROIDAL WAVE FUNCTIONS IN ELECTROMAGNETIC THEORY • *Le-Wei Li, Xiao-Kang Kang, and Mook-Seng Leong*

ARITHMETIC AND LOGIC IN COMPUTER SYSTEMS • *Mi Lu*

OPTICAL FILTER DESIGN AND ANALYSIS: A SIGNAL PROCESSING APPROACH • *Christi K. Madsen and Jian H. Zhao*

THEORY AND PRACTICE OF INFRARED TECHNOLOGY FOR NONDESTRUCTIVE TESTING • *Xavier P. V. Maldague*

METAMATERIALS WITH NEGATIVE PARAMETERS: THEORY, DESIGN, AND MICROWAVE APPLICATIONS • Ricardo Marqués, Ferran Martín, and Mario Sorolla

OPTOELECTRONIC PACKAGING • *A. R. Mickelson, N. R. Basavanhally, and Y. C. Lee (eds.)*

OPTICAL CHARACTER RECOGNITION • *Shunji Mori, Hirobumi Nishida, and Hiromitsu Yamada*

ANTENNAS FOR RADAR AND COMMUNICATIONS: A POLARIMETRIC APPROACH • *Harold Mott*

INTEGRATED ACTIVE ANTENNAS AND SPATIAL POWER COMBINING • *Julio A. Navarro and Kai Chang*

ANALYSIS METHODS FOR RF, MICROWAVE, AND MILLIMETER-WAVE PLANAR TRANSMISSION LINE STRUCTURES • *Cam Nguyen*

FREQUENCY CONTROL OF SEMICONDUCTOR LASERS • *Motoichi Ohtsu (ed.)*

WAVELETS IN ELECTROMAGNETICS AND DEVICE MODELING • *George W. Pan*

OPTICAL SWITCHING • Georgios Papadimitriou, Chrisoula Papazoglou, and Andreas S. Pomportsis

SOLAR CELLS AND THEIR APPLICATIONS • *Larry D. Partain (ed.)*

A complete list of the titles in this series appears at the end of this volume.

Advanced Integrated Communication Microsystems

JOY LASKAR
SUDIPTO CHAKRABORTY
MANOS TENTZERIS
FRANKLIN BIEN
ANH-VU PHAM

IEEE
IEEE PRESS

WILEY
A John Wiley & Sons, Inc., Publication

Published by John Wiley & Sons, Inc., Hoboken, New Jersey
Published simultaneously in Canada

Library of Congress Cataloging-in-Publication Data

Advanced integrated communication microsystems/Joy Laskar . . . [et al.].
p. cm.
Includes bibliographical references and index.
ISBN 978-0-471-70960-2 (cloth)
1. Radio–Transmitters and transmission. 2. Radio–Receivers and reception.
3. Radio frequency integrated circuits. 4. Wireless communication systems–Equipment and supplies. I. Laskar, Joy.
TK6560.A38 2008
621.384–dc22

2008021421

Printed in the United States of America

10 9 8 7 6 5 4 3 2 1

Contents

Preface

This book introduces readers to the implementation of miniaturized communication systems, which have matured significantly over 10 years. The GSM standard became popular around the early 1990s. Its implementation in silicon technology made it a mainstream focus of the semiconductor industry, and it remains an area of interest for major business even today. Over the years, semiconductor technologies have matured significantly, and slowly, integration of communication system blocks has reached maturity since their inception. Today, multiple radios are integrated on the same die, along with integrated circuit (IC) components for performance optimization and miniaturization. This process is motivated by our ever increasing need for improved mobile computing and connectivity. Wired connections are preferred for high bandwidth communications, and they carry much of the backbone traffic in communication systems. In practice, wired and wireless communication systems coexist in a synergistic manner to provide the overall communication system solution. Many communication standards have been developed, in both the wired and the wireless space to facilitate this coexisting aspect. In addition, there are several scenarios in which a wireless communication system is inevitable: (1) connectivity with the remote geographical areas, (2) satellite communication, and (3) implanted electronic devices.

One needs to appreciate the foundations of such systems. Although many communication systems are developed and deployed, very little has changed in the fundamentals of electromagnetic wave propagation, communication systems, and the basic functionality blocks. Integration of diverse functionality blocks in a system on chip and system on package are continuously evolving, which leads to innovative system solutions.

Between these two thought patterns, a gap exists. Although the foundations are a mature area of study led by early inventions in the 1900s, the integration and feasibility of miniaturized microsystems is only two decades old. Thus, while looking at the complexity of these miniaturized systems, "relating back" to the fundamentals becomes a difficult task for newcomers. At the same time, the reverse situation is true for experienced professionals and academicians. In this book, we make an attempt to provide a wholistic picture in the simplest possible manner to bridge the gap.

The book is organized into nine chapters, as outlined below.

Chapter 1 illustrates all the relevant fundamental concepts that need to be understood in order to appreciate various aspects of integrated microsystems. These aspects include (1) electromagnetics, (2) communication systems, (3) circuit fundamentals, and (4) semiconductor devices. These topics are all vast and very much mature areas today. We do not intend to make an attempt to perform a classic treatment of the individual disciplines. Our aim is to select a few principles in order to illustrate the different aspects of integrated communication systems. Hence, only a few selected basis functions are illustrated. Applications of these functions are shown in the chapters.

Chapter 2 illustrates wireless communication architectures. Wireless architectures are essentially the derivatives of Armstrong's original works related to superheterodyne architecture and the like. Because of the specific requirements of implementaion, one architecture is preferred over the others. This choice is also dependent on communication standard, semiconductor process, and level of integration. For example, one approach could include the implementation of superheterodyne radios using continuous-time signal processing in the front end. In this respect, one may enjoy the benefits of advanced semiconductor technologies (up to a certain extent). In the other extreme, one may also incorporate "mostly digital" schemes such as sampling architecture. These choices differ in the requirements of dynamic range, power, and form factor. In this chapter, we make an attempt to emphasize these tradeoffs.

Chapter 3 illustrates the various aspects of wired communication systems. Several architectural considerations are illustrated. Recently, there has been much discussion regarding the speed bottleneck of wired communication systems in the backplanes. To address these issues, development of equalizers demands specific attention. A fundamental aspect in the design of wired communication systems includes high-speed signal processing, which consumes significant power.

Chapter 4 illustrates the various basis functionalities in terms of circuit techniques. Both wired and wireless communication systems use similar building blocks or at least the same basis functionalities. We make an attempt to illustrate these similarities to the readers. The idea is to illustrate the versatility of the circuit blocks as they appear multiple times in any communication systems. We have covered various unique basic circuit topologies to illustrate this concept with specific implementation issues.

Chapter 5 provides practical examples of both wired and wireless communication systems. It illustrates the design methodology, building circuits, and a few architecture choices. One of these examples is in the second generation BiCMOS process, whereas the other is in deep submicron CMOS. Circuit designs, as well as layout considerations, have been illustrated.

Chapter 6 provides some advanced concepts. It includes the discussions from previous chapters to illustrate the developments of low-voltage, low-power circuits and systems. In this chapter, we focus on architecture, as well as on circuits. Critical aspects of low-power radios are illustrated, and the fundamental determinants of these requirements are also emphasized. Many modern applications such as medical electronics require ultra-low power in their implementations. Our approach, again, is to cover the few fundamental principles, and to put them in relevance, to build integrated systems.

Chapter 7 focuses on aspects that are inherent to passive components, packaging, and the like. Often, the growth of packaging technologies is underestimated; however, their importance needs to be kept in mind. No radios can even be feasible without integrating passive, discrete components. Such components include antenna, matching network, resonators, and so on. Although semiconductor scaling and extrapolation of Moore's law is often emphasized, this domain deserves special attention, and often it is a key to the success of smaller form factor, optimized, multichip solutions.

Chapter 8 illustrates various developments in the area of compact antennas. The success of any integrated communication system is involved in optimizing power and form factor. Although power consumption can be mostly related to judicious choices of architecture, circuits, available passive components, and so on, a major challenge lies in realizing small form factor antennas, which are mostly governed by electromagnetic principles. This chapter illustrates a few approaches in order to achieve small antennas.

Chapter 9 illustrates the simulation and test methodologies to build communication systems and to characterize them. Although understanding fundamental principles and their relevance to understand complex systems are important, it may fail to develop design confidence and enthusiasm in beginners. This chapter covers a few simulation techniques to analyze circuits in an intuitive manner. Various aspects of test calibrations are also covered in greater detail.

All in all, we have tried to maintain a good balance of theoretical foundation, design procedures, and practical implementation. Many textbooks are already available, and this attempt is, in no way, exhaustive. However, we hope that we can cover the seemingly complicated aspects in a simplistic manner. Since this discussion is an attempt to introduce the interdisciplinary approaches to realizing integrated systems, we have assumed a basic knowledge of electromagnetics, circuits, and architecture. Based on this assumption, we have provided the next level of details to the readers.

JOY LASKAR
Atlanta, GA
SUDIPTO CHAKRABORTY
Dallas, TX
MANOS M. TENTZERIS
Atlanta, GA
FRANKLIN BIEN
San Diego, CA
ANH-VU PHAM
Davis, CA
August 2007

Acknowledgments

This work is the culmination of many years of teaching, research, and practical development. We acknowledge our colleagues at the Microwave Applications Group at the Georgia Institute of Technology (Georgia Tech) and our research collaborators from GEDC/GTAC, including Prof. John Cressler, Prof. John Papapolymerou, Prof. Kevin Kornegay, as well as postdoctoral fellows, including Dr. Kyutae Lim and Dr. Stephane Pinel. This work would not have been possible if not for the creative freedom provided at Georgia Tech under the direction of Prof. Roger Webb. In addition, the authors would like to thank the support from the NSF Packaging Research Center under the direction of Prof. Rao Tummala and the Georgia Tech Microelectronics Research Center under the direction of Prof. Jim Meindl. We also deeply acknowledge the technical feedback received from students over numerous years that contributed toward refinement of the material presented in this book.

Most importantly, the authors thank their parents and family members for their continued patience and support during the many hard and strenuous phases of composition.

CHAPTER 1

Fundamental Concepts and Background

INTRODUCTION

In this chapter, we would like to illustrate a few fundamental concepts related to communication systems, circuits, devices, and electromagnetics to serve as a background for the materials to be illustrated in the later chapters. Any integrated system solution is a combination of the following functionalities: (1) data acquisition (sensor/analog interface), (2) signal processing, (3) communication (wireless or wired), and (4) power management. Irrespective of whether the end prototype is intended for wired or wireless communication applications, these four broad functionalities would be present in some form. Although each of these domains is diverse in nature, we illustrate only the fundamental concepts that are used in development of integrated communication microsystems. We start with communication systems, with an illustration of mathematical and physical tools that are necessary for understanding the principles of communication systems. Such tools can be used for design and analysis of systems architecture, circuits, and so on in an analytical, as well as intuitive manner.

1.1 COMMUNICATION SYSTEMS

Although diverse in their nature, wired and wireless communication systems work together to provide end-user services. Figure 1.1 illustrates this aspect. Let us consider the following situation: A user located in a cell in geographical area A needs to

Advanced Integrated Communication Microsystems, edited by Joy Laskar, Sudipto Chakraborty, Manos Tentzeris, Franklin Bien, and Anh-Vu Pham

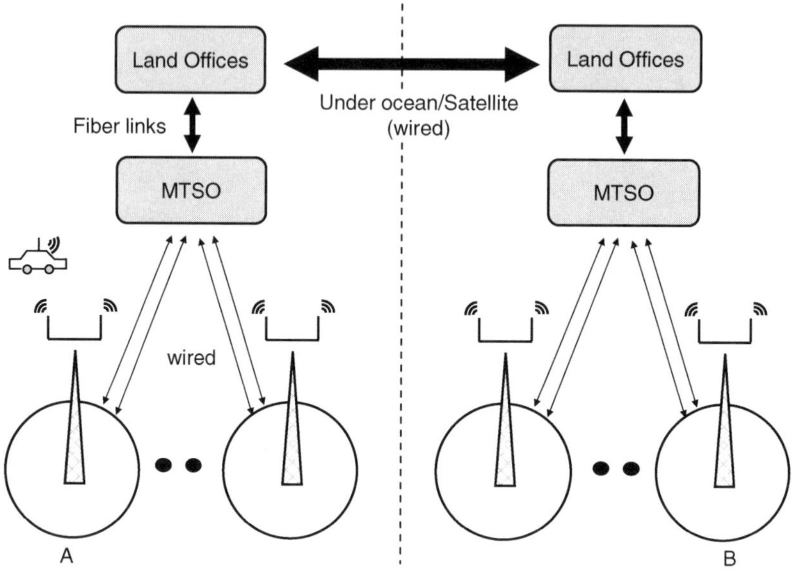

Figure 1.1. Coexistence of wired and wireless communication systems.

communicate to another user in a geographical area B while moving on the highway, at the end of work. Call from the mobile phone is accurately received by the base station in area A and communicated to the Mobile Terminal Switching Office (MTSO). Various MTSOs are connected to the central switching office by optical fiber backbone; they communicate with the central office, which communicates with its counterpart in B through optical fiber links laid underneath oceans or through geostationary satellites (with some communication delay). Modern communication systems mostly use optical fibers. As the message is received by the central office in B, it then diverts the traffic to a specific city, and the specific user gets the call from a telephone exchange. In case the end user is also mobile, the central office then communicates with another MTSO, which is responsible for delivering the message to the appropriate mobile user.

The entire process is complicated, in terms of its switching, traffic handling, and other network management issues. The above example has been used to illustrate the basic mechanism of a voice communication. Other types of high-data-rate communications are also feasible. For example, transferring large files, or multimedia movies, from one wireless device to another falls in the same category of high-speed wireless communications. Many times it is difficult to lay fiber optic cables because of geographical problems (rough terrains, mountains, etc.), and a direct line-of-sight wireless communication may be preferred. Our focus in this book is to provide an understanding of how to develop the physical-level hardware solution to enable such communication systems.

Our focus in this chapter is on the physical layer of these communication systems, in order to develop insight toward developing miniaturized hardware. A single chip, which can perform the functionalities of wireless communications at a desired data rate and frequency within a required power and area is the subject of this book. As the two communication systems are essentially diverse in nature, we focus on the various considerations toward wireless and wired communication systems. First we illustrate the nature of each of these communication systems and their fundamental aspects. Then we cover the key background needed for appreciation and design of such systems. This background is essential for developing any type of systems, wireless or wired.

1.2 HISTORY AND OVERVIEW OF WIRELESS COMMUNICATION SYSTEMS

The basic developments in the area of wireless communication date back to the early twentieth century. Since those early years, wireless engineering has come a long way. Most of the basic principles of the sophisticated radio architecture, as we see it today, were developed using vacuum tubes around 1930. Starting with the basic foundation provided by Maxwell (1883), and with subsequent inventions in wave propagation and wireless telegraphy by Hertz, Bose, Marconi, and others, wireless technology was born around 1900 in a very primitive form. Demonstration of a superheterodyne receiver by Armstrong dates back to as early as 1924. Various illustrations of Armstrong's superheterodyne receiver were reported during the 1920s and 1930s. At this time, radio pioneers considered the use of homodyne (/direct conversion) architectures for single vacuum tube receivers. For over two decades, the standard low-end consumer AM-tunable radio used a system of five vacuum tubes. A major milestone was set by the invention of the transistor by Bardeen, Brattain, and Schockley in 1948, which changed the world of vacuum tubes. However, implementing radios was a farsighted vision at that time. As semiconductor technologies became more mature, more circuit integration took place. Starting with small-scale integration in the standard integrated circuits, the trend moved toward more integration and high-speed microprocessors. With the tremendous growth in digital signal processing, very large-scale integration (VLSI), demands for ubiquitous computing and wireless applications increased.

During the 1990s, the maturity of digital electronics and signal processing hardwares led to the perception that a single-chip implementation of the front end could be feasible. This belief led to various developments of integrated filters, radio architectures based on frequency planning [super heterodyne to low intermediate frequency (IF) to direct conversion], and modulation techniques (such as DC-free spectrum) to combat known problems associated with direct conversion and so on.

Two fundamental operations of a receiver/transmitter include down/upconversion and demod(/mod)ulation. However, this is different in the case of coherent versus noncoherent radios. In the downconversion function, the desired signal is filtered and separated from the interferers, and it is converted from the carrier frequency to a frequency suitable for the demodulator for low signal processing power. Demodulation is performed at a lower frequency, either by a simple in-phase and quadrature phase (I/Q) demodulator or digitally sampled and performed by a digital signal processor (DSP). The latter allows for the use of complicated modulation schemes and complex demodulation algorithms. The demod(/mod)ulator and the other signal processing functionalities are usually performed using a digital signal processor, and its power consumption can be reduced by using advanced process technology nodes (which reduces the supply voltage and area). However, the down/upconversion functionality is not easily scalable, and the power consumption is a function of operating frequency, bandwidth, as well as intermediate frequency (which is dependent on blockers). Thus, numerous radio architectures are considered. Modern communication devices provide more and more integration on chip. The use of lower IF or elimination of IF from the frequency plan has many implications on the receiver/transmitter architecture. Low IF receivers combine the advantages of zero IF and IF architectures. It can achieve the performance advantages of an IF receiver, reaching the high level of integration as in a zero IF receiver.

1.3 HISTORY AND OVERVIEW OF WIRED COMMUNICATION SYSTEMS

Advanced wired communication systems today require transfer of multi-Gb/s data rate across bandlimited channels. Even computer hardware requires clock speeds of more than 2 GHz to be sent over motherboards. Overall, 10 Gb/s serial data have been transferred over FR-4-based backplanes, which were originally designed for 1-Gbps Ethernet applications. Advances in optical links and supporting electronics have dramatically increased the speed and amount of data traffic handled by a network system. Bandlimited channels continue to be a critical bottleneck for delivery of multigigabit serial data traffic.

The primary physical impediments to high data rates in legacy backplane channels are the frequency-dependent loss characteristics of copper channels. Above rates of 2 Gbit/s, the skin effect and dielectric loss in backplane copper channels distort the signal to such a degree that signal integrity is severely impaired. This dispersive forward channel characteristic contributes to the Inter-Symbol Interference (ISI).

Meanwhile, a major limiting factor to increasing transmission speeds and distances in fiber-optic communication links is modal dispersion causing ISI. Modal dispersion is caused as the numerous guided modes are transmitted in different paths in the multimode fiber (MMF) resulting in different receiving times at the receiver side of the fiber communication system. Modal dispersion becomes a severe factor as the length of the MMF is extended or the data rates are increased.

A brief comparison/contrast between wireless and wired systems can be represented as follows:

Electrical Characteristics	Wireless	Wired
Impact of channel	Mostly attenuation, and fading caused by path loss	Mostly dispersion caused by group delay variation
Bandwidth	Inherently narrowband	Inherently broadband
Effect of interferences	More interferers	Less interferers

(Continued)

Electrical Characteristics	Wireless	Wired
Synchronization problems	Very significant issue	A major issue to be considered
Modulation	A variety of modulation techniques are present starting from BPSK, QAM etc.	Mostly OOK, and some multilevel signaling in the electrical domain
Noise	Device noise plays a major role, as the signal is quite weak	Device noise is not an issue, as the signal levels are quite high
Architectures	Differ with each other in terms of frequency shift, up or down	Differ with each other in terms of synchronization schemes, half-rate/full-rate clock-data recovery systems, etc.
External components	Usually filter, balun, switch, duplexer (all electric/ electromagnetic in nature)	Usually photodiodes, VCSEL, other lasers, and electronic couplers

1.4 COMMUNICATION SYSTEM FUNDAMENTALS

In a wireless communication system, communication channel characteristics are defined by the environment in which we decide to operate, and this may vary among rural, urban, suburban, hilly area, and so on. In the case of wired communication, the choice of channel is dependent on the distance we want to communicate over, and the overall cost of the material (usually multimode or single-mode optical fiber). Once again, our target is on the channel capacity and the Signal-to-noise ratio (SNR) degradations associated with it.

1.4.1 Channel Capacity

The capacity of the channel is defined by Shannon–Hartley theorem, which is defined as

$$C = B\log_2\left(1+\frac{S}{N}\right) = B\log_2\left(1+\frac{E_b}{N_0}\right) \times \left(\frac{C}{W}\right)$$

where C is the channel capacity (bits/s), $\frac{S}{N}$ is the SNR obtained from average signal and noise powers, and $\frac{E_b}{N_0}$ is the energy ratio of the bit to noise energy, also known as "bit-energy per noise-density."

This theorem shows the achievable limit on the transmission bit rate, whereas the accuracy is given by whether the transmission bit rate, $R \leq C$. With this condition, the probability of error could be sufficiently small by using some channel coding, whereas in the region of $R > C$, no channel coding would lead to a sufficiently small error rate.

1.4.2 Bandwidth and Power Tradeoff

The above equation also leads to interesting consequences in terms of two aspects, a bandwidth-limited transmission scenario and a power-limited transmission scenario. In a bandwidth-limited situation, the transmission bandwidth is higher than the channel bandwidth, and we use symbols to represent several bits, along with some channel coding. This is possible, however, with a compromise in higher bit energy per noise density. It is certainly possible to consider a situation in which one may be interested in transmitting a lower bit rate through a higher capacity channel, while operating in the power-limited region of the capacity $\frac{E_b}{N_0}$ plane. This situation is illustrated in Figure 1.2, and it leads to fundamental considerations while determining radio architectures. For example, it a spectrally efficient modulation scheme, more bits would be packed in a symbol, which leads to the requirements of moderate to high accuracy for the signal processing necessary. To design such systems, a certain amount of power should be consumed to ensure the accuracy of the signal processing. In power-limited modulation techniques, low spectral efficiency modulation techniques are usually preferred. These techniques are used for low/moderate data rate systems, where battery longevity is the prime consideration.

One can also conclude from the above equation that, in the regime of low SNR communication systems, the logarithmic nature can be expressed as

$$W = C \times \left(\frac{S}{N} \right)$$

which implies that one can extend the bandwidth significantly while using a low SNR. This is particularly applicable to ultrawideband systems. Communication is performed by embedding information in the amplitude, frequency, and/or phase of

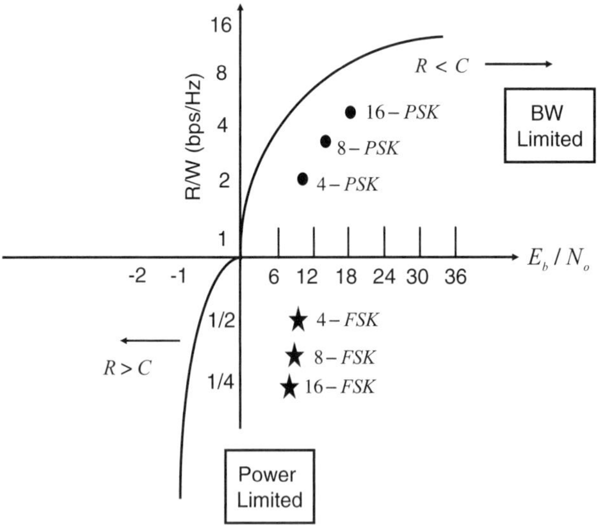

Figure 1.2. Bandwidth-efficiency plane.

the transmitted signal. Any communication system design is a tradeoff between the bandwidth and power usage. The task of the receiver is to recover the transmitted information successfully, which has discrete states w.r.t amplitude and time.

Radio architectures also evolve around these fundamentals. In a bandwidthlimited modulation scheme, spectral efficiency is a key factor, and the target is to "pack" a maximum number of bits into a symbol, in order to achieve high data rates. However, this process requires high signal processing accuracy. On the other hand, power efficient modulation schemes lead to low-power hardware, at the expense of low spectral efficiency. Depending on the application, each scheme should be chosen to fit the needs.

1.4.3 SNR as a Metric

Given a wireless environment, operating frequency, and distance, one can easily calculate the path loss, and the SNR degradations caused by multipath and shadowing effects. One can identify various geographical regions and map the associated SNR with them. Once these are determined, then, it is calculated how much SNR degradation is obtained from the RF/analog front end. After the signal processing at the RF/analog front end is performed, the demodulator obtains a specific SNR. We would then refer to the waterfall curve of the bit error rate in order to obtain a suitable modulation scheme. Thus, SNR is the major performance parameter that determines the choice of modulation scheme.

From an RF/analog perspective, a certain modulation scheme, bit error rate, and channel coding scheme defines the available bandwidth. The SNR degradation resulting from RF/analog blocks is contributed by the regular noise phenomena such as thermal noise, flicker noise, as well as intermodulation distortion product. These effects are further complicated in the case of wideband systems. The SNR improvements and, hence, the signal processing accuracies in the RF/analog front end are dependent on the power consumption.

The operating frequency is an important parameter in deciding the feasibility and cost of a communication system to be deployed. The propagation characteristics in a free space (path loss) is a function of frequency and the distance, and they are given by

$$L = \left(\frac{4\pi d}{\lambda}\right)^2$$

Thus, the path loss is lower at low frequencies, which leads to better signal propagation. However, the antenna size is inversely proportional to frequency of operation. However, it should be kept in mind that the above equation is a free space loss only. In an office environment or a home environment, the path loss assumes a much different profile, and the losses are usually much higher. In the case of mobile devices, the Doppler effect occurs between mobile devices, which needs estimation and compensation algorithms.

1.4.4 Operating Frequency

Although the above arguments hold good, choice of operating frequency is strongly motivated by licensed free frequency spectrum. Several frequency bands are dedicated for industrial, scientific, and medical applications under the FCC regulations. Medical applications usually operate in dedicated frequency bands because of the high reliability considerations of these devices. Such ISM bands are located in the 315M/433M/868M/915M/2400M bands. Associated with these center frequencies, there are various interference patterns from adjacent frequency bands. All ISM band devices have restrictions on maximum transmitter power as well. These restrictions, almost always determine the usable frequency band and the maximum distance achievable. The frequency allocation of various bands is shown in Figure 1.3 along with their applications.

Choice of frequency is a major decision point in the implementation of an integrated system. At lower frequencies, the data rate is lower, medium propagation is better, and a large antenna would be required. At higher frequencies, data rates are higher, medium propagation is worse, and a smaller antenna would be required for a low form factor solution. To trade off these constraints, most of the commercially available

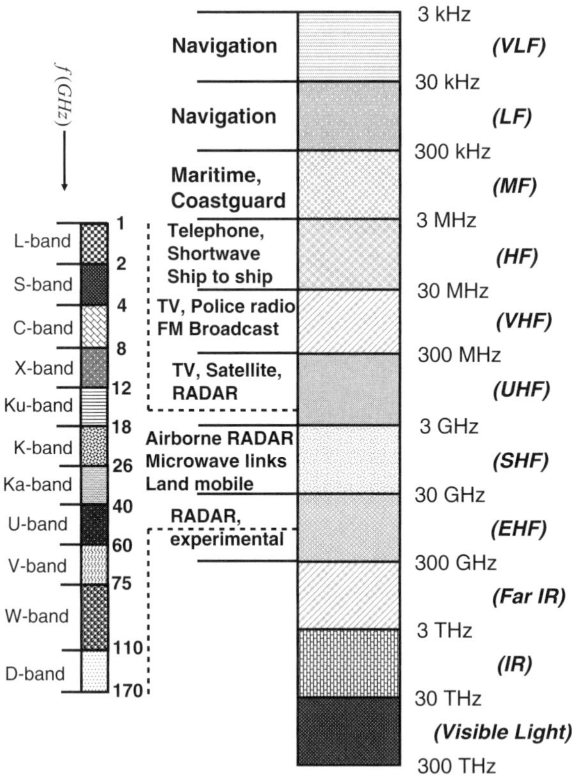

Figure 1.3. Frequency bands and allocations.

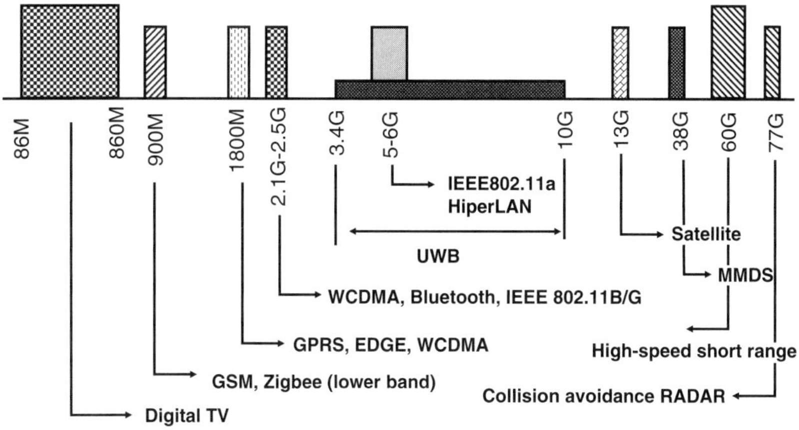

Figure 1.4. Commercial applications over frequency bands.

frequency bands range from 1 Ghz to 10 Ghz in present state-of-the-art cellular and wireless local area network (LAN) systems. Several emerging applications tend to operate in higher frequency bands as shown in Figure 1.4.

1.4.5 The Cellular Concept

Within the allocated frequency band of interest, multiple users can be accommodated by providing various frequency channels. It is the basis of the cellular radio system, as illustrated in Figure 1.5. A specific geographical area can be divided into multiple such cells, with frequency reuse planning. In practice, the individual cells are not

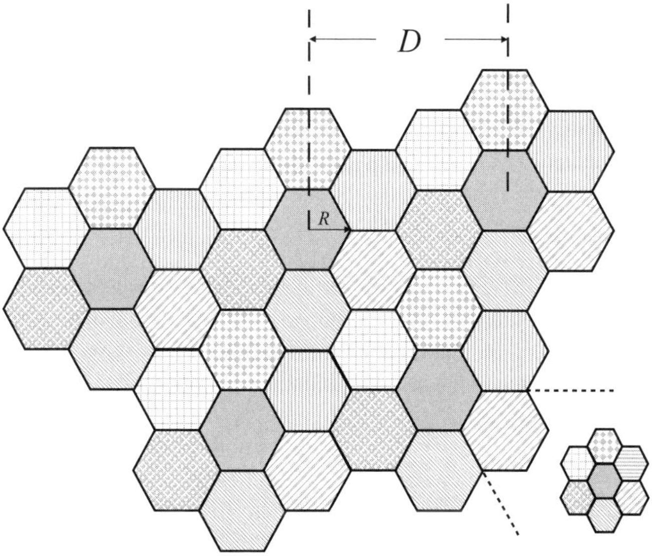

Figure 1.5. Cellular communication: frequency reuse, cell splitting.

hexagonal. In case of heavy traffic in an individual cell, it can be further divided into multiple smaller cells, and each one of the smaller cells operates at lower powers. Such allocations are dynamic in nature, which leads to increased flexibility of the cell-allocation scheme. This flexibility is illustrated in Figure 1.5, where different shades represent individual frequencies.

In analog communication systems, both the message signals as well as the time at which they are sent can assume continuous values. However, in a digital communication system, the information used and processed is discrete in time and amplitude. The fundamental aspects of any communication system design include (1) bandwidth, (2) power, and (3) error correction capability.

Obvious as it may seem, the bandwidth usage is specific to the FCC restrictions in specific countries under consideration. The task of the communication system designer is to select a specific frequency band and determine a modulation scheme such that the information transfer can be maximized at a given time. Depending on the situation, one can operate in the license-free ISM bands or the licensed bands with proper permissions from the governing agencies. Once the frequency is chosen, one should consider how much bandwidth is to be used, and what modulation technique is to be used.

1.4.6 Digital Communications

Our emphasis is on digital communication. In a digital communication system, information is arranged in a set of discrete amplitude as well as at discrete time instants. Such a signal, even being simply upsampled by a clock waveform, would lead to a digital waveform, which would lead to spectral spillover at the front end. At the same time, the antenna would need to be infinitely broadband in nature in order to accommodate all the useful information that is obtained. This would be highly inefficient, and we simply cannot transmit a digital waveform, however delicately it has been processed using careful techniques.

This issue is solved by using the concept of symbols. Symbols are formed from the sets of raw bits in the system, and as the information is discrete in amplitude and time, the symbols assume discrete states, and a diagram illustrating this is shown in Figure 1.6. Hence, one symbol may represent multiple bits at a time, and depending on the number of bits, the digital modulation is named. In binary notation, for an M ary communication, we obtain $\log_2 M$ symbols. Even symbols are digital in nature, so we have not really solved the spectral spillover problem.

The answer comes in constructing specific analog waveforms, which are bandlimited in nature. These waveforms can be obtained as a combination of three fundamental factors in information communication through waveforms by changing its (1) amplitude, (2) frequency, (3) phase, or a combination of them. Hence, we associate a specific analog waveform governed by our prespecific rule with each symbol. This process provides spectral containment. As the number M increases, we can associate more bits per symbol, and per analog waveform, or in other words, one waveform would contain the same information as so many raw bits. The nature of analog waveforms essentially determines the bandwidth, so the bandwidth in a digital communication is always associated with symbol duration.

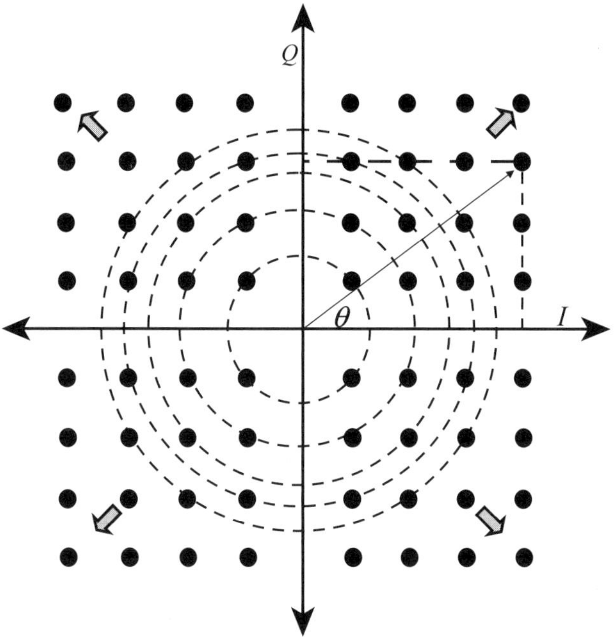

Figure 1.6. Signal constellation: illustration in polar/rectangular format.

1.4.7 Power Constraint

A communication system may be viewed as a combination of signal processing operations, which consume a certain amount of power. The signal processing can be continuous or discrete time in nature. If power consumption restriction is not present, the signal can be made arbitrarily large, and the individual symbol amplitudes can be made much differently from one another, which leads to their easy detection in case there is an error. However, we want to obtain the highest amount of information transfer in a given power budget. Hence, we need to obtain a proper choice of the modulation scheme. If the modulation order (*M*) is higher, the symbol constellation would become denser, and the distance between two symbols would reduce. In the case of a low order modulation scheme, the reverse is true. In the two cases, we have assumed the overall symbol power to be the same. Thus, for the same power, the higher *M* constellation symbols appear closer to one another, which leads to tolerance of the lower amount of impairment from noise, or require "higher SNR." Noise can appear from (1) quantization, (2) analog impairments, and (3) channel.

After the transmission is performed, the signal goes through several impairments in terms of channel characteristics and RF nonlinearity in the receiver. The digital receiver would then retrieve the correct state from the received impaired signal. There are several ways in which the transmitted signal can be distorted, including (1) intersymbol interference and (2) carrier offset.

Let us assume that the transmitted symbols are $x_1, x_2, x_3, \ldots x_n$. Assuming a linear superposition behavior, the received signal may obtain a value of $0.1 \times x_1 + 0.9 \times x_2$, instead of the second symbol x_2. This change may be caused by intersymbol interference, and in reality, it is a complex function of channel impulse response. At the same time, the carrier frequency generated from the VCO may provide a constant offset frequency, and the received symbols may appear to be as $x_1, x_2e^{j\phi}, x_3e^{2j\phi}, x_4e^{3j\phi}$ .... If the rotation of the symbols is larger, then symbols can get wound up in a manner in which the rotated symbols can significantly impact the detection of the received signal. A system designer always wants these degradations to be lowest, and comes up with the right selection of architectures and algorithms to mitigate these impairments.

1.4.8 Symbol Constellation

All of the impairments can be characterized in terms of symbol constellation, which is a graphical representation of symbols in a communication system with an orthogonal set of vectors (usually termed as in-phase and quadrature). Constellation represents the finite set of symbols in a digital communication system using a set of predefined states, using orthogonal axes. An example of orthogonal representation would be to represent symbols using amplitude and phase (I and Q), and it can be very well used in the case of BPSK, QPSK, QAM, etc. As the constellations become denser, which is the case of bandwidth-limited modulation, the SNR requirements become increasingly higher from RF/analog front ends. In the receiver, the digital demodulator delivers a reliable set of bits given distorted, quantized received signals at an oversampling factor N.

From the viewpoint of a communication system engineer, we would like to transmit the maximum possible bit rate while achieving the minimum probability of error with minimum available bandwidth and minimum SNR. We would also like to design the system for minimum complexity and maximize the number of users with a good quality of service in terms of delay and interference immunity. As these demands are contradictory to each other, a compromise needs to be obtained. Unlike the analog system, which works on reproducing original waveforms, digital communication sends waveforms to represent digits obtained by sampling the original waveform. Analog systems contain infinite energy, but finite power, whereas digital communication waveforms are of zero average power, with a finite energy. For this reason, the digital communication systems are better represented in terms of bit energy, with E_b/N_0 leading to bit error rate performance. From the above argument, it is clear why a communication system designer is always concerned with signal-to-noise degradations (SNR) in various signal processing blocks.

Currently, there have been various reports of communication system standards: architecture proposal. Key aspects of these standards include (1) communication channel under consideration; (2) center frequency, data rate, and distance; (3) modulation scheme; (4) connection protocol; and (5) targeted application.

1.4.9 Quadrature Basis and Sideband Combination

A specific way to understand transmitter and receiver architectures is by means of frequency translation and phase rotation. Let us consider direct conversion architecture as an example. In the transmitter, baseband signals are processed using a

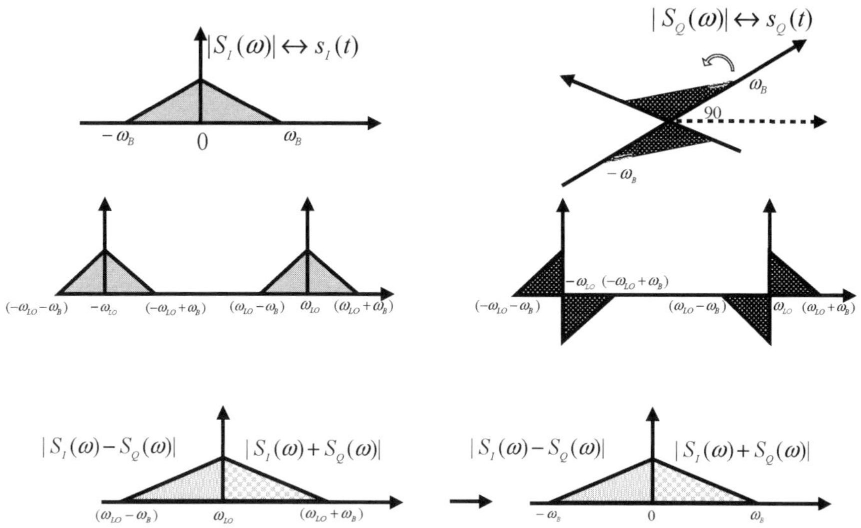

Figure 1.7. Frequency translation and negative frequency concept.

high-speed digital signal processor (depending on the data rate) to generate two streams of signals, in phase and quadrature. These streams are implemented by interleaving the original message sequence in in-phase and quadrature components and adjusting the delay between the two streams. This signal is then processed using a digital-to-analog converter (DAC), and finally up-converted by the LO frequency using quadrature phases, and combined at the output to obtain a single sideband. The mathematical synthesis can be represented as

$$S_{TX}(t) = A_{BB}\cos\omega_{BB}t \times \cos\omega_{LO}t - A_{BB}\sin\omega_{BB}t \times \sin\omega_{LO}t = A\cos(\omega_{LO} + \omega_{BB})t,$$

variations of this trigonometric formulation are also shown in Appendix A(1).

The frequency translation is illustrated in Figure 1.7.

1.4.10 Negative Frequency

In the frequency domain representation, the spectrum is symmetrically arranged around the LO frequency. The downconversion can be represented as a frequency translation in order to obtain the original signal centered around DC. Both sidebands contain a different amount of information. Both situations are consistent only if there is a "negative" frequency at the baseband. However, all along we are using real signals for illustration, and there is not a concept of negative frequency, we cannot generate it, and we cannot perceive it.

To understand this aspect, we represent signals as $S(t) = I(t) + jQ(t)$, where $I(t)$ and $Q(t)$ are real valued functions and "j" simply represents a "rotation" (or it could be thought about a transformation to construct a new variable). Thus, the frequency domain representation of this signal in frequency domain would contain $I(\omega) + Q(\omega)$

for positive ω and $I(\omega) - Q(\omega)$ for negative ω values. This is now upconverted at the transmitter and downconverted at the receiver. From $I(\omega) + Q(\omega)$ and $I(\omega) - Q(\omega)$, we can easily reconstruct $I(\omega)$, and $Q(\omega)$, and their time domain waveforms.

1.5 ELECTROMAGNETICS

Almost all developments in the area of communication systems, devices, and circuits can be correlated to some aspects of electromagnetics. Electromagnetic principles can well explain the propagation of waves, basis of wireless communication, skin effects of integrated inductors, Kirchoff's laws governing all areas of circuit design, standing waves formation, and the nature of electric field in scaled semiconductor devices.

1.5.1 Maxwell's Equations

Time-varying electrical and magnetic fields and their relationship with one another can be governed by Maxwell's equations, which is a generalized form of experimental results obtained by many researchers. In terms of generalized spatial coordinates, they can be stated as $E(x, y, z, t)$ for electric field and $B(x, y, z, t)$ for magnetic field. In free space, they are governed by the following four fundamental equations:

$$\begin{aligned} \nabla \cdot B &= 0 \\ \nabla \cdot D &= 0 \\ \nabla X H &= J + \frac{\partial D}{\partial t} \\ \nabla X E &= -\frac{\partial B}{\partial t} \end{aligned}$$

where $B = \mu_0 H$ and $D = \varepsilon_0 E$ in free space. These equations uniquely determine the nature of magnetic and electric fields at any spatial point as a function of time. The last two equations imply the inherent coupling between electric and magnetic fields (change in electric field produces change in magnetic field and vice versa).·Hence, two coupled first-order differential equations lead to formation of a second-order differential equation in individual variables, and they form the basis of a standing wave, which is used in the context of almost all electromagnetic phenomena. The first equation implies the absence of magnetic monopoles, and the second equation implies that the divergence of electric field is dependent on the net electric charge.

1.5.2 Application to Circuit Design

As a first application of the above equations, we consider the circuit design principles. The assumption is that there is no coupling between electric and magnetic fields, which is obtained by setting $\mu_0 = 0$, and $\varepsilon_0 = 0$, leading to

$$\begin{aligned} \nabla X H &= J \Rightarrow \nabla \cdot (\nabla X H) = \nabla \cdot J = 0 \\ \nabla X E &= 0 \Rightarrow \oint E \cdot dl = 0 \Rightarrow \oint -(\nabla \cdot V) . dl = 0 \end{aligned}$$

Divergence of curl of a vector is zero, and we use Stoke's theorem of line integral to obtain the second formulation. The first one implies that there is no divergence of current, implying that the sum of all the currents flowing to a node would be zero. The second one implies that in a loop, the sum of all the voltages along the loop would be zero. These form the basic theory of any circuit operation (Kirchoff's laws). The assumption of decoupled electric and magnetic fields is valid under the assumption that the lengths of the loop under consideration are much smaller than the wavelength under consideration. This is related to the physical dimension of the circuit under consideration, and in a semiconductor substrate, the wavelength is reduced by the relative dielectric constant.

1.5.3 Signal Propagation in Wireless Medium

In the above discussion, we have obtained the fundamentals of circuit theory under the assumption that electric and magnetic fields are uncoupled. In a coupled relationship, the wave propagation can be illustrated. A loop of wire carrying time-varying electric current causes a time-varying magnetic field around it. This changing magnetic field causes a continued time-varying electric field, and this happens in a three-dimensional fashion and with speed of light in the medium under consideration. Figure 1.8 illustrates this concept, which is the fundamental basis of radio propagation through air.

Maxwell's equations related to divergence the of electric field ($\nabla \cdot D = \rho$) provides fundamental equations related to device physics in the case of semiconductor junctions, maximum electric field, and so on, and they lead to a discipline well known as "electrostatics." Similarly the first equation related to magnetic fields (magnetostatics, also known as Biot–Savart's law) leads to understanding the nature of magnetic field lines resulting from inductors and so on. Hence, the generalized Maxwell's equations explain almost all the aspects, including circuit design, wave propagation, electromagnetic field lines, and so on. In modern integrated systems, these can be used to solve various problems, and in many cases, they are solved numerically in the case of practical problems in order to maintain computation speed and accuracy.

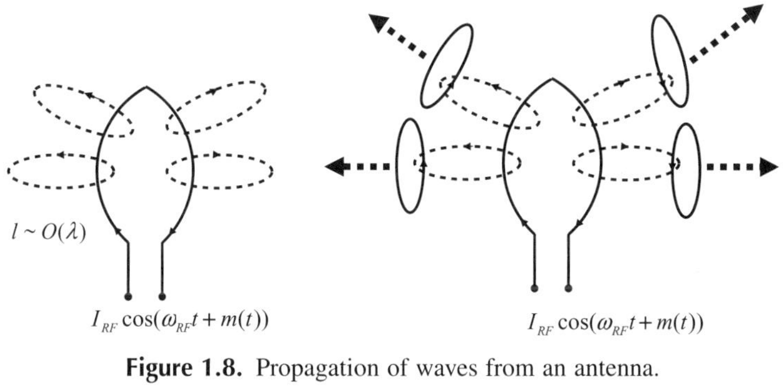

Figure 1.8. Propagation of waves from an antenna.

1.6 ANALYSIS OF CIRCUITS AND SYSTEMS

Several methods can be used to represent communication circuits and systems. We will illustrate a few of the analysis tools, which are used to obtain numerical efficiency, spectral information, as well as physical insights.

The first in this category of tools are a few transformations and signal processing components. The mathematical nature of signals in any communication systems can be represented in time or frequency domain. Transformations help with the conversion of complicated differential equations to linear equations, subject to the initial conditions, for simplified mathematical analysis. These processes were originally developed to analyze partial differential equations with boundary value problems, and later they were adopted in a variety of engineering disciplines.

1.6.1 Laplace Transformation

Laplace transformation is defined as $F(s) = \int_0^{\infty} e^{-st} f(t)dt$. This integral exists when $f(t)$ grows slower than e^{bt}, such that convergence is obtained. $f(t)$ need not be a continuous function, and it may be simply a piecewise linear function. If the transformation exists, it is uniquely determined. This transformation can be applied to convolution of two functions, which provides the multiplication of individual Laplace transforms. Laplace transformation can be used to provide impedances of inductance and capacitance, which are obtained under the conditions that the current through an inductor remains the same before and after an event occurred at time instant τ, while the voltage across a capacitor remains the same before and after a time instant τ. In the case of a simple example, we assume the initial current and charge values are 0, respectively. Since inductor and capacitors are represented by differential equations in the time domain, their voltage and current waveforms are represented as follows:

$$V_L(t) = L\frac{di}{dt}$$

$$V_C(t) = \frac{1}{C}\int \frac{di}{dt}$$

The Laplace transformation implies that $V_L(s) = (sL)I(s)$, and $V_C(s) = \frac{1}{sC}I(s)$; hence, inductor and capacitors are represented by frequency-dependent impedance values of sL, and $1/sC$, respectively.

1.6.2 Fourier Series

Analysis of periodic waveforms can be represented by Fourier series expansions. We first start with the analysis of periodic signals in the time domain. Periodic functions occur in numerous places, such as the LO drive of the mixer, which can assume various waveform types (sinusoidal, square, etc.). On the other hand, the input RF signal has a sinusoidal waveform shape. For a time domain waveform with period $p = 2T$, the

Fourier series can be represented as follows:

$$s(t) = a_0 + \sum_{n=1}^{\infty} \left(a_n \cos\frac{n\pi}{T} t + b_n \sin\frac{n\pi}{T} t \right)$$

and the Fourier coefficients are determined by

$$a_0 = \frac{1}{2T} \int_{-T}^{T} f(t)dt$$

$$a_n = \frac{1}{T} \int_{-T}^{T} f(t)\cos\frac{n\pi t}{T} dt, \quad n = 1, 2, 3 \ldots$$

$$b_n = \frac{1}{T} \int_{-T}^{T} f(t)\sin\frac{n\pi t}{T} dt, \quad n = 1, 2, 3 \ldots$$

For a different period, T can be replaced accordingly in order to obtain the desired Fourier series representation. As an example, a periodic square waveform, and a half-wave rectifier, can be considered, which are classical waveform shapes in electrical systems. The square waveform contains odd harmonics of the period, whereas the half-wave rectifier contains even harmonics of the period. The Fourier transform of the square waveform illustrated in Figure 1.9 is given by

$$s(t) = \frac{A}{2} + \frac{2A}{\pi T}\left[\cos\frac{\pi t}{2T} - \frac{1}{3}\cos\frac{3\pi t}{2T} + \frac{1}{5}\cos\frac{5\pi t}{2T} - \ldots\right]$$

which contains only the odd harmonics of the waveform frequency under consideration.

A phase shift (or delay) in the original signal $s(t)$ would lead to integral multiples of phase shift, depending on the harmonic tone under consideration, as phase is

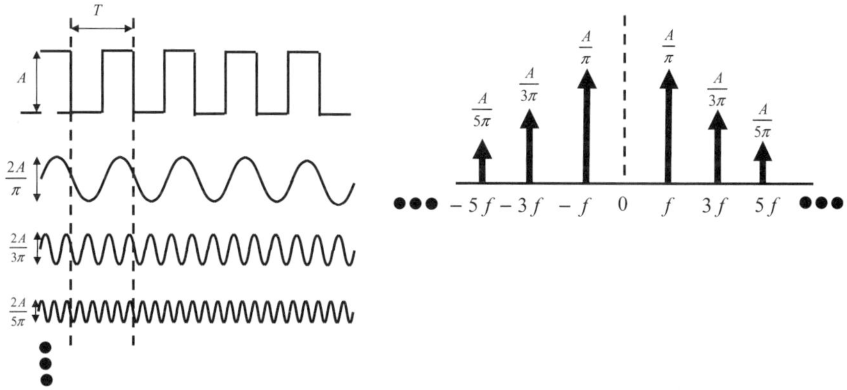

Figure 1.9. Spectral contents of a square waveform.

multiplied along with frequency. In the case of the half-wave rectifier waveform, the expansion is provided as follows:

$$s(t) = \frac{A}{\pi} + \frac{A}{2}\sin(\omega t) - \frac{2A}{\pi}\left[\frac{1}{3}\cos(2\omega t) + \frac{1}{15}\cos(4\omega t) + \ldots\right]$$

1.6.3 Fourier Transform

A more generalized case can be formulated by analyzing aperiodic signals, which can be formulated as signals limited in the time domain with a period of ∞. For a time-limited signal, we are interested in the equivalent frequency domain characteristics of the same signal, and the frequency spread is attributed to the bandwidth of the signal. When the signal is limited in time, the frequency spread increases and vice versa. For an aperiodic signal, the Fourier transform is given as follows:

$$S(f) = \int_{-\infty}^{\infty} s(t)e^{-j2\pi ft}\,dt$$

and the existence of Fourier transform requires: (1) $s(t)$ to have a minimum number of maximum and minimum and a single value, (2) an finite number of discontinuities, and (3) absolute integrability $\int_{-\infty}^{\infty} |s(t)dt| < \infty$.

Some frequently encountered signals can be evaluated in terms of the Fourier transform at this stage. A rectangular pulse, strictly limited in time, would lead to a "sinc" shape in the frequency domain;, similarly, a "sinc" shape in the time domain represents a rectangular waveform in the frequency domain. Various properties of a Fourier transform are illustrated in Appendix A(2), Figure 1.9. It can also be shown that for pulse signal families, the product of signals duration and the bandwidth is a constant. The energy contained in the signal can be evaluated from either frequency or in time domain representations, $E_s = \int_{-\infty}^{\infty} |S(f)|^2 df = \int_{-\infty}^{\infty} |s(t)|^2 dt$ (also known as Rayleigh's energy theorem). The concept of bandwidth can be also explained from the frequency spread of the signals, and it is determined as a frequency range within which maximum signal energy is contained. Common methods of indicating bandwidth include when the signal power is 3 dB below its peak value and can be used as a performance metric for low-pass, band-pass systems. In the case of "sinc" type pulses, "null-to-null" spacing in the frequency domain contains maximum energy (almost 92%), and it can be used as a measure of bandwidth.

Fourier transformation of standard functions are shown in Appendix A(2).

1.6.4 Time and Frequency Domain Duality

Analysis of convolution occurs in the same manner as illustrated before, in the case of Laplace transformations. Fourier transforms can be very powerful in analyzing complicated functionalities in the time domain. In nonlinear circuits and systems, often a square and cubic law characteristics are common, and they can be obtained as a result of time domain multiplication. A time domain multiplication leads to a convolution in frequency domain. The relationship is shown as follows:

$$s_1(t)s_2(t) \Longleftrightarrow \int_{-\infty}^{\infty} S_1(u)S_2(f-u)du$$

where $S_1(f)$ and $S_2(f)$ denote the Fourier transformation of $s_1(t)$ and $s_2(t)$, respectively. To evaluate the convolution of two signals in the frequency domain (the same procedure is true for the time domain as well), we first flip the frequency axis of one signal while keeping the other intact. Then the flipped axis variable is moved toward the right, and as it moves, integration of the product is performed.

In a practical case, consider the blocker scenario in a WCDMA standard, where we are to evaluate the impact of second-order nonlinearity upon the SNR requirement at the demodulator. The frequency domain representation of the blockers, as well as the desired tone, is shown in Figure 1.10. Instead of performing this complicated multiplication, we simply obtain the frequency domain representation of the input signals to the amplifier (a modulated signal, a continuous wave blocker, an amplitude modulated blocker). Then we perform a flip in the frequency axis and slide the "flipped" terms to the right of the frequency axis (starting from an infinite offset from the center),

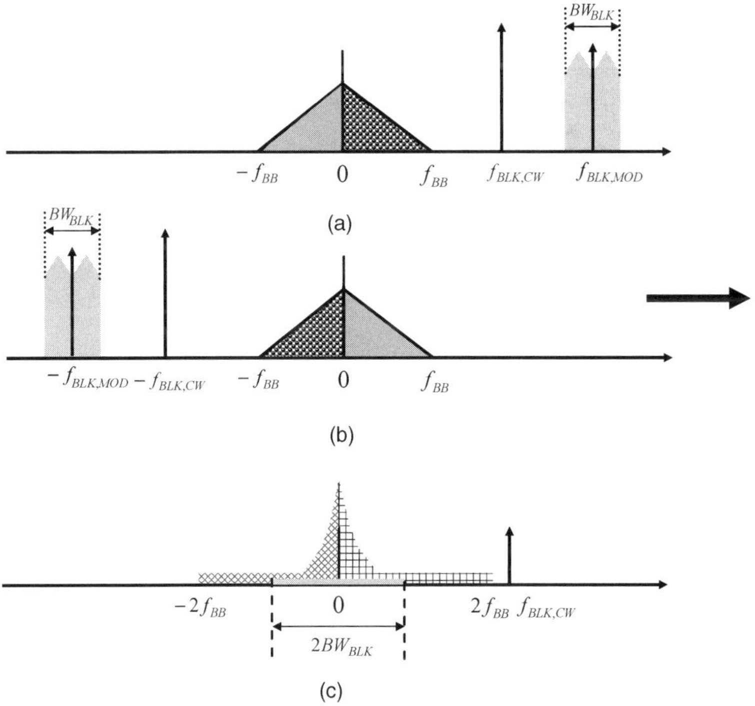

Figure 1.10. Analysis of modulated signal in the presence of blockers in nonlinear systems.

and we obtain the frequency domain representation of the nonlinear signal. As can be observed from the graphical illustration, the degradation is more when a modulated blocker with certain bandwidth is present, as opposed to a single tone. The convolution clearly illustrates the "spectral spreading" of the nonlinear terms in the desired bandwidth, and SNR degradation caused by the same. It can be easily observed that the evaluation in the time domain would be difficult, as the modulated blocker needs to be represented with time domain nonlinearity, leading to many terms, which are, hence, difficult to handle. At the same time, the graphical representation provides easily interpretable insights. For a cubic nonlinearity, convolution can be performed once more by flipping the frequency axis. Convolution-based evaluation becomes very effective in the case of multicarrier signals with uniform power distribution [a case for multicarrier orthogonal frequency devision multiplexing (OFDM) signals]. In this case, the frequency domain profile is rectangular and the time domain waveform is Gaussian in nature (because of presence of many carriers) with a high crest factor. Second-order nonlinearity leads to a cubic profile in the frequency domain, and cubic order nonlinearity leads to the parabolic shape of the frequency domain profile. When two sequences of bandwidth ω_1 and ω_2 are convolved with each other, the result would provide a frequency component up to $\omega_1 + \omega_2$. Hence, when a signal is convolved with itself, it "spreads" in frequency, which leads to SNR degradation throughout the bandwidth under consideration. In the case of unmodulated tone, no "spreading" is observed, providing a lower amount of SNR degradation.

1.6.5 Z Transform

In mixed signal systems, often *Z* transforms are useful in order to represent the operation of sampled signals. This is especially the case when the signal processing occurs at different time instants. It is true for a switched capacitor circuit, which stores charge at time τ_1 and transfers at time instant τ_2. These discrete time systems are well represented using *Z* transforms. *Z* transform is convenient in analyzing discrete time systems; e.g., digital filters and switched capacitor circuits. *Z* transforms are especially helpful in analyzing systems which are discrete in value (amplitude) and time (sampled). Analog/digital converters, especially sigma-delta type ones, are extensively analyzed using *Z* transforms. Analogous to the continuous time case, a–*Z* transform represents the frequency content and shaping function in the case of sampled data systems. They can be correlated to the continuous time counterparts with appropriate analog sampling frequencies. The transform can be represented by

$$F(Z) = \sum_{-\infty}^{\infty} f(n) z^{-n}$$

where *f*(*n*) is a discrete sequence, which can assume any amplitude values (usually determined by the quantization of the system). *Z* is usually represented as $Z = e^{-j\Omega_n}$. Most of the properties of *Z* transforms are similar to the Laplace and Fourier transforms discussed before.

1.6.6 Circuit Dynamics

Although the transformations illustrated above provide computation flexibility and speed, we need to be careful not to forget the true nature of the circuit dynamics. For example, a large signal charging/discharging of capacitor and time domain dependence of current/voltage waveforms can be easily forgotten by representing the capacitor by an impedance $1/\omega C$. We can use the transformation of impedances in frequency domain in order to relate voltage and currents through them in the steady state. We must use these transforms to solve complicated networks, but under a given situation, they must clarify them from circuit dynamics, which are captured well in the charging/discharging behavior of components.

1.6.7 Frequency Domain and Time Domain Simulators

Circuits and systems can be analyzed in the time or frequency domain. Both approaches are common, and for a given circuit complexity (number of components, feedback loops, etc.), the time domain proves efficient when many harmonics are involved in the waveform (e.g., a square waveform). This may be the situation in digital circuits, where mixers are driven with a large signal square waveform shape. However, a difficulty, which is often faced with time domain simulators, is the presence of multiple time constants that are varying by orders of magnitude from one another. This is often the case for integrated transceivers where a low-frequency signal is upconverted to an RF signal and an RF signal is downconverted to a low frequency signal. In both cases, the system would be allowed to settle within the limits of the time constant of the circuit. Frequency domain simulators prove efficient in these cases, as knowing the nature of harmonic tones and their placement along the frequency axis would require few iterations for simulation convergence. Such techniques are well known as "harmonic balance," and the signals are treated as a combination of DC and a finite number of harmonics of the signal. It solves for magnitudes and phases of all spectral lines in the frequency domain simultaneously. The frequency domain current and voltages are adjusted w.r.t. their amplitude and phase characteristics until their sum equals the input current and voltages according to Kirchoff's laws. Currently, frequency domain analyses are becoming computation efficient by incorporating "envelope simulation" techniques, whereasthe time domain simulators tend to analyze various parts of the circuits w.r.t different timesteps, and by correlating the results w.r.t. sampling techniques.

1.6.8 Matrix Representation of Circuits

In this section, we will discuss the various forms of circuit representations using matrix form. Usually such representations are generic in nature, and they are applicable to transistors also. Modern technologies use many parameters to represent a transistor model, and they may appear as a complicated circuit themselves. Most commonly used under this category are the Z (impedance), Y (admittance), S (small signal), and H(hybrid) Matrices. These representations assume a "black-box" representation of the circuit element, assuming a two terminal model, in which one is input (terminal 1) and another is output (terminal 2). The matrix relationships are represented as follows.

$$\begin{pmatrix} V_1 \\ V_2 \end{pmatrix} = \begin{pmatrix} Z_{11} & Z_{12} \\ Z_{21} & Z_{22} \end{pmatrix} \times \begin{pmatrix} I_1 \\ I_2 \end{pmatrix}$$

$$\begin{pmatrix} I_1 \\ I_2 \end{pmatrix} = \begin{pmatrix} Y_{11} & Y_{12} \\ Y_{21} & Y_{22} \end{pmatrix} \times \begin{pmatrix} V_1 \\ V_2 \end{pmatrix}$$

$$\begin{pmatrix} B_1 \\ B_2 \end{pmatrix} = \begin{pmatrix} S_{11} & S_{12} \\ S_{21} & S_{22} \end{pmatrix} \times \begin{pmatrix} A_1 \\ A_2 \end{pmatrix}$$

$$\begin{pmatrix} V_1 \\ I_2 \end{pmatrix} = \begin{pmatrix} H_{11} & H_{12} \\ H_{21} & H_{22} \end{pmatrix} \times \begin{pmatrix} I_1 \\ V_2 \end{pmatrix}$$

Each one of these representations is capable of describing the performance of two-port networks completely. Z and Y parameters are homogeneous, whereas H parameters combine voltage and currents. The fundamental difference in the case of S parameters is that they tend to use wave reflection methodology to represent a network. Each of these parameters can be interconverted as illustrated in Appendix A(3), and they can be used in appropriate scenarios. Although the other matrices do not require a standard reference impedance, any conversion from/to S parameters requires a reference impedance (50 Ω can be used assuming that characteristic impedance is not frequency dependent). All of these parameters capture the linear behavior of a network.

1.6.8.1 S Parameters. S parameters are widely used in microwave frequencies because of easy measurements (measurements are based on signal reflection and transmission), and convenience in using them for modeling purposes. In the power domain, they can be represented as follows:

$$\begin{pmatrix} |B_1|^2 \\ |B_2|^2 \end{pmatrix} = \begin{pmatrix} |S_{11}|^2 & |S_{12}|^2 \\ |S_{21}|^2 & |S_{22}|^2 \end{pmatrix} \times \begin{pmatrix} |A_1|^2 \\ |A_2|^2 \end{pmatrix}$$

where $|B_i|^2$ denotes the reflected power, $|A_i|^2$ denotes the incident power, $|S_{11}|^2$ denotes the reflected power at port 1, $|S_{21}|^2$ denotes the transmitted power from port 1 to 2, $|S_{12}|^2$ denotes reverse isolation, and $|S_{22}|^2$ denotes the output reflectance. To obtain the individual S parameters, the other terminal is terminated using characteristic impedance Z_0. To obtain S_{11}, port 2 is terminated. The accuracy of these parameters depends on the termination quality (how close Z_0 is to 50 Ω). The magnitudes of S_{11} and S_{22} are always less than 1, whereas S_{21} can have a magnitude greater than 1 (gain) and S_{12} is usually less than 1 (reverse isolation). S parameters also provide the phase shift information through the network, as they are essentially complex numbers. In the case of passive devices, reciprocity holds good, which leads to $S_{21} = S_{12}$. The magnitudes of the following cases can also be observed:

$S_{ii} = -1$: Amplitudes are inverted and reflected (0 Ω)

$S_{ii} = 0$: No reflections terminated at (50 Ω)

$S_{ii} = 1$: Voltage reflections without inversion

In the case of passive circuits, all values of S_{mn} are between −1 and 1, and in general, this implies that m is the output port and that n is the input port.

Impedance matching is an important consideration in designing integrated systems, and two critical examples occur in the termination of clock distribution networks in high-speed digital system and in the input of the low noise amplifier (LNA) in the case of wireless systems. Unterminated lines in digital systems result in reflections, which lead to significant distortions of the square waveform. A typical input matching of 15 dB can be adopted for narrowband wireless standards.

S parameters are important at a high frequency. At low frequencies, the voltage and current waveforms are the same at all points along the line. As frequency increases, line lengths are comparable with the wavelengths, and when the line is not terminated in Z_0, an entire signal is not absorbed by the load and reflected back to the source. In the case of a short-circuited termination, reflected and incident voltage waveforms would be equal in magnitude but oppositely phased. In the case of an open-circuit configuration, reflected and incident voltage waveforms are in phase, whereas the current waveforms are oppositely phased. In the case of a perfect termination, no standing wave is formed, and the energy flows from the source to the load in one direction. In the case of reflections, the ratio of maximum to minimum values of the RF envelope is termed the voltage standing wave ratio (VSWL). In the case of a perfect termination, VSWR = 1 and infinity for full reflection.

1.6.8.2 Smith Chart. A graphical representation of impedances can be made via a Smith chart, which can be effectively used in the case of designing matching networks. It starts with computing the reflection coefficient Γ, which is defined as $\Gamma = \frac{Z_L - Z_0}{Z_L + Z_0}$, and for load impedance variation in the range of $0 < Z_L < \infty$, $-1 < \Gamma < +1$. Hence, in the Smith chart, we can plot a set of impedances conforming to certain constraints, and the Smith chart essentially would contain the impedance states, which determines the reflection coefficient. This transformation is graphically illustrated in Figure 1.11. A polar plot can be represented as well.

From the impedance transformation, the rightmost point in the Smith chart denotes infinite impedance, and the leftmost point (diametrically opposite) denotes zero impedance. The center point of the Smith chart denotes the characteristic impedance Z_0, and for a perfect match, the impedance would coincide with the center. In practical cases, however, the matching levels are determined by the distance of the impedance from the center. The upper half of the Smith chart contains inductive impedance states, and the lower half contains capacitive impedance states. With increasing frequency, the impedances always traces clockwise.

Impedance states provide important graphical information for circuit designers, and they are traditionally used in designing high-performance stand-alone RF circuits such as LNA, power amplifier (PA), as well as matching networks. For example, a set of impedances can be plotted in a Smith chart, which optimizes the noise figure of the

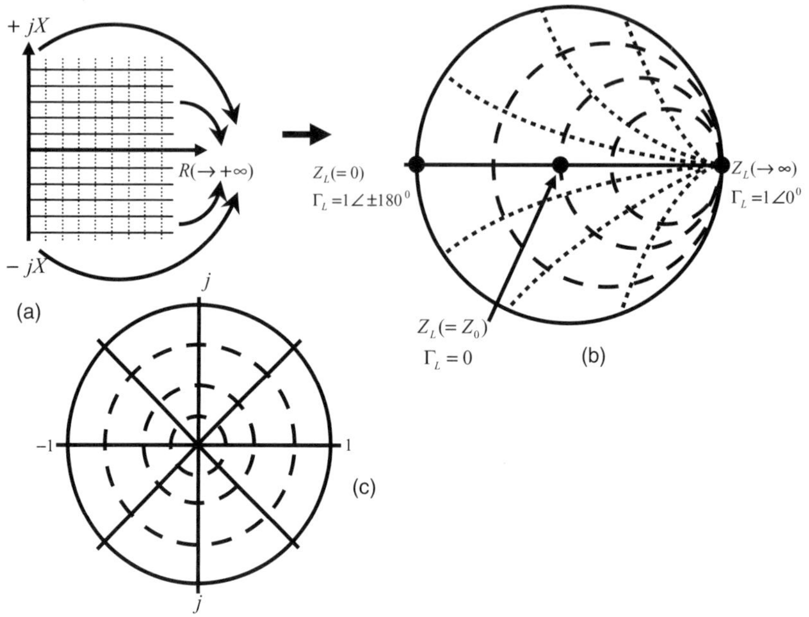

Figure 1.11. Rectangular, polar, and Smith chart representation of impedances.

amplifier, and simultaneously, the impedance states for maximum available gain can also be plotted. The locus of such impedances usually results in circles in the Smith chart (constant noise, gain, and stability circles can be plotted in the Smith chart as well) and the intersection of these circles would determine the optimum impedance states for circuit performance. In the case of power amplifiers, a load pull technique is commonly used to provide the designers realizable impedance states to maximize power transfer.

Impedance matching can also be graphically realized by Smith charts. The output impedance of a transistor is usually capacitive, and for maximum power transfer, we require an inductive impedance match. We first plot the device output impedance and its conjugate in the same Smith chart. Then we consider the 50 Ω load impedance and work backward. A series capacitor with the 50 Ω impedance provides an impedance state that is capacitive and is represented in the lower half of the Smith chart. This capacitive impedance is then considered in its admittance domain by flipping the Smith chart along its real axis. Finally, an inductive admittance takes it to the desired conjugate impedance. This simple graphical illustration suggests the use of an L-type matching network, which is commonly used at the output of the circuit. In practice, however, these components have finite Q, and the quality of matching is affected by the achievable Q from the components.

1.6.8.3 Practical Applications of S Parameters. S parameters are useful in device modeling. In the modeling step, the first part consists of an S parameter measurement, and then it converts to the appropriate parameters for better interpretation.

Critical performance parameters such as f_T and f_{MAX} can be easily interpreted from S parameters.

1.6.8.3.1 Amplifier Design. S parameters are commonly used in classic amplifier designs. A few illustrations include

$S'_{11} = S_{11} + \frac{S_{12}\times S_{21}\times\Gamma_L}{1-S_{22}\times\Gamma_L}$, input reflection coefficient with arbitrary load impedance Z_L

$S'_{22} = S_{22} + \frac{S_{12}\times S_{21}\times\Gamma_S}{1-S_{22}\times\Gamma_S}$, output reflection coefficient with arbitrary source impedance, Z_S

$A_v = \frac{S_{21}\times(1+\Gamma_S)}{(1-S_{22}\times\Gamma_L)\times(1+S'_{11})}$, voltage gain with arbitrary Z_S and Z_L

$K = \frac{1+|D|^2-|S_{11}|^2-|S_{22}|^2}{2\times|S_{12}\times S_{21}|}$, stability factor, where $D = S_{11}\times S_{22}-S_{12}\times S_{21}$

In the case of obtaining the unity current gain cutoff frequency f_T, we first obtain the S parameter measurement data and convert it to H parameters using the following relationship.

$$H_{21} = \frac{-2S_{21}}{(1-S_{11})\times(1+S_{22})+S_{12}\times S_{21}}$$

Usually transistors behave as a single-pole, low-pass filter, and f_T is determined by the frequency where $|H_{21}| = 1$.

1.6.8.3.2 Modeling of Passive Circuits. S parameters can be useful in constructing models of passive circuits. Since our target is to obtain the lumped element representation of such networks and to obtain the values of each of the lumped element components, a transformation to either Y or Z parameters should be applied. From these parameters, the individual lumped element components can be extracted. The networks are usually represented by a series and a parallel combination of lumped elements, which can be accurately extracted from the Y or Z parameters. This procedure is applicable in the case of package models and spiral inductor models.

Inductor models at high frequency can be obtained using S parameter measurement as well. In this case, the S parameters are first obtained using a two-port or one-port measurements. In RF circuits, often a differential inductor is employed for area efficiency, and a two port S parameter is most appropriate. From two-port S parameter measurement data, one-port S parameter data can be obtained (as shown in Appendix A) and subsequently converted to Z parameter data using

$$Z_{11,1p} = \frac{1+S_{11,1p}}{1-S_{11,1p}}\times Z_0$$

and the Q factor given as

$$Q = \frac{\text{imag}(Z_{11,1p})}{\text{real}(Z_{11,1p})}$$

w.r.t. Y parameters, it is given as

$$Q = \frac{-\text{imag}\,(Y_{11,1p})}{\text{real}\,(Y_{11,1p})}$$

A few commonly used networks, and their S, Y, and Z parameters, are provided in Appendix A(5). Many of these networks can be interpreted as a combination of T or π configurations.

1.7 BROADBAND, WIDEBAND, AND NARROWBAND SYSTEMS

The small signal response of any transistor-based circuit is usually broadband (determined by the f_T of the circuit). It is determined by the parasitic capacitance of the device as well as by the load capacitance. It forms a single-order pole at the output. Any signal processing operation in analog requires power, and the power consumption is proportional to the frequency of operation and square of the bandwidth under consideration. A broadband system is capable of operating from DC to its bandwidth. Wideband systems operate over bandwidths greater than the center frequency BW $\sim 1.5f_c$. Narrowband systems usually operate with BW $\sim 0.2f_c$, a commonly used scenario in wireless systems. Operating over narrow bandwidths reduces the power consumption in the front end. This result is quite intuitive, as we consume lower power and process smaller bandwidth signals. The characteristics of a semiconductor active device are usually broadband, limited by the parasitic device capacitance. To use such circuits, a narrowband signal processing element would need to use frequency-selective load impedances. Such components can be very easily implemented by a parallel combination of two frequency-dependent reactances, one of which grows with frequency (inductor) and another decreases with frequency (capacitor). Thus, we obtain a high impedance at a certain frequency (resonance) and a low impedance away from it. Broadband and wideband systems pose a severe group delay restriction on the circuits that perform signal processing, and they require a wideband antenna, which is challenging to implement. Narrowband systems can be implemented by selective peaking of wideband device characteristics using frequency-selective impedances.

1.7.1 LC Tank as a Narrowband Element

Each component in an LC resonant circuit can be represented in a series/parallel combination of the reactance and Q factor as illustrated in Figure 1.12. In a series representation, the Q factor is given by $Q_s = \omega L/r$ (reactance divided by resistance), where r is the series resistance. The parallel equivalent of this circuit consists of the same inductance, and a parallel resistance denoted by $R = Q_s^2 r$, and the Q factor is given by $Q_P = R/\omega L$ (resistance divided by reactance). Of course, we cannot obtain a different Q factor just by merely changing a series or a parallel combination of a component representation. The same is true for capacitors, where $Q_s = 1/\omega Cr$, and a parallel combination would provide $Q_P = \omega CR$, where $R = Q_s^2 r$.

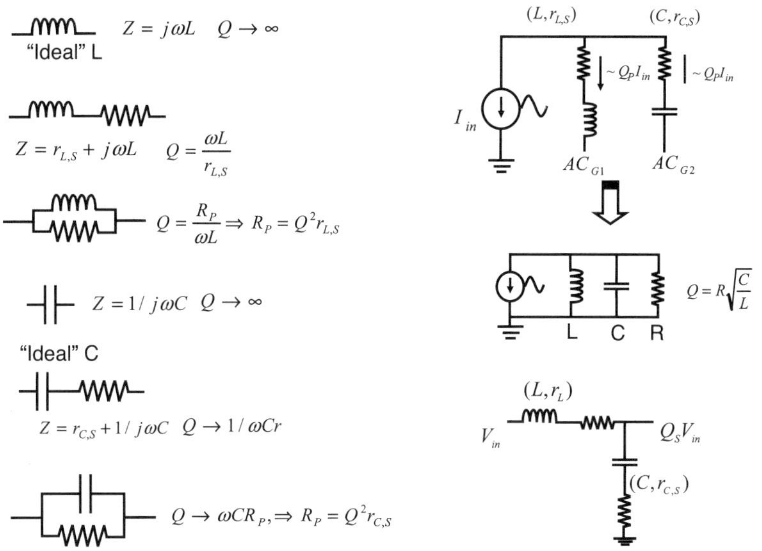

Figure 1.12. Series and parallel representation of LC circuits.

The above analysis can then be extended to *LC* resonant tanks. Each component can be transformed to its parallel equivalent, and at resonance, the inductive impedance will cancel the capacitive impedance. The resistive part from each of the components is connected in parallel, and the Q factor at resonance is determined by the ratio of the parallel resistance to the individual reactance.This implies that the Q factor of an *LC* resonant tank is obtained by the parallel combination of the individual Q factors. Usually on-chip capacitors provide $Q > 25$, and the overall Q is dominated by the inductor Q. Hence, we emphasize the unloaded Q factor of inductors in conjunction with *LC* resonant circuits. The circuit performance always depends on the "loaded Q" of the *LC* tank, which is governed by the parallel combination of the individual component Q factors.

1.7.2 LC Tank at Resonance

At resonance, the individual currents through the resonating components are increased by Q times the input current. However, this current is circulated through the inductor and the capacitors, and it does not flow anywhere else. The sum of the branch currents has to be the same as the input current to validate Kirchoff's laws.

Let us consider a transconductor stage loaded with an *LC* tank in the limits of a small signal operation. As shown in Figure 1.13, it can be observed that the voltage swing across the tank at resonance is given by (assuming high output impedance of the cascode pair)

$$V_{\text{TANK}} = g_m V_{in}(\omega L)Q$$

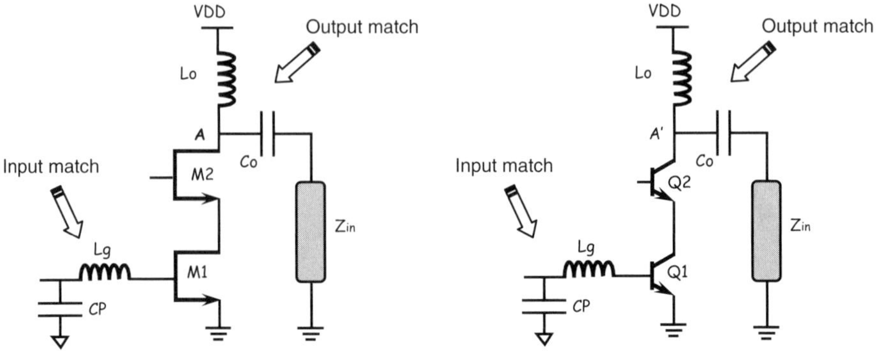

Figure 1.13. Inductively loaded MOS and bipolar stages.

1.7.3 Q Factor, Power, and Area Metrics

This result, is very important, as it implies that a Q factor of 3 can reduce the power dissipation by a factor of 3. We also assume here that the amplifier is operating at open loop. Q is the loaded quality factor associated with the tank ($= Q_L||Q_C||Q_{Cas}$). Assuming that we use the highest quality factor capacitors in tank load, and a high Q of the output parasitic capacitance from the cascode device, Q_L would dominate in the determination of loaded tank Q. At the same time, it can be observed that $g_m = \frac{I_C}{V_t}$ in the case of bipolar and $g_m = \sqrt{2\beta I_d}$ in the case of long-channel MOS devices. Thus, at a specific operating frequency (ω) and an operating input signal condition (V_{in}), the transconductance, hence, the DC current, and the power can be reduced proportionally with an increase in the LQ product. Hence, one can increase L or increase Q, or obtain an intermediate optimization stage. A higher value of L would lead to an increase in inductor area (given by the number of turns and the outer diameter), and to a reduction in capacitor area for operating at the same resonating frequency. The opposite would happen for a reduced L, and the capacitor area consumed would be more. If the Q is increased, then the circuit would provide a narrowband frequency response. Hence, two types of optimization domains would exist:

1. High L, low Q, low C, higher bandwidth, lower power
2. Low L, High Q, high C, lower bandwidth, lower power

1.7.4 Silicon-Specific Considerations

It would also be observed that as the Q is increased for the amplifier stages, it leads to increasingly higher signal swing and narrower bandwidths, which then leads to saturation of the subsequent stages and more susceptibility to process variation. As mentioned, the passive components do not consume any voltage headroom, and it is suitable to obtain signal swings beyond supply rails (assuming that the reliability considerations of the devices are satisfied for large swings). Thus, it leads to high dynamic range systems in the analog/RF domain. It can also be observed from the above argument that high Q is not always desirable. In fact, one can design circuits for low-power applications by using a Q factor of 4–5, which is fairly reasonable for silicon-based process technologies. Hence, ideally one does not have to look for technologies where a Q of 1000 is to be achieved. This may be desired in the case of oscillators for obtaining a very good phase noise at lower current consumption. However, oscillators always operate with a PLL, but the amplifiers usually operate in open loop. Hence, very high Q is attractive for VCOs, but not desirable for amplifiers. At the same time, it can be observed seen that the impedance at resonance is governed by the ratio of $(Q_L\sqrt{L/C})$. Various types of passive components are used in integrated systems, and we would explain the passive components a little later. While being useful for the power consumption perspective, they tend to consume large area on chip, and they generate electromagnetic cross-talk with other components present in the same substrate. It must be noted that passive components do not provide power gain (it would be wonderful if they did!). They can provide voltage gain or current gain.

1.7.5 Time Domain Behavior

The above illustration provided frequency domain representation of L-C resonators. We have observed the frequency-dependent nature of individual impedance components and have obtained a frequency where they cancel each other to provide high impedance. The transient domain analogy is also possible, and one can apply a current step at the input of the LC tank, where the voltage waveform would "ring" for Q cycles before reaching its steady state value. The frequency of oscillation determines the resonating frequency. It is caused by the fact that voltage across a capacitor is obtained by "integrating" the current over time, whereas the same for the inductor is obtained by "differentiating," thus leading to the formulation of a second-order differential equation.

1.7.6 Series/Parallel Resonance

The above illustrations are true for circuits at resonance, which can be a parallel resonance, or a series resonance, based on the circuit configurations and component arrangements, as illustrated in Figure 1.14. Series resonance can be used at the input of the LNA circuit to provide voltage gain, thereby improving the noise figure. Parallel resonance provides current gain at resonance, and this is true for an LC resonant tank or a matching network. At resonance, multiple reactance elements come in parallel to one another; the current through each element is multiplied Q times at resonance. In an L matched section, this current flows through the branch consisting of 50 Ω of impedance and responsible for the voltage swing at the output. As illustrated in Figure 1.14, both common-mode and differential-mode impedances must be taken into consideration.

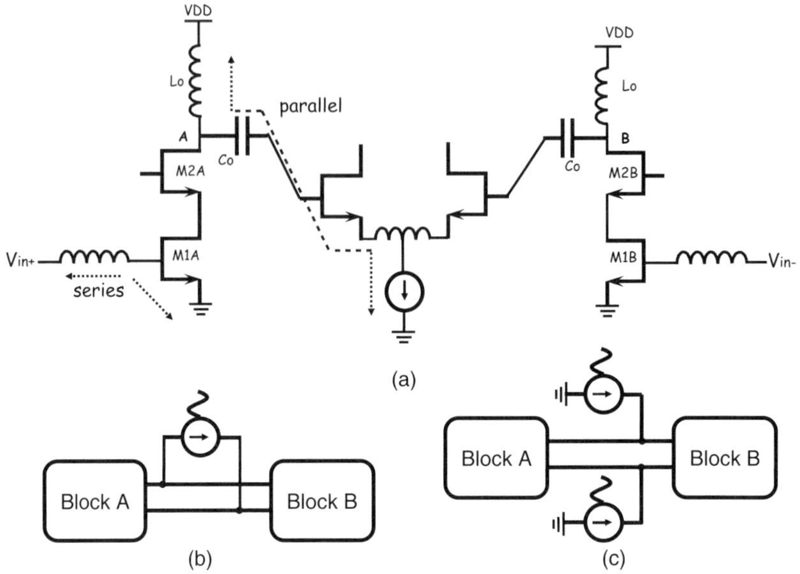

Figure 1.14. Two building blocks w.r.t. interfacing impedance and Q.

1.8 SEMICONDUCTOR TECHNOLOGY AND DEVICES

Because of the high mobility of III–V transistors, they were preferred in the early years for developing integrated radios. Most III–V transistors also provide a direct alignment of valence and conduction bands, leading to light-emitting behavior from these materials. Hence, they gained significant popularity: (1) high cutoff frequency to perform frequency translation and other high-speed operations, and (2) light emission capability, so that they can be integrated as a part of high-speed digital signal processing systems. However, cost was a major factor for these devices. Large wafers had manufacturing issues, whereas much complex functionality could not be easily integrated. At the same time, in many cases, the differences between the electron and the hole mobility were significant, such that feasibility of complementary circuit topologies was difficult. However, at the time of their popularity, radios used to be a combination of separate functional blocks, and were not viewed as complicated integrated systems. Many of these devices are still used in the high-speed industry, such as a defense electronics and RADARs, for their superior performance. However, in many of these cases, the quality of substrate was also a winning factor, as high Q and low parasitic passives, such as inductors, capacitors, and resistors could be easily realized and were accurate. Typical examples include GaAs MESFET, GaAs bipolar, InP bipolar, GaAs pHEMT, and so on. Each of these devices is different from another in terms of its physical construction and operation according to band-gap theory.

1.8.1 Silicon-Based Processes

Developments on silicon-based platforms started with the invention of bipolar transistors (1947). During the 1980's, various fabrication difficulties related to CMOS devices were solved and CMOS based digital circuit techniques became more and more popular. During the late 80's Germanium doped graded base profile bipolar transistors could achieve both the speed (f_T) and the RF performance (f_{MAX}) comparable with the III–V semiconductors. As the CMOS technology nodes were scaled successively, the high-frequency handling capability increased, and the minimum feature size was reduced. This improvement led to a lower voltage, lower geometry device. Although it became popular in the digital domain, there were several factors to consider:

1. Fineline CMOS is costly, because of the maskset and lithographic precision.
2. Silicon substrate provides moderate to low Q passives.
3. Process controllability is poor in fineline CMOS, which leads to worse component matching.
4. There may be incremental area advantages for high-density I/O circuits (circuits may be pad limited).

Although digital circuits enjoy scaling in terms of power and area (although leakage is a significant issue for the advanced nodes), RF/analog circuits may suffer because of lower breakdown voltages, poor matching, and lower drive strengths.

The proper choice of semiconductor platform is a critical decision in the implementation of integrated communication microsystems. The basic elements of any high-frequency communication circuit are transistors and high Q, high-density, low-parasitic passives. For the analog/RF building blocks, the transistor cut-off frequency f_T and the maximum oscillation frequency f_{MAX} are the key performance metrics, which usually increase with each technology node. This implies that one can realize progressively higher frequency circuits and systems using advanced silicon-based technologies. At the same time, the power consumption at a specific frequency of operation reduces with increased cutoff and oscillation frequencies. Advanced MOS transistors can be operated in a subthreshold region, which has been proven to be effective in terms of their low-power consumption. Advanced CMOS technologies tend to demonstrate subthreshold cutoff frequencies in the GHz range. To realize higher transconductance, the device sizes need to be significantly large, which leads to parasitic loading. The weak inversion region also has a worse noise performance when compared with the strong inversion region. As a MOSFET is driven into a subthreshold region from strong inversion (by reducing v_{GS}), NF_{min} increases sharply and then becomes saturated at a higher value [23].

1.8.2 Unity Current and Power Gain

The maximum device cutoff frequency (f_T), maximum oscillation frequency (f_{MAX}), broadband noise factor (NF_{min}), and flicker noise ($1/f$) profile become the determining

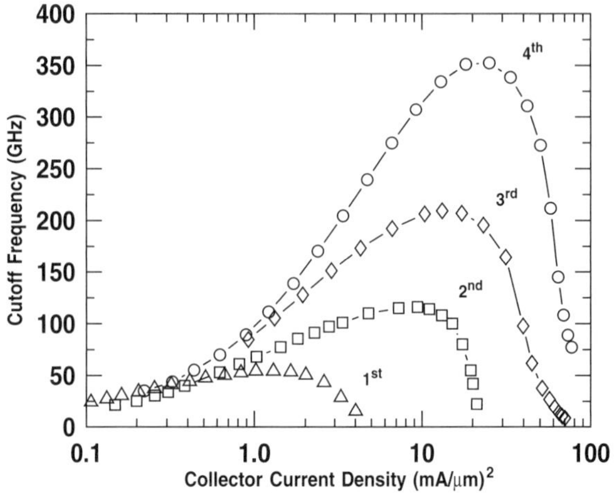

Figure 1.15. Cutoff frequency versus current density in bipolar.

factors for RF designs at nanometer geometries. These parameters play an important role, irrespective of their device technology (bipolar or MOS).

Figures 1.15 [24] illustrates the scaling impact on $f_T$across various generations of SiGe HBT transistors. As f_T increases, it is observed that a transconductor would require progressively lower power consumption at a certain frequency of operation (f_T GHz). In other words, increasing f_T enables the feasibility of RF designs at progressively higher frequency regime.

Figure 1.16 illustrates the scalability impacts on the cutoff frequency of MOS, across two technology nodes, 130 nm and 90 nm [25,26]. The cutoff frequency is given by

$$f_t = \frac{g_m}{2\pi(C_{gs} + C_{gd} + C_{gb})} \tag{1.1}$$

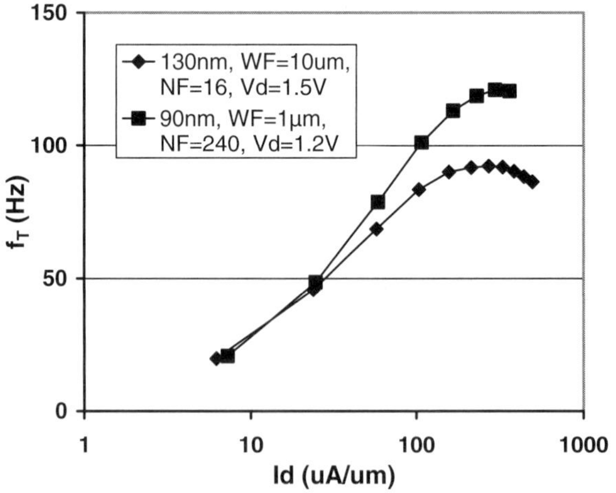

Figure 1.16. Cutoff frequency versus current density in CMOS.

where g_m is the transconductance and C_{gs}, C_{gd}, C_{gb} indicate the gate-source, gate-drain, and gate-bulk capacitances associated with the respective terminals. Using short channel approximations for g_m, and expressing the capacitances in terms of the area parameter,

$$\begin{aligned} f_t &= \frac{\mu_n C_{ox} W E_{sat}}{4\pi(C_{gs}W + C_{gd}W + C_{gb}WL)} \\ &= \frac{\mu_n C_{ox} E_{sat}}{4\pi(C_{gs} + C_{gd} + C_{gb}L)} \end{aligned} \tag{1.2}$$

In conjunction with f_T, f_{MAX} determines the frequency of unity power gain and is given by

$$f_{\max} = \frac{f_t}{\sqrt{8\pi C_{gd} R_G f_t + 4 g_{ds}(R_G + R_s)}} \tag{1.3}$$

It can be observed that, although f_T is a relatively straightforward expression in terms of forward current gain, f_{MAX} is a complicated function of device geometry and layout. The gate resistance plays a significant role in determining f_{MAX}, along with the substrate and gate resistances associated with the device.

In addition to their high-frequency behavior, two device noise mechanisms such as broadband noise (thermal noise) and flicker noise become important for nanometer MOSFETs. The broadband noise factor of MOSFET is given by

$$NF_{\min} = 1 + B\frac{f}{f_t}\sqrt{g_m(R_s + R_G)} \tag{1.4}$$

which implies that the minimum noise that can be obtained from these devices is dependent on the transconductance g_m, the gate resistance R_G, and the substrate resistance R_s. It can be observed that the improvement of broadband noise can be achieved by optimized layout, which also helps improve f_{MAX}. The substrate resistance directly impacts the noise factor, and hence, a reduction in the substrate resistivity would lead to minimization of the minimum noise factor.

1.8.3 Noise

The flicker noise contributed by the MOS transistors is given by

$$\overline{i_n^2} = \frac{K}{f} \cdot \frac{g_m^2}{WLC_{ox}^2} \cdot \Delta f \approx \frac{K}{f} \cdot \omega_T^2 \cdot A \cdot \Delta f \tag{1.5}$$

The flicker noise in MOSFET is usually much worse compared with bipolar transistors, as the fluctuation of the carriers in the channel occurs in the presence of traps in the oxide. For a given transconductance, a larger gate area and thicker oxide reduces the contribution of flicker noise. Since an increased area implies the loading for RF/analog circuits, careful design optimization needs to be performed.

Broadband noise is important for circuits such as low noise amplifiers, whereas flicker noise impacts the low-frequency circuits such as differential amplifiers and OP-Amps, in addition to the flicker noise upconversion in the cases of VCOs, frequency dividers, mixers, and other nonlinear frequency conversion circuits. The design in the RF/analog regime needs to be optimized in terms of power and area; hence, these formulations need to be used in specific cases, depending on the circuit under applications.

1.8.4 Bipolar vs. MOS

It is important to understand the differences between two devices in order to use them optimally in a transceiver architecture. Bipolars are vertical devices with regard to the current flow, whereas MOS current flow is lateral in nature and most of the action (electron transport) happens at the surface. From a circuit perspective, transconductance of a bipolar transistor is dependent only on the current it is biased at, and is not scaleable, whereas in the MOS device, the transconductance is a function of both the bias current and the device size. The transconductance is proportional to the bias current for a bipolar device (whereas the square root in MOS), which leads to the fact that bipolars provide superior transconductance performance. However, the collector and emitter terminals are asymmetric in nature in terms of electrical performances, whereas drain and source symmetrical in nature in a MOS device. Hence, MOS transistors can be used as excellent switches and are popular in digital circuits, sampling switches, and passive mixers. At deep submicron technology, bipolars tend to have superior output impedance performance compared with MOS. Area wise, construction of MOS device usually takes 15% more area compared to the bipolar device of the same transconductance performance. Being a surface device, MOS is susceptible to electron interaction with the trap states in the gate oxide, which contributes more flicker noise than a bipolar device.

A BiCMOS technology is optimum for integrated radios. Superior bipolar devices can be used to develop better low noise amplifiers, low $1/f$ noise VCO cores, and baseband amplifier stages, whereas MOS can be used for superior switching performance. In terms of fundamental device operations, bipolar is a minority carrier device, whereas MOS is a majority carrier device. Different parts of bipolar transistor characteristics are well modeled using exponential characteristics, whereas MOS is mostly a square law device and empirical modeling is used. However, at deep submicron MOS, this differs significantly. Bipolar modeling is complicated by the fact that the collector and emitter terminals are asymmetrical in nature, and the current distribution in the collector is difficult to model in advanced geometries. The difficulty in MOS modeling originates in order to construct a continuous model across all regions of operation. In their inherent nature, the operating regions of MOS devices are sometimes difficult to represent using a single mathematical equation, and usually, numerical fitting techniques are used. While modeling a device, accuracy over a wide operating range is often obtained using multiple parameters, which may imply intensive analysis time for the circuit simulators and so on. Although an accurate model can be obtained using fewer parameters for a fixed device geometry (and a large device can be an array of small devices) to obtain faster simulation of circuits, scalability is often, preferred by the designers. Moreover, in advanced CMOS technologies, all functional blocks can be integrated into one substrate, which leads to the popularity of CMOS technology nodes.

1.8.5 Device Characteristics

Without moving to much detail in modeling aspects, we intend to provide readers with basic intuition of the fundamental device parameters to develop circuits and systems. At the same time, a hand analysis is almost impossible using multiple model parameters. Key performance metrics are discussed in the following subsections.

1.8.5.1 DC Characteristics. I_D versus V_{GS} for MOS, I_C versus V_{BE} for bipolar MOS transistors operating in linear region $V_{DS} < V_{GS}-V_t$, for the triode region with a voltage variable resistor of $r_{ON} = \left[\mu C_{ox}\left(\frac{W}{L}\right)(V_{GS}-V_t)\right]^{-1}$

$$I_D = \frac{\mu C_{ox}}{2}\left(\frac{W}{L}\right)\left[2(V_{GS}-V_t)V_{DS}-{V_{DS}}^2\right]$$

MOS transistor in saturation region.
$V_{DS} > V_{GS}-V_t$ for saturation region, which leads to the square law characteristics,

$$I_D = \frac{\mu C_{ox}}{2}\left(\frac{W}{L}\right)(V_{GS}-V_t)^2$$

The threshold voltage V_t is given by, $V_t = V_{t0}+\gamma\left[\sqrt{2|\phi_F|+V_{SB}}-\sqrt{2|\phi_F|}\right]$.

In the case of bipolars, the corresponding relationships are given as follows:

$$I_C = I_S e^{(V_{BE}/V_T)}$$

1.8.5.2 Output Impedance. The real part of the output impedance is governed by the output DC characteristics, and the imaginary part is governed by a combination of one or more capacitances associated with the device. At RF frequencies, the capacitance usually plays a dominating role in determining the output impedance. At low frequencies and DC, the real part becomes important for construction of current sources and so on. Both the magnitude and the Q factor of this impedance are important for circuit design.

In the case of bipolar transistors, $r_O = \frac{V_A}{I_C}$, where V_A is the "Early" voltage and I_C determines the bias current. In the case of MOS, this is given as $r_O = \frac{1}{\lambda I_D}$, where λ denotes the channel length modulation parameter.

1.8.5.3 Capacitive Elements. Two types of capacitors are associated with transistors: (1) The geometry-dependent capacitor and (2) the bias-dependent capacitor. These capacitors are associated with MOS and bipolar.

In the case of MOS transistors, the source-drain regions form a diode structure with the substrate, and the capacitance is voltage dependent, which is given by

$$C_{SB} = \frac{C_{SB0}}{\sqrt{1 + V_{SB}/\phi_0}}, \quad \text{and} \quad C_{DB} = \frac{C_{DB0}}{\sqrt{1 + V_{DB}/\phi_0}}$$

The input capacitance referred to as the gate terminal is geometry dependent, and it is given by $C_G = C_{OX}(W.L)$. This capacitance is essentially divided between the drain and source terminals, depending on the geometrical shape of the formed channel, which differs from a triode region to a saturation region.

Triode/Linear region:

$$C_{GS} = \frac{1}{2}C_{OX}(W.L), \; C_{GD} = \frac{1}{2}C_{OX}(W.L)$$

Saturation region:

$$C_{GS} = \frac{2}{3}C_{OX}(W.L), \; C_{GD} = \frac{1}{3}C_{OX}(W.L)$$

In the case of bipolar transistors, the $C-B$, and $C-S$ junction capacitances are given by

$$C_{EB} = \frac{C_{EB0}}{(1 + V_{EB}/\phi_0)^n}, \; C_{CB} = \frac{C_{CB0}}{(1 + V_{CB}/\phi_0)^n}$$

The input capacitance C_B is bias dependent and given as $C_B = \tau g_m$, with $C_{BE} = C_B + C_{jE}$.

Voltage variable capacitors provide distortions to the signal waveforms. They also provide AM–PM conversion in large signal swings.

1.8.5.4 Device Noise. Device noise determines the fundamental limits to available signal-to-noise ratio in any circuit. Mainly three types of noise occur in devices:

Thermal noise: This is associated with random flow of electrons, and not associated with any bias current; it is present in all devices. The spectral density id given by

$$\langle i_n^2 \rangle = \frac{4kT}{R}\Delta f$$

Shot noise: This is associated with DC current flow, and it is always associated with a junction and independent of frequency. The spectral density is given by

$$\langle i_n^2 \rangle = 2qI\Delta f$$

Flicker noise: This is associated with DC current and present in all active devices; the spectral density is given by

$$\langle i_n^2 \rangle = K\frac{I^a}{f^b}\Delta f$$

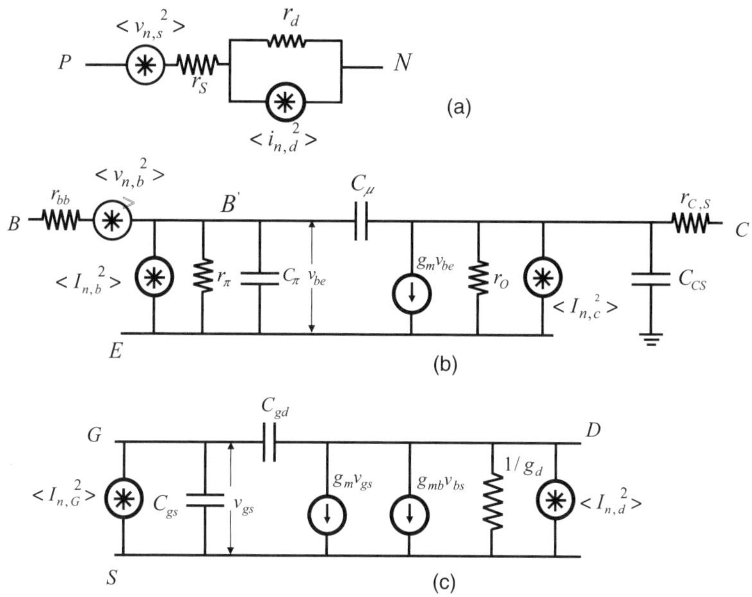

Figure 1.17. Noise models for diode, bipolar, and MOS transistors.

These equations can be applied to diode, MOS, and bipolar devices, respectively, and the noise model can be obtained as illustrated in Figure 1.17. Noise can be referred to the output as well as to the input, and both methods can be used in the circuit design process. Both are related to each other by the ratio of transconductances. Flicker noise is an important consideration in MOS transistors, and it is physically related to the number of surface states and to the change of threshold voltage from the gate oxide capacitance C_{OX}. The input-referred flicker noise spectral density is represented by

$$\langle v_{f,G}^2 \rangle = \frac{K_f}{WLC_{OX}f} \Delta f$$

which is independent of bias current. However, when referred by drain, this is given by

$$\langle i_{f,D}^2 \rangle = g_m^2 \langle v_{f,G}^2 \rangle = \frac{K_{f,D} I_D}{L^2 f} \Delta f$$

which is dependent on the bias current and the channel length. This result bears very important conclusions in the case of circuit designs: (1) Circuits, which do not consume any current (such as passive mixers, etc.), are inherently quiet in terms of flicker noise performance. (2) The output noise current is inversely proportional to the square of channel length, and migration to smaller channel lengths would contribute more flicker noise.

On the other hand, the spectral density for drain-referred thermal noise is given by $\langle i_{n,D}^2 \rangle = 4kT\beta g_m \Delta f$, where β varies between 2/3 and 2 from the saturation region to the

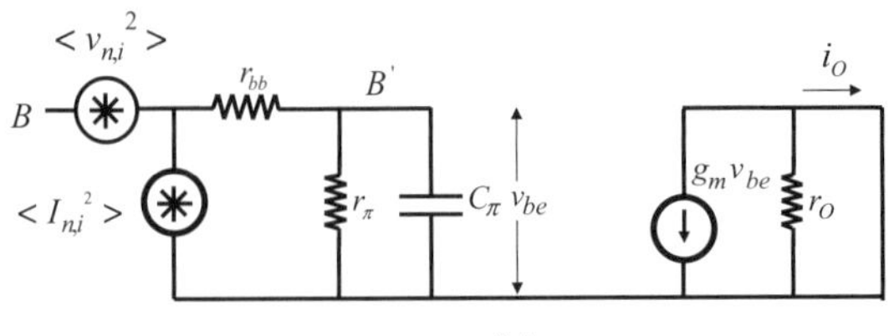

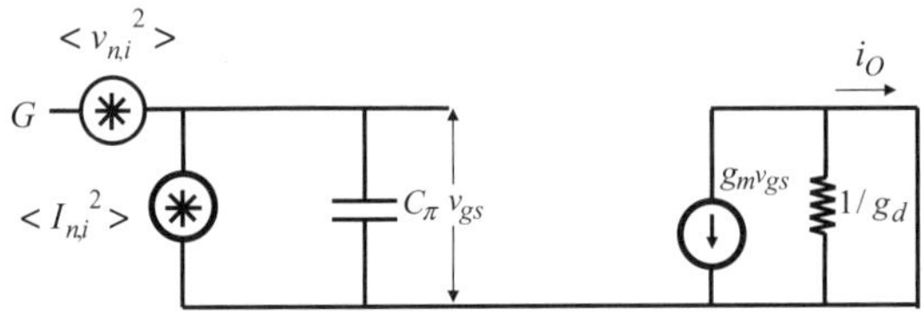

Figure 1.18. Equivalent noise models for circuit analysis.

subthreshold. Drain current caused by the input signal is given by $i_D^2 = g_m^2 V_{gs}^2$; thus, increasing g_m leads to a better SNR w.r.t thermal noise.

Gate leakage in MOS leads to shot noise, which can be represented as $\langle i_{n,G}^2 \rangle = 2qI_G \Delta f$. This is uncorrelated from the drain noise terms described above. Bipolar transistors can be analyzed in the same manner, starting with fundamental noise equations.

Using these noise models, important insight can be developed to construct optimum impedance for low noise circuits. Representation of bipolar and MOS circuits are illustrated in Figure 1.18. Noise models can be viewed as various noise sources in conjunction with the small signal models. In the case of bipolars, both the input referred voltage and the current noise terms are present. The voltage noise becomes dominant when being driven from a low impedance source, and the current noise becomes dominant when being driven from a high impedance source. Thus, an optimum noise figure is obtained in between the two impedances. In the case of MOS stages, the input noise current source is dominant, which leads to a high driving impedance for optimum noise performance.

Noise has an important implication in designing circuits; therefore, we will discuss a few typical cases. Often in communication systems, signals are sampled w.r.t. a clock waveform. To prevent aliasing, the clock frequency is chosen to be an integral multiple of the message signal. Although this process ensures signal reconstruction, it also downconverts noise from various clock harmonics, and places them in a band of interest. This action is typical of a sampling switch. In a simple $R-C$ stage, when resistor increases, the magnitude of the noise associated with it increases, but the bandwidth reduces. Similarly, when resistance is reduced, thermal noise reduces, but bandwidth is increased. The integrated noise is the same in both cases

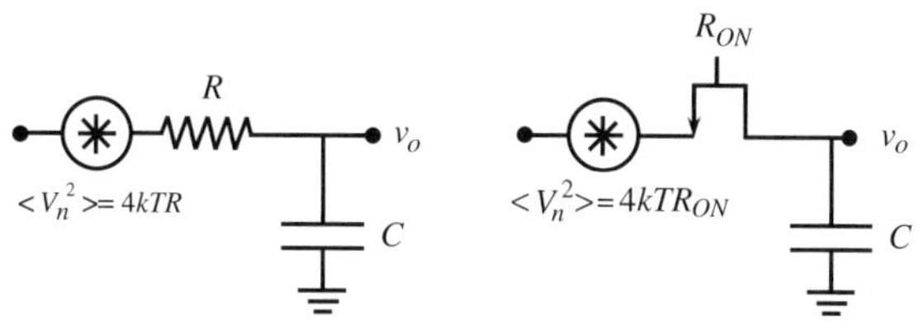

Figure 1.19. Noise in R–C stages.

and is given by

$$\langle V_{o,n}^2 \rangle = \int_0^\infty \frac{4kTR}{[1+(2\pi fRC)^2]} df = \frac{kT}{C}$$

However, it is also true for a bandlimited system. In a circuit, where multiple poles and zeros are present and the frequency response extends to infinity (a pole followed by a zero), the integration bandwidth does matter in the integrated noise consideration. Noise is contributed by the resistance only, but the capacitance comes into picture because of the band limitation of the white noise. This process is illustrated in Figure 1.19.

As this is dependent on the capacitor size, the output noise is invariant even when a MOS switch is present instead of a resistor. Thus, to suppress this noise, a large capacitor is to be used. To account for integrated noise and thermal noise spectral density, periodic steady-state noise and small-signal noise analysis can be used using circuit simulators.

1.8.5.5 Breakdown Voltage. From a device design perspective, device speed and breakdown voltages are important. They are related by the fundamental relationship, known as Johnson's limit $BV \times f_T = K$. Hence, the speed of a device cannot be increased arbitrarily without compromising the breakdown voltage. Hence, breakdown plays a critical role in submicron geometries, when designed for a high RF frequency.

Breakdown can fundamentally occur because of (1) application of an electric field across a semiconductor junction, which is more than the maximum electric field to be sustained at that junction, and (2) damage to the oxide caused by electrons moving in high velocity (caused by a high-input electric field). In the first case, the device can still be recoverable, but in the second case, it is permanent damage to the device.

In a bipolar device, two critical voltages associated with breakdown are V_{CBO} and V_{CEO}, which are called the "collector to base breakdown with emitter open" and "collector to emitter breakdown with base open," respectively. They are related to each other by

$$BV_{CEO} = \frac{BV_{CBO}}{\sqrt[n]{\beta_F}}$$

In a MOS transistor, similar quantities are referred to as "gate to drain breakdown" and "drain to source breakdown." The gate breakdown voltage is significantly lower compared with the drain-to-source breakdown voltage.

In large signal circuits and systems, breakdown is a critical consideration, and it depends on operating temperature as well. To obtain a higher breakdown voltage, thick gate oxide transistors can be used in MOS technologies with the compromise of lower transconductance. In a bipolar transistor, collector current has a positive temperature coefficient, and circuits are susceptible to thermal runaway. This is a major consideration in PA design, and resistors in the emitter are used to counter this effect, as RF power devices comprises arrays of small bipolar junction transistor (BJTs). On the other hand, the drain current for a MOS device has a negative temperature coefficient, which prevents thermal runaway, and multiple MOS transistors can be connected in parallel without ballasting requirements. Because of the inherent switching functionality of MOS devices, they are popular for switching the mode operation of PAs.

1.8.5.6 Technology Scaling. As semiconductor technologies are scaled, the physical device dimensions (W,L,t_{OX}) get reduced by the scaling factor α, supply voltage also drops by the same factor, delay of digital gates reduces by α (delay $=CV/I$), and so on. However, the wiring delay remains the same (the resistance of wires increases and the capacitors reduces). Hence, at scaled geometries, interconnecting delays play a critical role in gate delays. Power dissipation in digital gates reduces, and so is the power-delay product, which contributes positively to the performance of digital gates. In terms of RF/analog performance, f_T and f_{MAX} play a significant role in determining power consumption at a specific center frequency.

The impacts of scaling can be categorized in terms of analog and digital circuits as well. In a digital circuit, the number of gates would be an important parameter, and a technology scaling would significantly reduce the area of the digital part of the chip. Although there are several advantages to scaled CMOS geometries, they tend to provide significant leakage currents.

At deep submicron CMOS technology nodes, gate leakage contributes to a significant fraction of the power consumption of large digital chips. In RF circuits, it may lead to noise figure degradation. Leakage mechanisms can be categorized in the following major categories: (1) drain-induced barrier lowering (DIBL), (2) gate-induced drain leakage, and (3) hot carrier effect. Figure 1.20 illustrates the various leakage mechanisms in deep submicron MOS devices. In addition to these effects, process variation plays a critical role in analog circuits. As the lithographic geometries are extremely small, a little variation in geometry may lead a to a large variation of threshold voltage, transconductance, and so on. Broadband noise performance improves with scaling, but the flicker noise performance usually gets worse. Hence, the impact in terms of a continuous-time signal processing block (analog/RF circuit) includes (1) reduced supply voltage and dynamic range limitation; (2) increased component variation, which leads to the need of calibration circuits; (3) increased leakage; and (4) increased mask cost. However, sometimes the use of analog/RF blocks along with digital in the same substrate is encouraged to obtain a single die solution (system on a chip).

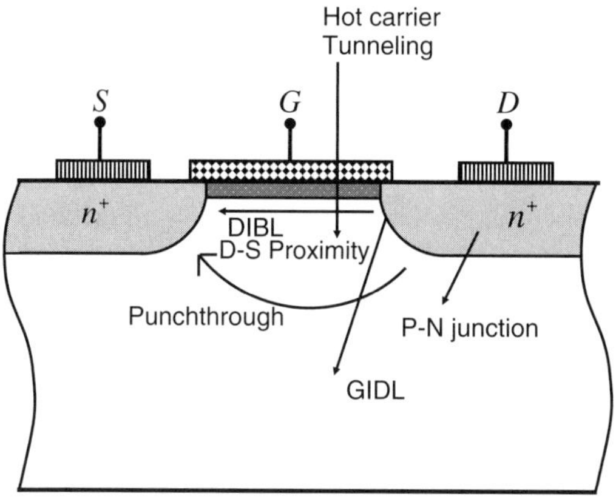

Figure 1.20. Leakage mechanisms in deep submicron CMOS technologies.

With this basic introduction, we will now consider a few testbenches in order to evaluate semiconductor technology platforms. These testbenches are generic in nature and routinely used by circuit designers. They can be used toward any technology platform at hand: Silicon CMOS, BiCMOS, GaAs, and so on.

1.8.6 Passive Components

Passive components are a key aspect to the development of analog/RF circuits. Commonly used passive components include:

1. Resistors
2. Capacitors
3. Inductors
4. Transformers

1.8.6.1 Resistors. Resistors are used mostly for the following functionalities:

1. Load impedance of the circuit (usually an amplifier/current mode logic etc.)
2. Biasing

Key considerations in using resistors include:

1. Parasitic capacitance
2. Component matching accuracy
3. Component variation with process and temperature
4. Voltage variation
5. Sheet resistancc

Any resistively loaded circuit is essentially a broadband circuit. These circuits consume voltage headroom whenever they are used in the signal path, as part of amplifiers, mixers, and so on. The bandwidth is determined by the *RC* product, where *C* denotes the output capacitance (consists of resistor's parasitic capacitance and the output capacitance of transistors). Component matching is a critical performance in determining I/Q accuracy, asymmetry in balanced amplifiers, and so on. Component matching improves with a large component area. However, parasitic capacitance increases with area and nonlinear capacitance from reverse-biased diodes increases with an increase in area.

Resistors are categorized by their sheet resistance, which indicates their to area, and in fineline CMOS technologies, process and temperature variations of resistors may lead to as large as $\pm 40\%$ in terms of their values. Often, "dummy" resistors are placed alongside the main resistors to improve component matching performance. Although one can ignore the process variations of DC biasing resistors, variations in the load impedance may cause transistors to run out of voltage headroom.

1.8.6.2 Capacitors. Capacitors are the second-most area-consuming elements in ICs following inductors. The main usage of capacitors includes:

1. *LC* resonating tank
2. Coupling RF signals
3. Decoupling for power supply

Capacitors are categorized by the following performance aspects:

1. Capacitance density
2. Voltage variation
3. Bottom plate parasitics
4. *Q* of main capacitor as well as its bottom plate
5. Process variation
6. Leakage (in the case of MOSCAPs)
7. Breakdown voltage

Capacitance density directly implies the area consumption on chip. Capacitive impedance decreases with frequency. Since driving higher impedances provides improvement in power consumption, we prefer lower capacitance values with a high *Q* factor (to reduce tank loading). However, a lower value of the capacitor may lead to more fringing capacitance and poor component matching. Bottom plate capacitance and its *Q* factor are major considerations, as it would lead to signal shunting to ground. Bottom plate capacitance is proportional to the area. In large signal circuits such as the PA driver, the voltage variation of capacitance is important. It is determined from the $C - V$ characteristics of a capacitor. In the case of AC coupling capacitors, a large capacitance value with low bottom plate capacitor value is desired. It is also desired that the voltage variation of coupling capacitor be as small as possible. The Q of the bottom plate capacitor would directly impact the Q of the input impedance.

A large variety of capacitors are used in mixed signal systems. Three common types of capacitors are used frequently in communication circuits:

1. FLUXCAPs
2. MOSCAPs
3. Varactors

MOSCAPs are used in the case of supply bypassing capacitors, and they may be used as coupling capacitors. Whenever a MOSCAP is used, DC leakage current may alter the DC operating point, and one must be very careful to ensure that it is modeled. MOSCAPs require proper biasing, and the capacitance value depends on the AC voltage swing. FLUXCAP, on the other hand, is made in a comb-like structure using metal structures, and it provides the lowest voltage variation. Varactors (voltage variable capacitors), on the other hand, are accumulation-type capacitors, and they are intended for large voltage variation, as they directly impact the analog tuning characteristics of VCOs. One needs to be extremely careful when using voltage variation of capacitors, as any variation in the AC swing (amplitude noise) would lead to a frequency shift. In the case of varactors, the ratio of maximum-to-minimum capacitance provides an important design guideline, and it is optimized around $4 \sim 5$ for most cases.

Capacitive impedance is 160 ohm per pF at 1 GHz, and it can be calculated at commonly used wireless frequencies such as 0.9 GHz, 2.4 GHz, 5.2 GHz, and so on.

1.8.6.3 Inductors. Inductors are essential to the development of high-frequency circuits. They add a "zero" in the transfer function of the circuit when loaded at the output, thereby boosting the high-frequency response of the circuits. Narrowband tuning and filtering is essential for the RF front end, and this is manifested by "tuning out" the capacitive load of the transistors. The current gain is dependent on the Q factor of the inductor (a high Q factor provides more gain at resonance). A fundamental advantage of using inductors is the achievable swing beyond the supply voltage, which is essential for high dynamic range, low-voltage circuits. Inductors can well extend the bandwidth of the circuit, in the case of both high-speed digital or analog/RF blocks. In summary, usage of an inductor in circuit blocks can be illustrated as follows:

1. Resonating tank
2. RF chokes
3. Lossless feedback
4. Matching networks

The Q factor of on-chip inductors is of the order of 8–10, and it depends on the area of the components. A larger area usually provides more inductance and Q, but it leads to more electromagnetic cross-talk (as the number of flux lines is proportional to the area).

Assuming that one can obtain a high Q from the on-chip capacitors (depends on the frequency of operation and on layout of the capacitors), the Q factor of the LC tank would be limited by the Q factor of the inductor. This fundamental issue has motivated various developments of area-efficient high Q inductors in digital CMOS technologies. In rest of this section we will pay close attention to various types of inductor topology and to key performance issues.

The following parameters can be used to benchmark inductor performances:

1. Inductance value
2. Q factor
3. Parasitic capacitance and Q factor of parasitic capacitance
4. Self-resonating frequency
5. Area

1.8.6.3.1 Inductor Geometry. Inductors are categorized by the following key geometrical parameters:

1. Number of turns (n)
2. Turn width (W)
3. Turn spacing (S)
4. Outer diameter (OD) (determines the area)

Qualitatively, when the inductor turns are closer , mutual coupling increases, as well as the interwinding capacitance. Hence, the inductance increases (total inductance self-inductance of each turn + mutual inductance effects between two turns), and self resonating frequency decreases. As the width W increases, the series resistance of the turns drops, and the Q factor improves. The inductance also drops because of the current flowing through the edges of the conductor. The physical construction of inductors is illustrated in Figure 1.21. The inner and outer diameters are related to each other by $OD = ID + 2n \times (W + S)$.

The inductance of an inductor is given by $L \sim n^2 \times OD$, where n is number of turns and OD is the outer diameter of the inductor. The series resistance is given by $R_S \sim n^2$. An optimum Q factor is usually obtained in the case of $ID \sim 0.5 \times OD$. While using inductors from a standard technology library, various combinations of the geometry parameters can be used and the right dependence can be obtained.

1.8.6.3.2 Self-Resonating Frequency. Let us now consider the self-resonating frequency of inductors. The self-resonating frequency determines the usable limits of an inductor. A simplified electrical model of inductor is shown in Figure 1.22.

In a center-tapped differential inductor, the parasitic capacitance can be obtained by connecting all the terminals, and by observing the overall capacitance by injecting a current source at the common terminal. The combined capacitance (C_P) can be distributed by placing two capacitors of $C_P/4$ at each end and a capacitor of $C_P/2$ at the center terminal. The self-resonating frequency is determined as $1/2\pi\sqrt{L_h C_t}$, where L_h

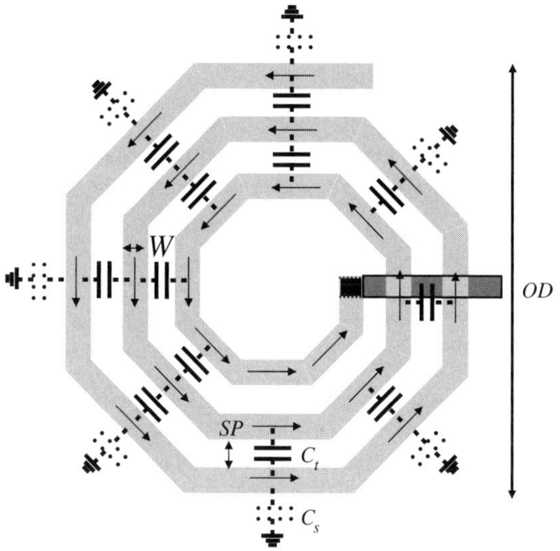

Figure 1.21. A single-ended inductor and its electrial components.

is the inductance of the half section of the inductor, and $C_t = C_P/4$. Physically, this capacitance is attributed by the interwinding capacitance and the capacitance to substrate. If the self-resonating frequency is too low, then one can reduce the outer diameter (smaller area reduces capacitance to substrate) or use a larger turn spacing S. Reducing the diameter by a factor of two would lead to a factor of 4 reduction in area, reducing the parasitic capacitance to substrate. Thus, larger inductors tend to exhibit lower self resonating frequencies. This phenomenon can also be illustrated due to the fact that, $L \sim ID$ and $C \sim ID^2$, hence, the rate of increase in capacitance is higher than the inductor, leading to lowered self resonating frequency.

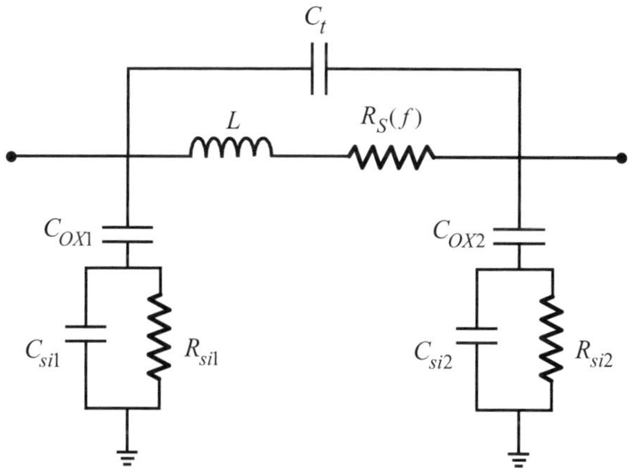

Figure 1.22. Inductor model.

1.8.6.3.3 Geometry and the Q Factor. The Q factor of inductors is dependent on the ratio of inductive impedance to its series resistance, and it is given by the ratio of area to the conductor length. For the same area of circular cross section provides the highest area-to-length ratio $(2/\sqrt{\pi})$, which leads to a high Q factor. However, it can be observed that an octagonal geometry is often a compromise, as it is easier to fabricate. Inductors generate electromagnetic fields, which propagate to adjacent circuit components, and create electromagnetic cross-talk. To prevent this action from occurring two approaches can be taken:

1. Isolate the inductor structure by providing a high resistivity exclusion zone around it.
2. Provide a low impedance substrate shield (patterned ground shield).

Both approaches are effective, depending on substrate resistivity. However, in the first approach, the electromagnetic field decays as $1/r$, where r is the distance from the excitation to the point of observation. However, the exclusion area uses much additional overhead in terms of area and processing step. In the second approach, however, the shield uses a patterned ground shield, which leads to an $1/r^2$ decay of the electromagnetic field and then to less cross-talk to the adjacent circuit components. This result can be explained with the help of image theory, which implies the conceptual formulation of image components in terms of current and charge in the substrate. For the sake of simplicity and understanding, we assume that the ground plane can be "perfect" and that the created image components would have the same strength as the original excitation element.

1.8.6.3.4 Single-Ended and Differential Inductors. We will now consider two types of inductor topologies: (1) the single-ended inductor and (2) the differential inductor. Figure 1.23 illustrates differential inductor configurations.

A differential inductor is compact in size compared with the single-ended inductor. In an inductor structure, two current flow paths exist: (1) a direct path

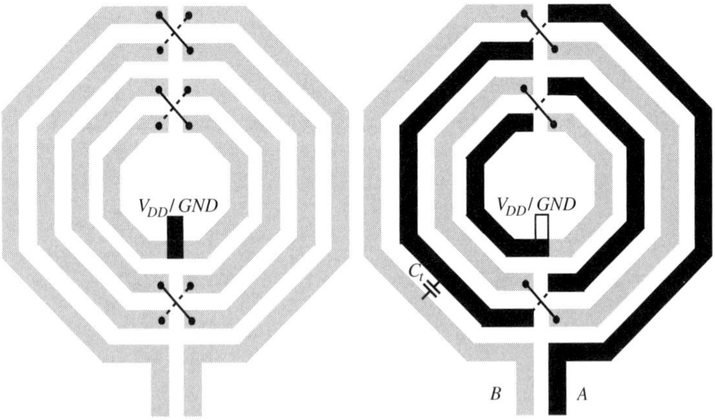

Figure 1.23. Differential inductor configuration.

through inductive component of the structure and (2) a secondary path through the capacitances or the interwinding or substrate parasitics. Current in the direct path flows along the length of the inductor, whereas it flows laterally in the secondary path. Our aim is to enhance the current flow through the direct path and to minimize the current flow through the secondary path, in order to obtain more inductive behavior. Let us assume that the inductors are used in differential circuits and that the terminal peak AC swing is V_P (i.e., differential $2V_P$). In a single-ended inductor, this voltage gradually drops across the inductor turns because of impedance, so we should simply assume that the turns get nodal voltages such as V_P, $0.8V_P$, $0.6V_P$, $0.4V_P$, and $0.2V_P$ (we assume equal drop for a five-turn inductor). Ultimately, the inductor terminals go to AC ground, no matter whether they are connected to circuit supply or circuit ground. In this case, the voltage difference across adjacent spatial turns is $0.2V_P$, which is responsible for current flow through the lateral interwinding capacitor. The inductive path is a single turn length, and the capacitive path is determined by the lateral separation of two turns (spacing). However, in the case of a differential inductor, to traverse from one turn to its adjacent turn, the inductive current flows through all the turns (a much longer path compared with the single-ended inductor structure and, hence, a larger impedance). The voltage difference is $2V_P$, for the outermost spatial turns, which leads to much higher current through the capacitive path. Hence, the effect of capacitance is more dominant in the case of a differential inductor, and their self resonating frequencies are lower. Thus, the differential inductors are used in the 3–5 GHz range. In a differential inductor, the center tap must be designed to carry twice the current limit for each of the single-ended segments of the inductor. The central tap of a differential inductor can be used to provide bias to the circuit under consideration. Differential inductors provide symmetric loading to the circuits at each port. In the case of single ended inductors, there are two terminals: (1) AC terminal and (2) underpass. In circuit implementation, the AC terminal is driven, as they usually provide a higher Q factor. Usually, in the construction of inductors, several metal layers are strapped to obtain a lower series resistance for Q factor enhancement.

1.8.6.3.5 Q Factor versus Frequency. The frequency dependence of the inductor Q factor is an important aspect for RF circuit design, and at very low frequencies, inductor loss is usually a fixed quantity, determined by its resistivity. As frequency increases, the inductive impedance ($2\pi fL$) increases and hence the Q factor leads to $Q \propto f$. As frequency increases even more, skin effect comes into the picture, and the current tends to be more concentrated toward the outer periphery of the inductor turns. The skin effect impacts the inductor metallization from all directions, and as the thickness of the AC current flow is given by $\delta = 1/\sqrt{\pi f \mu \sigma}$, the resistance increases with a square root dependency ($r \propto \sqrt{f}$). As the inductive impedance grows linearly, the Q factor grows in a square root dependence $Q \propto \sqrt{f}$. At even higher frequencies, various capacitive coupling to lossy silicon substrate occurs, and Q falls as an inverse square law dependence $Q \propto f^{-2}$. These dependencies can also be observed by plotting the Q factor versus frequency in a log scale plot. In between the transition region, Q peaks at a certain frequency, and this peak Q frequency is a very important parameter for circuit designers, as this is the optimum performance point of an inductor given a certain inductance, and area considerations. Usually it is desired that peak Q is maintained over a broad range of frequencies. The circuit performances are usually dependent on the inductors as a factor of Q^2 (or Q, depending on the circuit); hence, a humble 5% improvement in Q may improve circuit performance by 10% (1 dB).

Use of an inductor does not have to be restricted to the circuits and systems based on amplifiers only (hence, implying wireless type systems). Inductors are used in many digital circuits and systems, where the bandwidth enhancement is obtained using inductors. However, it must be observed that compared with the wireless systems, digital circuits and systems are wideband in nature, need to include all harmonics of the clock, and require a relatively lower Q. Otherwise, they would selectively amplify a specific frequency content. Such is the case for inductively loaded inverters, multiplexers, and selector circuits operating at the 20–40 Gbps range [19, 20]. Fundamentally, inductors help realize higher impedance at a specific frequency of interest, and hence, they reduce the power consumption of these circuits. Any use of inductor in high speed digital system should be strictly observed for area considerations. However, in practice, the resonating impedance cannot be increased arbitrarily. The inductor has its own parasitic components itself, and usually the impedance realized is somewhere in the 200–400-Ω range.

1.8.6.3.6 Mathematical Analysis of Inductors. The search for a solution to the electromagnetic fields resulting from an inductor has been an interesting area of research for a long time. The analysis becomes complicated because of the wide variation in inductor geometries. Current generation inductors use turn spacing to be much smaller than the turn thickness and width, and the solution can be obtained in a two-dimensional (2D) current distribution, and a mesh can be obtained using Kirchoff's laws to solve them. To compute the various components of an inductor in a lumped element representation, the current and charge distributions need to be computed. These distributions vary across the cross section of the inductor turns, and they must be obtained by "mesh"ing the inductor using a minimum grid. Using numerical computation techniques (contributions from individual mesh points with appropriate weight factor), the charge and current distributions can be shown to have peaks at the edges of turn cross sections. Once these steady-state current and charge distributions are obtained, they can be used in conjunction with some standard inductor solver configuration such as Greenhouse, and so on in order to obtain a full solution of the inductor. Although the computation is interesting in nature, because of the coverage and focus of the book, we encourage readers to refer to [27–30].

Inductor Q, however, needs to be observed at the desired frequency. Inductors do not include any voltage variable component in any of their subcomponents, and hence, they are extremely linear in nature. However, because of lithographic limitations, the metal resistances may vary significantly, which leads to changes in the Q factor. It should be noted at this stage that the inductance does not vary w.r.t. process corners, as it is related to flux linkage, which, in turn is dependent on the number of turns and outer diameter.

Inductive impedance is usually given by 6 Ω per nH per GHz. This number can be used in calculating the impedance at commonly used wireless center frequencies such as 0.9 G, 2.4 G, and 5.2 G.

At lower GHz ranges, on-chip spiral inductors provide much lower Q because of low-frequency operations (and often a bondwire inductance is preferred). To realize the same impedance, a lower frequency inductor also needs to have a higher inductance for the same Q factor. This high inductance leads to significant area consumption in the case of on-chip inductors. Such a large inductor would also exhibit more parasitic capacitance values, which leads to tuning range limitations in the case of tank circuits in amplifiers and VCOs. At the same time, they are susceptible to creating electromagnetic cross-talk.

1.8.6.3.7 Active Inductors. An alternative arrangement can be obtained using analog circuit techniques using a "gyrator-C" approach, as shown in Figure 1.24. A capacitance connected at the interfacing node of two antiparallel transconductance stages provides an inductive impedance at the other end [shown in Figure 1.24(a)]. A similar arrangement is possible in a feedback topology, where g_m of transistor M_P is multiplied by the gain of the amplifier. The inductance can be properly controlled, and it must be observed that the inductance in all of these configurations is a ratio of capacitance to transconductance; hence, the lower the transconductance (implying lower power), the higher is the inductance. Active inductors are attractive at lower RF frequencies, as a passive counterpart would not only consume more area, but it would also exhibit a poorer Q factor because of the lower frequency of operation.

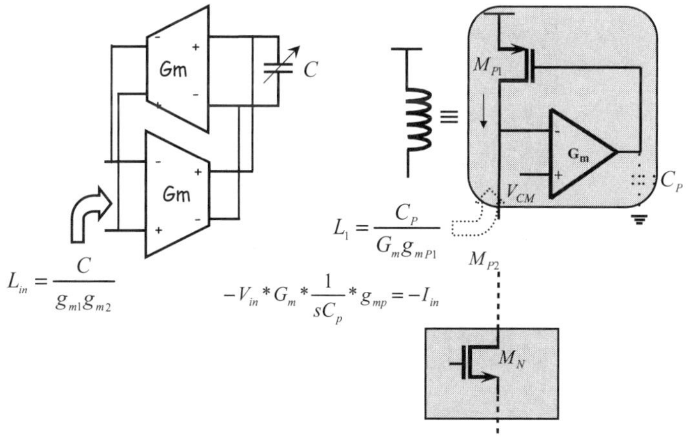

(a) (b)

Figure 1.24. (a) Gyrator-C configuration and (b) active inductor implementation.

Fundamental advantages of active inductors over the passive inductors are as follows:

1. Lower area consumption
2. No electromagnetic cross-talk

However, active inductors provide the following performance degradations as well:

1. Additional power consumption
2. Linearity degradations caused by more active components
3. Noise degradations
4. May require higher supply voltages for operation, depending on the configuration

However, as the inductance is a ratio of two dissimilar quantities, it would vary over process corners, unlike the passive inductors. Some calibration circuits would need to be used to guarantee the process-invariant behavior of the inductance.

1.8.6.4 Transformers. With the illustrations on inductors, we now illustrate the usage of transformers. A transformer is commonly used in an RF circuit to obtain either current or voltage gains. Transformers are DC isolated, and several configurations are possible as shown in Figure 1.25. The secondary terminals can use a different DC bias, and they are suitable in interfacing two fully differential circuits operating at different DC common mode levels. The DC isolation also makes them attractive for use in a feedback network feedback network without the need of DC blocking capacitor [32]. Voltage and current gains of a transformer are related by the

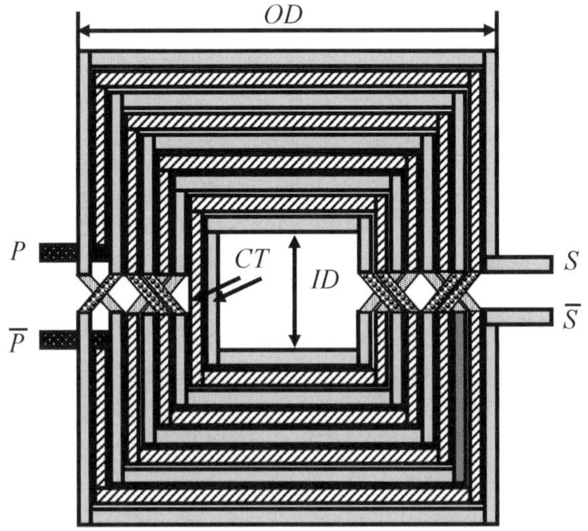

Figure 1.25. Physical construction of a transformer.

turns ratio (N) of transformers, whereas the impedance is related by N^2. Thus, it can act as an impedance buffer to reduce the effect of capacitive parasitics. The primary and secondary terminals of transformers provide inductive impedance, which can be resonated with various capacitances, such as device output capacitance, and so on. Another use of transformer is in the single-ended to differential transformation (balun) with current or voltage gains, and this operation can be performed with minimum imbalance between the two differential terminals. This implementation is important in the front end, as the LNA takes an input single-ended signal, which should be transformed to differential as early as possible with lowest amount of imbalance. Circuit elements based on active components tend to provide more imbalance than their passive counterparts.

The key element to successful implementation of transformer is the coupling coefficient of flux linkage from primary to secondary, denoted by $K(<1)$. Also, they are area inefficient. An efficient way of implementing transformers is to use an autotransformer configuration. A differentiator in this configuration is the fact that the terminals are not isolated in terms of DC voltages. However, this can be easily obtained from a simple inductor structure, which takes advantage of mutual coupling between the terminals.

At further higher frequencies (well into the mmW regime), inductors assume a distributed component model, rather than the standard lumped element model that we have been illustrating until now. In that situation, inductors are mostly microstrip line components. If we were to unwind the spiral inductor, it would certainly be a long line. Spiral winding helps in improving mutual coupling and increases inductance in a given area under considerations. At the same time, the inductance per unit length is lower because of interwinding capacitances. However, at frequencies comparable with the wavelengths of the operating frequencies, these capacitances would be prohibitive, and the physical dimension of the inductor would be comparable with the wavelength, so we can obtain larger inductance per unit length without the area penalty.

Another interesting component at mmW frequency range is called a "stub." A stub is essentially a transmission line used to realize "short" or "open" impedances in a frequency dependent manner. They also provide characteristic impedances, which is repeatable with frequency, implying that a stub providing an "open" circuit impedance would provide the same impedance (theoretically) at frequencies f, $2f$, $3f$, and so on, while providing a "short" circuit impedance at $1.5f$, $2.5f$, and so on.

1.8.7 Evaluation Testbenches

Prior to designing circuits, it is important to obtain some useful information about the technology components. These testbenches provide useful information about the technology to be used and its performance metrics. The following testbenches are absolutely essential for circuit design:

1. Transistor f_T, f_{MAX}, ON resistance, OFF capacitance, minimum noise factor
2. Parasitic capacitance of passive components (resistance, capacitance, inductance)

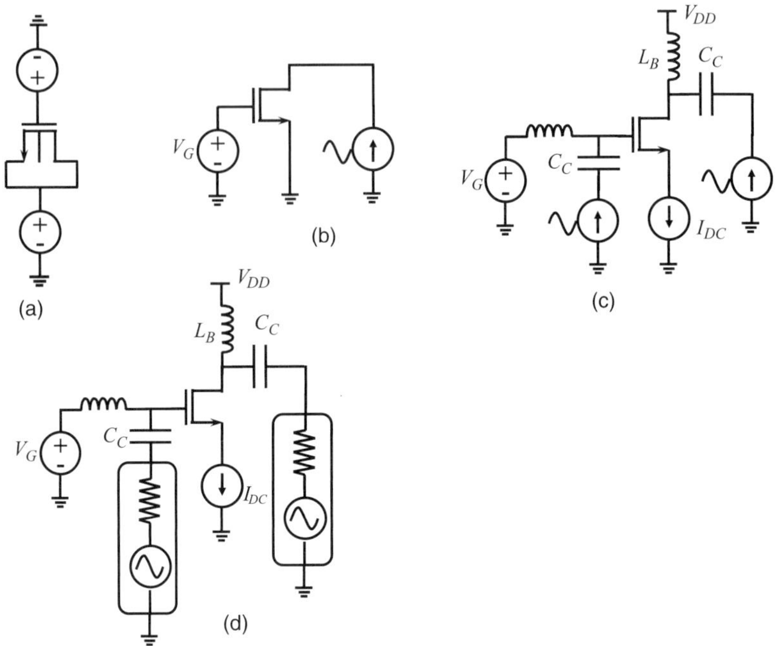

Figure 1.26. Various testbenches to characterize transistors.

3. Impedance testbenches
4. Nonlinearity testbenches

Figure 1.26 illustrates the various testbenches used to characterize the transistor. In fact various types of transistors in the technology include high speed transistor, high breakdown voltage transistor, lightly doped transistor, thick gate oxide transistor, unadjusted threshold voltage (native) transistor, and so on. f_T, f_{MAX} values should be obtained for the actual dimension of the transistor used in the circuit. The ON resistance of the MOS transistor and the OFF capacitance of the MOS transistors are important circuit considerations. ON resistance is pertinent in the cases of (1) passive mixer and (2) MOS switches in a binary capacitor array. In the case of a passive mixer circuit, ON resistance indicates the signal loss and the achievable Q factor of the mixer connected with a coupling capacitor. In the case of a switched capacitor binary array, it implies the Q factor of switchable components. ON resistance depends on the gate to source voltage, threshold voltage, and the transistor geometry. As the voltage at source terminal changes, the ON resistance would also vary accordingly, and signal-dependent variation of the ON resistance would provide important design insight. The OFF capacitance of a MOS transistor depends on the transistor geometry and also varies with voltage values. The OFF capacitance is a measure of the junction capacitances (drain to bulk or source to bulk). It is illustrated in the testbench (a)–(b). One can set the input voltage to different values to obtain the on resistance and off capacitance.

Testbench (c) can be used to set the gate voltage and bias currents separately, and the transistor's current gain, output impedance (g_m, g_{ds}), and so on, can be obtained. The biasing inductors and capacitors are of large values in order to provide high impedance to the DC biasing source and low impedance to the AC, respectively. The magnitude as well as the Q factor of these impedances is important for circuit design. It is usually obtained by separating real and imaginary parts of the input impedance obtained by simulating it with a 1-A current source. Testbench (d) can be used to estimate noise factor, and so on using 50 Ω ports at the input and output. This may provide insight to various types of microwave parameters for designing amplifiers and so on. Current sources are used to measure impedance, as current sources provide high output impedance, and minimum loading to the circuit, whereas voltage sources load the circuit by providing zero impedance. The minimum noise factor indicates the minimum achievable noise figure of a MOS transistor (with appropriate geometry and current consumption). An amplifier's noise figure is given by a parabolic curve w. r.t. impedance states. The minimum noise factor provides the lowest possible noise figure for specific transistor geometry at a given bias current. In reality, the minimum noise factor is never achieved because of the consideration of simultaneous noise and input return loss for LNAs.

Figure 1.27 illustrates component-level testbenches, which includes resistance, capacitance, inductance, and so on. The parasitic element of a resistance determines bandwidth shrinkage, whereas the parasitic capacitance of a capacitor (known as the "bottom plate capacitor") indicates a shrinkage in tuning range of integrated narrow-band circuits (VCOs, filters). Assuming parasitic capacitance to be evenly distributed across the device terminals, a two-terminal device can be shorted across the functional terminals (the two terminals of a resistor, inductor, or capacitor). Parasitic

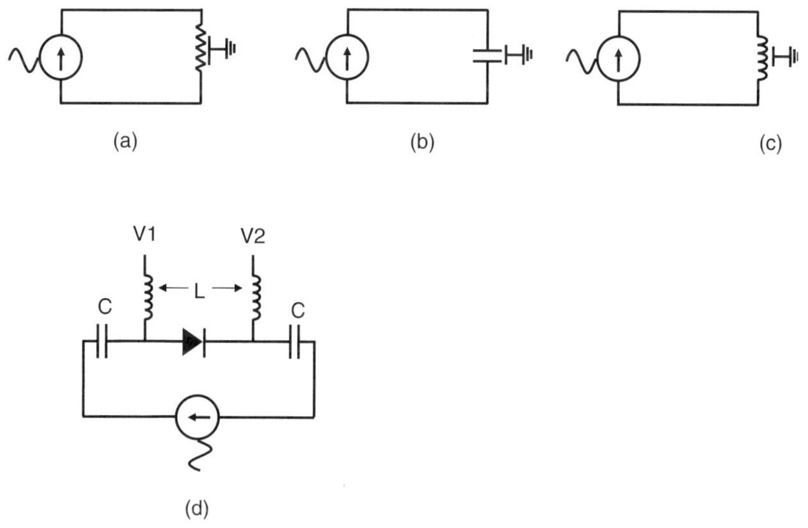

Figure 1.27. Various testbenches to characterize technology components as well as parasitic elements of (a) resistor, (b) capacitor, (c) inductor, and (d) diode.

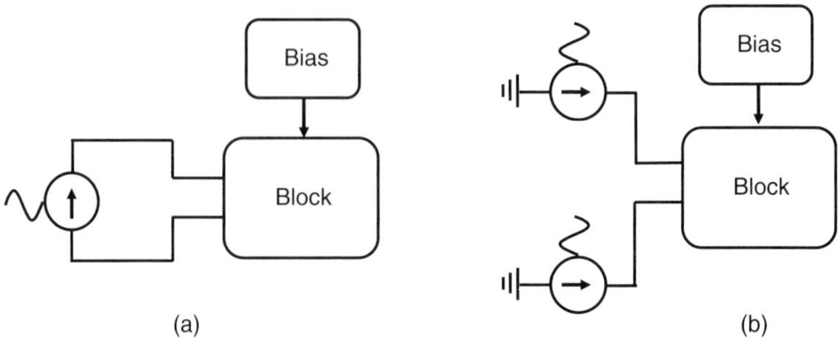

Figure 1.28. Various testbenches to characterize impedance of building blocks: (a) differential mode and (b) common mode.

capacitances indicate loading at RF, and they determines the maximum usable frequency for a component (self-resonating frequency for an inductor). Like any component, along with the parasitic capacitance value, the Q factor of the capacitance is important for circuit design considerations. In the case of varactors, capacitance versus voltage characteristics (known as C–V curves) are important to obtain information on the maximum and minimum capacitance values, the Q factor, as well as on potential AM-to-PM conversion. Figure 1.27(a)–(c) illustrates the resistance, capacitance, inductance, and so on, whereas (d) illustrates the input impedance determination of a diode with bias voltage applied as (V_1–V_2). To obtain the impedance, this voltage can be swept from negative to positive, and the voltage variation of the impedance, as well as its Q factor, can be obtained. In the figure, both the inductor and the capacitor are of large values to provide DC bias voltage and AC short without loading each other. This is a fully differential characterization, which is more accurate in terms of the voltage variation of parasitic components at each port. Similar approach can be taken for the other voltage variable components (e.g., varactors).

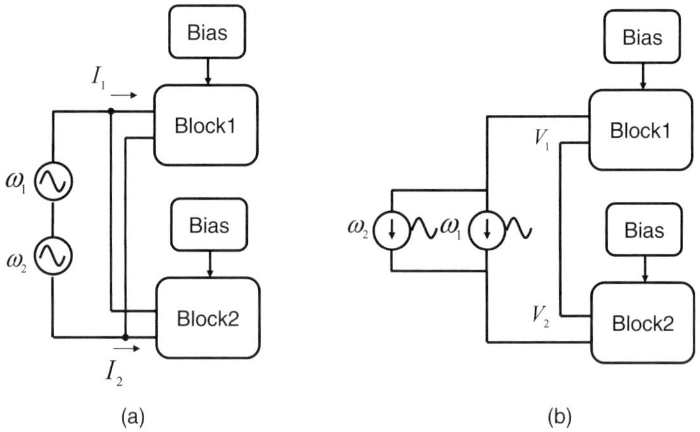

Figure 1.29. Testbenches to characterize nonlinearity: (a) with a two-tone voltage input and (b) with a two-tone current input.

In integrated circuit implementations, signal paths, as well as signal generation paths are usually differential in nature to reduce the common mode noise perturbation. Hence, the common mode and differential mode input impedances, as well as their Q factors, would need to be obtained for a circuit design. For an active circuit, proper bias condition must be provided to characterize the input and output impedances. In an integrated mixed signal environment, our aim is to amplify differential signals and reject common mode signals as much as possible. In this regard, the circuit interfaces are designed in a way such that the common mode path has lowest possible Q. If this is not the case, any finite common mode current resulting from mismatched differential circuit would lead to common mode instability. Figure 1.28 shows two testbenches to obtain differential and common mode impedances.

In a series/parallel combination of two or more components, it is imperative to know which component provides linearity limitation. Examples of these would be a parallel combination of three functional blocks in a highly linear system and so on. Such information can be obtained by providing two tones in the voltage domain (a series combination of two voltage sources) across the parallel network, and observing intermodulation terms in the individual branch currents. Two estimate linearity, two mechanisms can be used: (1) provide a two-tone voltage input (series combination) to a parallel combination of two blocks, and (2) provide a two-tone current input (parallel combination) to a series combination of two blocks. Usually, for intermodulation tests, the circuits are characterized at their linear operation range. Nonlinearities are embedded as part of signal-dependent behavior of the impedance (input/output). With two-tone voltage input as stimuli, the discrete Fourier transform (DFT) of input currents are performed to understand the block that is most limiting in terms of linearity. While performing a linearity simulation, the simulator options must be specified for the desired accuracy. It is usually obtained by providing a specific timestep of analysis in a time domain simulation. Figure 1.29 shows testbench to evaluate linearity performance using two tone voltage and current stimuli.

1.9 KEY CIRCUIT TOPOLOGIES

In this section, we will illustrate the fundamental circuit basis functions. These cores can be implemented using any type of transistors and can be used anywhere in a complicated system.

1.9.1 Differential Circuits

Differential circuits consume twice the area compared with single-ended counterparts, but they are used extensively in integrated circuits because of several key advantages such as follows:

1. Rejection of common mode signals (substrate noise, even order terms).
2. No additional hardware to create 180° phase-shifted version (inversion).

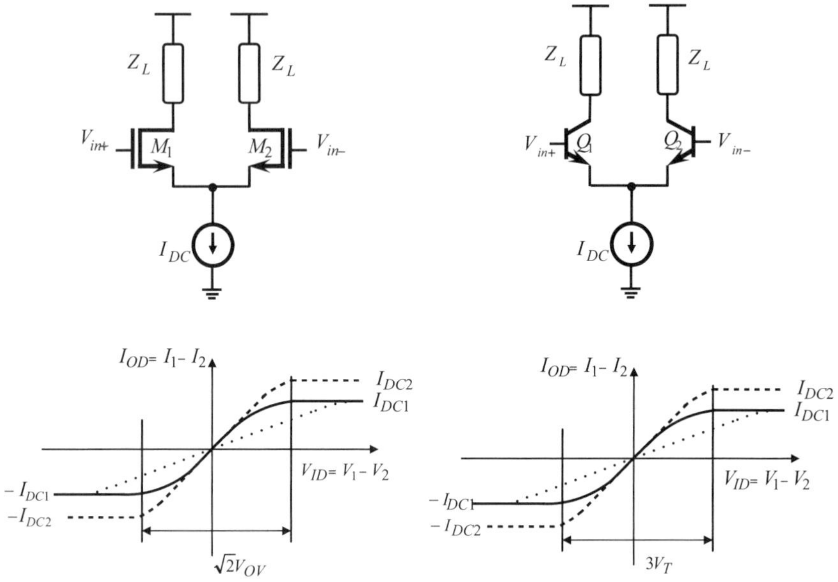

Figure 1.30. Basic differential pair topology.

3. Insensitivity to package parasitics connected at common mode terminals.
4. Increasing speed of digital circuits, and possibility of using lower supply voltage in I/O rails.

First in this category is the differential input transconductor stage, which may use bipolar or MOS devices at its input, as shown in Figure 1.30. We assume that the tail current is kept constant, and there is no voltage headroom problem across the transistors (they operate in active and saturation regions in the cases of bipolar and MOS, respectively, $V_{CE} > V_{CE,SAT}$, and $V_{DS} > V_{DS,SAT}$). Such configurations are also referred to as "source coupled" stages. In terms of large signal characteristics, the current/voltage relationships are given as follows:

$$\frac{I_{c1}}{I_{c2}} = e^{V_{id}/V_T}$$

$$I_{TAIL} = \frac{1}{\alpha_F}(I_{c1} + I_{c2})$$

$$V_{od} = V_{c1} - V_{c2} = \alpha_F I_{TAIL} R_L \tanh\left(\frac{-V_{id}}{2V_T}\right)$$

Thus, the input–output DC transfer characteristics in the case of a bipolar diff-pair is dependent on the thermal voltage $V_T = \frac{kT}{q}$. As can be observed from the plot shown in Figure 1.30, this stage is "hard-switched" when the differential input voltage exceeds $3V_T = 78$ mV. The slope around the zero point in the transfer curve plane determines the small signal amplification. The amount of deviations of this slope from a constant value (a perfectly linear curve would provide a constant derivative) determines small signal linearity (observed by providing two tones at the input). Large signal linearity is usually determined by signal clipping (essentially any arm of the differential pair running out of current). The derivative of the I–V characteristics determines the transconductance of the differential pair stage.

Hard-switched differential pairs behave as limiters in the case of driving mixers and so on, in order to reduce process and temperature variation of the input signal waveform. However, they are usually nonlinear in nature, and they use a small linear range. These problems can be alleviated by providing resistive feedback to linearize the input stage without sacrificing some headroom. A resistive degeneration linearizes the input stage.

A similar configuration can be performed using any type of transistors (MOS/MES FET, etc.). For the sake of simplicity, we will assume square law I–V characteristics to obtain the trends and insights in circuit design. In deep submicron technologies, the I–V curves are much different from a square law representation. From the individual gate-source voltages, we obtain

$$V_{id} = \frac{\sqrt{I_{d1}} - \sqrt{I_{d2}}}{\sqrt{\frac{K_N}{2}\left(\frac{W}{L}\right)}}$$

$$I_{TAIL} = I_{d1} + I_{d2}$$

$$I_{d1,2} = \frac{I_{TAIL}}{2} \pm \frac{K_N}{4}\left(\frac{W}{L}\right) V_{id} \sqrt{\frac{4I_{TAIL}}{K_N(W/L)} - V_{id}^2}$$

Both transistors operate in the saturation region when $|V_{id}| \leq \sqrt{\frac{2I_{TAIL}}{K_N(W/L)}} \Rightarrow |V_{id}| \leq \sqrt{2}V_{OD}$. Various curve families are shown in Figure 1.30. Once again, the linear range can be extended by using resistors at the sources of the transistors. When resistors are prohibitive in terms of headroom and noise, inductors can be used. As the differential input voltage range for linear operation is higher, MOS differential stages usually offer higher linearity compared with their bipolar counterparts with no degeneration added. With degeneration, these considerations would differ. In terms of a "hard-switched" differential pair (driving mixer switches, etc.), it can be observed that the bipolar transistors would require only a 78-mV differential signal swing in order to switch, whereas MOS can easily require 100–200 mV. Lowering of this voltage leads to larger transistor dimension and to increased loading on the signal generation network. Thus, bipolars prove to be attractive in these cases.

Apart from headroom, one needs to be careful about the pole created by the RC equivalent circuit designed by the degeneration resistor and the source capacitor. In high-frequency circuits, inductive degeneration is used, and this may resonate with the source capacitance as well. Careful attention should be provided in order to avoid such scenarios.

Tail impedance plays a critical role in differential circuits, as the common mode rejection performance is heavily dependent on its value. The extent of rejection experienced by unwanted signals is determined by the common-mode rejection ratio

(CMRR), and it is given by CMRR $= 1 + 2g_m R_T$. In reality, this is obtained by a transistor (part of a current mirror) or a small resistor. In a fully differential amplifier configuration, the output capacitance is not important, as it experiences an AC ground due to input signal symmetry. However, in reality, imbalance exists because of asymmetry in the device layout, input duty cycle error, and it is desired that the output capacitance be as small as possible.

Voltage offset for differential amplifiers is always referred to the input terminals. This is similar to small signal noise, and the overall mismatch is computed in terms of components in the amplifier stages, and then it is divided by the amplifier gain. In computation of offset voltages, we assume the components to be slightly mismatched and apply a differential voltage at the input to ensure that the output voltage is zero. Under these conditions, a MOS differential stage can be solved to obtain

$$V_{OS} = \Delta V_t - \frac{V_{OD}}{2}\left(\frac{\Delta R_L}{R_L} + \frac{\Delta W/L}{W/L}\right)$$

1.9.2 Translinear Circuits

The second widely used variety is translinear circuits, which are illustrated in Figure 1.31. A simple illustration of this principle can be provided by considering bipolar transistors, in which an equal number of clockwise and anti-clockwise transistors is connected from a reference point to the ground.

As the base-emitter voltages of bipolar transistors are given as a logarithmic function of the collector currents, summation of the clockwise and the anti-clockwise path V_{be}s can be equaled to provide

$$\prod_{CCW} I_i I_j = \prod_{CW} I_i I_j$$

In actual implementations, these currents would be a sum of DC biasing current and an AC current, providing versatile functions such as multiplication, division, and so on. Translinear circuits can also be realized using MOS transistors, and in modern technologies, the subthreshold operation of MOS transistors becomes analogous to bipolar transistors because of their exponential characteristics.

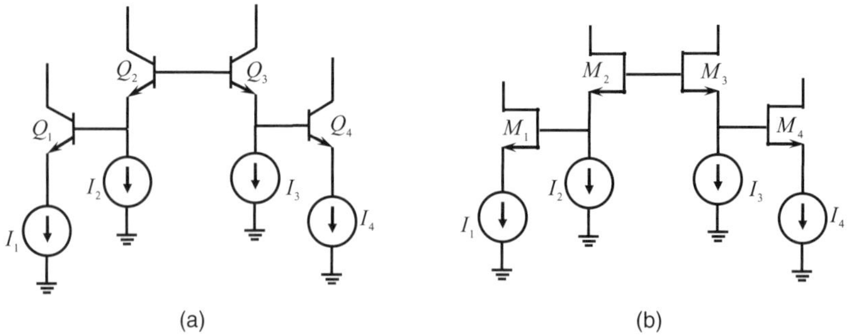

Figure 1.31. Translinear circuits using (a) bipolar and (b) MOS.

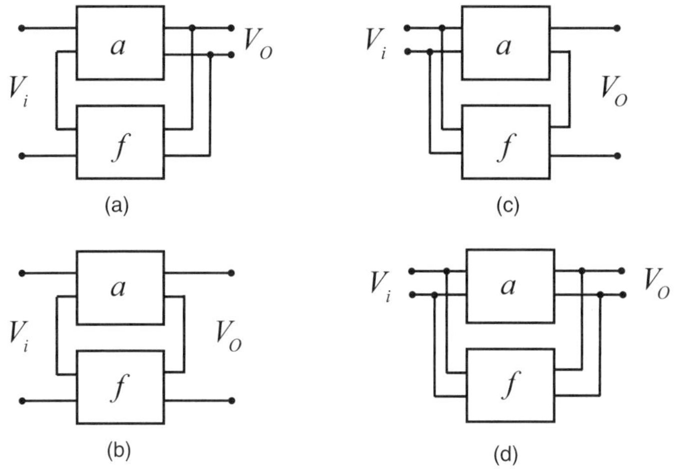

Figure 1.32. Feedback configurations: (a) series-shunt, (b) series-series, (c) Shunt-series, and (d) Shunt–shunt.

1.9.3 Feedback Circuits

Feedback plays an important role in integrated circuits and systems. Negative feedback is commonly used in linearizing amplifiers, reducing input impedance value, with the compromise of noise addition in the circuit. Positive feedback is essential in determining oscillation start-up conditions. Depending on the placement of the feedback component, their nomenclature is followed as follows: (1) Series–shunt, (2) shunt–shunt, (3) shunt–series, and (4) series–series. These topologies use voltage or current as sampling and feedback variables as appropriate. These configurations are illustrated in Figure 1.32. Negative feedback is used extensively in circuit linearization, however, they always contribute noise to the main circuit. Positive feedback has been used extensively in oscillator circuits.

1.9.3.1 Feedback in OP-AMPs. We will pay close attention to feedback loop of an OP-Amp, which is a shunt–shunt topology. In this case, the output current is sampled through the feedback resistor R_F and injected to the input terminal. The feedback resistor reduces the gain of the OP-Amp, and an open loop gain of 60 dB reflects to a close-loop gain of R_F/R_s, where R_S is the source resistance. It can be observed that although the OP-Amp consumes DC current, the gain is set by the resistor ratios! The DC current in the OP-Amp branches ensures that it can provide the large signal current swings. The input impedance is given by $Z_{in} = \frac{R_F}{1-A_v}$, where A_v denotes the open-loop voltage gain of the OP-Amp. If we were to assume the open-loop response of the OP-Amp to have a single pole response, then $A_v = \frac{A_{v0}}{1+s/\omega_P}$, which leads to inductive behavior in the input impedance of the OP-Amp.

1.9.3.2 Virtual Ground. The presence of low input impedance is referred to as "virtual ground." A low impedance implies a "short" between the two terminals in their electrical signal equivalence but not physically. The OP-Amp is capable of sinking any amount of current as required. In reality, the current does not flow to ground, but it flows through the feedback impedance path. Thus, we have the name "virtual ground."

1.9.3.3 Miller's Theorem. Analysis of electrical circuits, where input and output terminals are coupled through an impedance element, can be performed using Miller's theorem. It is a generalized analysis, and it can be performed with any impedance connected in the feedback path. Using Miller's theorem, we establish one-to-one equivalence between the two circuits, as illustrated in Figure 1.33. Using Miller's theorem, we aim to decouple the input output–connection by using mathematical representation. This approach helps in the analysis of the circuit's input and output nodes. We first solve the two systems w.r.t. nodal equations, and we use voltage equivalence to obtain

$$Z_{O,M} = \frac{Z_f - Z_O/|A_v|}{1 + 1/|A_v|} \approx \frac{Z_f}{1 + 1/|A_v|}$$

$$Z_{O,M} = \frac{1 + Z_O/Z_f}{1 + |A_v| + Z_O/Z_f(1 - |A_v|)} \approx \frac{Z_f}{1 + |A_v|}$$

The details of this derivation are provided in Appendix A(7).

In the above discussion, we have assumed that the feedback impedance is larger compared with the series output impedance, which is a reasonable assumption. Conceptually, Norton's equivalent circuit can also be obtained, and the amplifier

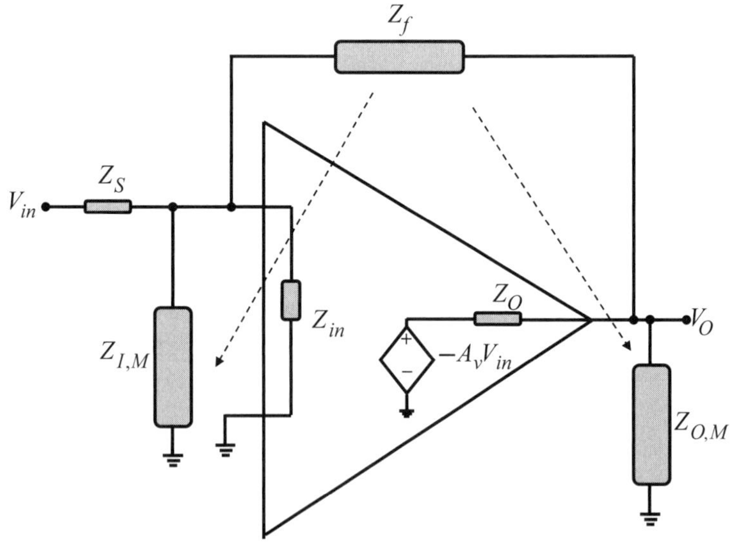

Figure 1.33. Illustration of virtual ground and Miller's impedance.

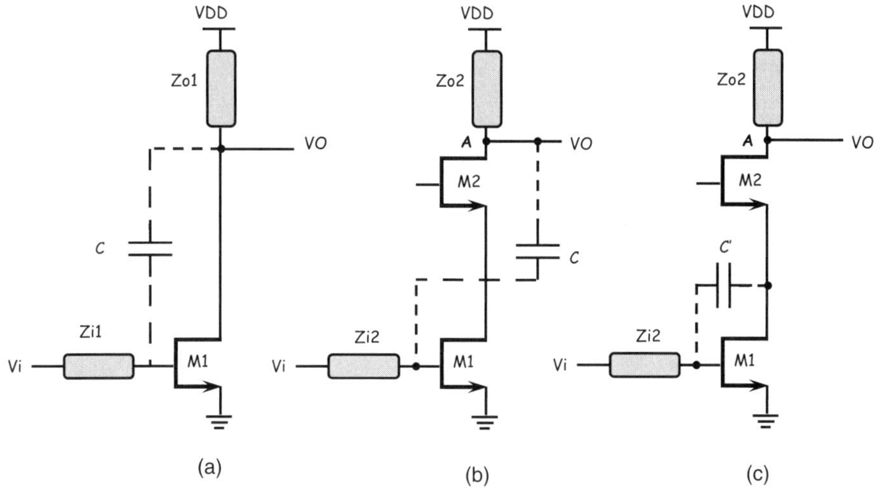

Figure 1.34. Miller effect consideration leading to cascode topology.

can be represented by the current gain. Miller's theorem provides important insights into several feedback circuits. The previous section on OP-Amp's virtual ground formulation was also an illustration of Miller's theorem. It can be observed that, if the feedback impedance is capacitive, it would lead to an equivalent input capacitance equal to the original capacitor multiplied by the open-loop gain of the amplifier. At RF frequencies, this leads to significant bandwidth reduction. Even if a capacitor is not connected deliberately, the parasitic capacitance of the device would provide the same effect. This is shown in Figure 1.34.

1.9.4 Cascode Circuits

To alleviate this problem, a cascode topology is often adopted. The purpose of a cascode transistor is to isolate the input and output networks, as shown in Figure 1.34. Assuming a voltage gain of 24 dB (linear factor of 16), without using cascode, the effective capacitance seen at the input is 16C. Assuming the same geometry of the main and the cascode devices, the voltage gain resulting from the main device is close to unity (ratio of transconductances). According to Miller's theorem, the input referred capacitance is 2C, which reduces the loading by 8 times. For this reason, cascoding is almost always used in RF circuits. Several other advantages exist apart from (1) bandwidth enhancement and (2) reverse isolation. As the input and output terminals are decoupled, cascoding helps in separate optimization of input and output matching, and it achieves high output impedance to provide high voltage gain. However, compared with a non-cascoded variant, a cascode device consumes more headroom, which leads to reduced signal swing. In the non-cascoded variant, the Miller capacitor reduces the output impedance, but the output can swing higher at the expense of increase current consumption.

1.9.5 Common Source, Common Gate, and Common Drain Stages

We will now cover a few basic circuit topologies, which can be easily analyzed in the analog domain and have been used as part of circuits in any building blocks. These topologies are known as (1) common source, (2) common gate, and (3) common drain (or source follower). They are named according to the terminal that is common in the input and the output networks. For the case of the common source amplifier, the source terminal is "common" between input and output and so on. Important performance parameters associated with this stage are as follows:

1. Gain, linearity, and noise
2. Input common mode range
3. Output common mode range
4. Current consumption and bandwidth
5. Signal handling capability

Input common mode implies the amount of signal swing at the input. It also implies the possibility of direct coupling among various blocks. A low common mode ($<V_{DD}/2$) is suitable for a PMOS input stage, whereas a high common mode is suitable for the NMOS-type input stage. This consideration is important at low frequency, as the stages cannot be capacitively coupled because of the area constraints. To interface two stages, we either require a level shifter or complementary stages (NMOS-based stage interfacing PMOS and so on). Gain bandwidth product is an important performance criteria for such amplifiers and is dependent on the bias current and load impedance. The signal handling capability is denoted by $V_{DSAT}$and by the bias current of the amplifier. Higher V_{DSAT} implies higher linearity. Linearity can also be improved by using degeneration resistors (or inductors at high frequencies), which provide negative feedback around the transconductor stage, reducing the overall gain. However, any capacitive component in the degeneration path leads to formation of negative impedance at the input, which leads to stability issues. The multiplication factor is $\frac{g_m}{j\omega C_{gs}}$, and an inductor leads to real input impedance, a resistor leads to capacitive input impedance, and a capacitor leads to negative impedance (causing stability concerns!). Hence, any parasitic capacitance at the source terminals needs careful attention in the design phase.

An advantage of using common source type topology is the amplification in both voltage and current domain. Other topologies, such as common gate, provide voltage amplification at the output (dependent on the load) but not current amplification. The input impedance of the common gate stage illustrated in Figure 1.35 is given by $Z_{in} = \frac{r_{dsL} + r_{ds1}}{1 + g_{m1} r_{ds1}}$. Thus, $Z_{in} \sim \frac{1}{g_{m1}}$, in the case of low load impedance, which is usually the case for RF circuits.

At lower frequencies, active loads are commonly used, and the input impedance can be $Z_{in} \sim \frac{2}{g_{m1}}$, assuming equal output impedance from the core transistor and the load. As the input impedance is lower, the voltage swing at the input is also reduced (leading to lower voltage variation of input capacitance). Current transfer is linear, which causes high linearity at this stage. The input impedance is wideband in nature, and this

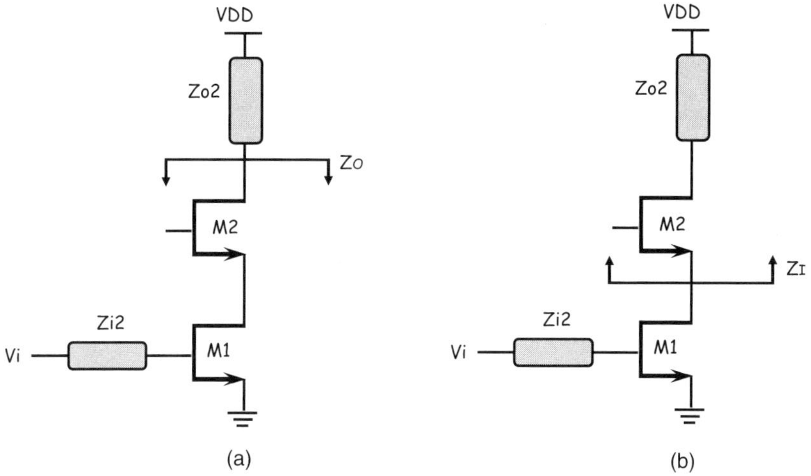

Figure 1.35. Impedance calculations.

can be optimized to 50 Ω in order to provide a power match. As the driving impedance is also 50 Ω, the minimum achievable noise figure is ~3 dB. Common source stages are usually preferred for LNAs, with simultaneous noise and power matching design considerations, as will be illustrated later.

Common drain or source/emitter followers (shown in Figure 1.36) are primiarily used as an impedance buffering stage, and they are capable of providing a voltage shift. In these cases, the source/emitter terminal voltage follows the gate/base, and no phase change occurs in terms of voltage transfer. However, the maximum voltage amplification out of this stage is usually lower than 1, and it is usually dependent on the g_m and the source/emitter impedances Z_s. It is desired that $g_m Z_s >> 1$. To meet this

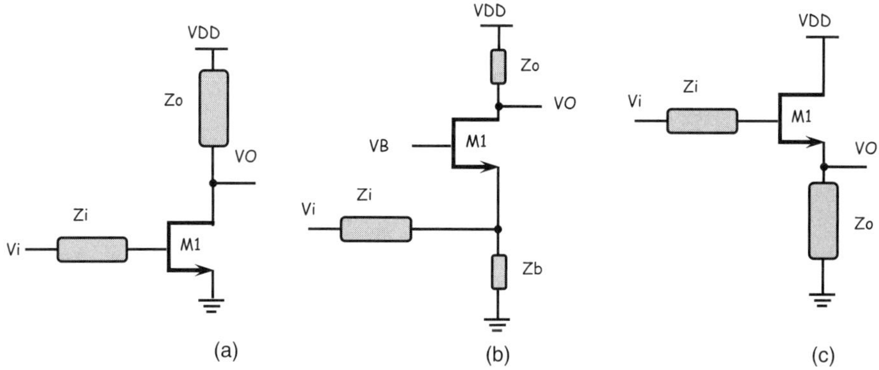

Figure 1.36. Common single transistor amplifier topologies: (a) CS, (b) CG, and (c) CD.

requirement, these stages can consume significant current. Also, because of body bias and other nonlinearity factors in V_t, the voltage gain is usually slightly lower than unity (usually 0.7–0.8) in UDSM CMOS nodes. These stages provide no voltage amplification, but they add noise, which leads to dynamic range degradations. However, they are effective in impedance buffering. As the voltage gain $A_v \approx 1$, the output capacitance is reflected back to the input by an amount $C_L(1-A_v)$, which causes a small impedance referred to at the input. The input impedance of these amplifiers is high, so they are perfectly suitable for driving large loads (off-chip) at low frequencies for measurement purposes. In this situation, enough gain is placed before them, so the resulting noise would not have much impact on system performance, and at the same time, the output capacitance of 5 pF would appear as 1 pF (with a voltage gain of 0.8) at about 1 Mhz, providing 160 kΩ, which would imply much reduced levels of loading. Similarly, their low output impedance suggests almost perfect voltage transfer to the subsequent driven networks.

1.9.6 Folded Cascode Topology

In low-voltage circuits, a configuration often known as "folded-cascode" is used to meet headroom requirements, as shown in Figure 1.37. In this configuration, current is "folded" through a high impedance to the delivering impedance. DC currents in these branches should be able to withstand the AC current swing.

Any cascode stage boosts the output impedance by the factor $g_m r_{ds}$, which leads to an output impedance of $g_m r_{ds} R$. In a stacked transistor configuration, the output

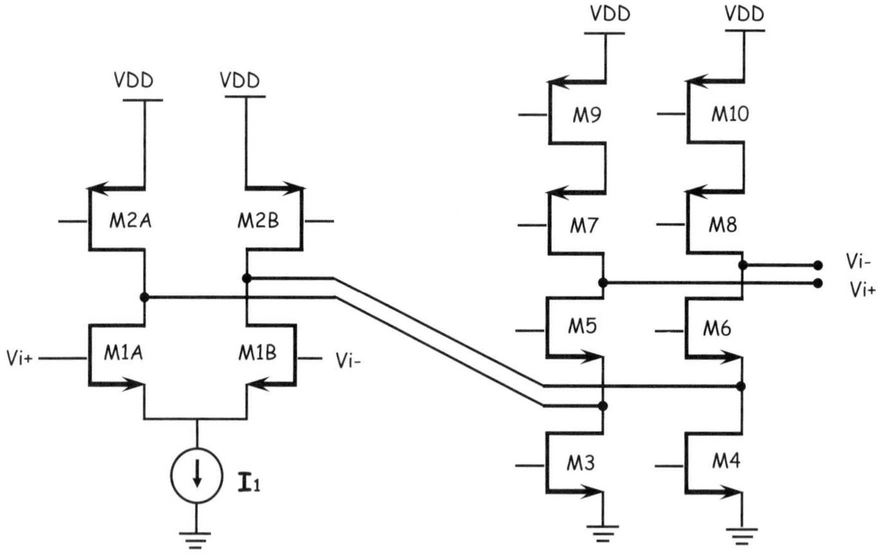

(a)

Figure 1.37. NMOS input folded cascode circuit configuration.

impedance of a transistor can be boosted as well, which results in the to $g_m r_{ds} r_{ds2}$. In the case of a bipolar transistor stack, the output impedance is βr_O. However, these impedances are only valid in the low-frequency regime of operation. At RF, these are shunted through output capacitances.

1.10 GAIN/LINEARITY/NOISE

Any building block in a mixed signal communication system can be attributed in terms of its gain, linearity, and noise contribution. Any circuit design for a certain functionality evolves around a systematic design compromise among these variables. Noise contribution can be considered in terms of a linear and nonlinear operation, as appropriate. These parameters are always considered in a cascaded system. The first block in the chain dominates in terms of noise contribution, whereas the subsequent stages determine the linearity. Hence, the linearity is usually associated with the last stage in a cascaded chain. Two types of linearity can be attributed: (1) small signal and (2) large signal. Small-signal, linearity is determined by the slope of the DC transfer characteristics near the origin whereas large-signal behavior is usually associated with the clipping behavior of signals, in an amplifier stage. Small-signal linearity terms are usually attributed in terms of two-tone inputs with closely spaced frequencies, and they are attributed as IIP3, IIP2, and so on. IIP3 relates to nonlinear terms in the output because of cubic nonlinearity, whereas IIP2 results from second-order nonlinearity effects, and is a function of component mismatches, duty cycle errors, and so on. In practical systems, output power levels of third-order intermodulation products are obtained and plotted as a function of input power level. In a double logarithmic scale plot, IM3 has a slope of 3, whereas the fundamental power has a slope of 1. IIP3 is given as the value of input power where these two cross each other. Physically, nonlinearity implies the redistribution of total available power from fundamental component to spectral harmonics. IM2 has a slope of 2, and the corresponding intersection with fundamental power is attributed as IIP2. Linearity is also related to the out-of-band blockers, and they need to be filtered using resonating tanks or passive filtering techniques as the signal propagates through the receiver chain. It must also be noted that while referring to IIP2 or IIP3 or any IIP products in (dBm), the referring impedance should be $50\,\Omega$ (or $75\,\Omega$ in the case of video standards). Otherwise they should be represented in terms of voltage or current (dBV or dBI) as appropriate.

1.10.1 Noise and Intermodulation Tradeoff

It can be observed that the linearity of cascaded blocks depends heavily on the phase of the signals traversing through them. Because of the phase shifts in the cascaded building blocks, it is possible to obtain cancellation of intermodulation terms. They also experience phase rotations because of various phase-shifting combinations such as *RC*, *CR*, and so on. Any filtering of out-of-band blockers relaxes the out-of-band linearity requirements.

Although phase is important for intermodulation, in terms of addition/subtraction, noise always adds in an uncorrelated fashion. The noise of cascaded systems is determined by the individual noise factors (linear scale) and the gain of the preceeding stages. As can be observed, providing more gain in the first few stages reduces the noise figure, but the increased level of signal causes linearity degradation. Thus, for a given power consumption, noise and linearity always pose tradeoffs in system design.

1.10.2 Narrowband and Wideband Systems

In narrowband communication systems, noise contributes more in order to degrade SNR, compared with nonlinearity. However, this scenario changes in the case of wideband communication systems, especially using multicarrier modulation techniques. Instead of evaluating nonlinearity terms based on two input tones, now we can use three or more, which leads to formation of triple order beats [terms located at $(f_1 + f_2 - f_3)$, $(f_1 - f_2 + f_3)$, etc., in addition to formation of $(2f_1 - f_2)$ and $(2f_2 - f_1)$]. It can be easily observed [shown in Appendix A(8)] that the magnitudes of triple beats is 6 dB higher than the original IM3 terms. Hence, in wideband systems, the intermodulation floor rises faster compared with narrowband systems, which causes higher power consumption in the building blocks.

CONCLUSION

In this chapter, we have provided the readers with the fundamental concepts to understand the basic principles of communication systems and circuit design. Key analysis methods have been illustrated along with circuit simulators and system design parameters. We have tried to focus on the basis functions that occur in all communication circuits and systems. Although the technology platforms keep changing, and various communication technology standards keep evolving, fundamentals are applicable everywhere, and they can provide a basic tool for understanding the principles of design. In the subsequent chapters, we will apply these concepts to solve complicated systems.

REFERENCES

Communication Systems

[1] S. Haykins, *Communication Systems*, John Wiley and Sons, 1995.

[2] J.G. Proakis, *Digital Communications*, McGraw-Hill, 2nd edition, 1989.

[3] G.L. Stuber, *Principles of Mobile Communication*, 2nd edition, Kluwer Academic Publisher, 1996.

[4] B. Sklar, "A structured overview of digital communications-A tutorial review-part I," *IEEE Communications Magazine*, Vol. 21, No. 5, Aug 1983, pp. 4–17.

[5] B. Sklar, "A structured overview of digital communications-A tutorial review-part II," *IEEE Communications Magazine*, Vol. 21, No. 7, Oct 1983, pp. 6–21.

[6] William C.Y. Lee, *Mobile Cellular Telecommunications*, 2nd edition, McGraw-Hill, 1995.

Electromagnetics

[7] R.E. Collin, *Foundations for Microwave Engineering*, 2nd edition, John Wiley and Sons, 1992.

[8] D.M. Pozar, *Microwave Engineering*, 2nd edition, John Wiley and Sons, 1998.

[9] D.J. Griffiths, *Introduction to Electrodynamics*, 2nd edition, Prentice Hall, 1989.

[10] J.R. Reitz, F.J. Milford, and R.W. Christy, *Foundations of Electromagnetic Theory*, 3rd edition, Addison Wesley, 1980.

[11] R.F. Harrington, *Field Computation by Moment Methods*, Oxford University Press, 1993.

[12] S. Ramo, J.R. Whinnery, and T. Van Duzer, *Fields and Waves in Communication Electronics*, 3rd edition, John Wiley and Sons, 1994.

Basic Mathematical Analysis

[13] E. Kreyszig, *Advanced Engineering Mathematics*, 8th edition, John Wiley and Sons, 2001.

Digital/Mixed-Signal Circuit Design

[14] H. Taub and D. Schilling, *Digital Integrated Electronics*, McGraw-Hill, 1977.

[15] J.P. Uyemura, *Digital MOS Integrated Circuits*, Kluwer Academic Publisher, 1999.

[16] R.J. Baker, H.W. Li, and D.E. Boyce, *CMOS Circuit Design, Layout, and Simulation* volumes 1 and 2, Prentice Hall, 2002.

[17] P.E. Allen and D.R. Holdberg, *CMOS Analog Circuit Design*, 2nd edition, Oxford University Press, 2004.

[18] D.A. Johns and K. Martin, *Analog Integrated Circuit Design*, John Wiley and Sons, 1997.

[19] P.R. Gray, P.J. Hurst, S.H. Lewis, and R.G. Meyer, *Analysis and Design of Analog Integrated Circuits*, 4th edition, John Wiley and Sons, 2001.

[20] S. Gondi, J. Lee, D. Takeuchi, and B. Razavi, "A 10Gb/s CMOS adaptive equalizer for backplane applications," *IEEE International Solid State Circuits Conference*, Vol. 1, Feb 2005, pp. 328–601.

[21] T. Dickson, E. Laskin, I. Khalid, R. Beerkens, J. Xie, B. Karjica, and S. Voinigescu, "A 72Gb/s 2^{31}-1 PRBS generator in SiGe BiCMOS technology," *IEEE International Solid State Circuits Conference*, Vol. 1, Feb 2005, pp. 342–602.

Electronic Devices

[22] B.G. Streetman and S. Banerjee, *Solid State Electronic Devices*, 5th edition, Prentice Hall, 2000.

[23] E.O. Johnson, "Physical limitations on frequency and power parameters of transistors," *IRE International Convention Record*, Vol. 13, Part 5, Mar 1965, 27–34.

[24] K.-H. To, Y.-B. Park, T. Rainer, W. Brown, and M.W. Huang, "High frequency noise characteristics of RF MOSFETs in subthreshold region," *IEEE Radio Frequency Integrated Circuits (RFIC) Symposium*, June 2003, pp. 163–166.

[25] J.D. Cressler (editor), *Silicon Heterostructure Handbook – Materials, Fabrication, Devices, Circuits, and Applications of SiGe and Si Strained-Layer Epitaxy*, CRC Press, 2005.

[26] G. Baldwin, et al., "90 nm CMOS RF technology with 9.0V I/O capability for single chip radio," *IEEE VLSI Symposium*, 2003.

[27] J.Y. Yang, et al., "0.1 um RFCMOS on high resistivity substrates for system on chip applications," *IEEE IEDM*, 2002.

Inductors

[28] H.M. Greenhouse, "Design of planar rectangular microelectronic inductors," *IEEE Transactions on Parts, Hybrids, and Packaging*, Vol. PHP-10, No. 2, Jun 1974, pp. 101–109.

[29] E. Pettenpaul, H. Kapusta, A. Weisgerber, H. Mampe, J. Luginsland, and I. Wolff, "CAD models of lumped elements on GaAs up to 18 Ghz," *IEEE Transactions on Microwave Theory and Techniques*, Vol. 36, No. 2, Feb 1988.

[30] A.M. Niknejad and R.G. Meyer, "Analysis, design, and optimization of spiral inductors and transformers for Si RF ICs," *IEEE Journal of Solid State Circuits*, Vol. 33, No. 10, Oct 1998.

[31] A.M. Niknejad and R.G. Meyer, "Analysis of Eddy-current losses over conductive substrates with applications to monolithic inductors and transformers," *IEEE Transactions on Microwave Theory and Techniques*, Vol. 49, No. 3, Jan 2001.

[32] J.R. Long and M.A. Copeland, "The modeling, characterization and design of monolithic inductors for Silicon RF ICs," *IEEE Journal of Solid State Circuits*, Vol. 32, Mar 1997, pp. 357–369.

[33] D.J. Cassan and J.R. Long, "A 1-V transformer-feedback low noise amplifier for 5-Ghz wireless LAN in 0.18-um CMOS," *IEEE Journal of Solid State Circuits*, Vol. 38, No. 3, Mar 2003, pp. 427–435.

CHAPTER 2

Wireless Communication System Architectures

INTRODUCTION

In this chapter, we will illustrate the system architecture considerations for wireless communication systems. Decisions related to system architecture play a major role in power consumption, form factor, sensitivity, and selectivity of integrated wireless systems. Given a specific performance requirement, a system architect tries to optimize cost (form factor, etc.) and power consumption. These considerations also differ depending on whether the system is narrowband versus wideband. In some extreme cases, even noncoherent radios are used where simplicity is the prime requirement or ultra-low power is needed. Usually more emphasis is placed on receivers, as in the transmitter, one has good control over various parameters, such as modulation, out-of-band rejection, and so on. Sensitivity and selectivity are the key considerations for any wireless system architecture.

Radio architectures, and high-speed semiconductor processes, have attained a state of maturity over the years. Superheterodyne radios usually offer the highest fidelity, with the necessity of multiple off-chip components, which leads to high form factor. Architecture decisions are usually based on (1) frequency of operation, (2) modulation technology, and (3) semiconductor technology under consideration. As briefly discussed, the tasks of a radio front end consist of (1) selective low noise amplification, (2) rejection of unwanted blocker signals, and (3) frequency shifting (up/down), to enable complex signal processing at the lowest possible signal frequencies. Performing complicated signal processing at lower frequencies leads to signal

Advanced Integrated Communication Microsystems, edited by Joy Laskar, Sudipto Chakraborty, Manos Tentzeris, Franklin Bien, and Anh-Vu Pham

processing accuracy and lower power. Although such functionalities are classic in nature, this chapter aims to illustrate various practical considerations related to radio architecture design. The miniaturaization and integrability with complicated digital signal processing blocks have led to low intermediate frequency (IF) and direct conversion architectures. Current state-of-the-art consists of single-chip multifunction radios in a standard digital CMOS substrate. In general, wireless front ends are targeted to be lower power to enhance the battery life and talk time significantly.

2.1 FUNDAMENTAL CONSIDERATIONS

A system designer/architect's task begins with the understanding of frequency planning in a specific country under consideration. Every country has an allocated set of frequencies, which are designated for industrial, scientific, and medical usage and termed as "license-free" bands. An example of "band-allocation" is shown in Figure 2.1, which illustrates the spectrum allocation in the United States within the 1–10- GHz frequency band. Our discussion in this chapter is heavily focused on the coherent receiver schemes; i.e. a local clock (oscillators) is used in the receiver to process the received signal. Theoretically, noncoherent schemes can be used, and prove to be significantly lower power, with the compromise on sensitivity. In the context of high data rate systems, coherent systems are almost always used to provide superior sensitivity and a larger data rate. A category of wireless devices known as "short-range devices" (SRDs) operate at a very low data rate and low power.

2.1.1 Center Frequency, Modulation, and Process Technology

Selection of any radio architecture starts with the choice of center frequency and modulation technology (along with data rate). Any radio architecture implementation is a compromise between area (form factor) and power, which is determined by the

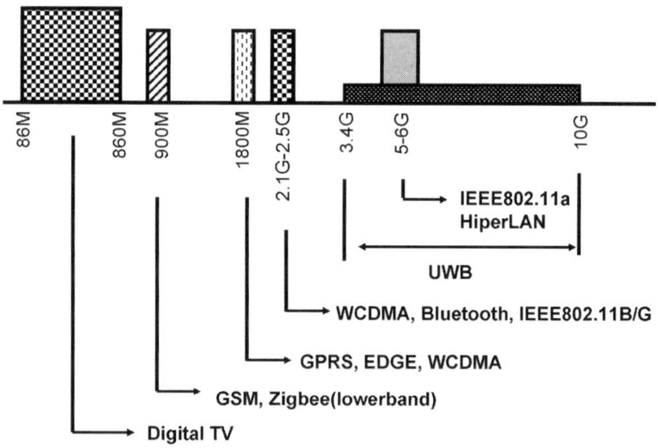

Figure 2.1. Wireless applications in the 1–10-GHz range and blockers.

semiconductor technology under consideration. Typically III–V semiconductors were preferred in the past because of their high electron mobility. However, the maturity of mainstream silicon-based technologies (BiCMOS and CMOS), and their scaling at deep submicron geometries, have attracted researchers and developers to consider a single-chip solution for the integrated system. Such integrations, however, tend to be heavily dominated by digital content and the roadmap to place all functionalities of a radio in a single chip.

The power consumption of a building block is proportional to the center frequency of operation. Fundamentally, high frequencies are suitable for (1) higher bandwidth, hence data rates, and (2) lower antenna size, hence smaller form factor. The obvious drawbacks of using a higher frequency are as follows: (1) reduced medium penetration, and (2) higher power consumption per signal processing function. In comparison, the lower frequencies are better suited for lower power and better medium penetration. The choice of center frequency also determines the blocker signals, which leads to dynamic range considerations in the front-end building blocks.

Bandwidth plays a major role in determining the power consumption and accuracy of continuous-time signal processing in the radio front ends. Bandwidth is determined by the symbol rate, which is determined by modulation, and raw data rate (QPSK maps two raw bits to one symbol and so on). A spectrally efficient modulation scheme maps as many bits as possible for one symbol, thus leading to maximum data transmission in the available bandwidth. The modulation scheme also determines the peak-to-average ratio (crest factor) of time domain signals, which directly implies the linearity and power consumption of the specific building block. A denser symbol constellation (more symbols in a given signal energy) will require more accurate signal processing in the front end. Often it is desirable that the modulated spectrum be "DC-free." This is preferred for most direct conversion architectures, which is made suitable by multi-carrier-type systems such as OFDM. In these systems, omission of information around DC tone does not lead to significant signal-to-noise ratio (SNR) degradation of the wireless system. Another DC-free modulation scheme is FSK. Direct sequence spread spectrum can also provide good immunity to DC impairments such as DC offset and 1/f noise, as will be discussed later.

Mainstream silicon processes (BiCMOS, CMOS) are commonly used to develop radio circuits. Silicon-based technologies are favored for their high integration and potential for containing digital, analog, radio frequency (RF), and power management building blocks in the same die. Although suitable for high integration, such technologies tend to suffer from component mismatch and lossy substrate for required passive performance.

2.1.2 Frequency Planning

The design of a frequency plan has a direct dependence on the transceiver topology, the number of downconversions, and the modes of operation (full duplex or half duplex). This section will focus on superheterodyne topology operating in a full-duplex system. The block diagram of such a receiver is shown in Figure 2.2. There are only two downconversions, one from RF to IF and the other from IF to baseband I and Q.

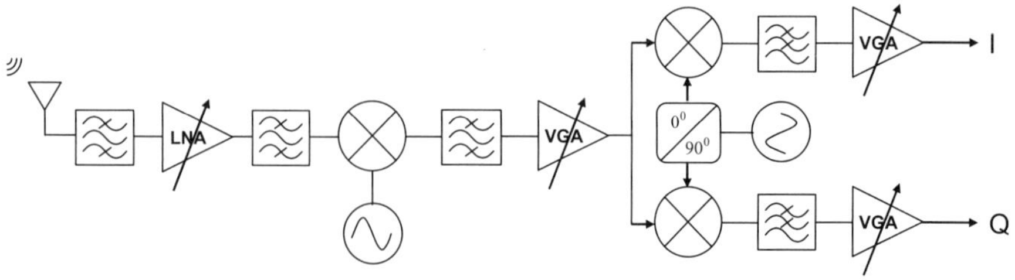

Figure 2.2. Architecture of a superheterodyne receiver.

Thorough frequency planning involves study of blockers, spurs, and image frequencies. Blockers, spurs, and image interferers are RF signals that are transmitted into the air by other wireless devices and can degrade the SNR performance of the desired receiver. Because of multiple stages of amplification, downconversion, and filtering, superheterodyne radios offer the best fidelity performance. However, as the architecture needs to employ signal processing blocks at various frequencies, it tends to be bulky and power hungry.

2.1.3 Blockers

Understanding the wireless applications that coexist in the frequency spectrum surrounding the band of interest is one of the most important steps in determining a workable frequency plan. Applications that use high-power transmitters can create problems by saturating the receiver front end and potentially damaging such components. Applications such as mobile phone services that are widely deployed and operate with handset output powers in excess of 1 W are especially troublesome. Operating close to such frequency bands places great demands on front-end filter selectivity. Many satellite receivers use L-band frequencies for IF and avoid using the bands occupied by the mobile phones as an IF frequency to avoid interference.

In addition to the ISM band applications as illustrated in [1], additional government and military bands exist, especially at 10 GHz that operate high-power radars and have to be considered. Although some of these higher frequency blockers experience higher medium loss, they typically operate at higher output power. Radar applications can easily exceed a few Watts of output power.

Blockers play a critical role in determining system power. The presence of high-amplitude blockers leads to high dynamic range requirements from the building blocks, which then essentially leads to more power consumption. Finite intermodulation characteristics of the front-end circuits also lead to in-band SNR degradation caused by any modulation existing in the blocker signal. This will be demonstrated.

Figure 2.3 illustrates the propagation of a blocker signal through the front end of a receiver for three different cases depending on the relative frequency of the blocker to the receive band. In case (1), the frequency of the blocker is very close to the receiver

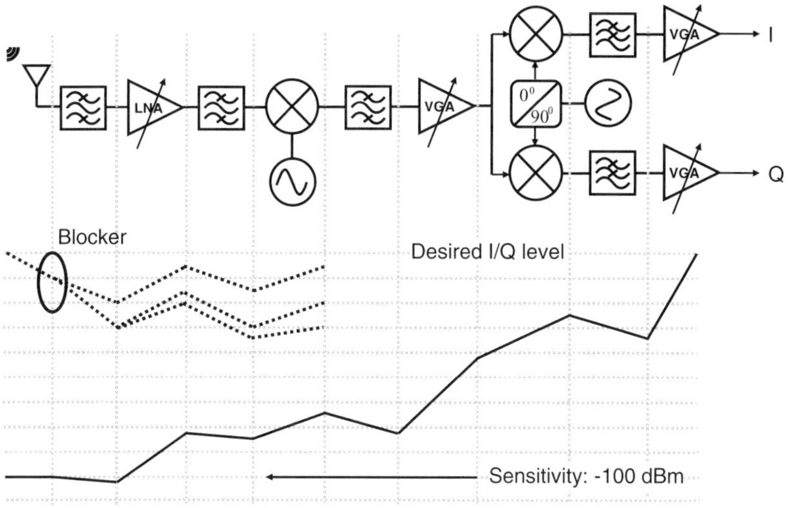

Figure 2.3. Blocker propagating through the signal path in a superheterodyne receiver.

band of interest and it experiences little filtering in the band-pass filter. This places stringent linearity requirements on the low noise amplifier (LNA) and RF mixer, which has a subsequent impact on the power consumption of these components. In case (2), the blocker frequency is significantly lower than the receive band and it experiences adequate rejection through the filters. In case (3), the blocker frequency is separated from the receive band as in case (2), but this time it is located at the higher side of the band. Assuming the same filter rejection for this higher frequency blocker, the blocker level is further attenuated by the high-frequency gain role-off in the active components in the front end such as the LNA and the RF mixer.

The designer needs to perform this analysis for every potential blocker in the spectrum and make adequate corrections either to the frequency plan or to the filter specification. Unfortunately, the frequency plan may not provide much flexibility for alleviating the impact of blockers on the receiver RF front end based on a specific power budget. The location of the receive band is typically predetermined by standards and cannot easily be shifted, leaving the filtering as the only means for addressing the blockers in the RF front end. The only limitation for selection of an IF frequency is the availability of IF surface-acoustic wave (SAW) filters at the chosen frequency.

The formula below describes the rejection required for the front-end band-pass filters (BPFs). A margin (MRG) is typically needed to determine the blocker back-off from the input 1-dB compression point (IP1dB) of the receiver and its individual components. The input blocker level (IBL) is the blocker strength at the antenna output. This number has already been adjusted for path loss and antenna selectivity.

$$\begin{aligned} \mathrm{BPF_{rej}} = {} & (\mathrm{IBL} + \mathrm{MRG} - \mathrm{IP1dB}_{LNA@f(blocker)}) \\ & + (\mathrm{IP1dB}_{LNA} + \mathrm{G}_{LNA@f(blocker)} - \mathrm{IP1dB}_{Mixer@f(blocker)}) \end{aligned} \tag{2.1}$$

Example of Figure 2.3:

$$\text{IBL} = -10\ \text{dBm} \qquad \text{IP1dB}_{LNA@f(blocker)} = -20\ \text{dBm}$$

$$\text{MRG} = 10\ \text{dB} \qquad \text{IP1dB}_{Mixer@f(blocker)} = -20\ \text{dBm}$$

$$\text{G}_{LNA@f(blocker)} = 25\ \text{dB}$$

$$\text{BPF}_{\text{rej}} = (0 + 10 + 20) + (-20 + 25 + 20) = 45\ \text{dB}$$

2.1.4 Spurs and Desensing

Spur analysis is an extension of analysis performed for blockers. Here we study not only the impact of other transmitters operating in the surrounding frequency bands but also the unwanted spurious frequencies that are generated by interaction between various components of our own transceiver. This includes the interactions between the low-frequency crystal oscillator used for the synthesizer, RF and IF local oscillators (LOs), as well as transmit and receiver signals. This analysis is performed to identify spurs that appear in either RF, IF, or LO frequency bands. The spurs that appear at high-power levels relative to the signal of interest in these bands can be very troublesome and have to be addressed early in the frequency plan. Typically, the spurs that interfere with the LO frequencies are less problematic since the LO signals are significantly higher in power when compared with the spurs. On the contrary, the IF and especially the RF signal, which are typically low in power, are very susceptible to spur interference. It is usually best to design the frequency plan to avoid spurs that fall directly on the RF or IF bands. Desensing is one of the outcomes of such interference where a spur with a higher power level than the desired RF or IF signal lands either directly in the band or adjacent to the band and saturates the transceiver.

2.1.5 Transmitter Leakage

Leakage from the transmitter is a major concern for any advanced RF subsystem, especially in full-duplex systems. The transmitter, operating at a high power level, requires stringent filtering to avoid interference with the adjacent bands, especially the receiver. Transmitter leakage into the receiver can result in desensing of the receiver by saturating the receiver front end.

The transmit signal can leak into the receiver input and cause an oscillation by finding a leakage path after the front-end gain. This is one major issues, which makes it very difficult to integrate both transmit and receive functions of a duplex system on a single chip.

2.1.6 LO Leakage and Interference

Local oscillator signals and their harmonics are major sources of spurious interference. As shown in Figure 2.4, many potential paths are available for LO leakage and interference. These leakage paths are created either through the integrated circuit (IC) substrate and package or through the board on which the IC is mounted. Often, it is very

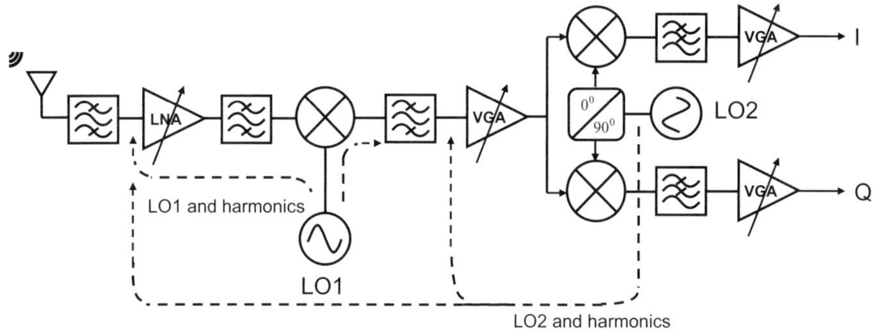

Figure 2.4. Illustration of LO leakage in a superheterodyne receiver.

difficult to identify and address every leakage path that may exist in the system. Therefore, the best method for avoiding interference is to devise a frequency plan where all LO frequencies and their harmonics, and even frequencies resulting from higher order mixing of these signals, do not fall in the RF or IF bands. Since LO power levels are relatively higher than RF and IF levels, it is very likely to have a higher order term involving an LO signal to create significant interference or overpower the RF input signal and desense the receiver.

For the cases where LO1 reaches the input of the LNA, the interfering signal is amplified by the LNA, making it even a larger interferer. This typically results in saturation of the RF mixer or any other active element that follows the LNA. Since the frequency selectivity of the typical LNA is not very high and no protection is provided from the first BPF, this type of interference heavily depends on the amount of LO or LO harmonic coupling either through the substrate or the board.

Again, the frequency plan and careful selection of IF and LO frequencies can play an important role in alleviating this problem. A good way of looking at the extent of this problem is to estimate the level of desired isolation using the inequality shown in the example below. Once the needed isolation is calculated, the designer can determine whether this level of isolation is practical in the IC process, packaging, and board characteristics under consideration. Local oscillator power at the mixer input (PLO), input 1-dB compression point (IP1dB) of the blocks, and a safety margin (MRG) are used in the calculations.

$$\text{Isolation} > G_{\text{LNA@f(LO)}} + G_{\text{BPF2@f(LO)}} + \text{MRG} + P_{\text{LO}}\text{-IP}_{\text{1dBMixer}} \tag{2.2}$$

Example:

$$\text{PLO} = -5\ \text{dBm} \qquad \text{IP1dB}_{Mixer@f(LO)} = -20\ \text{dBm}$$

$$\text{MRG} = 10\ \text{dB} \qquad G_{BPF2@f(LO)} = -20\ \text{dB}$$

$$G_{LNA@f(LO)} = 25\ \text{dB}$$

Isolation $> 25 - 20 + 10 - 5 + 20 < 30$ dB (very practical)

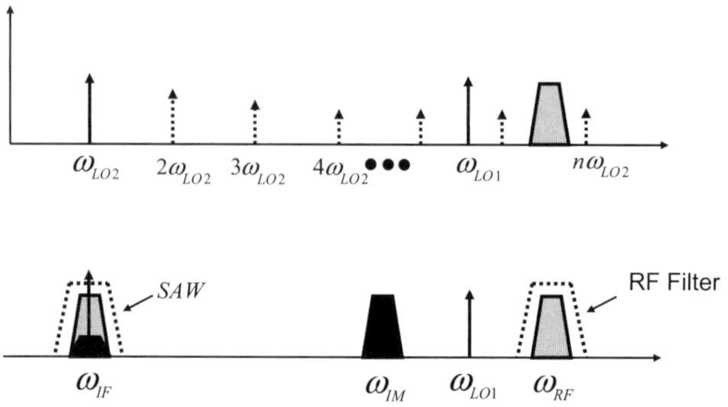

Figure 2.5. Harmonics of LO tone in a superheterodyne receiver.

Figure 2.5 demonstrates how a second LO signal can generate harmonics that can appear in or close to the RF band of frequency. There can also be potential problems if the harmonics of the second LO interfere with the first LO signal. To avoid these two undesirable scenarios, the frequencies allocated for LO1, LO2, and RF bands should be selected so that they are not integer multiples of one another. Allocation of the LO frequencies should be done in such a way that their harmonics completely clear the entire RF band with a reasonable margin.

2.1.7 Image

Image frequency is one of the most problematic issues with designing traditional superheterodyne receivers. With a receive frequency ω_{RF}, and an LO frequency ω_{LO}, the IF frequency and image frequencies are given by $\omega_{IF} = \pm\omega_{RF} \mp \omega_{LO}$, and $\omega_{IM} = \omega_{\mathrm{RF}} \pm \omega_{\mathrm{LO}}$, which corresponds to this choice of IF frequency.

Pictorially, the image signal is located on the opposite side of the LO frequency and folds on top of the IF band as the signal is downconverted in a mixer. This process creates a serious interference issue that needs to be addressed by either filtering or image-reject mixing topologies. In a semiconductor technology platform, image rejection is limited by component mismatches and so on.

2.1.8 Half-IF Interference

Interference of half-IF frequencies is another issue that plagues most receiver topologies. As shown in Figure 2.6, the half-IF frequency is located directly between the LO and the RF. This half-IF signal can create a second harmonic in the LNA or RF amplifiers in the front end and get downconverted into the IF band by mixing with the second harmonic of the LO signal. This problem can be avoided with adequate filtering in the front end or low-distortion LNA and RF amplifier designs. By using an active front-end component with low even-order distortion products, the half-IF frequency will no longer produce a significant second harmonic eliminating the concern for interference.

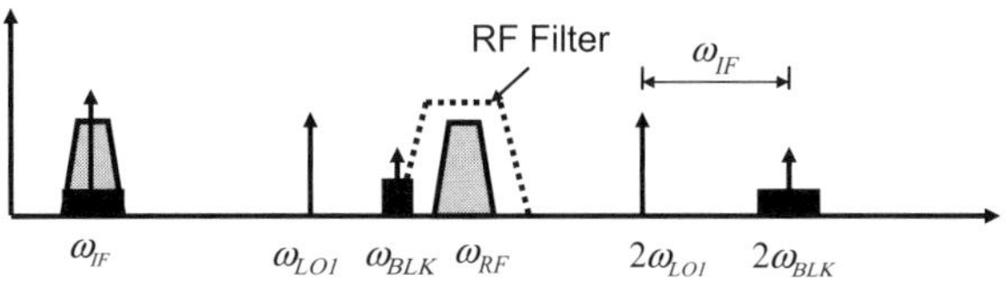

Figure 2.6. Illustration of half-IF interference.

Equations (2.3) and (2.4) describe the interfering component that develops from the interaction between half-IF and LO harmonics for the case of a low-side injection mixer.

$$\mathrm{F}_{\frac{1}{2}IF\ Interference} = \mathrm{F}_{\mathrm{RF}} - \left(\tfrac{1}{2}\right) \times \mathrm{F}_{IF} \tag{2.3}$$

$$F_{IF} = [2 \times F_{\mathrm{LO}}] \times [2 \times F_{\frac{1}{2}IF\ Interference}] = Interference \tag{2.4}$$

2.2 LINK BUDGET ANALYSIS

The purpose of a link budget analysis is to determine the system level specifications of individual transceiver blocks. This analysis is dependent on several key system parameters such as sensitivity, dynamic range, and input signal range required for the analog-to-digital converter (A/D) or limiting amplifier terminating the back end of the receiver.

The basic purpose of a modern receiver is to detect and deliver an RF signal from an antenna to an A/D while maintaining the desired signal quality. The sensitivity and dynamic range of a receiver are the two main parameters that define the range of input RF power that must be received. Bit error rate (BER) and symbol error rate are the performance metrics that define the acceptable quality of the received signal. Sensitivity defines the lowest input RF signal that must be detected and distinguished by the receiver with acceptable quality, and the dynamic range defines the entire range of input RF power from the sensitivity threshold up to the maximum detected signal. A link budget analysis uses these given criteria to determine the receiver lineup and the requirements of various receiver blocks. This typically involves calculations for gain, noise figure, filtering, intermodulation products (IMs), and input 1-dB compression (P1dB). In this section, we will identify these components and illustrate the common methods used to quantify them.

2.2.1 Linearity

Linearity is the criteria that define the upper limit for detectable RF input power level and set the dynamic range (ratio of maximum allowable signal to the minimum detectable signal) of the receiver. Linearity in a receiver is mostly determined by two-tone intermodulation products: (1) third-order intermodulation (IM3) and (2) second-order intermodulation product (IM2). As the circuits and systems in the modern front ends are inherently differential, IM3 is usually affected by the amount of power consumption, and IM2 is usually affected by the amount of even order impedance (common mode impedance, etc.). These two parameters are the result of a two-tone analysis, where two in-band signals are subjected to the receiver or one of its components. These tones mix with each other because of the nonlinear elements in the receiver, and generate products that can be used to characterize the extent of the nonlinearity. These intermodulation terms can be extended to define IM4, IM5, and so on.

Equations (2.5) and (2.6) describe the generation of IM2 and IM3 and the input intercept points for both of these types of nonlinearity behavior. Although the intermodulation products (IM2 and IM3) are dependent on the signal power and cannot be used independently to describe performance, the intercept points are indeed independent parameters that can be used to quantify the linearity.

$$\mathrm{IM2} = \mathrm{A2} \times \mathrm{RFin}^2 \tag{2.5}$$

$$\mathrm{IM3} = \mathrm{A3} \times \mathrm{RFin}^3 \tag{2.6}$$

where

A2: measure of device second-order nonlinearity
A3: measure of device third-order nonlinearity

The relationship described above dictates a 2 : 1 slope for the IM2 and 3 : 1 slope for the IM3 products as shown in Figure 2.7. This relationship can be used in Eqs. (2.7) and (2.8) to determine the input intercept points for the second-order (IIP2) and third-order (IIP3) products. Output intercept points for the second-order (IP2) and third-order

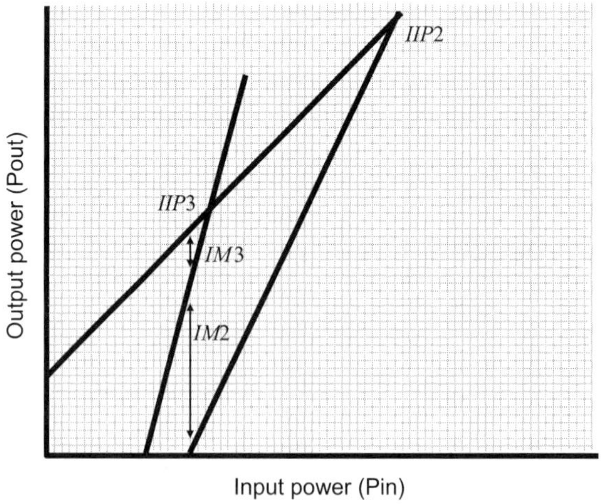

Figure 2.7. Input power sweep to illustrate IM2 and IM3.

(IP3) products can also be calculated easily by adding the gain of the cascaded blocks to the appropriate input intercept points.

$$\text{IIP2[dBm]} = \text{RFin[dBm]} + \Delta\text{IM2[dB]} \tag{2.7}$$

$$\text{IIP3[dBm]} = \text{RFin[dBm]} + \Delta\text{IM3}/2\text{[dB]} \tag{2.8}$$

Figure 2.7 illustrates the result of a two-tone power sweep with a fundamental signal and its intermodulation products plotted as a function of input RF power. Intercept points are extrapolated using the plotted data. This graph can be generated by either simulation or measurement to determine a measure of linearity for a receiver or any of its components. For a link budget analysis, these numbers are then used in an architecture described by Eq. (2.9) in order to determine the impact of individual components on the linearity of the overall receiver.

$$1/\text{IP3}_{overall} = 1/\text{IP3}_1 + \text{G}_1^2/\text{IP3}_2 + \ldots \tag{2.9}$$

By understanding the formula above, one realizes that the overall linearity is highly dependent on the linearity of later stages in a cascade. For example, in a receiver front end, the mixer typically becomes the limiting component. This can be addressed by designing a passive mixer or by consuming higher current in the mixer to meet the linearity requirements.

It must be remembered that these terms only indicate the circuit's performance. Real linearity of the circuits occurs with the clipping of the input signal and with the input power where the gain drops by 1dB for the front-end amplifier. Using the cubic transfer characteristics of amplifiers, IP3 and P1dB are related by

$$\text{IP3} = \text{P1dB} + 10$$

Linearity terms can also be interpreted from the DC I–V transfer characteristics. Cubic order linearity is related to the slope of I–V characteristics around the origin of the DC I–V characteristics. It should also be kept in mind that while determining these small-signal linearity terms, IM2 and IM3, the input power range must be kept low. In other words, these terms must be evaluated in the "linear" operating range of the amplifier, and the results should be extrapolated along with the fundamental terms to obtain IP2 and IP3 terms.

Often, gain switching is used in the front end of the receiver to ease the linearity constraint and improve intermodulation performance. When the input RF signal is increased, the receiver no longer needs to amplify the input signal as much. Thus, it can reduce the gain of the LNA or any other front-end amplifiers.

In the transmitter systems, the linearity performance is determined by the harmonics of the baseband signal. The IM2 and IM3 terms are relevant in the case of a receiver, where these terms indicate the in-band SNR degradation. In the transmitter, the SNR degradation is related to the harmonics of the baseband signal. If ω_{BB} is the baseband frequency, then the SNR degradation in the transmitter is related to the extent of the $2\omega_{BB}$, $3\omega_{BB}$ terms, which appear inside the transmitter band. Intermodulation terms resulting from baseband signal are also important to consider.

2.2.2 Noise

The noise performance of a receiver defines its sensitivity by limiting the lowest input RF signal that can be detected by the receiver. In this section, we will discuss various noise sources and methods of calculations.

2.2.2.1 Thermal Noise. Thermal noise is a function of random movement of electrons in circuits, and it is present irrespective of the power consumption of the block. The topic of thermal noise has been covered extensively in many other references, so we will only describe it briefly and highlight the relevant formulas required for a receiver system analysis. As shown in Eq. (2.10), the thermal noise power is dependent on the signal bandwidth and on the temperature of the medium. Naturally, the noise power increases with increasing temperature and bandwidth.

$$P_n = kTB \tag{2.10}$$

where

$k =$ Boltzman's constant $= 1.38 \times 10^{-23}$ [J/K]
$T =$ Temperature [K]
$B =$ Bandwidth [Hz]

The thermal noise of a receiver is typically referred to the input of the chain in the form of either an overall system noise temperature or noise figure. Frii's formula for calculation of receiver noise figure is illustrated as follows:

$$\begin{aligned} F = F_{BPF1} + (F_{LNA}-1)/G_{BPF1} + (F_{BPF2}-1)/(G_{BPF1}G_{LNA}) \\ + (F_{MIX}-1)/(G_{BPF1}G_{LNA}G_{BPF2}) + \end{aligned} \tag{2.11}$$

where

$G =$ Gain [dimensionless]
$F =$ Noise Figure [dimensionless]

It is very critical to note that the noise figure is dominated by the noise figure of the first stage.

2.2.2.2 Transmitter Noise. In modern communication microsystems, multiple receivers and transmitters are integrated in the same die. As shown in Figure 2.8, the broadband noise generated by the power amplifier (PA) can dominate the thermal noise of the receiver and result in a significant increase in the noise floor as well as consequently limit the sensitivity of the receiver. Apart from filtering or use of a better

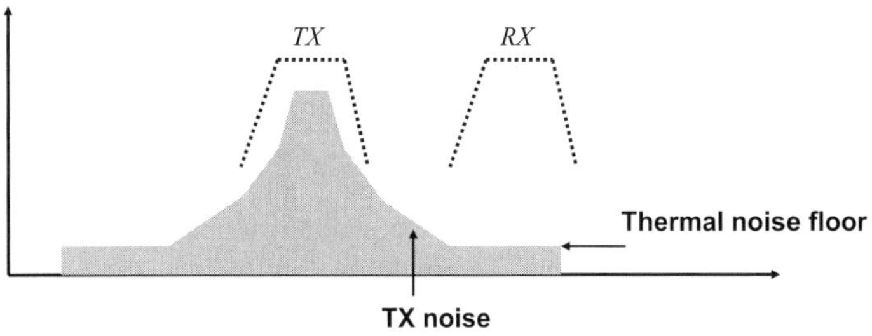

Figure 2.8. Impact of transmitter noise on receiver.

power amplifier, the only other choice for addressing this problem is moving the receive band further away from the transmit band.

This situation can occur if the transmitters and receivers are located very closely from one another. From a practical standpoint, let us assume that a Bluetooth and a WCDMA device are integrated in the same die, and they are a few centimeters apart from one another. The Bluetooth transmitter, operating at 2.4 GHz, would provide a broadband noise, which would be coupled through the respective antennas, and to the upper receive band of WCDMA (2.1 GHz). This would lead to desensitization of the WCDMA receiver.

2.2.2.3 Phase Noise. The LO signal generator provides a noise profile, known as "phase noise." When the signal is downconverted in the mixer, the phase noise of the LO is added to the existing noise of the incoming RF signal, which is downconverted into the IF band along with the RF signal. Phase noise is described as a relative measure of the difference between the peak LO power and the noise floor as a function of frequency offset. The phase noise contribution of the LO is calculated by integrating the LO noise power over the RF signal bandwidth. As shown in Figure 2.9, the close-in phase noise is much more adverse. The phase noise typically flattens out further from

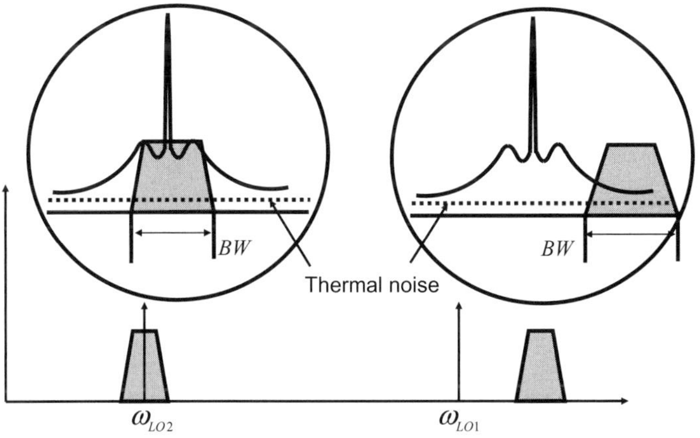

Figure 2.9. Phase noise and its impact on the transmitter.

the LO frequency. The close-in phase noise of the LO is typically dependent on the loop response and on the phase detector performance of the phase lock loop (PLL) synthesizer, whereas the far-out phase noise is dependent on the phase noise performance of the voltage-controlled oscillator (VCO). This generates a different set of requirements for different local oscillator sources. As shown in Figure 2.9, the critical component of the LO1 source will be the VCO, whereas the critical component of the LO2 source will include the PLL phase detector and loop filter in addition to the VCO.

2.2.3 Signal-to-Noise Ratio

The SNR and BER are the key parameters that define the performance of the receiver. As shown in Eqs. (2.12) and (2.13), the SNR is a simple performance metric that describes the difference between the signal power and the noise floor. This measure is sometimes modified to include the interference and described as the signal-to-noise and interference ratio (S/N + I). The SNR is used to define the energy-to-noise (Eb/No) parameter needed to predict the BER performance of a receiver. The relationship between SNR and Eb/No is dependent on the modulation scheme of the received signal and is described in detail in the literature [2,3].

$$\text{SNR} = \text{Signal}/\text{Noise} \tag{2.12}$$

$$\text{SNR}[\text{dB}] = \text{Signal}[\text{dBm}] - \text{Noise}[\text{dBm}] \tag{2.13}$$

2.2.4 Receiver Gain

One of the most important parameters in the receiver is the overall receiver gain and the range for gain variation. Typically, a user can be located at a cell edge, or very close to the base station. In these two cases, the received power can vary from extremely small to a much higher level. In between these two extremes, signal strengths would vary continuously depending on the position, and the gain is usually decided by the receive signal strength indicator (RSSI). As shown in Figure 2.10, most modern receiver chains start with the antenna and end with an analog-to-digital converter (A/D). Therefore, it becomes very important for a receiver designer to not only understand the limitations of the input power and noise at the receiver input, but also to consider the requirements and limitation of the A/D to accommodate the large gain variation. The dynamic range of the A/D is the key criterion that defines the range

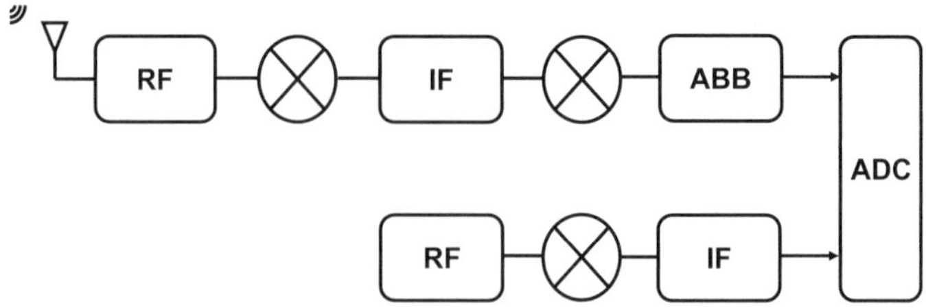

Figure 2.10. Receiver chains with an A/D interface.

of voltages that can be digitized with acceptable quality. The dynamic range is defined by the number of bits and by the maximum input voltage swing of the A/D. The number of bits defines the lowest voltage that can be detected by the A/D and sets the maximum gain required by the receiver to boost the input RF signal from the antenna to this minimum voltage level at the A/D input. The maximum input voltage swing allowed for the A/D sets the minimum gain required by the receiver. This relationship is described as follows:

$$G_L = \text{Minimum Receiver Gain}[\text{dB}] = Pin_{A/D,MIN} - RS \tag{2.14}$$

$$G_H = \text{Maximum Receiver Gain}[\text{dB}] = Pin_{A/D,MAX} - RS - DR \tag{2.15}$$

where

RS = Receiver Sensitivity [dBV] (w.r.t. reference impedance, usually 50 Ω)
DR = Receiver Dynamic Range [dB]
$Pin_{A/D,\ MAX}$ = Maximum Input Signal to A/D [dBV]
$Pin_{A/D,\ MIN}$ = Minimum Input Signal to A/D [dBV]

In the above calculation, the impedance levels must be adjusted. The input impedance of the RF front end is significantly different from the ADC input impedance. This gain variation range of the overall receiver needs to spread across various components in the receiver chain. This gain variation is typically delegated to a switched-gain LNA or RF amplifier in the receiver front end and one or two low-frequency variable gain amplifiers (VGAs) in the IF or baseband chain. For ease of design and implementation, most of the gain variation is set for the low-frequency VGAs.

2.3 PROPAGATION EFFECTS

Although signal propagation is an external phenomenon that does not occur in the receiver, its effects have significant impact on receiver signal integrity. In our discussion, we will consider an air medium and study the impact of air and other stationary and moving objects in the path of the signal. Although air is the most commonly used propagation media, high dielectric constant materials can also be mediums, such as water (for underwater communications) and muscle tissues (for implantable radios).

2.3.1 Path Loss

When a source transmits in all directions from an isotropic antenna, the signal is propagated in a spherical pattern as shown in Figure 2.11.

This result implies that a given electromagnetic energy is spread over a surface of a sphere that grows in radius as the signal moves further away from the antenna. It causes

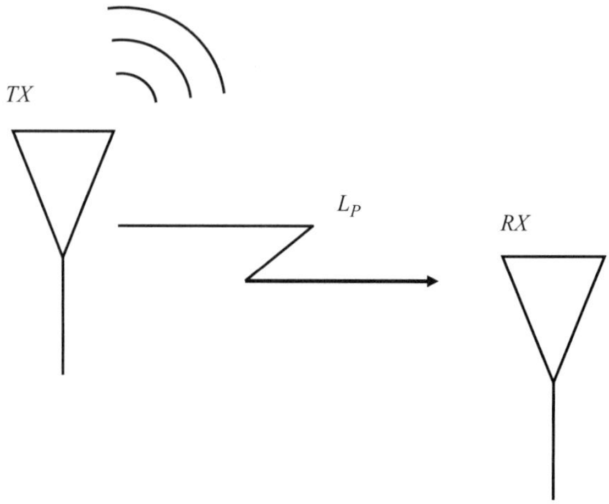

Figure 2.11. Communication link illustrating path loss.

the signal density to attenuate over distance resulting in a signal power loss, which is referred to as path loss (Lp). Path loss can be calculated as follows:

$$L_P = 20\log(4\pi R/\lambda)[dB] \quad (2.16)$$

where

R = Path Length [m]
λ = Wavelength [m]

In addition to regular path loss through an air medium, the signal can also be attenuated by rain or high water vapor concentration in the air. The water molecules absorb the electromagnetic energy resulting in a frequency-dependent attenuation through the air. Water vapor attenuation peaks at around 2 GHz, which is the resonant frequency of the water molecules. This information has been experimentally calculated and is available in the literature [2,4] .

The relationship that describes the propagation of a signal from transmitter, through the air and to a receiver is described in the formula below. This relation also accounts for antenna gain in case the antenna is not isotropic.

$$P_R = P_T + G_{T,ANT} - L_P + G_{R,ANT}[dBm] \quad (2.17)$$

where

P_T = Transmitter Output Power [dBm]
P_R = Receiver Input Power [dBm]
EIRP = $P_T + G_{T,ANT}$ = Radiated Transmitter Power [dBm]
$G_{R,ANT}$ = Receive Antenna Gain [dB]

$G_{T,\ ANT}$ = Transmit Antenna Gain [dB]
L_P = Path Loss [dB]

2.3.2 Multipath and Fading

In our increasingly mobile lifestyles, multipath and fading have become important challenges faced by most of today's wireless applications. Fading refers to fluctuation of the RF signal amplitude at the receiver antenna over a small period of time. Fading is caused by interference among various versions of the same RF signal that arrive at the receiver antenna at different times. These versions of the RF signal, which are called multipath waves, take different paths during their propagation and are subject to different phase shift and amplitude attenuation. They may even be the subject of a Doppler shift caused by a mobile object or antenna. Figure 2.12 illustrates such a scenario where multiple reflections of the same signal arrive at the receiving antenna at different phase, amplitude, and Doppler frequency shift. A simple formula describing the impact of a moving receiver or a transmitter is shown as follows:

$$F_D = \text{Doppler Shift} = V/\lambda\,[\text{Hz}] \tag{2.18}$$

where

F_R = Received Frequency [Hz]
F_T = Transmitted Frequency [Hz]
V = Relative Velocity [m/s]

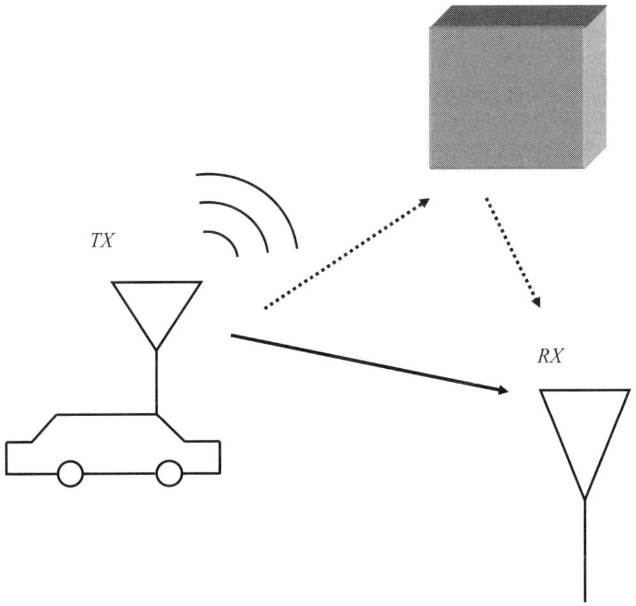

Figure 2.12. Multiple reflections leading to multipath and fading at the receiver.

Formulating fading and multipath is somewhat more involved, and it is covered in detail in the literature [4].

Several techniques have been developed to help combat fading and to improve the general link performance of a system in a hostile environment.

2.3.3 Equalization

Equalization is used to combat intersymbol interference (ISI) caused by the multipath within the channels. This form of interference occurs when the radio channel bandwidth truncates the signal modulation bandwidth, resulting in time spreading of modulation pulses.

To reduce ISI in a mobile environment, adaptive equalization is used to track the time-varying characteristics of the channel. Typically a known training sequence is transmitted to characterize the channel. This information is then used to calculate and set the proper filter coefficients for equalization in the receiver back end. The data are transmitted after the training sequence, received and corrected by the equalizer. In an adaptive equalizer, the filter coefficients of the equalizer are constantly optimized to compensate for the changing radio channel [4].

2.3.4 Diversity

Diversity is another method that can help reduce the severity of fading. Diversity can be provided by using polarization, time, and frequency. The most common diversity technique is spatial diversity. In this technique, multiple receiver antennas are strategically placed at different locations, which allows the antennas to receive different versions of the transmitted signal, thus providing the receiver with a choice on which version to use.

As shown in Figure 2.13, there are two different methods to implement a spatial diversity receiver. In the first method, one receiver path can switch between

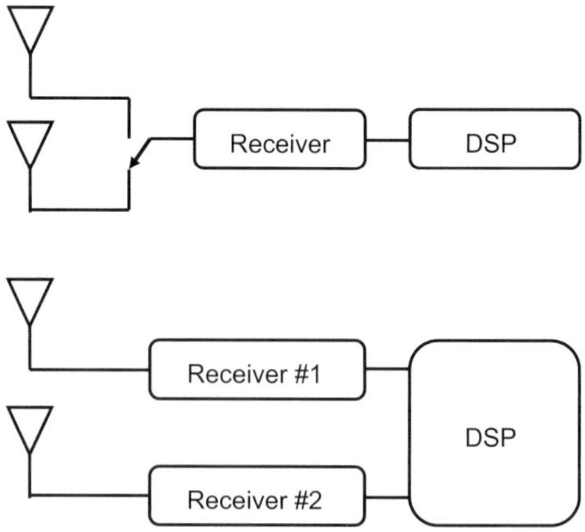

Figure 2.13. Block diagram of a receiver using antenna diversity and another one using a complete receive chain diversity.

multiple antennas, which requires the receiver to first test each antenna and then make a decision. In the second method, multiple independent receiver paths with their own antenna are used. In this case, the receiver back end can have both signals available simultaneously while choosing the better one. This process allows for a more dynamic operation, but the cost of the receiver is significantly increased.

2.3.5 Coding

Another method used to improve the performance of the communications link is channel coding. In this technique, redundant data bits are added to the original message prior to modulation and transmission of the signal. These added bits follow a specific code sequence that helps the receiver to detect and correct some or all of the error created by the radio channel. The addition of coding bits does reduce the overall throughput, but it is very effective in reducing errors.

2.4 INTERFACE PLANNING

For a working communication link, a receiver should be able to interface effectively with its environment and other components in the link. This should be addressed by interface planning during the system-level design of the receiver prior to determining the block-level specification. Interface planning helps to determine the number of ICs in a chipset and their input and output characteristics. The perceived nature of a signal at different points in the receiver path and at input and output pins of the ICs is also determined.

In most modern receivers, the level of on-chip integration is continuously growing to include components that operate at RF, IF, and baseband frequencies. The reference impedance used for the RF components is 50 Ω, but this impedance typically changes to hundreds of ohms at IF and thousands of ohms at baseband. In designs that cover receiver blocks from RF to baseband, one may use voltage as a variable for tracking power, noise, and intermodulations throughout the system. Because of the change in impedance levels, the numbers in "dBm" may be misleading.

With the introduction of fundamental functionalities of basic receiver and transmitters, we now proceed to various radio architectures. They are discussed in the order of evolution: (1) superheterodyne, (2) very low IF, and (3) direct conversion. In addition, we will also discuss the subsampling-type architecture, which is suitable for more "digitally" controllable receivers. All of these architectures are discussed with the frequency planning, along with advantage and disadvantages.

2.5 SUPERHETERODYNE ARCHITECTURE

"Heterodyne" means to mix down, and "super" is used for "super audio" frequency. Superheterodyne radios have been present for a long time, and often they provide the highest fidelity. These radios use multiple stages of amplification, filtering, and frequency shifting, in order to provide sufficient image suppression. A two-stage superheterodyne radio is common, using off-chip high rejection filters and so on. Because of the requirements of off-chip filters and matching to 50 Ω at various places, super heterodyne radios are often implemented as multiple-chip solutions integrated as a module, and use a higher form factor compared with to the other architecture variants.

2.5.1 Frequency Domain Representation

Figure 2.14 illustrates the frequency planning of the first stage of the superheterodyne receiver. Multiplication in the time domain is illustrated in terms of frequency domain convolution, and the output at the first mixer stage is spectrally located at ω_{IF}, with the image rejection performance determined by the band-pass filter's rejection characteristics. After the first stage, image rejection is employed by using quadrature phase-shifted signals as illustrated in Figure 2.15. This addition can be implemented simply by the addition of output currents from two output stages (similar to an wired-AND connection).

Two options can be employed to obtain phase shift, as illustrated in Figure 2.16. For practical implementations, the 90° phase shift in Hartley architecture is usually split into 45° in both branches using a lag-lead network. This helps reduce mismatch between the two paths and balance loading to the previous blocks. In Weaver's architecture, phase shift is realized by multiplication of same frequency signals and added later. It is a more desirable solution, as the two phase shifts are derived synchronously from the same signal generator.

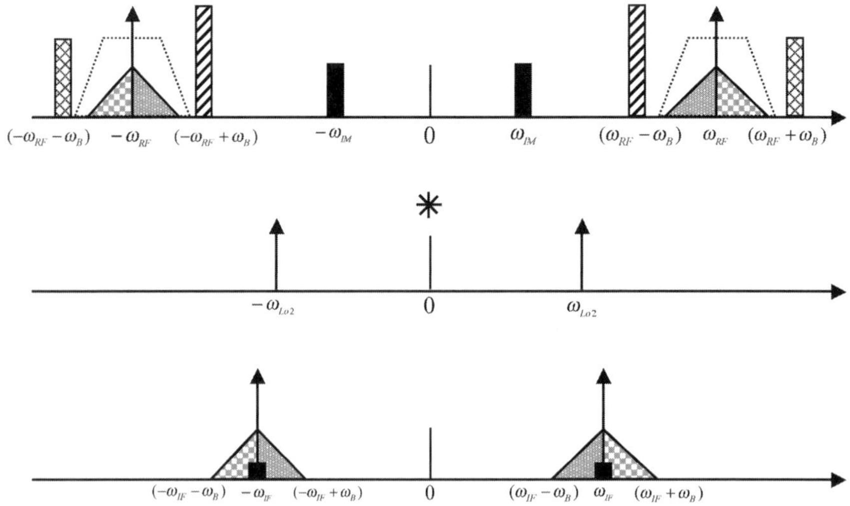

Figure 2.14. Frequency planning of a superheterodyne receiver.

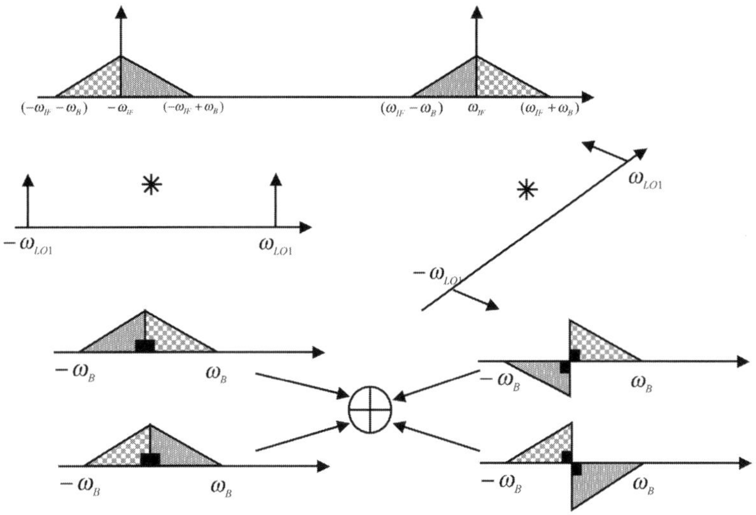

Figure 2.15. The frequency planning of a superheterodyne receiver.

2.5.2 Phase Shift and Image Rejection

Phase shifting is an important functionality of any radio, and two options can be used to provide phase shift, as illustrated in Figure 2.17. Passive phase shifters usually provide signal loss, which leads to degradations in sensitivity. The amplitude and phase mismatch of the two paths directly impact the image rejection performance of a

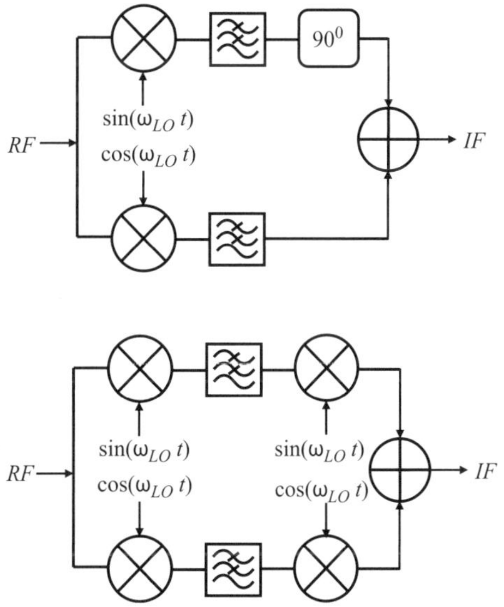

Figure 2.16. (a) Hartley and (b) Weaver image reject architectures.

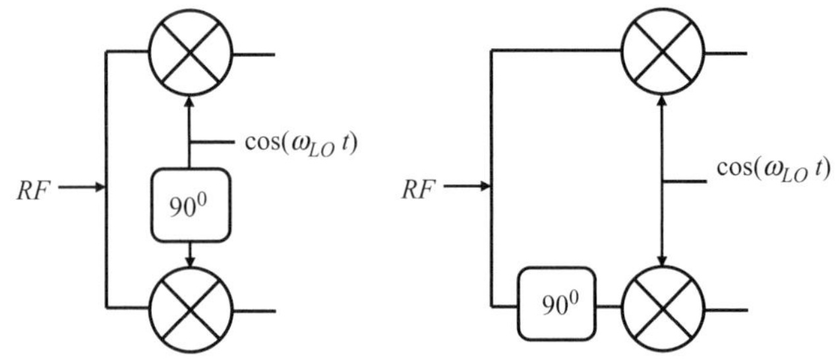

Figure 2.17. Two options for a phase shifter.

superheterodyne radio. Usually, the phase shift in the signal generation path is commonly employed, as this does not degrade sensitivity, and can be implemented accurately. The image rejection ratio is given by

$$\text{IRR} = \tfrac{1}{4} \times [(\Delta A/A)^2 + \theta^2] \tag{2.19}$$

where

$\Delta A/A$: relative gain mismatch
θ: relative phase mismatch in radians

Using the relationship described above, it can be seen that an amplitude mismatch of 0.1 dB and a phase mismatch of 1° yields an approximate 40 dB of IRR.

To maintain an acceptable receiver signal quality, most modern wireless standards require 60 to 90 dB of image rejection. The traditional method of image rejection uses filters designed with a stop-band at the image frequency. However, because of the stringent requirements, image rejection is typically performed through a combination of filtering and the use of image rejection mixing techniques.

To employ the highest selectivity, SAW filters are commonly used in superheterodyne receivers. In between the circuit blocks, a 50-Ω impedance match may or may not be provided, depending on the system requirements, as well as on the integration requirements (on-chip vs. off-chip).

2.5.3 Transmitter and Receiver

The functionality of transmitters is similar to receivers, in the inverse order according to the frequency scheme. However, a few considerations are different. For example, the second harmonic of a 2.4-GHz receive band is located at 4.8-GHz, which is filtered by the interstage filter. While in the transmitter with 1-GHz bandwidth, the second harmonic of 300-MHz is inside the bandwidth and causes degradations. Differential architecture with adequate common-mode rejection would reduce the second harmonic, whereas the magnitude of an in-band third harmonic would be dependent on the power consumption of the amplifier block.

2.5.4 Imbalance and Harmonics

In a superheterodyne radio, any imbalance in the two quadrature phases in terms of amplitude and phase leads to image rejection performance. Any imbalance in the two individual differential paths in the same signal chain (either I or Q) leads to LO leakage at the output of the transmitter and, with the phase noise profile of the on-chip signal generator (VCO/PLL), provides significant degradations to the EVM performance of the transmitter. In a superheterodyne radio, this can be comfortably filtered out using a high out-of-band rejection filter. It can be alleviated using deliberate mismatch and calibration at the baseband.

In the transmitter, a finite amount of harmonic linearity of the baseband amplifier appears as an unwanted signal on the opposite side of the LO signal in the spectrum of the desired band. For example, let us assume that we have a bandwidth of 400 MHz, and the first LO frequency is 3.0 GHz; then the highest frequency component output signal would be at 3.2 GHz, and the third harmonic would be located at 2.4 GHz ($=3.0-3\times0.2$)G. As can be seen, separation of this harmonic component from the desired fundamental depends on the center frequency and on the bandwidth of the signal under consideration.

It must be kept in mind that the power consumption of any signal processing is dependent on the center frequency, bandwidth, noise, and linearity requirements. Superheterodyne radios employ multiple blocks operating at a higher frequency, leading to higher power consumption. Also, the area increases as a result of multiple interstage filters. If these filters are implemented on-chip, the impedance matching consideration is not critical, and higher impedance can be used to reduce power. For modern handheld applications, low IF and direct conversion architectures are being used more and more. However, where high performance and reliability is critical, such as a satellite uplink/downlink, the form factor is usually not much of a consideration, and superheterodyne radios are preferred.

2.6 LOW IF ARCHITECTURE

Low IF architectures evolved as a reasonable compromise between super-heterodyne and direct conversion architectures. In this approach, the input RF signal is down-converted to a low IF (IF frequency being at least half of channel bandwidth) using quadrature signal paths. It is illustrated in Figure 2.18.

These quadrature signals are subsequently filtered using a complex filter. Complex filters are a cross-coupled combination of two real filters using a quadrature phase shift. This approach became attractive, as from the low IF frequency, on-chip filters

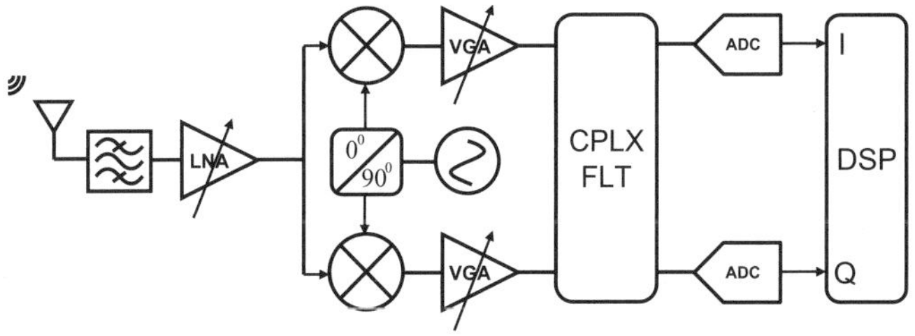

Figure 2.18. Architecture of a low IF receiver.

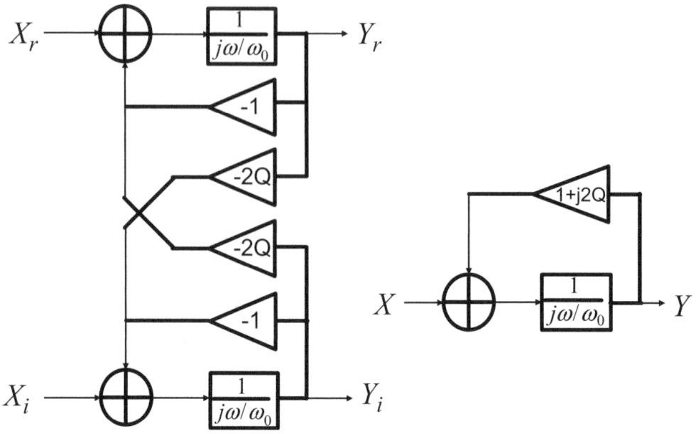

Figure 2.19. Signal processing in a complex filter.

could be integrated using low Q filtering stages (hence, lower power consumption), and most of the external filters could be avoided. The output of the complex signal is a band-pass combination of I and Q signals obtained from the mixer, and hence, at the output, a single ADC would suffice. It reduces the number of building blocks in the transceiver architecture, compared with their direct conversion counterpart. However, the building blocks need to process signals of higher frequency content, leading to higher power consumption.

Figure 2.19 illustrates an arrangement for a complex filter. They can provide sufficient rejection ($\sim$60 dB) to the image signals located at negative frequencies. To meet the requirements of stringent blocker specifications for modern wireless standards, it employs multiple filtering stages (integrators). In the low IF receiver, a polyphase filter takes two differential inputs in I and Q phases, and it provides a passband at positive frequencies, while rejecting negative frequencies. The transfer function of a band-pass polyphase filter is given by shifting the pole of a low-pass filter. Hence, from low-pass transfer function $H_{lp}(j\omega) = \frac{1}{1+j\omega/\omega_0}$, the realized band-pass transfer function is given by $H_{bp}(j\omega) = \frac{1}{1+j(\omega-\omega_c)/\omega_0}$. For a higher order system, multiple stages can be cascaded. The realization of the block diagram illustrated in Figure 2.19 can be easily performed by connecting resistors from the output of the I phase to the virtual ground of the Q phase and vice versa. The input and output signals are fully differential and quadrature in phase. The core active component can be an OP-Amp-based stage or a gm-C stage. The frequency planning of a low IF receiver is illustrated in Figure 2.20.

2.7 DIRECT CONVERSION ARCHITECTURE

In a direct conversion architecture, RF signals are directly converted to DC, and I/Q signals can be available at the ADC input. As the RF and LO frequencies are very close to one another, it is also called the "homodyne" architecture.

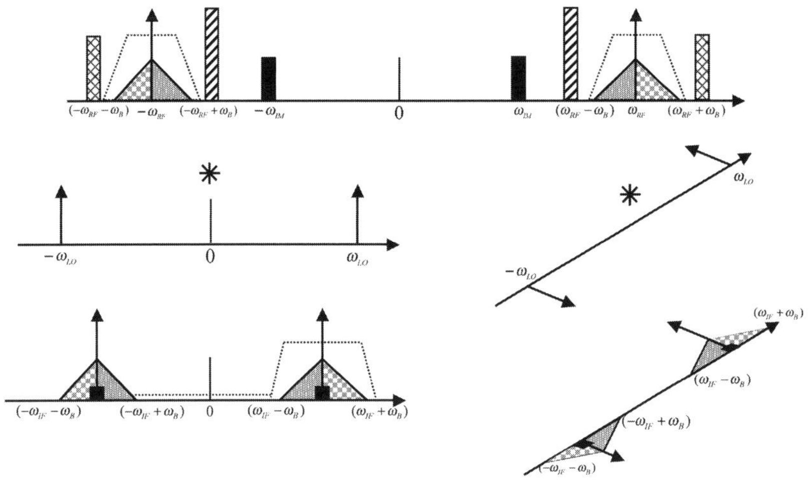

Figure 2.20. Frequency planning of a low IF receiver.

2.7.1 Advantages

This architecture uses a minimum amount of high-frequency hardware and a minimum extent of filtering, which leads to lower power and a smaller form factor. In addition, the baseband circuits need to process signals up to half the signal bandwidth. At the baseband, I/Q signals are directly available for any calibration and compensation. For these reasons, direct conversion architecture has been adopted for many wireless solutions. Downconverting RF signals to DC provides flexibility in frequency planning in transceivers. It makes system design much easier w.r.t. frequency planning and integration of multiple oscillators, and so on.

2.7.2 Modulation

In a direct conversion radio, impairments around DC cause significant performance degradations. Hence, suitable modulation techniques must be used in designing direct conversion architectures. Impairments around DC can be filtered using a high-pass filter. However, decisions about the cutoff frequency of these filters are based on the SNR and settling times. An example of DC-free modulation includes (1) FSK modulation and (2) OFDM modulation with elimination of tones around DC. Direct sequence spread spectrum (DSSS) modulation can be used, and this can provide immunity to impairments around DC.

2.7.3 Architecture and Frequency Planning

Figure 2.21 illustrates the receiver architecture for direct conversion. The filter following VGA can be a fully integrated low-pass filter. The order of this filter is dependent on the rejection profile required for the blockers at the receiver.

At the output of the ADCs, I and Q signals are available in digital form, which can be used for compensation and calibration of I/Q mismatches, and so on. In a low-IF

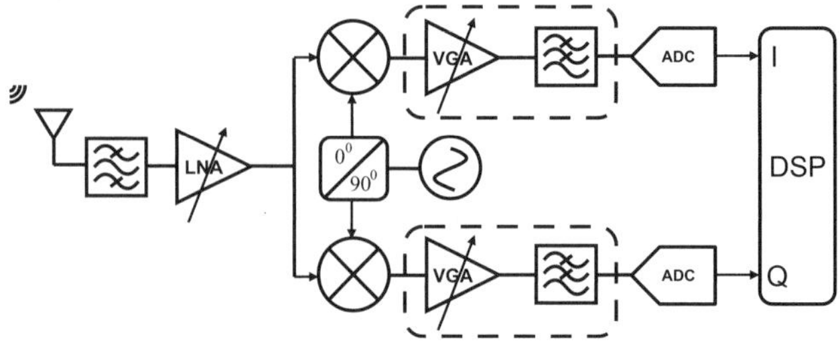

Figure 2.21. Single-stage direct conversion receiver.

architecture, two IF amplifiers and one ADC (operating at twice the speed of direct conversion ADC) are used. In a direct conversion radio, the signal processing speeds are usually half of the low-IF processing speeds, reducing the power consumption. At the same time, I and Q signals are available explicitly at baseband, which can be used for additional signal processing.

Frequency planning of a single-stage direct conversion receiver is illustrated in Figure 2.22. As illustrated before, the quadrature phase shift is provided from the signal generator. Figure 2.23 illustrates implementation of a direct conversion receiver for FSK modulation systems.

This simplistic implementation requires no image rejection stages, which leads to very compact and low power solutions used in pagers. However, the decision on demodulated waveform is significantly affected because of any spikes on the clock or data signal lines.

2.7.4 Challenges in the Direct Conversion Receiver

2.7.4.1 Finite IIP2, IIP3. Direct downconversion receivers are vulnerable to the even and odd orders of intermodulation terms. Equations (2.20) through (2.26)

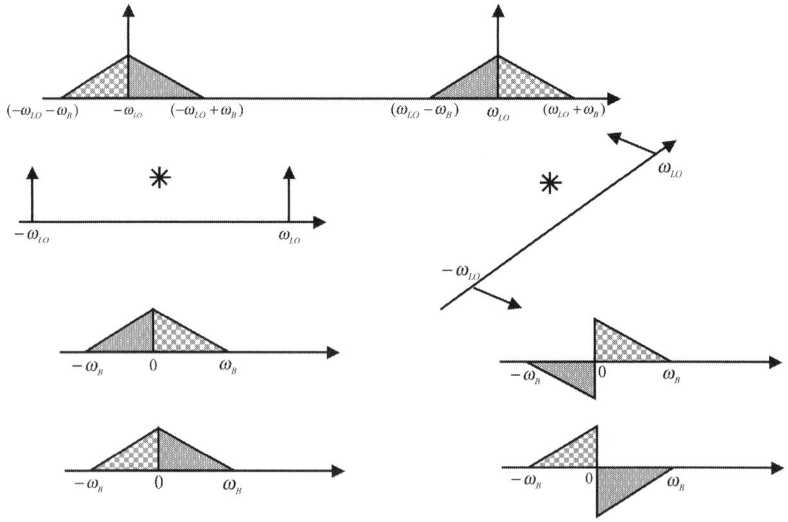

Figure 2.22. Frequency planning of a zero IF receiver.

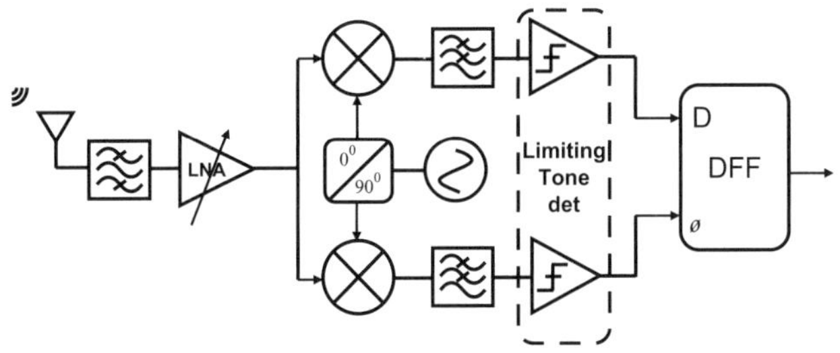

Figure 2.23. Direct conversion receiver for FSK demodulation.

incorporate two-tone mixing and generation of IM2 and IM3 signals in addition to the DC offsets. These equations show demodulation of two baseband tones ω_a and ω_b that have been modulated on an RF carrier labeled as ω_1 and ω_2.

$$\omega_1 = \omega_{RF} + \omega_a \tag{2.20}$$

$$\omega_2 = \omega_{RF} + \omega_b \tag{2.21}$$

$$\omega_{LO} = \omega_{RF}(\text{for direct conversion}) \tag{2.22}$$

$$\begin{aligned}\text{RF}_{\text{input}} = {} & C_1\cos(\omega t - \omega_1 t) + C_2\cos(\omega t - \omega_2 t) + C_{\text{IM3}}\cos[\omega t - (2\omega_2 - \omega_1)t] \\ & C_{\text{IM3}}\cos[2\omega t - (2\omega_1 - \omega_2)t] + C_{L+R}\cos(\omega t - \omega_{LO} t) + \ldots\end{aligned} \tag{2.23}$$

$$\text{LO} = C_{\text{LO}}\cos(\omega t - \omega_{LO} t) \tag{2.24}$$

$$\text{IF} = \text{RF}_{\text{input}} \times \text{LO} \tag{2.25}$$

$$\text{IF} = 1/2C_{\text{mixer}}C_{\text{LO}}\left[\begin{array}{l} C_1\cos(\omega t - \omega_a t) + C_2\cos(\omega t - \omega_b t) + \ldots \\ C_{\text{IM3}}\cos[\omega t - (2\omega_a - \omega_b)t] + C_{IM3}\cos[\omega t - (2\omega_b - \omega_a)t] + \ldots \\ C_{\text{IM2}}\cos[\omega t - (\omega_a - \omega_b)t] + \ldots \\ C_{L+R} + \ldots \end{array}\right] \tag{2.26}$$

Several mechanisms can contribute to intermodulation distortions in a DDC receiver. As demonstrated in Eq. (2.26), the major portion of the even-order intermodulations are generated by the nonlinearities in the mixer where two adjacent RF tones or interference signals are mixed together to generate a low-frequency beat at the mixer output. It occurs because of mismatches in mixer components as well as because of the presence of even-order effects in the mixer switching waveform (duty cycle error). The second-order intermodulation generated from the LNA is not important, as it gets upconverted by the mixer, and eventually filtered out. Even-order intermodulation causes SNR degradation from an amplitude-modulated interferer. As shown in Figure 2.24, the AM modulation on such an interferer can be demodulated in the mixer, resulting in AM noise in the baseband output. Using the same analogy of time and frequency domain representations, even-order nonlinearity leads to signal spreading that occupies twice the modulated bandwidth of the

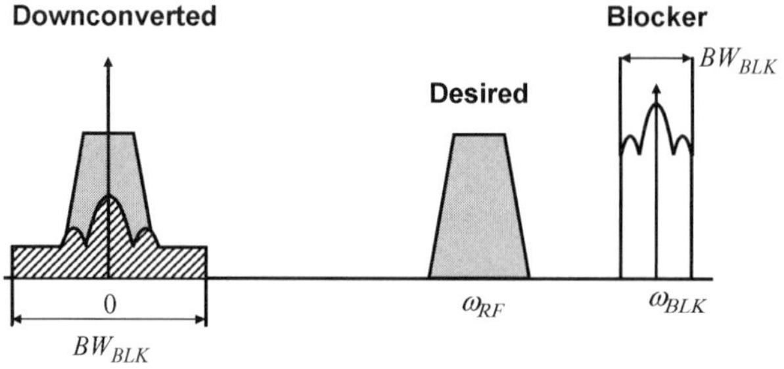

Figure 2.24. The effect of finite IIP2.

interferer. Because of its frequency spreading nature, a modulated blocker may provide more SNR degradation compared to a single-tone blocker. When two blockers are present, one single tone and another modulated, separated by the half bandwidth of a desired signal, SNR degradations occur and the distortion terms appear in-band from the modulated blocker. These scenarios can be easily evaluated in terms of mathematical convolution, as illustrated before.

Also as demonstrated in Eq. (2.26), the RF signal carries third-order distortions that are generated by the nonlinearities in both the LNA and the mixer. These odd-order distortions are then demodulated by the LO signal and downconverted to the baseband, adding to the additional odd-order products generated in the mixer, further corrupting the desired signal.

Two methods can be used to alleviate the problems of finite IIP2. In the first approach, the common-mode impedances can be increased, in order to reduce the generated IM2 product. In the second method, an auxiliary clock can be used to provide chopper action and the IM2 products can be translated outside the bandwidth under consideration. IIP2 limitation results from the mismatches of transistors, and so on, which are significantly enhanced in the second approach.

Figure 2.25 illustrates a technique in order to enhance IIP2 of receivers. IIP2 is mostly determined by the mismatches of the components in the mixer, delay mismatches in the signal generation path, and any mismatches in the load network. Even with the best possible layout, some mismatch is inevitable. Improvements can be made by using dynamic matching techniques. It is used in precision OP-Amps (chopper stabilized amps) and linear multibit DACs. It can be used to alleviate component and device mismatches. The overall operation in the receive path is given by

$$BB(t) = [\phi_i \times \cos(\omega_{\text{LO}} t) \times \phi_o]\cos(\omega_{\text{RF}} t)$$

In the actual implementation, ϕ_i, ϕ_o are square wave signals, and they are implemented using FET switches in conjunction with the actual mixer. These two signals are synchronized with each other, and their product forms the downconverted signal. Even-order products generated at low frequencies are frequency translated out of band, and they are eventually filtered out by the LPF at baseband. A PN sequence can be used to spread the imperfections. Usually, a periodic waveform can be used for the ease of implementation.

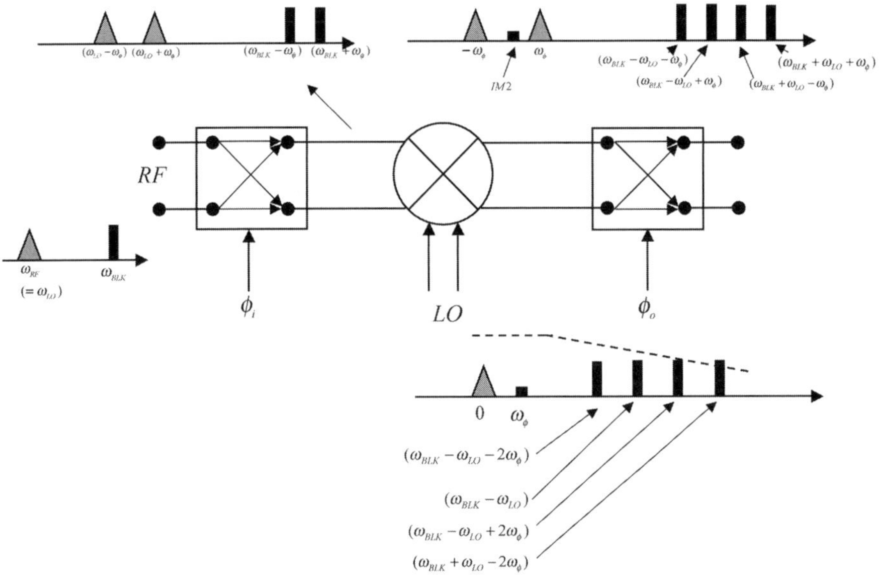

Figure 2.25. IIP2 enhancement approach.

The input switches modulate the bandlimited input RF signal to frequencies $\omega_{\mathrm{RF}} \pm \omega_\phi$. The LO switches translate these frequencies to $\omega_{\mathrm{RF}} \pm \omega_\phi \pm \omega_{\mathrm{LO}}$. The next set of switches operate at ω_ϕ, which leads to the baseband signal at $\omega_{\mathrm{RF}} \pm \omega_{\mathrm{LO}}$. All imperfections near DC caused by IIP2 (resulting from the RF input stage transconductors) as well as flicker noise gets upconverted to ω_ϕ, where ω_ϕ is a frequency outside the bandwidth of the baseband filter. A reasonable choice can be made by selecting ω_ϕ to be twice the bandwidth of the baseband filter. The spurious responses from ω_ϕ would be filtered by the mixer's filtering networks.

In the above discussion, we have considered the intermodulation terms caused by in-band signals only. However, the same consideration is true for out-of-band signals as well. Finite intermodulation terms are applicable for two out-of-band signals, the cross product of which falls inside the bandwidth of interest.

2.7.4.2 DC Offset. DC voltages in the demodulated spectrum of a DDC receiver not only corrupt the signal spectrum, but they can also propagate through the baseband circuitry and saturate the subsequent stages. These DC offsets are mostly generated through self-mixing of LO signal and mismatches in the mixers. To avoid the frequency coupling issues, the VCO is usually operated at a frequency different from the incoming RF frequency. Because of the large signal operation, VCOs can couple to other parts operating at the same frequency, resulting in DC offsets.

In direct conversion receivers, the mixer is immediately followed by LPFs and a chain of high-gain direct-coupled amplifiers that can amplify small levels of DC offset and saturate the stages that follow (Figure 2.26). Consequently, the sensitivity of the receiver can be directly limited by the DC offset component of the mixer output. The

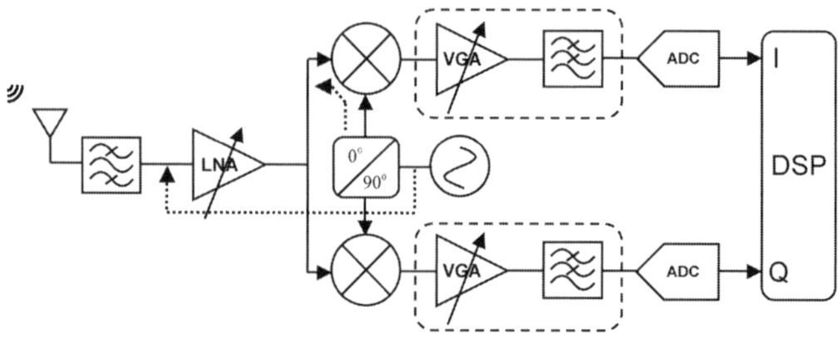

Figure 2.26. A direct conversion receiver with a leakage mechanism.

DC offset of a mixer can be separated into two components, a constant and a time-varying offset. The constant DC offset can be attributed to the mismatch between mixer components, whereas the time-varying DC offset is generated by self-mixing of the LO. As demonstrated in Figure 2.27, a finite amount of LO leakage into the RF port persists because of the imperfect isolation between the LO and the RF ports of the mixer. In addition, LO leakage, and even the LO radiation, can reach the LNA or other stages prior to the mixer and propagate through the front end. DC offset levels can also vary depending on the time-varying load of the antenna.

Representing LO leakage and radiation level by C_{L+R}, mixer conversion gain by C_{mixer}, and LO level by C_{LO}, Eqs. (2.27) through (2.31) demonstrate the recovery of a baseband tone (ω_a) by the LO frequency (ω_{LO}). As observed in Eq. (2.31), the amplitude of the DC offsets generated from the self-mixing is mainly proportional to the LO leakage and radiation C_{L+R}, mixer gain, and the LO power level C_{LO}.

$$\omega_1 = \omega_{\text{RF}} + \omega_a(\text{desired signal}) \tag{2.27}$$

$$\omega_{\text{LO}} = \omega_{\text{RF}}(\text{fordirect conversion}) \tag{2.28}$$

$$RF_{\text{input}} = \cos(\omega t - \omega_1 t) + C_{L+R}\cos(\omega t - \omega_{\text{LO}} t) + \ldots \tag{2.29}$$

$$\text{LO} = C_{\text{LO}}\cos(\omega t - \omega_{\text{LO}} t) \tag{2.30}$$

$$\text{IF} = \text{RF}_{\text{input}} \times \text{LO} = 1/2 C_{\text{mixer}} C_{\text{LO}}[\cos(\omega t - \omega_a t) + C_{\text{LO}} C_{L+R} + \ldots] \tag{2.31}$$

A possible solution for removing the DC offset is AC coupling of the mixer output. This process will not only remove the unwanted DC offsets, but at the same time, it will

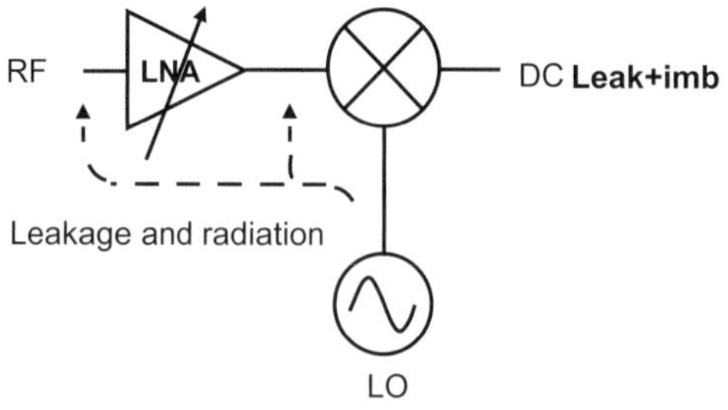

Figure 2.27. LO Leakage and radiation.

corrupt the downconverted signal by attenuating the components near DC. At the same time, this is not acceptable for demodulating most random binary modulation schemes that exhibit a DC peak in their signal spectrum. Use of "DC-free" modulation schemes such as binary-frequency shift keying (BFSK) can reduce susceptibility to DC offsets while taking away spectrum efficiency and other advantages of popular and mainstream digital modulation schemes. Therefore, DC offset cancellation techniques are necessary to accommodate the use of direct conversion topology in today's wireless applications. The signal distortion would be lower and could be of an acceptable level, if the time constant is longer ($\sim$1 s), but this long time constant makes the settling time of the receivers too long, and it cannot be integrated as a part of analog circuits.

Many DC offset cancellation techniques have been reported over the past few years, which can be divided into baseband analog and digital techniques. In the DSP, a complex nonlinear scheme can be used toward estimation of DC offset in a dynamic manner. In many cases, the reported techniques only provide solutions for specific system topologies.

The static part of the DC offset is caused by systematic mismatch of components, and it can be canceled by the baseband at power-up. The dynamic component of the DC offset can be caused by a change in an antenna's reflection characteristics and LO leakage.

DC offset correction can be performed in two ways: (1) DAC-based calibration and (2) downconversion-based calibration. In a DAC-based calibration, DC currents are programmed from high-impedance current sources and are fed to the baseband amplifier's input node. The DAC should be optimized for a low noise injection, and the extent of DC offset cancellation would be dependent on the DAC resolution. In a servo loop configuration, there may be phase perturbation of the filter's response near DC.

2.7.4.3 LO Leakage. Component mismatches in the receiver lead to DC offset, and LO leakage in the transmitter. In direct conversion, the LO tone is situated at the center of the transmitted band. The level of this LO w.r.t. transmitted signal can increase when we are operating at the lowest power level of the transmitter. For this reason, the maximum amount of transmitter gain steps are provided in the RF and not in the baseband.

Figure 2.28 illustrates a single-stage direct conversion transmitter. It should be noted that the signal generation path consists of VCO, divider, and phase locked loop

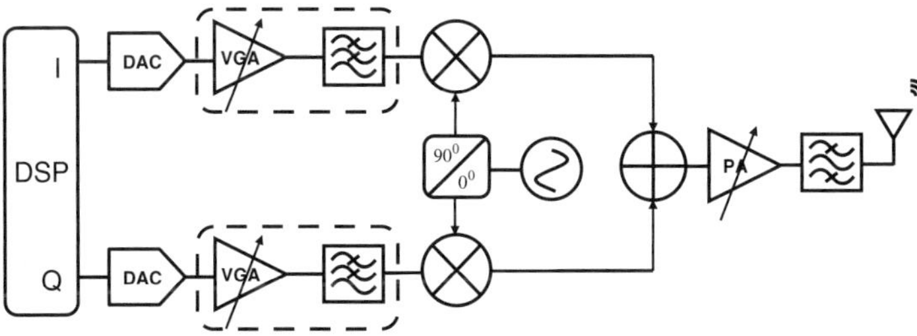

Figure 2.28. A single-stage direct conversion transmitter.

(PLL). The filtering needed at the baseband section is fairly relaxed, and it can be realized using a single-stage filtering. Modulators are responsible for single sideband combination, which is performed by combining the output currents of I and Q upconverted paths. As I/Q signals are available at baseband, compensating for the mismatches would allow complete LO cancellation. It can be performed in two ways: (1) DAC-based calibration and (2) downconversion-based calibration.

In a DAC-based correction scheme, the output signal from the PA is coupled, and envelope detection is performed in order to detect the signal level. Depending on this level, the DAC would output DC currents from high-output impedance current sources in order to cancel the effects of LO leakage. In a downconverted scheme, the output signal from the PA is coupled, downconverted to I/Q, and properly weighted for a correction of initial I/Q signals.

2.7.4.4 I/Q Imbalance. Figure 2.29 illustrates the frequency planning and I/Q combination at the transmitter. Also, any imbalance in the I and Q paths results in the SNR degradations, as the third harmonic of the baseband signal (ω_{BB}) is located in the message spectrum on the opposite side ($-3\omega_{BB}$) of the LO frequency. This appears as an in-band spurious signal and degrades the SNR in the transmitter. Since this is related to the third harmonic distortion of the baseband signal, the transconductor part of the transmit modulator may consume a significant amount of current to enhance linearity.

In the receiver, I/Q imbalance distorts the received symbol constellation. A phase imbalance rotates the constellation, whereas an amplitude imbalance compresses and expands the constellation. The combination of both effects can be attributed by error vector magnitude (EVM). Once again, as the I and Q signals are directly available at baseband, these impairments can be corrected using DSP.

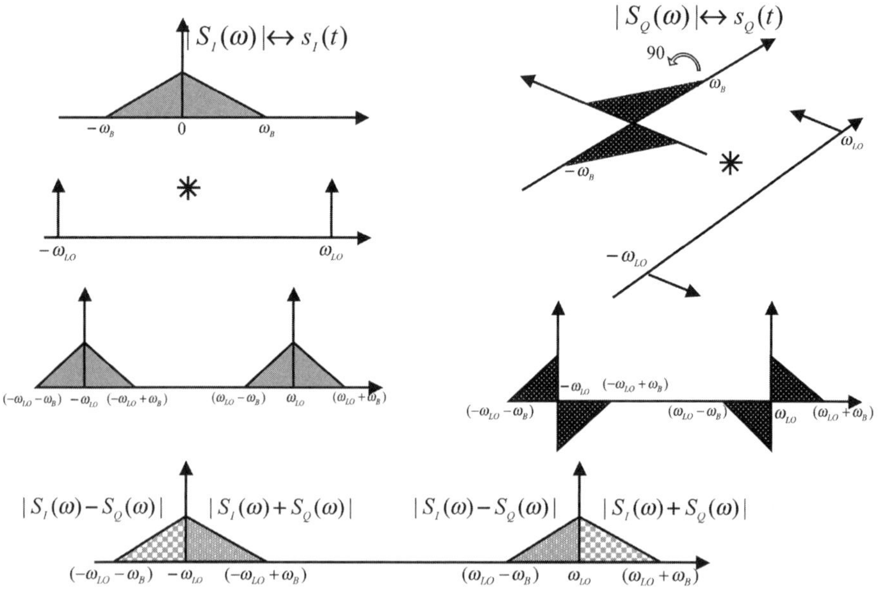

Figure 2.29. Frequency planning of a zero IF transmitter.

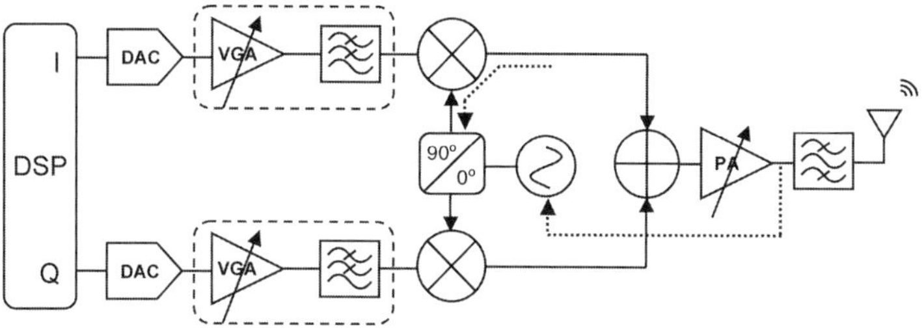

Figure 2.30. A direct conversion transmitter with a leakage mechanism.

2.7.4.5 LO Pulling. Lo pulling can also happen in direct conversion transmitters. Figure 2.30 illustrates a situation where the VCO is operating at twice the transmitter frequency.

Because of finite mismatches in the PA circuitry and/or load components, it can generate a second harmonic, which is the same as the VCO frequency. If this is integrated in the same die (which is usually the case for modulator and PA driver), the output of a PA driver can interact with VCO as both are at the same frequency. This leakage can occur because of the finite resistance of the substrate, as well as electromagnetic cross-talk through the supply lines and inductive loads.

2.7.4.6 TX-RX Crosstalk. Figure 2.31 illustrates an integrated direct conversion transmitter and receiver. The cross-talk performance is dependent on the number of blocks operating at the same frequency. Let us consider the case where a single transceiver is present in a die, where both transmitter and receiver are operating at the same time but at different frequencies. This is commonly known as the frequency

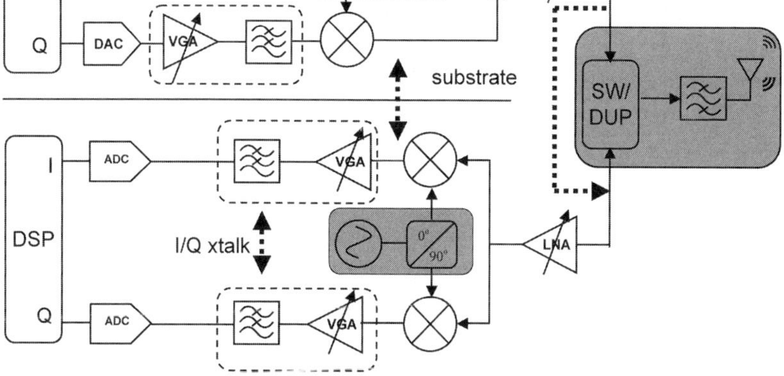

Figure 2.31. Direct conversion transceiver with various cross-talk mechanisms.

division duplex (FDD) architecture. In this scenario, finite leakage from a duplexer will lead to leakage of the transmitter signal into the receiver, and it may cause severe performance degradations. This situation is apparently not present in the case of a time division duplex (TDD), as the receiver and transmitter do not operate at the same time. However, in recent times, multiple transceivers are being integrated in the same die, and it is possible that two radios may operate simultaneously. In this situation, finite leakage from the "ON" transmitter to the "ON" receiver will cause severe cross-talk.

Inside the same transceiver chain, cross-talk may occur through substrate coupling. Let us assume that the LO frequency is twice the RF frequency, and that the architecture uses a direct conversion scheme. In this case, the PA will cause a finite amount of second harmonic of RF frequency, which will be the same as the VCO operating frequency. In this case, cross-talk will occur between the two blocks in the transmitter. Some of it is associated with the electromagnetic cross-talk occurring through inductors and antenna. The situation becomes worse when multiple radios are integrated in the same die, and many radios may operate simultaneously.

2.7.4.7 Flicker Noise. In the transmitter, the 1/f noise is produced by various noise sources and the noise is "up-converted" through the switching action of the oscillator transistors. In the transmitter, this 1/f noise appears around the LO tone and degrades SNR, whereas in the receiver, it appears around DC and degrades SNR. There could be many ways to mitigate this effect:

1. Consume significant current in the oscillator: Phase noise relates to the current consumed in the VCO, and in the current limited region, with every doubling of current in the VCO core, phase noise improves by 3 dB.
2. System techniques: The choice of modulation technique is important for direct conversion radio. Ideally, a modulation scheme with high spectral efficiency would be suitable, with no energy at DC. Multicarrier modulation schemes, such as OFDM proves to be efficient in this aspect, as dropping a few carriers near DC does not cause much SNR degradation. Another example of DC-free modulation would be FSK, although the spectral efficiency is poor. Direct sequence spread spectrum can be very suitable, as it is mostly immune to the noise and, with a large spreading gain, can be very suitable for such applications.

2.8 TWO-STAGE DIRECT CONVERSION

So far, we have discussed single-stage direct conversion transmitter and receivers. However, a direct conversion architecture may use two stages of up/down conversion. Figure 2.32 illustrates a two-stage direct conversion, where two frequency translations are involved, and a direct access of I/Q signals is possible at the baseband. Although this approach involves more hardware and power dissipation, often it is preferred in order to improve blocker performances through the radio signal chain.

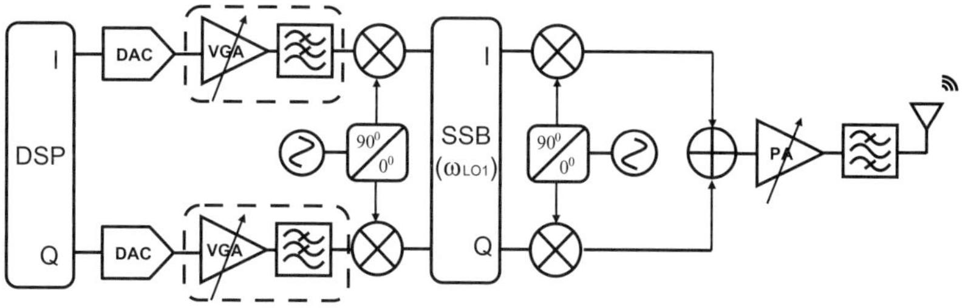

Figure 2.32. Dual-stage direct conversion transmitter.

Although it is commonly perceived that the "downconversion" is performed using an LO frequency lower than RF frequency, fundamentally, the opposite could be true as well. The choice of frequencies depends on the blockers, placement of image, and on how many radios are integrated in the same die. In the latter situation, the area and power consumption considerations should be performed, considering all radios, and not a single radio itself.

2.9 CURRENT-MODE ARCHITECTURE

Any radio system thrives on dynamic range, which is the ratio of maximum signal handling capability to the noise floor and distortions. Often the ratio of "signal to noise and distortion" or SNDR is used in terms of system design. The intermodulation products are a function of the system bandwidth and linearity of the building blocks. Most front-end nonlinearity occurs in the voltage to current-mode conversion. Once the signal is available in the current mode, signal processing becomes linear. In the case of narrowband wireless systems, noise is the dominant degradation term in SNDR, whereas in the wideband systems, intermodulations become the dominating term. This current is again converted to voltage using transconductance stages, and we must allow sufficient headroom for the current processing elements. Figure 2.33 illustrates this situation.

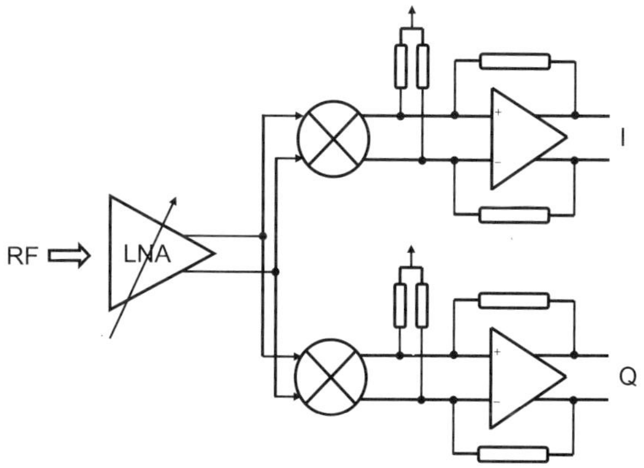

Figure 2.33. A current-mode output mixer.

The LNA converts voltage to current, which is processed using a switching mixer, and the current is input to the virtual ground of the baseband filter. As it is at virtual ground, the signal swing is reduced, and hence, the distortions are minimized throughout. However, to minimize linearity degradations, a minimum number of active stages (or V–I conversion stages) should be used. However, use of minimum active stages achieves lower gain (as much gain in a single stage is avoided for stability reasons). It provides a direct trade-off between noise and linearity in any receiver architecture. In a cascade of multiple active stages, the current consumption increases subsequently in order to obtain good linearity.

2.10 SUBSAMPLING ARCHITECTURE

Radio architectures evolve from many diverse disciplines, and sometimes, they are motivated by thoughts from the signal processing domain. A judicious utilization of continuous time and discrete time signal processing can be used in the front end. Subsampling architectures are motivated by the faster processing speeds of semiconductor processes, and they are illustrated in Figure 2.34.

From a message reception and demodulation perspective, we are only interested in the message bandwidth, and thus, the incoming signals, after being amplified by an LNA, can be digitized by the faster sampling clock at the ADC, which leads to discrete samples at the output. The sampling rate should ideally satisfy the Nyquist criterion for the message bandwidth. Thus, the sampling rate is a fraction of the incoming RF signal and, hence, the name "subsampling." However, as the sampling frequency is low, noise contributions from all the harmonics of the clock frequency appear in the message bandwidth and degrade the overall noise figure of the receiver. Thus, although this architecture is suitable for extensive digital processing and flexibility, it suffers from the low sensitivity in the front end. Figure 2.35 illustrates the frequency planning of the subsampling receiver.

In a subsampling radio, the desired frequency content at $n\omega_s$, $n = \pm K$, is down-converted along with the thermal noise from other clock harmonics. A simple filtering can be used to reject the blockers. However, the thermal noise aliasing from multiple harmonics of the sampling clock cause major degradation of the SNR.

Figure 2.36 illustrates a circuit implementation as well as the sampling instants for a subsampling receiver. The input RF current is sampled using the sampling switches, and fed to the baseband filter, which rejects the out-of-band blockers.

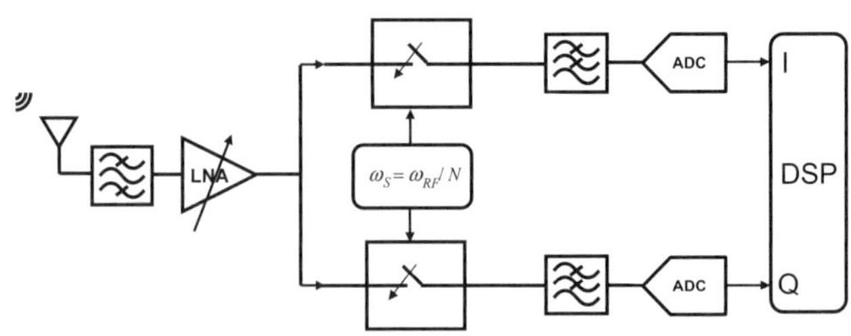

Figure 2.34. Subsampling receiver architecture.

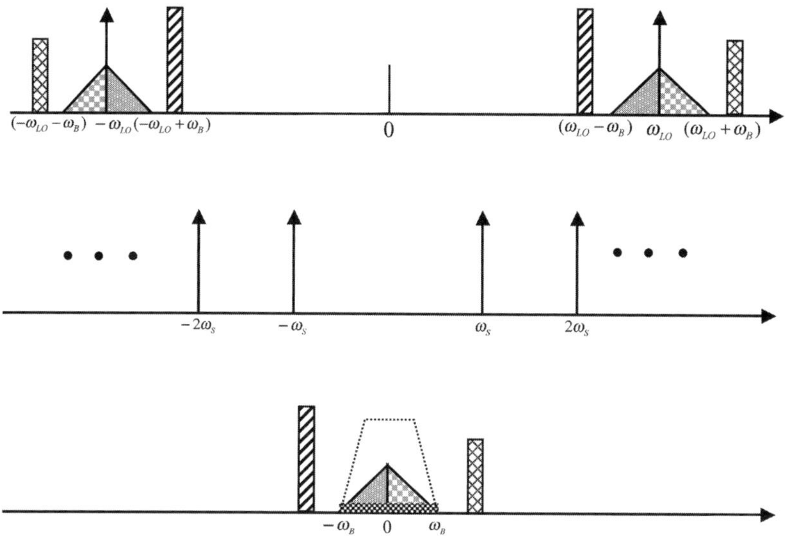

Figure 2.35. Frequency planning of the subsampling architecture.

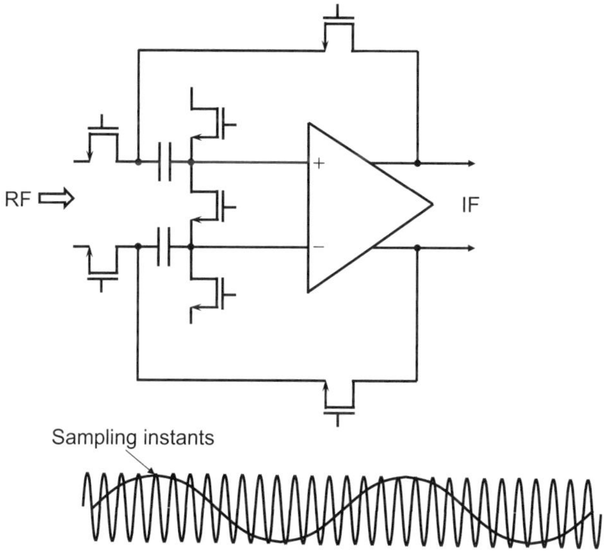

Figure 2.36. Subsampling receiver architecture.

2.11 MULTIBAND DIRECT CONVERSION RADIO

Because of many emerging applications to support high data rates for video and other multimedia applications, supporting multiband front ends becomes essential. In modern times, many high-data-rate applications, such as UWB, have become attractive. In UWB applications, transmission occurs at different center frequencies, and

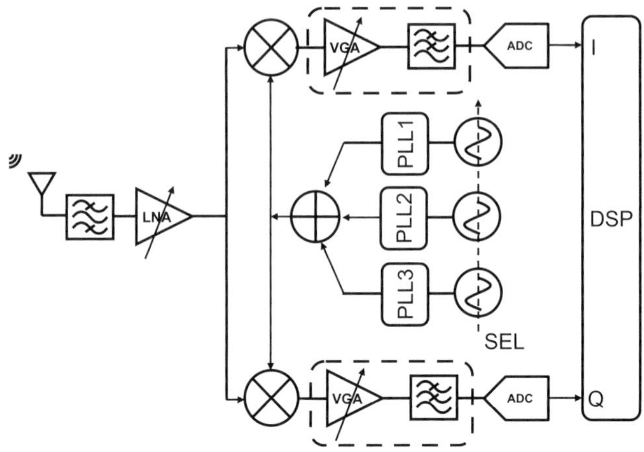

Figure 2.37. The direct conversion receiver architecture for multiband applications.

they last for brief periods of time. In these cases, PLL locking times become important system design aspects. The multiband UWB standard is a classic example of a frequency-hopped, high-frequency direct conversion radio. As this approach requires fast frequency hopping, it poses a significant challenge to the signal generation scheme. As conventional PLL-based approaches require a long time to settle, the center frequencies can be generated beforehand, and a frequency selector employing multiplexing logic can select the desired frequencies. Two approaches can be taken in order to achieve the fast switching: (1) offset mixing technique and (2) multiple PLL technique. In the offset mixing approach, a single VCO–PLL combination is used, and multiple frequencies are generated by frequency dividers, and they are combined using a single sideband mixer to produce various center frequencies. However, a significant disadvantage of this technique is the presence of spurious tones at the output of these offset mixers, which causes SNR degradations because of unwanted downconversion. To alleviate this problem, multiple PLLs can be used, which are individually locked, and the outputs can be selected using a multiplexer block. Although it consumes area, it provides much superior performance compared with the offset mixing approach. Figure 2.37 illustrates frequency-hopped, multiband direct conversion architecture.

Although the above approach is used to illustrate multiband OFDM applications, it can be used for other multiband direct conversion radios as well. One advantage in using a direct conversion radio for multiple frequency bands is its use of single LO for the downconversion scheme, which is suitable for frequency planning. An implementation of such architecture using ring oscillators has been illustrated in [10].

2.12 POLAR MODULATOR

In the previous sections, we have discussed the signal processing aspects in terms of a Cartesian modulator approach. In a Cartesian approach, a band-pass signal can be constructed from its orthogonal basis as $X(t) = I(t)\cos(2\pi f_c t) + Q(t)\sin(2\pi f_c t)$. Because of this formulation, the baseband signals are translated up in frequency in the case of a transmitter and combined to provide the transmitted signal. At the receiver, this band-pass signal is "downconverted" by the same quadrature basis function in terms of carrier at frequency f_c. Depending on the specific architecture under consideration, this band-pass signal can be obtained in a single-stage frequency upconversion or multistage upconversion, followed by interstage filtering. The individual signal paths I and Q experiences homogeneous signal processing operation in terms of frequency shift, amplification, and filtering. To ensure the quality of the band-pass signal, close attention needs to be paid to the matching of the components. Hence, in a typical direct conversion transmitter, the baseband signals are filtered, upconverted, and added in order to provide a single sideband transmitted signal. One or more VCO–PLL is necessary in the signal generation path.

The band-pass signal in the time domain can be represented in polar format, and it can be represented as $X(t) = r(t)\cos(2\pi f_c t + \theta(t))$. The phase modulation can be implemented in a VCO–PLL combination (fractional PLL), and the amplitude modulation can be implemented by varying the supply voltage or bias current in the driver amplifier following the VCO and buffer. Using this technique, I–Q switching mixers can be eliminated, and the architecture can be significantly simplified, which leads to a low-power transmitter implementation. However, because of the PLL stability, the data rate is limited by the loop bandwidth. In the case of a constant envelope modulation, this approach is suitable, as the amplitude can be held at a constant value while VCO–PLL changes in frequency. Figure 2.38 illustrates the polar modulator architectures.

As illustrated here, the baseband part provides the two quantities, amplitude (r) and phase (θ). Information in phase changes the center frequency of the VCO–PLL combination by adjusting the inputs to Σ-Δ modulator, whereas the amplitude signal is filtered before providing it to the power amplifier output. The supply voltage or the programming current of the PA driver can be changed to reflect this amplitude.

In a practical implementation, an I-Q modulated signal can be represented by, IF $(t) = A(t)\cos(\omega_{IF} t + \theta(t))$. The phase information is obtained by limiting this signal

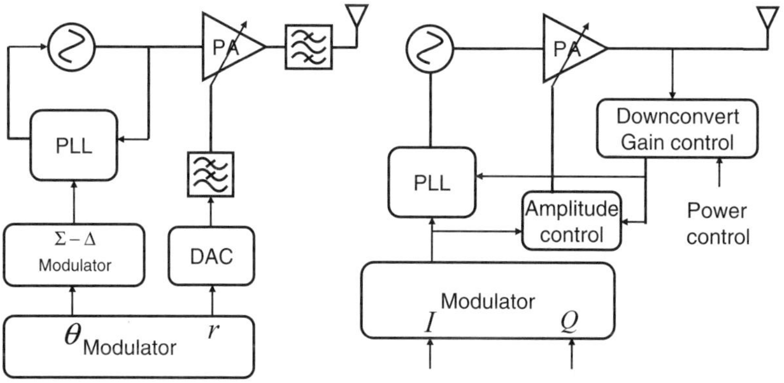

Figure 2.38. Polar modulator: (a) open loop and (b) polar loop transmitter.

and by removing the amplitude information. The amplitude information is obtained by envelope detection of the IF signal, which is obtained by mixing the If signal with itself and low-pass filtering the resultant output (filtering removes the $2\omega_{IF}$ component). The phase modulation path is then applied to the VCO control signal path, and the amplitude is applied to the PA driver circuitry to control its supply voltage or bias current. Thus, the phase and amplitude information are obtained through separate paths, and their individual effects are combined at the RF domain. In the case of a constant envelope modulation, the amplitude path can be eliminated, which leads to additional saving in power and efficiency.

Although attractive in nature, polar modulators provide their own challenges. The difference in delay between the amplitude and phase paths provide significant degradation of the transmitter EVM and, hence, the constellation. Inadequate bandwidths in the AM and PM paths result in spectral regrowth, which can violate the transmitter spectral mask. The time constants of amplitude and phase paths are fundamentally different in nature. VCOs operate in a positive feedback loop, and PLLs provide phase/frequency feedback in addition to this. Hence, the response time corresponding to any change in the frequency path is significantly different from the response time of an amplifier. This delay mismatch leads to significant degradations in transmit EVM. The group delay of the AM path must match the group delay of the PM path to maintain the transmitter spectral mask requirement. Other impairments include (1) DC offset in the amplitude path and (2) finite isolation between the PM signal and the transmitter output. However, for a suitable application at hand, polar modulators provide the best possible power efficiency in a transmitter.

2.13 HARMONIC REJECT ARCHITECTURE

The coexistence of various wireless standards in ISM bands has become a common practice in modern wireless communication systems. It results in the downconversion of unwanted RF signals operating at a harmonic of the receiver clock frequency. Let us assume that a "hard-switched" mixer is operating at a frequency f GHz. This will downconvert the signals present at $3f$, $5f$, and so on, as they are part of a spectrally decomposed square waveshape. Hence, a receiver designed at 800 M would downconvert Bluetooth signals at third harmonic (2400 M); a center frequency of 1100 M would downconvert a 5.5-GHz WLAN signal using its fifth harmonic. This leads to the requirements of extensive filtering in the front end.

Figure 2.39 illustrates the harmonic reject architecture. It can be adopted in the transmitter as well as the receiver. The LO waveform, a square wave, is delayed and advanced by 45° in phase with an amplitude of 0.707 times the amplitude of the original waveform. Frequency multiplication also implies phase multiplication, which leads to a delay of 135° at third harmonic and of 225° at the fifth harmonic component. The vector sum of the output currents result in cancellation of third and fifth harmonic terms. However, it can be seen that the seventh and ninth harmonics are boosted up, although their relative levels are small compared with the fundamental amplitude.

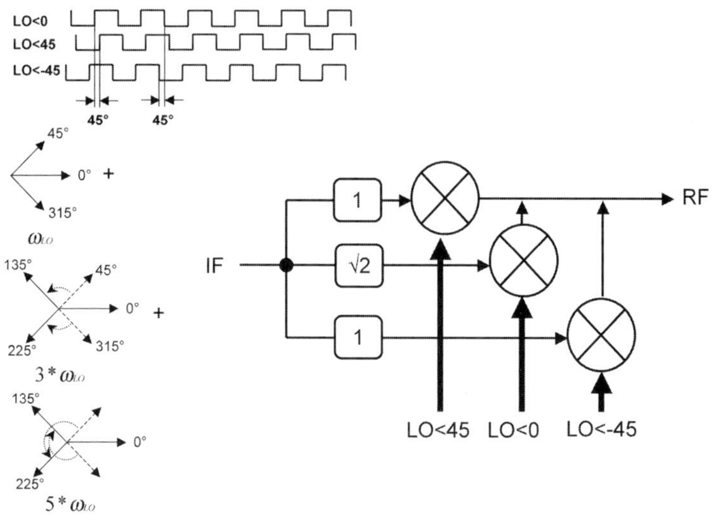

Figure 2.39. Harmonic reject architectures and operating principles.

The extent to which harmonic reject architecture can be successful depends on the matching of the different paths of the LO waveform. Any mismatch leading to the deviation from 45° phase would lead to imperfect suppression of third and fifth harmonic terms.

2.14 PRACTICAL CONSIDERATIONS FOR TRANSCEIVER INTEGRATION

2.14.1 Transmitter Considerations

The output of the TX modulator drives a PA driver, and the input impedance of the PA driver determines the signal swing at the interface of the modulator and PA driver. Two scenarios can be considered: (1) power-efficient architectures, which employ mostly a constant envelope modulation; and (2) bandwidth-efficient architecture, where linearity and performance are key considerations. In terms of circuit implementations, current-mode and voltage-mode interfaces should be used as appropriate. In voltage-mode interfaces, progressively higher impedance levels are used, whereas for current-mode interfaces, progressively lower impedances are used. Current sourcing and sinking from the baseband stage should be minimized to consume overall lower DC current in the transmitter. Inductively loaded circuits provide current gain by the loaded tank quality factor, and they lead to a lower power in the circuits. Harmonic components of the transmitter are obtained by Fourier transform of the output waveform over a complete cycle of baseband frequency. Linearization techniques at baseband are used to provide low levels of in-band degradation as shown in Figure 2.40.

Gain control in a transmitter is an important consideration, and it can be performed by using any of the techniques that were mentioned earlier, or changing the bias

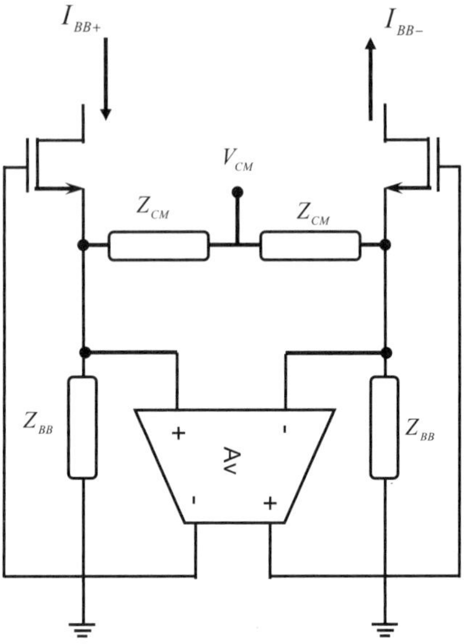

Figure 2.40. Circuit used to implement baseband linearization functionality.

current, as at lower power transmit, we also consume lower currents in the transmitter. However, depending on the topology of the PA driver, a change in bias current may lead to a change in output matching levels. Gain control in the transmitter is usually implemented at RF; otherwise, the transmitter performance would degrade w.r.t. LO leakage at the output of the transmitter. PA drivers should meet the linearity requirements in order to comply with the spectral mask and adjacent channel leakage requirements.

2.14.2 Receiver Considerations

Most receiver functionality and considerations have been presented in terms of building blocks in this chapter. Once again, two fundamentally diverse applications are prevalent: (1) power/energy efficient radios and (2) bandwidth-limited radios. Impedance plays a significant role in determining the circuit interfaces in terms of current mode or voltage mode. Use of active stages provides the necessary gain in the signal path, while degrading the intermodulation distortion (and hence linearity) performance. However, a reduced number of active stages leads to lower sensitivity of the receiver.

A signal strength indicator provides a key functionality to any receiver implementation. Fundamentally, as receivers move from the edge of the cell to the proximity of the base station, the distance changes, and the power consumption needs to be dynamically scalable to obtain highest battery efficiency. In next-generation standards, the modulation scheme (QPSK, QAM, etc.) can be adaptively used depending on application in order to provide energy efficiency for signal processing.

CONCLUSION

In this chapter, we have discussed various wireless architectures from a system perspective. Fundamentally, wireless signal processing can be visualized as a combination of amplification, selective rejection, frequency translation, sampling, and detection functionalities. The choice of a specific architecture for a certain application is based on the characteristics of desired and undesired signals. In many cases, the power consumptions of radios are dictated by blockers and enhanced linearity requirements. Desired signals influence the selection of center frequency, bandwidth, and modulation techniques (dictates the crest factor). Undesired signals (blockers) influence the architecture decision in terms of their relative strengths compared with desired signals, offset from the desired frequency of considerations, and the modulation on them. A system designer optimizes the architecture based on all of these factors, and the end result of the optimization is a power and area-efficient integrated radio. One needs to pay close attention to the situations where multiple radios are integrated on the same die. In such cases, frequency planning and finite isolation of substrate would be considered in a practical manner. Design of radio architecture is a mature area at present, and developments in all areas, including modulation techniques, coding scheme, integrated RF architectures, and so on, are in progress in order to develop compact multiradio, single-die CMOS solutions.

REFERENCES

Frequency Band Allocations

[1] http://www.ntia.doc.gov/osmhome/allochrt.pdf.

[2] T. Pratt and C.W. Bostian, *Satellite Communications*, John Wiley and Sons, 1986.

[3] L.W. Couch II, *Digital and Analog Communication Systems*, Macmillan, 1993.

[4] T.S. Rappaport, *Wireless Communications Principles and Practice*, Prentice-Hall, 1996.

Architecture

[5] A.A. Abidi, “Low power radio-frequency IC’s for portable communications,” *Proceedings of the IEEE*, Vol. 83, No. 4, April 1995, pp. 544–569.

[6] A.A. Abidi, Direct conversion radio transceivers for digital communications, “*IEEE Journal of Solid State Circuits*, Vol. 30,” No. 12, Dec 1995, pp. 1399–1410.

[7] J.L. Mehta, Transceiver architectures for wireless ICs. www.rfdesign.com.

[8] B. Razavi, “Design considerations for direct conversion receivers,” *IEEE Transactions on Circuits and Systems*, Vol. 44,No. 6, June 1997, pp. 428–435.

[9] J. Crols and M.J. Steyart, “Low-IF topologies for high performance analog front-ends of fully integrated receivers,” *IEEE Transactions on Circuits and Systems – II: Analog and Digital Signal Processing*, Vol. 45, No. 3, Mar 1998, pp. 269–282.

[10] B. Razavi, “A UWB CMOS transceiver, ” *IEEE Journal of Solid State Circuits*, Vol. 40, No. 12, Dec 2005, pp. 2555–2562.

Low IF

[11] J. Crols and M. Steyart, "An analog polyphase filter for high performance low-IF filter," *International VLSI Symposium*, 1995, pp. 87–88.

[12] M.J. Gingell, "Single sideband modulation using sequence asymmetric polyphase networks," *Electrical Communications*, Vol. 48, 1973, pp. 21–25.

IIP2 Enhancements

[13] E. Bautista et al., "A high IIP2 downconversion mixer using dynamic matching," *IEEE Journal of Solid State Circuits*, Vol. 35, No. 12, Dec 2000, pp. 1934–1941.

[14] K. Kivekas et al., Characterization of IIP2 and DC offsets in transconductance mixers, *IEEE Transactions on Circuit and Systems-II: Analog and Digital Signal Processing*, Vol. 48,No. 11, Nov 2001, pp. 1028–1038.

Polar Modulators

[15] T. Sowlati et al., "Quad-band GSM/GPRS/EDGE polar loop transmitter," *IEEE Journal of Solid State Circuits*, Vol. 39,No. 12, Dec 2004, pp. 2179–2189.

[16] M.R. Eliott et al., "A polar modulator transmitter for GSM/EDGE," *IEEE Journal of Solid State Circuits*, Vol. 39, No. 12, Dec. 2004, pp. 2190–2199.

Harmonic Reject Architecture

[17] J.A. Weldon et al., "A 1.75-GHz highly integrated narrow-band CMOS transmitter with harmonic-rejection mixers," *IEEE Journal of Solid State Circuits*, Vol. 36, No. 12, Dec. 2001, pp. 2003–2015.

Subsampling

[18] D. Shen et al., "A 900-MHz RF front-end with integrated discrete time filtering," *IEEE Journal of Solid State Circuits*, Vol. 31, No. 12, 1996.

CHAPTER 3

System Architecture for High-Speed Wired Communications

INTRODUCTION

The demand for higher data throughput has been increasing tremendously over the past decade. Figure 3.1 illustrates bandwidth growth from the 1980s. The time interval between one order of magnitude growth has been decreasing, indicating that the need for 10-Gb/s data transmission is now imminent over existing bandlimited channels such as copper-based cables, backplanes, and fibers. As of 2006, the IEEE 802.3ae, the IEEE 802.3ak, the IEEE 802.3an, and the IEEE 802.3aq standards for 10 Gigabit Ethernet over fiber, 10 Gigabit Ethernet over twin-axial cable, 10 Gigabit Ethernet over unshielded twisted-pair, and 10 Gigabit Ethernet over multimode fiber (MMF) have been approved. The IEEE 802.3ap standard for 10 Gigabit Backplane Ethernet was approved in 2007.

Moreover, the increasing demand on multimedia contents usage such as streaming video over Internet and satellite broadcasting over handheld devices, as illustrated in Figure 3.2 and Figure 3.3, has further increased the requirements for higher data-rate processing over the existing infrastructures that was originally designed to handle lower data throughputs.

Meanwhile, the advances in optical links and the supporting electronics have dramatically increased the speed and amount of data traffic handled by a network system. However, the bandlimited links are not keeping pace with these technical

Advanced Integrated Communication Microsystems, edited by Joy Laskar, Sudipto Chakraborty, Manos Tentzeris, Franklin Bien, and Anh-Vu Pham

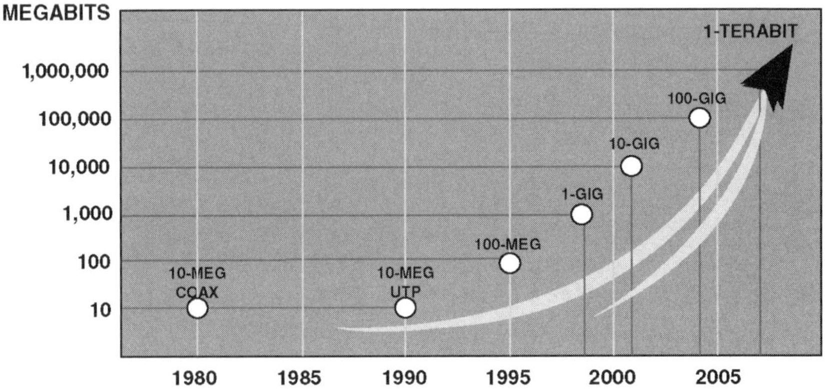

Figure 3.1. Bandwidth growth timeline.

improvements for multi-Gbit serial data communication and are becoming a critical bottleneck.

The primary physical impediments to high data rates in legacy backplane channels are the frequency-dependent loss characteristics of copper channels. Above rates of 2 Gb/s, the skin effect and dielectric loss in backplane copper channels distort the signal to such a degree that signal integrity is severely impaired. This dispersive forward-channel characteristic contributes to the Inter-Symbol Interference (ISI).

Meanwhile, a major limiting factor to increasing transmission speeds and distances in fiber-optic communication links is modal dispersion that causes ISI. Modal dispersion results when the numerous guided modes are transmitted with different paths in the MMF, resulting in different receiving times at the receiver side of the fiber communication system. Modal dispersion becomes a severe factor as the length of the MMF is extended or the data rates are increased.

Figure 3.2. Satellite-based digital multimedia broadcasting handhelds that requires broader bandwidth capabilities of the copper-based PCBs inside.

Figure 3.3. Increasing demand on streaming multimedia contents over the Internet, pushing the overall data throughput higher in the network.

The channel bandwidth limitation and modal dispersion can be addressed by using a channel-compensation technique, namely, equalization at the transmitter and/or receiver side. An equalization technique compensates the frequency-dependent channel loss characteristics. The bandlimited channel has a low-pass frequency response, as shown in Figure 3.4(a). The larger loss in high-frequency range causes the signal power to smear into the neighboring symbols. The equalization technique restores the high-frequency component of the original transmitted signal. Thus, the frequency response of the equalizer has larger gain values for the high frequencies compared with the low frequencies around DC, as shown in Figure 3.4(b).

Digital equalization techniques have traditionally been used to reduce ISI in bandlimited wire-line applications, but such techniques require high-resolution analog-to-digital converters with sampling rates at or above the symbol rate. The increased circuit complexity and power consumption required to apply these techniques to high-speed serial data transmission are prohibitive at the considered data rates. Hence, analog or mixed-signal equalization techniques are attractive alternatives for multi-Gb/s serial transmission.

The most common type of analog equalizer used in practice to compensate for ISI is a linear finite impulse response (FIR) filter with adjustable tap-coefficients, as shown in Figure 3.5. Each tap coefficient is updated through certain equalization algorithms. With the FIR structure implementation, there are several equalization algorithm criteria to reduce the ISI.

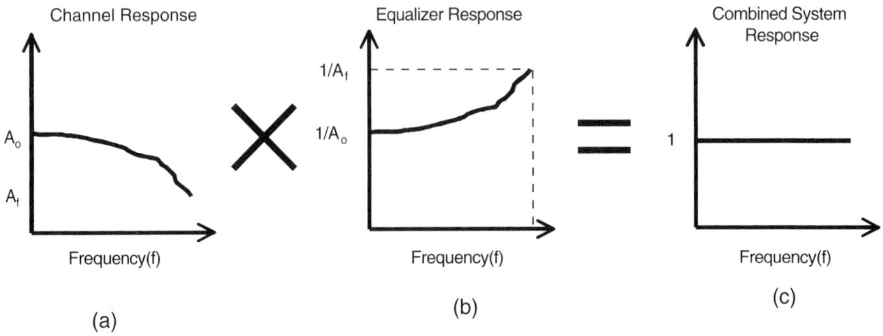

Figure 3.4. Conceptual illustration of equalization: (a) channel response, (b) equalizer response, and (c) equalized response in the frequency domain.

Meanwhile, if the channel frequency-dependent loss characteristics are time invariant, the channel can be measured and the tap coefficients for the equalization can be extracted from the measured channel characteristics. As the channel is time invariant, once the tap coefficients are set, the data can be transmitted without further adjusting of the tap values.

Another popular equalization technique is pre-emphasis, which was suggested to realize transmit-side equalization. This equalization scheme predistorts transmit signal waveforms to enhance the data transition. As channel loss increases as a result of longer trace geometry, this technique needs to increase the amount of predistortion. However, the maximum voltage swing is limited by the system constraints as well as by the voltage headroom issue in integrated circuit (IC) implementation. Thus, the resulting decreased average signal level thereby leads to reducing the overall signal-to-noise ratio. Furthermore, this equalization technique may increase the amount of near-end cross talk, which is another major signal impairment factor in backplane applications. Thus, the receive-side equalization technique is considered suitable for multi-gigabit data transfer.

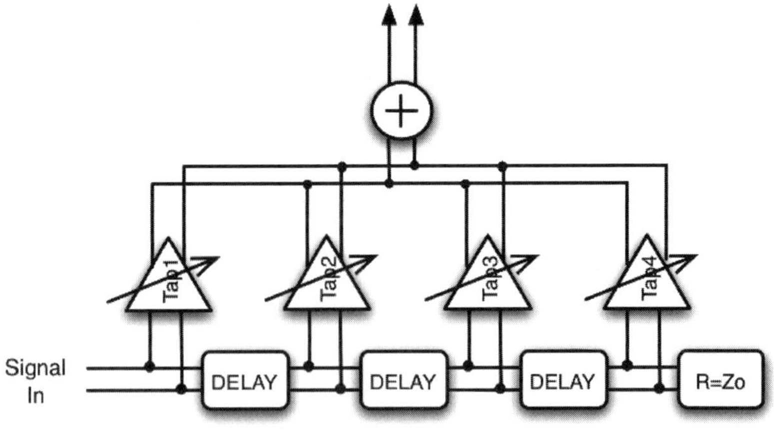

Figure 3.5. Functional block diagram of an FIR-based linear equalizer.

The receive-side equalization technique using a FIR filter structure was reported for a 10-Gb/s backplane application. On-chip passive components were adopted to offer a bandwidth benefit necessary for 10-Gb/s equalization. Because of the intrinsic loss problem of this passive delay line, this analog equalizer has the limitation of the maximum number of taps in the FIR structure. Therefore, the development of a novel delay line structure is still requested. Furthermore, this passive component-based equalizer cannot provide adjustable compensation to diverse channel configurations.

Figures 3.6(a) and (b) show a typical legacy backplane configuration and the corresponding channel loss characteristics, respectively. The loss characteristics are different depending on the trace length and on the board material. Therefore, the

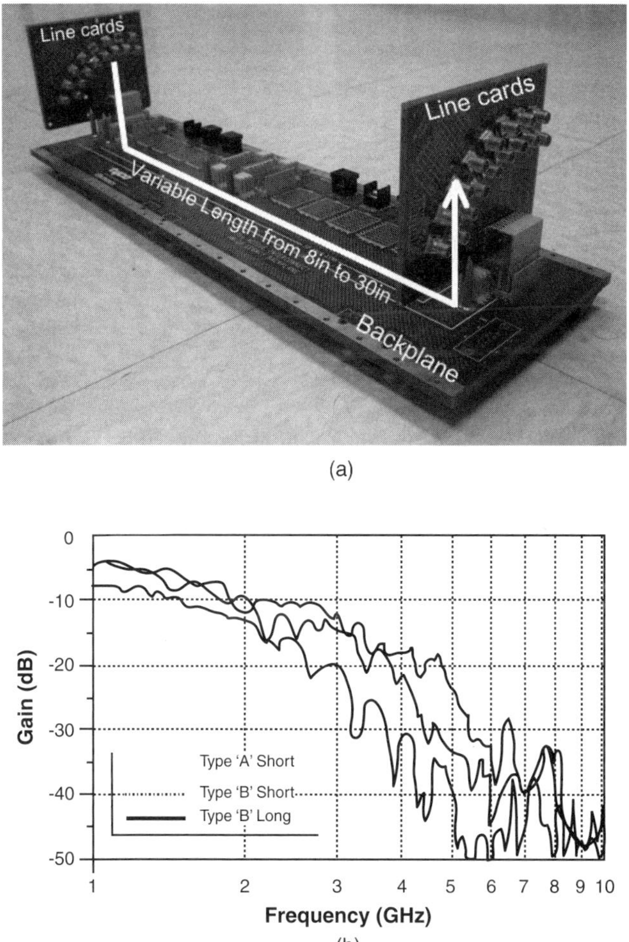

Figure 3.6. (a) Backplane channel configuration including 8-in and 20-in FR-4 trace length and (b) the corresponding channel frequency response.

amount of channel equalization needs to be adjusted and reconfigured to reflect each channel configuration.

3.1 BANDLIMITED CHANNEL

When a signal goes through a bandlimited dispersive channel with an impulse response illustrated in Figure 3.7(a) or Figure (b), its output signal power spreads in time. This spreading of signal power causes ISI. In other words, transmission of a square pulse through such a dispersive channel results in a widened and flattened pulse at the far end, which implies that each data bit of information overlaps with its adjacent bits. This overlap can cause major distortions of the signal. At high data rates and in long channels, the ISI can be so severe that it becomes impossible to recover the original transmitted data. This major phenomenon limits data transmission and must be addressed for multi-Gb/s serial data communications over bandlimited channels.

Therefore, it is necessary to analyze the impact of channel characteristics on signal integrity in order to compensate for the degradation caused by each channel. In this section, optical fiber links and backplane channels are investigated in more detail. The major cause of signal dispersion in both fiber optic channels and backplane channels is analyzed with the suggested solution to recover signal integrity. After that is the summary of different channel compensation techniques to improve the signal integrity over serial data links. Finally, channel compensation with a settable equalizer is shown to demonstrate the channel compensation.

3.1.1 Fiber Optical Link

The major concern in fiber optic communications is pulse dispersion resulting in ISI. The ISI becomes more severe as data rate and distances are increased. In this section, three different types of dispersion are briefly reviewed.

In MMF, the numerous guided modes travel at different speeds, resulting in pulse dispersion at the receiver. This is called differential-modal delay (DMD) and results in

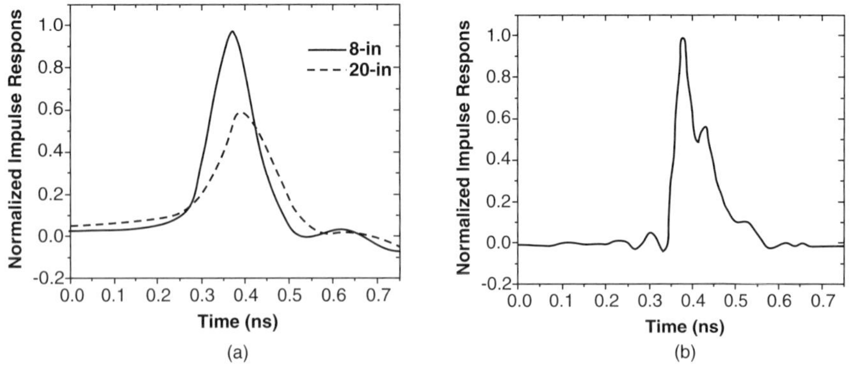

Figure 3.7. Impulse response of (a) 8-in and 20-in backplane channels and (b) 500-m MMF.

ISI. Because of DMD and the resulting ISI, MMF usage is limited to short-haul applications at 10-Gb/s up to 300 m with non-return-to-zero (NRZ) serial data.

In a single-mode fiber (SMF), data links as illustrated in Figure 3.8(a), polarization-mode dispersion (PMD), and chromatic dispersion (CD) cause ISI. PMD is created when two polarization modes experience slightly different conditions, as a result of a generic imperfect circular symmetry of fibers and other external stress on the fibers, and they travel along the fibers at different speeds. CD is created by the variation of the speed of light through the fiber depending on a wavelength. The CD is the sum of two quantities, dispersion inherent to the material and dispersion originating from the structure of the waveguide. PMD and CD are the main dispersion factors in SMF.

An optical system for the characterization of 25-km SMF is shown in Figure 3.8(a). The optical signal is transmitted with a continuous-wave laser module, and it is received with a pin diode, forming a two-port network. The corresponding impulse response of the optical link is plotted in Figure 3.8(b). As expected, the channel is dispersive. As shown in Figure 3.8(c), the signal integrity of the transmitted 10-Gb/s signal has been severely degraded and the original information is unrecoverable without compensation.

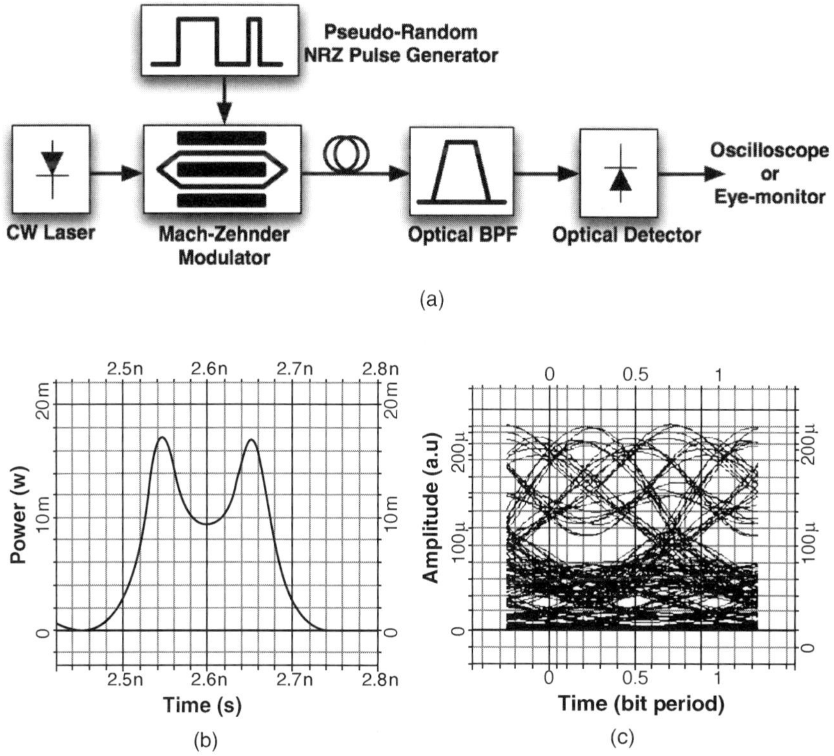

Figure 3.8. Optical link system simulation: (a) schematic, (b) channel response, and (c) eye diagram after 25-km SMF with 10-Gb/s NRZ signal input.

As illustrated in Figure 3.8, fiber optical links introduce dispersion in signals, which results in degraded signal integrity. As data rate and/or link distance increases, dispersion in the fiber optical links becomes more severe and contributes to ISI. Thus, it is necessary to compensate this degradation. Moreover, a fixed compensation for degraded signals cannot cover different types of channels such as MMF and SMF that has different channel impulse responses as illustrated before. Hence, it is necessary to include a method to adjust flexibly to variations in data rate, types of fiber, and link distances for optimum channel compensation. In this chapter, a settable equalizer is introduced with variable tap weights that can address the channel dispersion to a certain extent. In the next chapter, a reconfigurable equalizer is introduced with variable tap spacing in addition to the variable tap weights to improve the channel compensation.

3.1.2 Dispersion in Fibers

It is necessary to review the different types of fibers with their characteristics before covering the fiber dispersion, because the dispersion in the fibers depends on the fiber types. In this chapter, three major types of fibers are reviewed as follows:

1. Step-index MMF
2. Graded-index MMF
3. SMF

MMF has been used mainly with light sources such as light emitting diode (LED) for short-haul application. A large core diameter of the MMF has the merit of collecting light efficiently from inexpensive light sources. However, MMFs can generate multimodes of light that lay in the fiber, which is an undesirable effect from a communication perspective. Multimode generation depends on the core diameter, a numerical aperture, and light launch conditions.

As multimode properties introduce modal dispersion, which limits the MMF usage in optical communication, a fiber with a gradual refractive index profile is developed (graded-index MMF). The faster light speed in the low refractive index compensates the differential modal delay effects, which are severe in the step-index MMF. The MMFs with core sizes of 50 μm and 62.5 μm are standard for short-distance fiber communication. The main application of the multimode fibers today is in systems where connections must be made inexpensively and transmission distances and data speeds are modest. However, the MMF still is not an ideal candidate for long-haul optical communication.

The SMF has a small core size, which is small enough to restrict transmission to a single mode. Because the single-mode transmission avoids modal dispersion, modal noise, and all other effects that come with multimode transmission, SMFs can carry signals at much higher speeds and longer distances than MMFs. As a result, the SMF is used widely over long-haul, fiber-optic communication systems. Figure 3.9 shows the core size of the fibers and conceptually describes how the light is transmitting over the MMFs and SMFs.

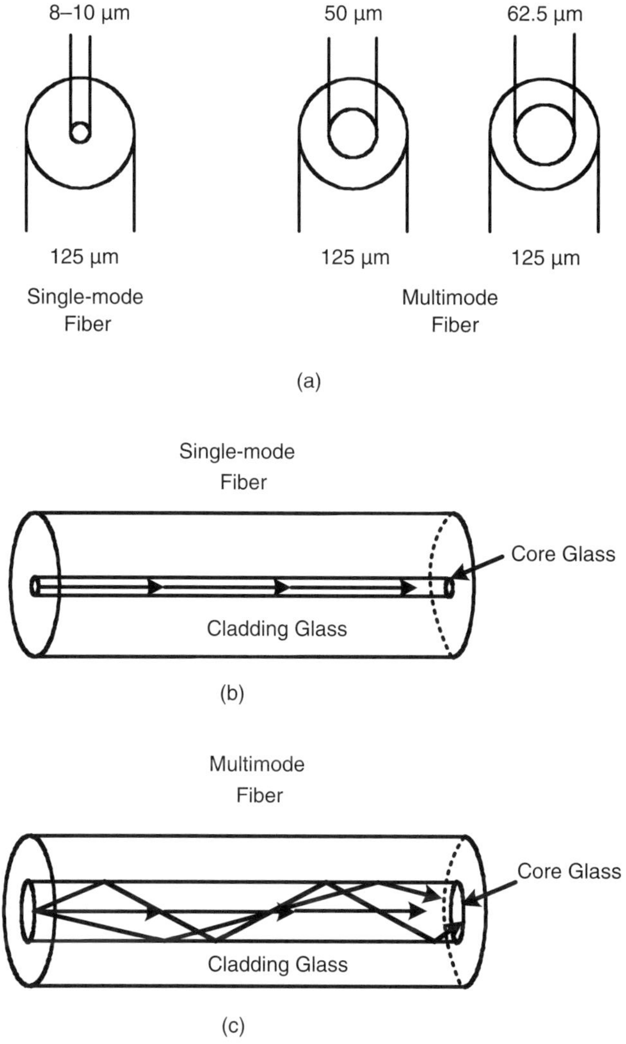

Figure 3.9. Light transmission over SMF and MMF.

The major concerns in fiber communication are the dispersion, which generates ISI and becomes severe as the data speed and distance are increased. In this chapter, three different types of dispersions are briefly reviewed. First, DMD is the dispersion as the numerous guided modes are transmitted with different paths in the MMF resulting in different receiving times at the receiver side of the fiber communication system. The DMD becomes a severe factor as the length of the MMF is extended or the data rates are increased. For these reasons, the MMF is specified by the bandwidth-distance product. Figure 3.10 shows the ISI penalty versus distance plot in different bandwidth-distance product MMFs assuming 10 Gbps of data throughput. Because of the DMD, MMF usage is limited to short-haul applications.

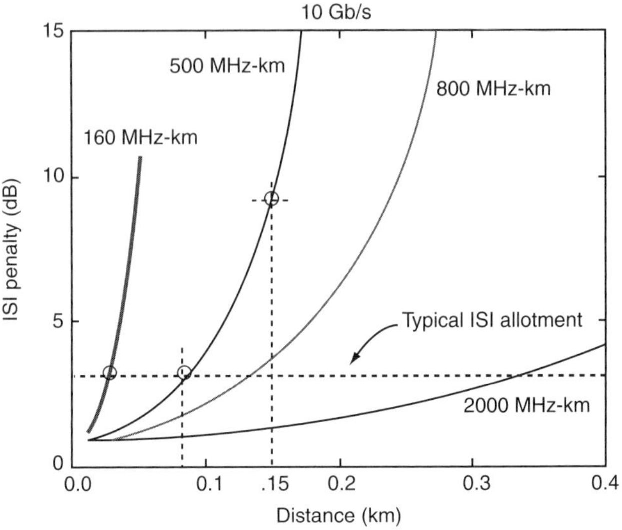

Figure 3.10. ISI versus distance plot with different bandwidth-distance product.

The second type is PMD, which is generated when the two polarization modes in the SMF experience slightly different conditions and travel along the fibers at different speeds as a result of the generic imperfect circular symmetry of the fibers and other external stress on the fibers. Finally, CD is caused by the variation in the speed of light through the fiber depending on a wavelength. The CD is the sum of two quantities, dispersion inherent to the material and dispersion originating from the structure of the waveguide. With the PMD, the CD is the main dispersion factor in an SMF. However, for the MMF, DMD is the main dispersion factors and PMD and CD can be neglected. Figure 3.11 shows the eye diagram of the received signal from a photoreceiver through the MMF with VCSEL as light sources. As the data rate increases, the eye diagram experiences more dispersion resulting in increased ISI penalties.

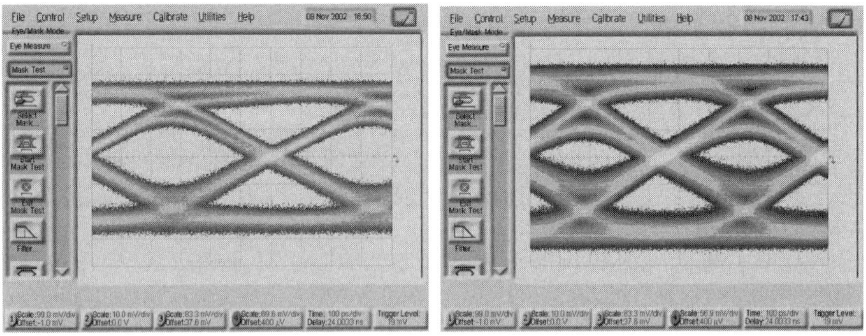

(a) (b)

Figure 3.11. Eye diagram dispersion caused by DMD in MMF: (a) 1.5-Gbps PRBS after 300-m MMF and (b) 1.85-Gbps PRBS after 300-m MMF.

3.1.3 Backplane Multi-Gb/s Data Interface

Channel loss is an important electrical parameter that affects the channel response and influences the design of various components in a backplane link. The channel loss is composed of conductor loss and dielectric loss. Both conductor loss and dielectric loss are directly proportional to frequency and thus become severe in the microwave frequency range (i.e., beyond 1 GHz) for FR-4 dielectric-based components such as backplanes. This channel loss induces dispersion and degrades signal integrity severely. The channel loss is a major impediment in multi-Gb/s backplane signaling.

At low frequency around DC, the conductor loss depends on the resistivity of the conductor and total area over which current is flowing. Since the dielectric material in printed circuit boards (PCBs) is not a perfect insulator, DC loss is associated with current flow through the dielectric material between a signal conductor and a reference plane. However, the conductor loss at DC for commercial PCB substrates is usually very negligible and can be ignored. However, as frequency increases, the skin effect comes into play. The skin effect is a physical phenomenon in which current flowing in a conductor migrates toward the periphery of the "skin" of the conductor as frequency increases. With increasing frequency, the nonuniform current distribution in the transmission line causes the resistance of a conductor to increase with the square root of frequency. Thus, high-frequency components experience more loss than low-frequency components.

Figure 3.12(a) shows the system setup to characterize backplane channels. Two line cards are connected by transmission lines on a backplane, forming a two-port network, and S21 of the network has been measured for 8-in and 20-in channels. The line card can be inserted at different separation length via connectors resulting in various overall trace lengths. As can be expected from the impulse responses, they behave like low-pass filters, depressing high-frequency components, thus causing dispersion for longer trace length. Figure 3.12(b) shows the resulting eye diagram of a 10-Gb/s NRZ signal at the output of a 20-in FR-4 backplane channel. As shown in Figure 3.12(b), the output signal is severely degraded for the 20-in case such that the signal cannot be recovered. It clearly illustrates the need for compensation to maximize the link distance while maintaining signal integrity. Furthermore, different board materials with unique dielectric constants show different characteristics, which implies that the compensation should be adjustable or reconfigurable to cover these variations.

The backplane channel loss characteristics are frequency dependent. Specifically, high-frequency components of the input signal experience larger loss than the lower frequency components around DC. This high-frequency loss becomes worse in the longer backplane channel environment, as shown in Figure 3.13. Figure 3.13 shows that a 20-in FR-4 backplane has a much larger attenuation or loss compared with an 8-in FR-4 backplane observed in the frequency domain. The resulting impulse response of the 20-in FR-4 backplane has more DC signal power loss and more widened pulse shape compared with the 8-in channel. In the next section, the loss mechanisms for backplanes are elaborated in detail.

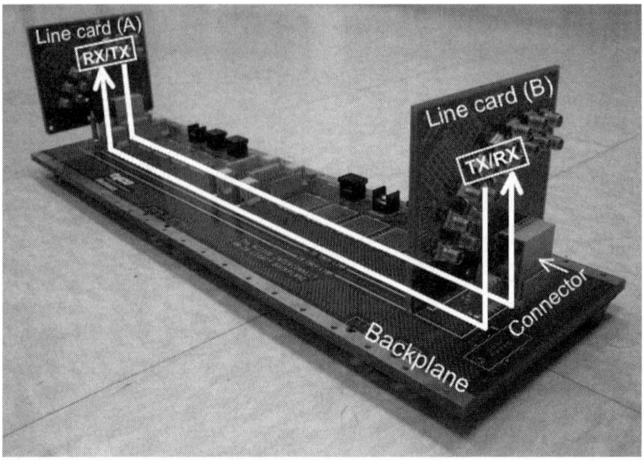

(a)

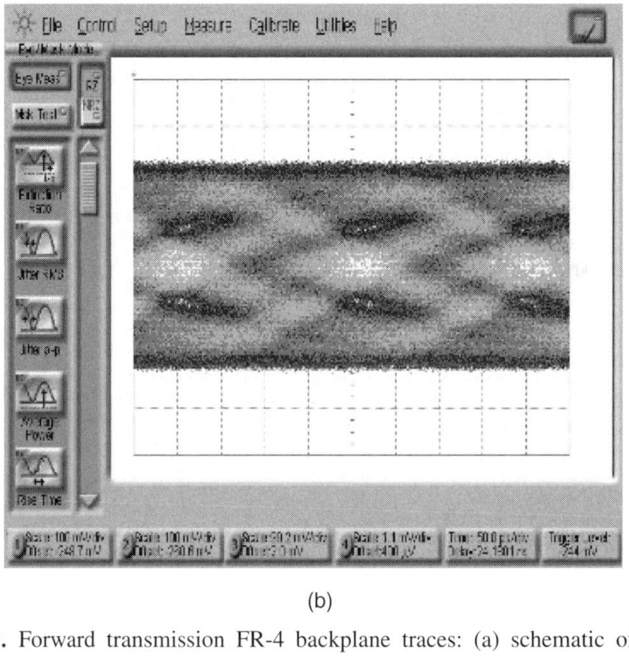

(b)

Figure 3.12. Forward transmission FR-4 backplane traces: (a) schematic of the system configuration and (b) eye diagram at the receiver input of a 10-Gb/s NRZ signal after a 20-in backplane trace.

3.1.4 Backplane Channel Loss

As high-speed input/output (I/O) interface technology evolves, resistive loss affects the link performance by decreasing the signal amplitude and slowing the edge rates. The primary origins of this loss are DC loss; the skin effect and dielectric loss are described in the subsequent sections.

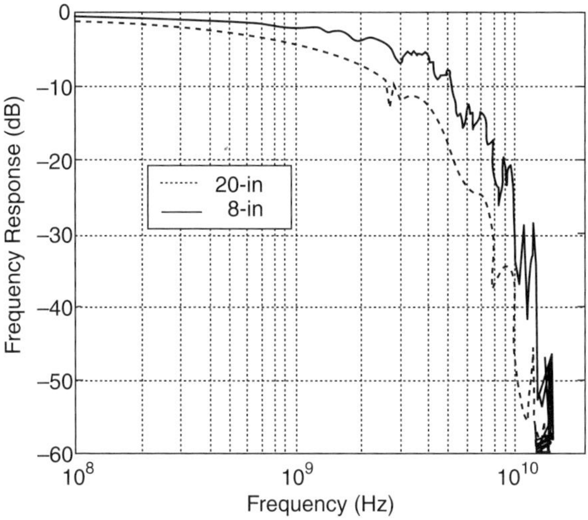

Figure 3.13. Forward transmission frequency response of a 8-in and 20-in FR4 backplane traces.

3.1.4.1 DC Loss. A DC loss depends on the resistivity of the conductor and on the total area in which the current is flowing. The resistive loss of the channel shown in Figure 3.14 can be calculated as

$$R = \frac{\rho L}{A} = \frac{\rho L}{Wt} \tag{3.1}$$

where R is the total resistance of the line, ρ is the resistivity of the conductor material in ohm-meters (the inverse of conductivity), L is the length of the line, W is the conductor width, t is the conductor thickness, and A is the cross-sectional area of the signal conductor.

Since the dielectric materials used in PCBs are not perfect insulators, a DC loss is associated with the resistive drop across the dielectric material between the signal

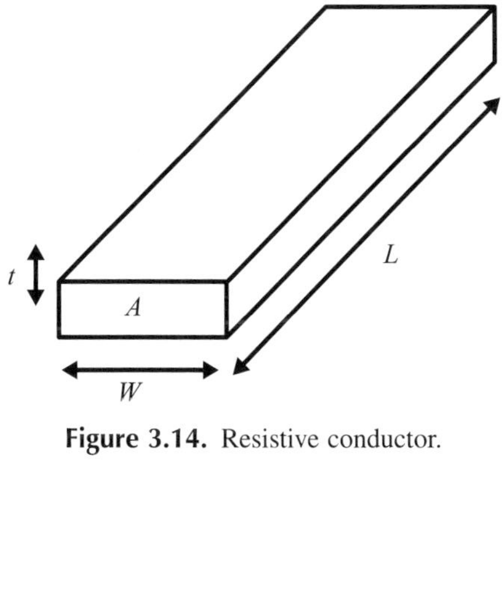

Figure 3.14. Resistive conductor.

conductor and the reference plane. The dielectric losses at DC (it is not just at DC but throughout the frequency span) for commercial PCB substrate are usually very negligible and can be ignored.

3.1.4.2 The Skin Effect. The skin effect is a physical phenomenon related to high-frequency transmission on a wire. Beyond tens of megahertz, the electromagnetic field of the wire causes most of the electrical current to become crowded at the edges of the wire. This phenomenon alters the distribution of the signal current throughout the wire and changes the effective resistance of the wire. The current flowing in a conductor will migrate toward the periphery or "skin" of the conductor. This is the origin of the name "skin effect". The resulting effect is the increased signal attenuation at higher frequencies.

The skin effect manifests itself primarily as resistance and inductance variations. As frequency increases, the nonuniform current distribution in the transmission line causes the resistance to increase with the square root of frequency and the total inductance to fall asymptotically toward a static value called the external inductance.

In the microstrip transmission line, the electric and magnetic fields intersect the signal trace or the ground plane conductor. They will penetrate the metal, and their amplitudes will be attenuated. The amount of attenuation will depend on the resistivity ρ of the metal and on the frequency content of the signal. The amount of penetration into the metal, known as the skin depth, is shown as follows:

$$\sigma = \sqrt{\frac{2\rho}{\varpi\mu}} = \sqrt{\frac{\rho}{\pi f\mu}} \tag{3.2}$$

where ϖ and μ are the angular frequency and the permeability of free space, respectively.

3.1.4.3 Dielectric Loss. As frequency increases over 1 GHz, dielectric loss becomes another dominant loss factor in the legacy backplane applications. When dielectric losses are accounted for, the dielectric constant of the material becomes a complex value shown as follows:

$$\varepsilon = \varepsilon' - j\varepsilon'' \tag{3.3}$$

where the imaginary portion represents the losses and the real portion is the typical value of the dielectric constant. Since the imaginary portion of Eq. (3.3) represents the losses, it is convenient to think of it as the effective conductivity of the lossy dielectric.

Subsequently, $1/\rho = 2\pi f\varepsilon$ becomes the equivalent loss mechanism, where ρ is the effective resistivity of the dielectric material and f is the frequency. The typical method of loss characterization in dielectrics is by the loss tangent shown as follows:

$$\tan|\delta_d| = \frac{1}{2\rho\pi f\varepsilon} = \frac{\varepsilon''}{\varepsilon'} \tag{3.4}$$

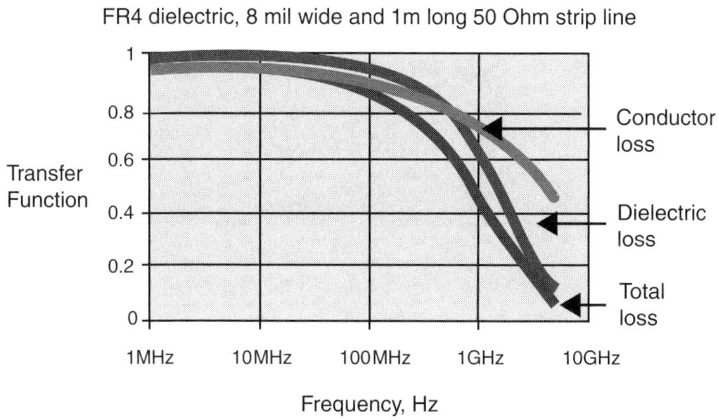

Figure 3.15. Loss transfer functions of the FR4 legacy backplane.

Figure 3.15 shows the overall loss transfer function of the FR4 legacy backplane channel. The dielectric loss increases and becomes the major loss factor as well as the conductor loss beyond 1 GHz. The resulting overall channel loss drastically increases, and the corresponding frequency response is similar to that of a typical low-pass filter. Thus, the multi-Gbit/s signal experiences the loss of the high-frequency components through the backplane. The next section describes the impacts of this channel loss on the signal integrity performances.

3.1.4.4 Impacts of Channel Loss on the Signal Integrity. The Fourier series represents the spectral contents of the periodic time-domain signal. Wideband digital signals can be approximated to a square-wave pulse train. The Fourier series expansion of this square-wave pulse train contains many frequency components shown as follows:

$$f(x) = \frac{2}{\pi} \sum_{n=1,3,5,\ldots} \frac{1}{n} \sin(2\pi n f x) \tag{3.5}$$

where f is the frequency and x is the time. The backplane channel loss characteristics are frequency-dependent. Specifically, high-frequency components of the input signal experience larger loss than the lower frequency components around DC.

This high-frequency loss becomes worse in the longer backplane channel environment. When the signal goes through the dispersive backplane channel, its output signal power spreads in time. This spreading of signal power causes ISI. In other words, transmitting a square pulse through such a dispersive channel results in a widened and flattened pulse at the far end, which implies that each data bit of information overlaps with its adjacent bits. This overlap can cause major distortions of the signal. At high data rates and in long channels, the ISI can be so severe that it becomes impossible to recover the original transmitted data. This major phenomenon

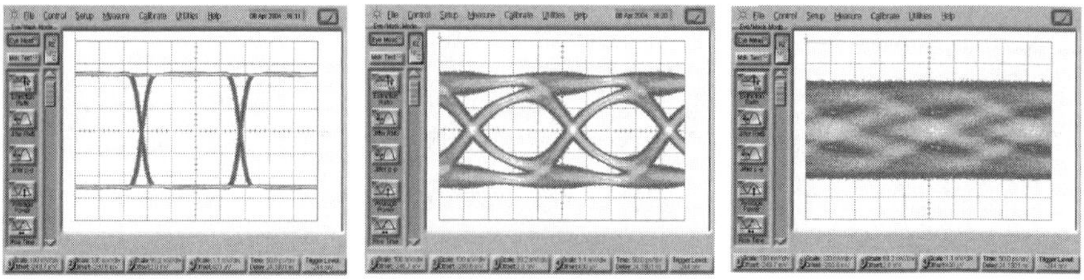

(a) (b) (c)

Figure 3.16. Eye diagrams of 5-Gbit/s NRZ signals of (a) clear input signal, (b) output signal of 8-in backplane trace, and (c) output signal of 20-in backplane trace.

limits data transmission and must be addressed beyond 5-Gb/s data rates in the legacy backplane channels longer than 20 in.

Figure 3.16(a–c) show how ISI affects the signal integrity performance based on transmission distance with a 5-Gb/s NRZ signal. When a clean 5-Gb/s NRZ signal, as illustrated in Figure 3.16(a), propagates through 8-in and 20-in backplane board traces, its eye becomes smaller and almost closed because of ISI, as shown in Figure 3.16(b) and (c), respectively.

Figure 3.17(a–c) show how ISI affects the signal integrity performance based on data rate with 2.5-, 5-, and 10-Gb/s NRZ signals over a 20-in FR4 backplane channel. The 2.5-Gb/s NRZ signal has a large eye opening at the backplane output that is large enough to provide a reliable link performance. Meanwhile, beyond 5 Gb/s, the resulting backplane output signal becomes severely impaired, as shown in Figures 3.17(b) and (c).

As shown in Figures 3.16 and 3.17, the NRZ signal transmission beyond 5 Gb/s turns out to be very challenging over the 20-in FR4 backplane channel. Thus, a channel loss compensation technique is essential to increase the maximum data throughput in the legacy backplane signaling environment. Different system architectures for high-speed wired communication links are discussed in the next section.

As illustrated in Figures 3.16 and 3.17, data communications through both fiber and backplane copper channels distort the transmitted signal, which causes considerable ISI. As a result, it becomes impossible to communicate at high speeds beyond a certain distance using existing infrastructure.

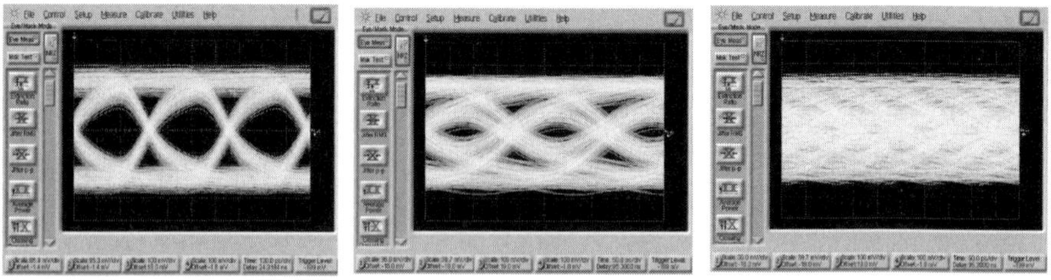

(a) (b) (c)

Figure 3.17. Eye diagrams of a 20-in FR4 backplane channel output signal of (a) 2.5 Gbit/s, (b) 5 Gbit/s, and (c) 10 Gbit/s.

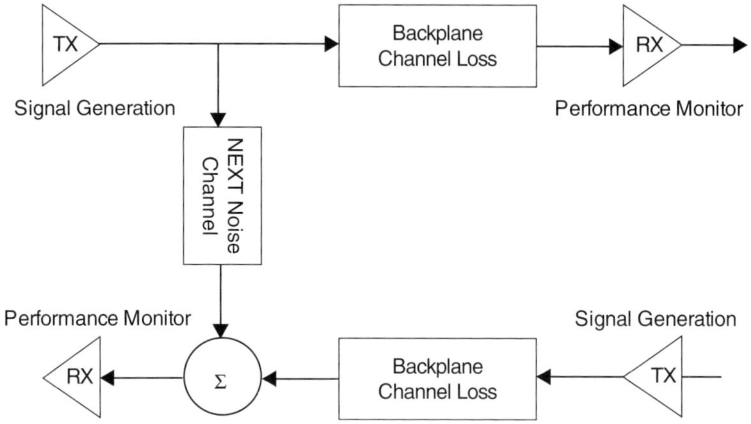

Figure 3.18. A backplane signaling system model.

3.2 EQUALIZER SYSTEM STUDY

Backplanes are bandlimited channels with severe loss at higher frequencies, which becomes worse as the channel length increases. This low-pass dispersive channel characteristic obstructs the high-speed data transition and leads to ISI. To alleviate this ISI effect, a channel-compensation technique, e.g., equalization, is essential beyond 5 Gb/s in a legacy FR4 backplane channel longer than 20 in. Meanwhile, as the data rate increases, coupling noise becomes another major noise component. Specifically, NEXT noise is the dominant factor to deteriorate signal integrity beyond 6 Gbit/s in the legacy backplane channel. The noise cancellation technique is a promising solution to achieve reliable multi-Gb/s transmission in legacy backplane channels.

Before implementing the channel-effect mitigation techniques, system simulation is performed to investigate the optimum system architecture and the corresponding building block requirements. Based on the system model, shown in Figure 3.18, the system trade-offs are investigated and the system specification and the function of building blocks are defined. These resulting specifications are used as the initial design goals in the IC implementation.

3.2.1 Equalization Overview

An equalization technique compensates the frequency-dependent channel loss characteristics by applying the inverse of the channel transfer function as shown in Figure 3.19. The bandlimited channel has a low-pass frequency response, as shown in

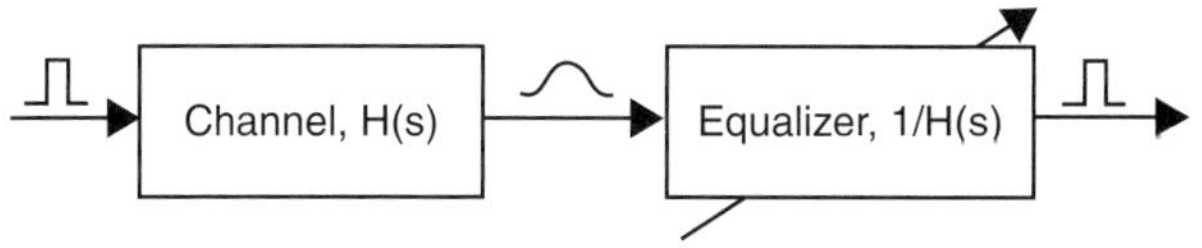

Figure 3.19. A system model of equalization for a dispersive channel.

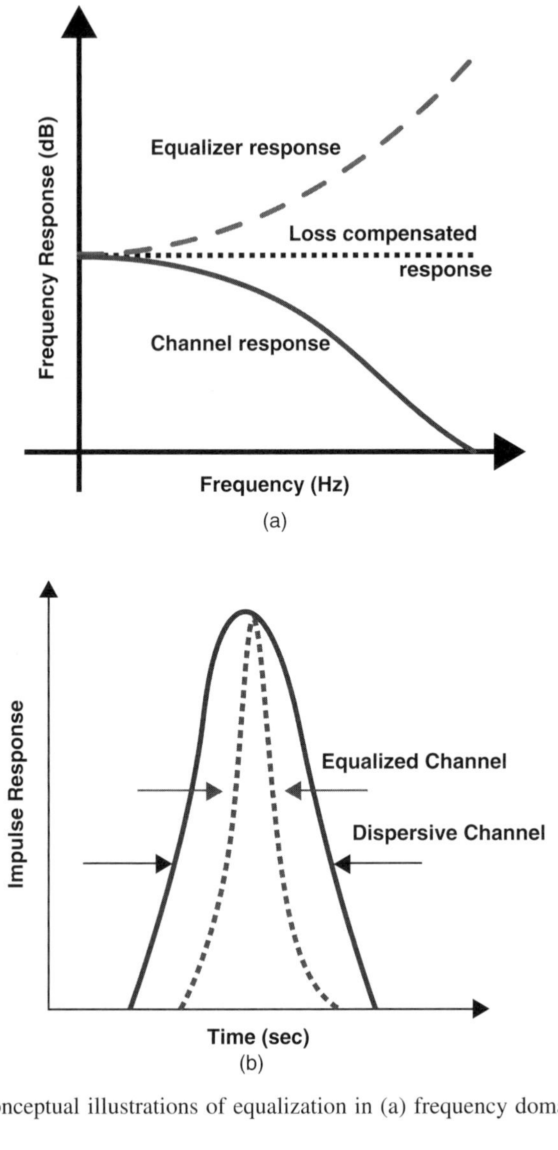

Figure 3.20. Conceptual illustrations of equalization in (a) frequency domain and (b) time domain.

Figure 3.20(a). The larger loss in high-frequency range causes the signal power to smear into the neighboring symbols. The equalization technique restores the high-frequency component of the original transmitted signal.

Thus, the frequency response of the equalizer has larger gain values for the high frequencies compared with the low frequencies around DC, as shown in Figure 3.20(a). Meanwhile, the equalization can be interpreted as a process to sharpen the channel impulse response, as shown in Figure 3.20(b). The width of the channel impulse response means the degree of signal power dispersion in the time domain for a given pulse width. Therefore, an equalizer can be regarded as a spectrum-shaping filter to shorten the channel impulse response to bring it back to its original transmission width. The subsequent sections introduce the historical background and various topologies used in several applications.

3.2.2 Historical Background

The equalization technique has been widely used to alleviate the ISI effects in several dispersive channels such as the magnetic storage channel, untwisted pair network cable, coaxial cable, backplane PCB trace, and optical fiber channel. At first, an analog cable equalizer was introduced for coaxial cables. The coaxial cable channel has attenuation characteristics proportional to the cable length as shown in Eq. (3.6). To compensate for this frequency-dependent channel loss, the equalizer transfer function was determined to have the reciprocal of the channel response, i.e., $H^{-1}(f)$, and it was approximated to the linear superposition of 1 and $b \cdot Y(f)$, as shown in Equation (3.7). The block diagram and frequency responses of the cable equalizer are shown in Figure 3.21(a) and (b), respectively:

$$H(f) = e^{-aL\sqrt{f}} \tag{3.6}$$

$$H^{-1}(f) = e^{aL\sqrt{f}} \cong 1 + b \cdot Y(f) \tag{3.7}$$

The IEEE standard 802.ab 1000BASE-T specified the physical layer for Gigabit Ethernet (GbE) over CAT-5 cabling system. Since the widely deployed CAT-5 cabling systems had been used for 100BASE-TX, the 1000BASE-T was supposed to provide a smooth way to increase the data rate by 10 times over 100BASE-TX. However, CAT-5 cable was not designed to offer enough channel capacity for 250-Mbit/s data transmission per each cable pair.

Meanwhile, the DSP-based equalizer technique was also suggested and implemented with 0.18-μm CMOS process technology. This digital equalizer solution had an intrinsic problem of power consumption. Thus, a mixed-signal IC solution was proposed to overcome the power consumption problem of the digital solution. This work adopted the sample-and-hold (S/H) based transversal equalizer with rotating tap weights. The sign–sign least-mean-squared-error (SS-LMS) adaptation algorithm was implemented using digital-to-analog converters (DACs) for the tap weights of the suggested equalizer.

Along with the equalization technique for copper cables, the equalization technique for fibers was also developed. MMF supports multiple modes of light propagation, each with a different velocity resulting in many received pulses of light with different amplitudes, as shown in Figure 3.22. Bandwidth limitations of the receiver front end smear together pulses into one Gaussian electrical pulse. To address this ISI problem in the MMF channel, the distribution network by the LC ladder type of the artificial transmission line was implemented, which supported continuous-time signal delay for a transversal filter-type equalizer. All seven taps

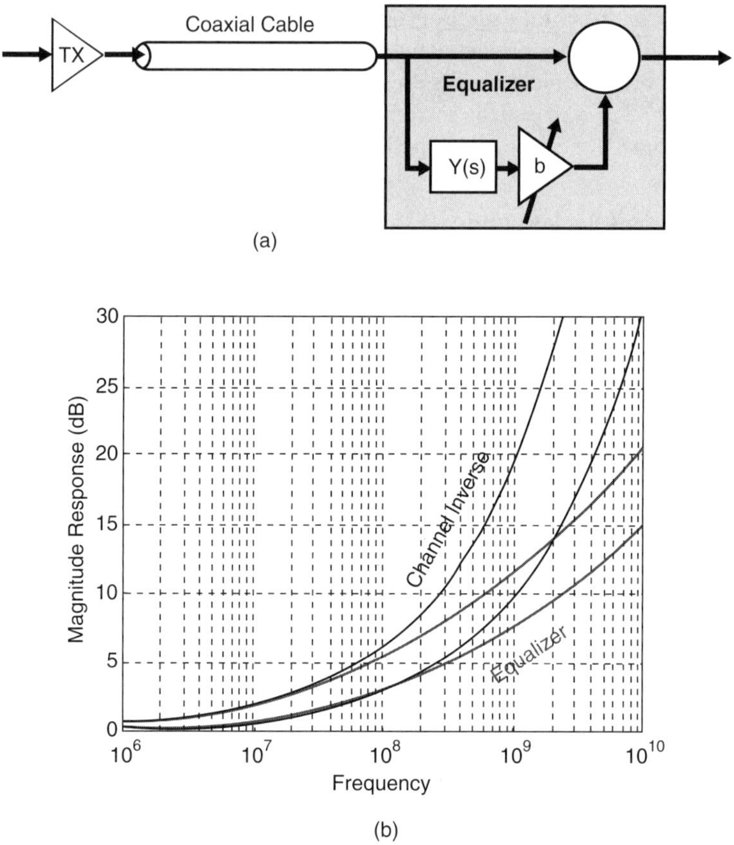

Figure 3.21. (a) Block diagram and (b) frequency responses of the coaxial equalizer.

with tap coefficient multipliers were used to compensate the DMD for 10-Gbit/s NRZ data transmission over 800-m MMF.

Furthermore, the adaptive transmit pre-emphasis equalizer IC for a backplane application was developed. The equalizer IC was implemented with 0.25-μm CMOS process technology for 5-Gb/s transmission over typical FR4 backplanes. This equalizer predistorts the transmission data waveform to combat the channel's dispersive feature. The resulting pre-emphasized waveform was able to compensate the

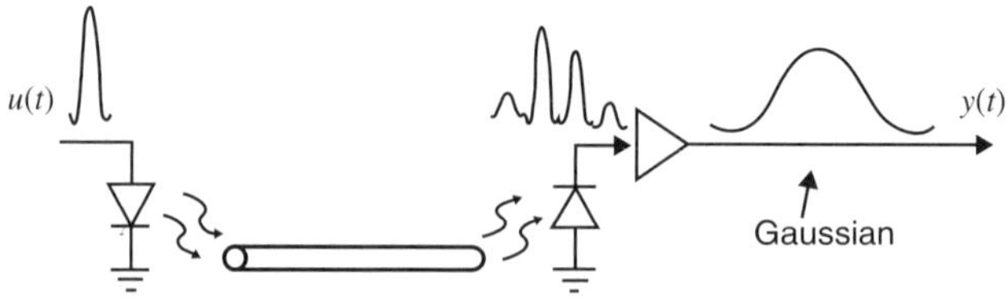

Figure 3.22. Signal pulse dispersion caused by the differential modal dispersion effect of the multimode fiber channel.

channel loss effect successfully. However, the boosted high-frequency components of the transmitted signal waveform induced the increase of coupling noise between the connector pins. Moreover, as channel loss increases because of longer trace geometry, this technique needs to increase the amount of predistortion. Meanwhile, the maximum voltage swing is limited by the system constraints as well as by the voltage headroom issue in IC implementation. Thus, the resulting decreased average signal level thereby leads to reducing the overall signal-to-noise ratio. Thus, novel system architecture is still needed to handle the ISI and coupling effects efficiently.

The subsequent sections introduce the background of the various topologies of equalizers widely used in the dispersive channels. Then, the most adequate equalizer topology is selected and optimized for the legacy backplane applications.

3.2.3 Equalizer Topology Study

The basic function of the equalizer is to compensate the channel loss. A simple linear equalizer has the equivalent mathematical transfer function, shown as follows:

$$G_E(f) = \frac{1}{C(f)} = \frac{1}{|C(f)|} e^{-j\theta_c(f)} \tag{3.8}$$

where $C(f)$ is the channel characteristic and $G_E(f)$ is the equalizer transfer function characteristic. Therefore, the amplitude response of the equalizer is $|G_E(f)| = 1/|C(f)|$, and its phase response is $\theta_E(f) = -\theta_c(f)$. As the equalizer transfer function is the inverse form of the channel, the equalizer completely eliminates the ISI in theory. This equalizer is called the zero-forcing equalizer. For example, the copper channels such as telephone line or twisted-cable have low-pass filter characteristics resulting in increased rising time and falling time of the transmitted signal. The increased rising time and falling time causes the ISI, in other words, the dispersion in the channel impulse response. The ISI is the main source of the signal distortion in digital communication systems. Figure 3.23 shows the conceptual view of the signal dispersion in the lossy channel.

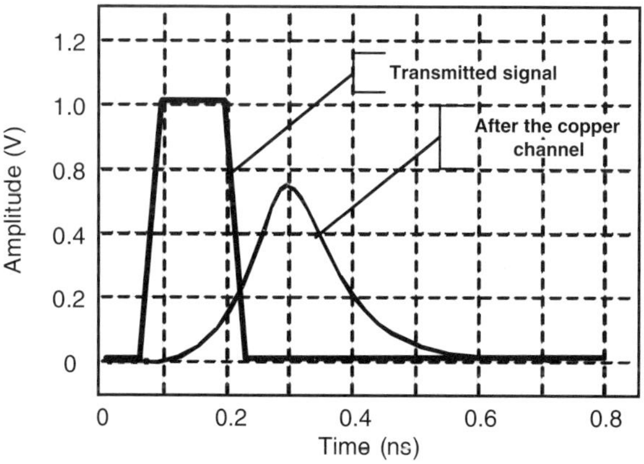

Figure 3.23. Channel impulse response dispersion in a copper channel.

Through the equalization at the receiver side, the dispersive channel effect can be compensated at the receiver front end. This work can be done in the receiver side as explained above, or the signal can be transmitted with some intended signal distortion at the transmitter side. This is called the pre-emphasis technique. This section will touch on the background knowledge of equalization, various types of equalizations, and the pros and cons of each equalization technique.

3.2.3.1 Linear Equalizer. One of the most common equalizer types is a linear FIR filter with adjustable tap-coefficients, as shown in Figure 3.24. Each tap-coefficient is updated through the certain equalization algorithms. With the FIR structure implementation, there are several equalization algorithm criteria to reduce the ISI.

Depending on the tap-coefficient extraction algorithms, this linear equalizer is classified to a zero-forcing equalizer and a minimum-mean-squared-error (MMSE) linear equalizer. The zero-forcing equalizer has the transfer function characteristics as described in Eq. (3.3). A time delay element, shown in Figure 3.24, is called tap delay. The tap delay can be as large as a symbol interval, and the delayed version of the signal is $x(t - k\tau)$ (where $\tau = \text{T}$, T is the symbol period of the signal, and $k = 1, \ldots, n$). Also τ can be smaller than T; in this case, it is called a fractional tap-spaced equalizer.

The fractional tap-spaced equalizer can reduce the aliasing problem in a symbol-spaced equalizer and improve the performance assuming the delay is implemented by sampling. As the zero-forcing equalizer has an inverse channel transfer function characteristic, it can significantly increase the additive noise in the channel. An alternative solution to ameliorate this problem is the MMSE algorithm, where the tap value is optimized to minimize the power in the residual ISI and the additive noise in the channel.

If the channel frequency-dependent loss characteristics are time-invariant, the channel can be measured and the tap coefficients for the equalization can be extracted from the measured channel characteristics. As the channel is time-invariant, once the tap coefficients are set, the data can be equalized without further adjusting the tap values. However,

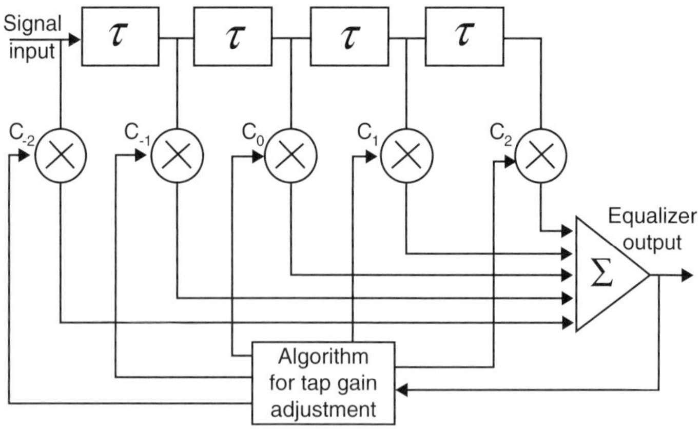

Figure 3.24. Linear FIR equalizer.

if the channel is time-variant such as the wireless channel, the equalizer tap values should be updated periodically based on the real-time channel frequency characteristics. The equalizer that can update the tap coefficient by tracking the channel characteristics is called an adaptive equalizer. The most commonly used adaptive equalization algorithm is the least-mean-squared (LMS) error algorithm. The tap coefficients updated by the LMS algorithm are shown as follows:

$$p(k+1) = p(k) - \mu \frac{\partial E[e^2]}{\partial p} \quad \text{or} \quad p(k+1) = p(k) + 2\mu \cdot e(k) \cdot f(k) \qquad (3.9)$$

where $p(k)$ is the tap coefficient, μ is the parameter controlling the adaptation rate, $e(k)$ is the error signal between the desired signal and received signal, and $\phi(k)$ is the derivative form of the received signal (i.e., $\partial y/\partial p$, where y is the signal after the adaptive equalization). Figure 3.25 shows the one example of adaptive equalization. In this example, the transmitted signal is required at the receiver side (i.e., training sequence) as shown in Figure 3.25(a), or the desired signal can be extracted from the receiver's decision block as shown in Figure 3.25(b).

The practical implementation of the LMS algorithm induces hardware complexity. So there are several alternative simplified algorithms to reduce the burden in

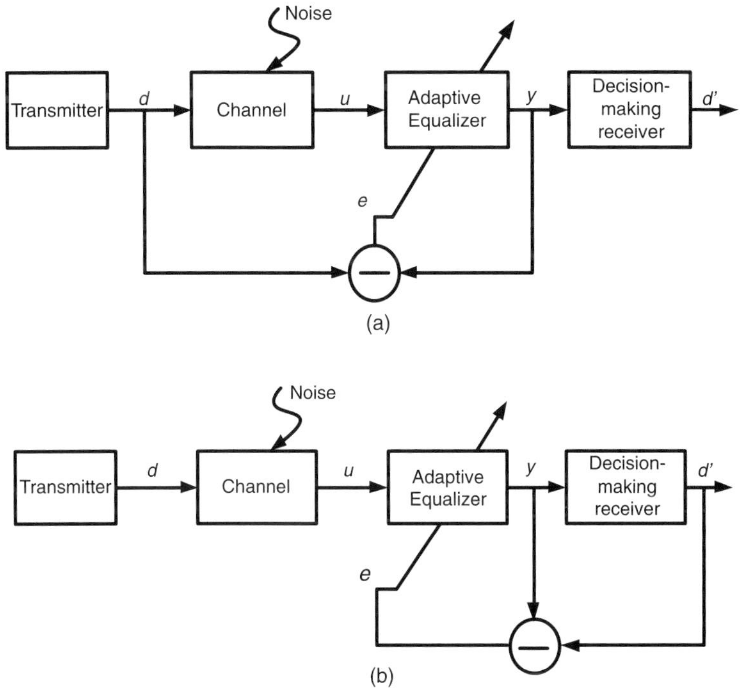

Figure 3.25. Adaptive equalization: (a) using the training sequence and (b) using the decision signal at the receiver as the desired signal.

hardware implementation. The simplified version of LMS algorithms is shown as follows:

$$\text{Sign-data LMS}: \quad p(k+1) = p(k) + 2\mu \cdot e(k) \cdot sgn(\phi(k)) \tag{3.10}$$

$$\text{Sign-error LMS}: \quad p(k+1) = p(k) + 2\mu \cdot sgn(e(k)) \cdot \phi(k) \tag{3.11}$$

$$\text{Sign-sign LMS}: \quad p(k+1) = p(k) + 2\mu \cdot sgn(e(k)) \cdot sgn(\phi(k)) \tag{3.12}$$

Even with the advantage in hardware implementation, these simplified algorithms may not converge or may have more iterations than the original algorithm.

3.2.3.2 Nonlinear Equalizers. The linear equalizers described in the previous section are very effective on channels such as wire-line and telephone channels, where the ISI is not so severe. However, in some channel environments having spectrum nulls, the linear equalizer will introduce a large amount of gain to compensate for the spectrum null. Thus, the noise in the channel will be enhanced severely. Such channels are often encountered in a mobile radio channel, such as those used for cellular radio communications.

A decision feedback equalizer (DFE) is a nonlinear equalizer that employs previous decisions to eliminate the ISI caused by the previously detected symbols on the current symbol to be detected. The block diagram for the DFE is shown in Figure 3.26.

The DFE is typically used with the conjunction of a linear FFE as shown in Figure 3.26. Even though the linear FFE alone can be used to cancel the ISI, the combination of the linear FFE and DFE has better performance. The principal

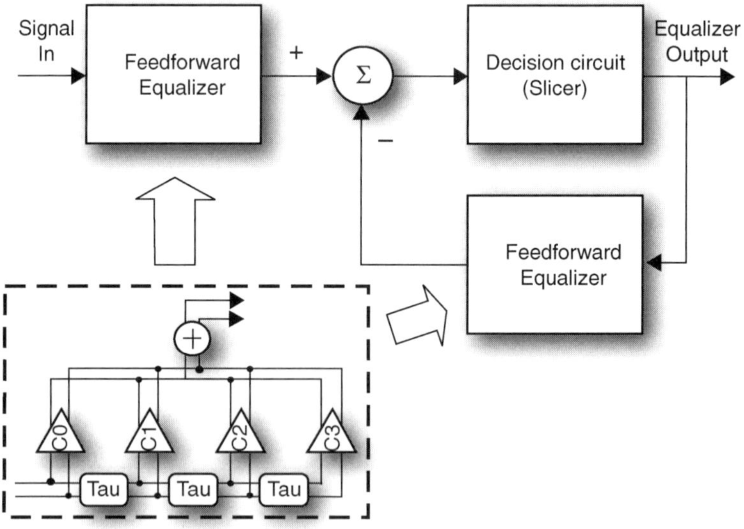

Figure 3.26. Block diagram of a DFE.

reason for this improvement is that the DFE uses the linear combination of the noiseless binary decisions to eliminate some ISI and does not add noise at the input of the decision circuit. The linear FFE amplifies the high-frequency portion of the signal and the noise to cancel the ISI, which is not compensated by DFE. So the noise enhancement of the linear FFE in conjunction with DFE is less than the one when the linear FFE alone is used. Recently, the DFE is reported as a good candidate for backplane channel equalization, where the NEXT noise is severe. Otherwise, the FFE alone will significantly amplify the NEXT noise because the NEXT noise channel frequency response is similar to the high-pass filter response. One potential problem with a DFE is the error propagation. If the DFE provides an incorrect decision, the error will propagate through the feedback filter and increase the probability that another incorrect decision will be made.

Consequently, another algorithm was suggested, which finds the sequence that maximizes the joint probability of the received sequence conditioned on the desired sequence. This sequence is called the maximum-likelihood sequence detector. An algorithm that realizes maximum-likelihood sequence detection (MLSD) is the Viterbi algorithm. Partial-response maximum-likelihood (PRML) detectors using various implementations of the Viterbi algorithms have been widely adopted for the hard disk drive read channel. Meanwhile, the major drawback of MLSD is the exponential behavior of the computational complexity, which is a function of the ISI span. Thus, the MLSD is practical for the channel where the ISI spans only a few symbols.

3.2.3.3 Cable Equalizer (Bode Equalizer). In this section, one typical form of the equalizer specifically for the cable channel will be covered. As the cable channel can be modeled with a simple low-pass filter transfer function, the cable equalizer can be implemented with the combination of the high-pass filter with several poles as design parameters and the variable gain controller as shown in Figure 3.27 (a-b).

The variable gain can be controlled via an LMS or other algorithm for adaptation. The cable equalizer is a practical solution to implement by analog continuous-time signal processing. The continuous-time equalization techniques have some advantages over discrete-time solutions. For example, the continuous-time equalizer does not need any sampling-phase recovery block, so that the equalizer adaptation can be realized independently with the timing recovery function. Also the continuous-time equalization is well fitted for high-speed operation over the discrete-time counterpart as it does not need any high-speed sampling function. Despite these advantages, the cable equalizer has some potential problem that it can boost up the high-frequency noise, which is the similar phenomenon in the linear FFE.

3.2.3.4 Transmitter- and Receiver-Side Equalizer. As mentioned, the equalizer can be installed at the transmitter side or the receiver side. The conceptual block diagram is shown in Figure 3.28. The transmitter-side equalizer, which is called pre-emphasis, is easily implemented by the FIR filter type with digital control. However, the pre-emphasis technique boosts up the high-frequency portion

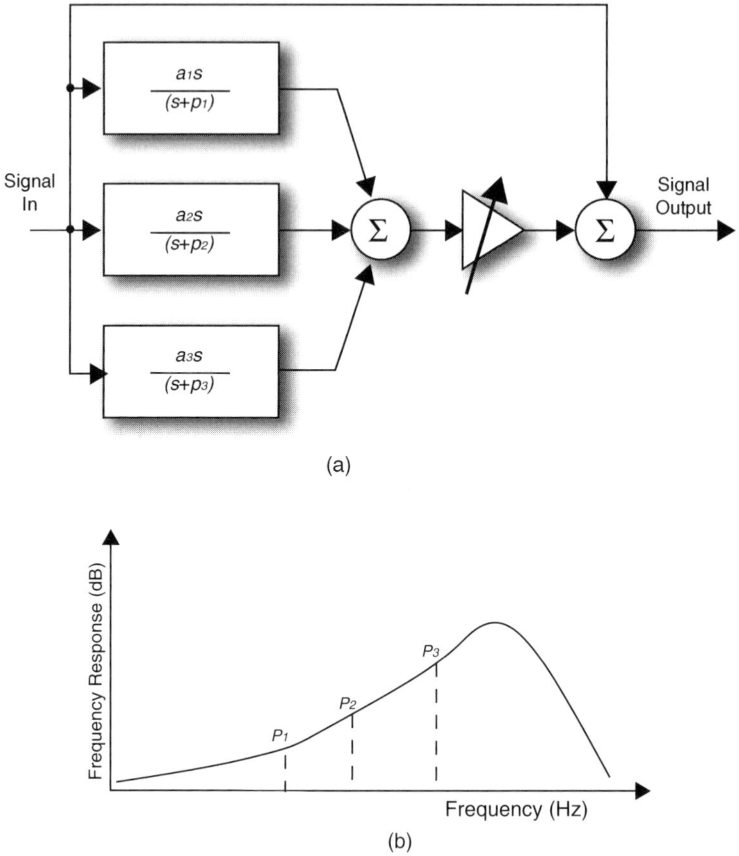

Figure 3.27. (a) Block diagram for the simple cable equalizer: (b) the corresponding equalizer frequency response.

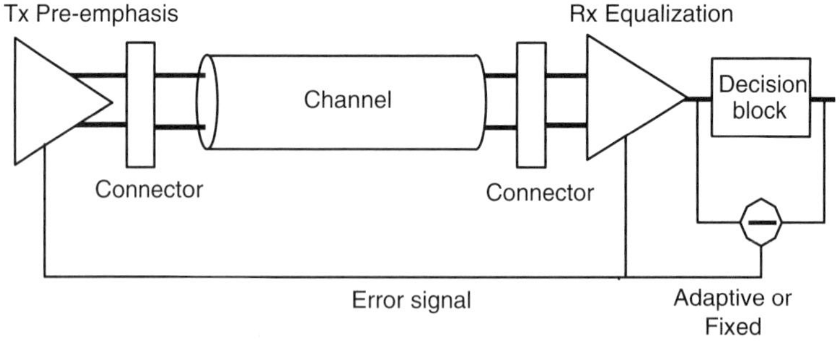

Figure 3.28. Equalization at the transmitter side, receiver side, and both sides.

on the transmitter side increasing the NEXT noise for high-speed chip-to-chip interconnections. Additionally, the pre-emphasis requires the information sent from the receiver side for dynamic or fine-tuned tap-coefficients updates. Furthermore, as channel loss increases, the pre-emphasis needs to apply more gain to boost the high-frequency components of the transmit signal. Since the maximum signal swing is limited by the system constraints and the IC process technology, the average signal swing level at the transmitter side needs to be decreased, thereby requiring additional gain at the receiver side. For these reasons, the equalizer at the receiver side is a better candidate over the pre-emphasis for adaptive or fine-tuned equalization. However, as mentioned, the FIR-type equalizer alone at the receiver side enhances the noise at high-frequency ranges, whereas it compensates for the channel loss to reduce the ISI.

It is also possible to use the combination of the pre-emphasis and receiver-side equalization to increase the overall bit error rate (BER) of the high-speed interconnections. The resulting gain-boosting requirement for specific spectral loss can be relaxed for each equalizer. However, it needs more complex hardware implementation increasing the overall system cost.

For the digital communication systems, the equalizers have been implemented with digital circuitry below Gb/s. For example, the wireless communication system requires the equalization to compensate for the multipath fading effects. However, as the data rate is increased over multi-Gb/s, the conventional digital approaches are no longer a proper solution. For this reason, several I/O standards such as XAUI, PCI-express, and UXPI have emerged to address any high-speed interconnection problems in system and packaging level. Moreover, from an IC implementation point of view, have been made several efforts to implement the equalizer by the continuous-time analog signal processing, the mixed-signal circuit, or the RF/microwave techniques over conventional digital circuit approaches.

3.2.4 Equalizer System Simulation

The backplane channel has coupling noise effects from the connector pins. This coupling effect must be considered to select the optimum equalizer architecture for the backplane channel. To select the optimum equalizer type, transmit-side and receiver-side equalizers are examined for the backplane channel environment. Then, the equalizer architecture and the corresponding signal processing algorithm are investigated for optimum performance.

A transmit equalizer pre-emphasizes the transmit signal waveform to combat ISI. As channel loss increases, this transmit equalizer needs to apply more pre-emphasis to boost the high-frequency components of the transmit signal. Since the maximum voltage swing is limited by system constraints and the CMOS voltage headroom, the average signal level needs to be decreased, thereby requiring additional gain at the receiver. Moreover, the transmit equalizer uses channel information fed back from the receiver to adjust the tap gains for compensating the channel loss, which means that the transmit equalizer has to be controlled by both the transmitter and the receiver. Meanwhile, a receiver equalizer does not change the transmit signal level. Therefore,

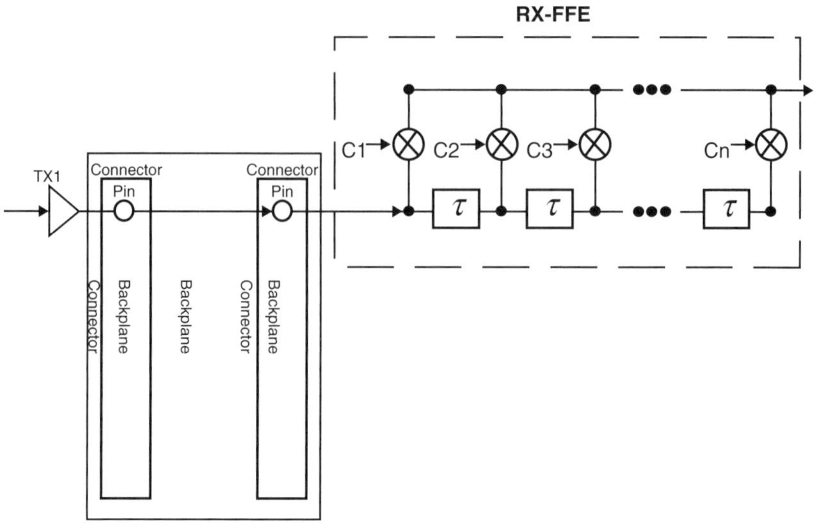

Figure 3.29. Block diagram of a receiver FFE with an FIR filter structure.

the receiver does not need the additional gain that is required in the transmit equalizer scheme. The receive side equalizer is simply adjusted using the channel information obtained by an eye-monitoring unit at the receiver. Additionally, this channel information is reused by the NEXT noise canceler. As a result, the receiver equalizer integrated with a NEXT noise canceler can be configured in a more efficient way than a transmit equalizer.

A receiver-side FFE is implemented using an FIR filter structure. The FIR filter consists of variable gain amplifiers for tap gains and a tapped delay line (TDL), as shown in Figure 3.29. These tap coefficients are derived from the measured impulse response data set.

Equalizer tap coefficient values are calculated based on the measured backplane channel impulse responses using signal processing algorithms such as the zero forcing-linear equalizer (ZF-LE) and the minimum-mean-squared-error-linear equalizer (MMSE-LE). As the transfer function of ZF-LE is the reciprocal of the channel transfer function, it can remove ISI completely but does so neglecting the impact of high-frequency crosstalk, which is also amplified. In contrast, MMSE-LE can ameliorate this noise enhancement problem, since its tap coefficients are calculated to minimize overall signal degradation from both ISI and crosstalk noise. Figure 3.30 shows the frequency response for the ZF-LE and MMSE-LE solution to a 20-in FR-4 backplane channel.

To obtain the optimum FFE configuration and values of the tap coefficients, the performances of the equalizer with different tap-spacings and number of taps were simulated. Figure 3.31 shows the eye diagrams and histogram plots of Ts/2 and Ts/3-spaced FFE output signals for 3- and 4-taps, respectively, without an aggressor source, where Ts is the symbol duration (i.e., 100 ps) of 20 Gbit/s 4-PAM signal.

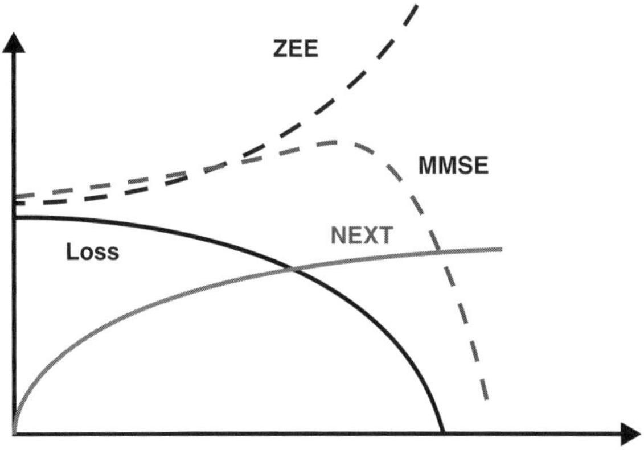

Figure 3.30. Transfer function of the ZF-LE and the MMSE-LE for a 20-in FR-4 backplane channel.

The Ts/3-spaced, 4-tap equalizer has the largest voltage margin and eye-opening size compared with other configurations, as shown in Figure 3.31(d).

In this section, the Ts/3-spaced, 4-tap receiver-side FFE was determined by the optimum equalizer structure. The optimum tap-coefficient values were calculated with the MMSE algorithm. The corresponding building blocks are a tap delay line with 33ps second tap-spacing and variable gain block with bipolar gain value within $-1 \sim +1$. These basic definitions and specifications of the building block function will be used to design and implement the corresponding building block ICs.

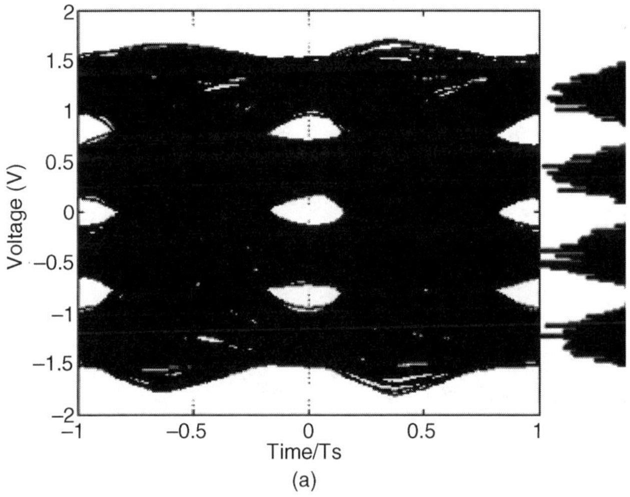

Figure 3.31. Eye diagrams for the proposed FFE output: (a) 3-tap Ts/2-spaced FFE, (b) 4-tap Ts/2-spaced FFE, (c) 3-tap Ts/3-spaced FFE, and (d) 4-tap Ts/3-spaced FFE.

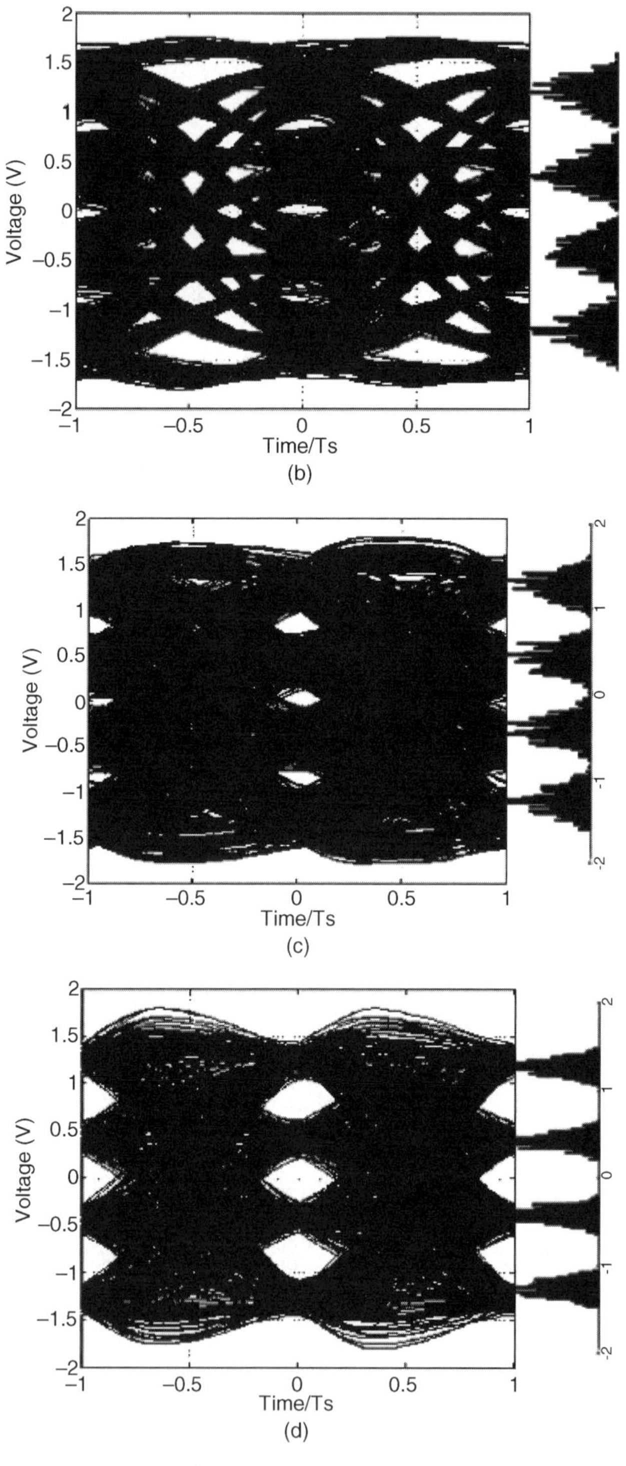

Figure 3.31. (*Continued*).

CONCLUSION

In this chapter, band-limited channels, such as backplanes and multi-mode fiber are studied and its impact on signal integrity is analyzed. With the understanding of the respective channel environment, various equalization system architectures have been reviewed that enables serial data transmission achieving signal integrity beyond their disigned specifications. Based on these principles, electrical equalization can be implemented in an integrated circuit (IC) form to successfully compensate for various band-limited channels at the targeted data rate with a single equalizer IC.

REFERENCES

[1] M. Maeng, F. Bien, Y. Hur, S. Chandramouli, H. Kim, Y. Kumar, C. Chun, E. Gebara, and J. Laskar, "A 0.18μm CMOS equalizer with an improved multiplier for 4-PAM/20Gbps throughput over 20-in FR-4 backplane channels," *IEEE International Microwave Symposium*, Vol. 1, June 2004, pp. 105–108.

[2] C. Pelard, E. Gebara, A. J. Kim, M. Vrazel, F. Bien, Y. Hur, M. Maeng, S. Chandramouli, C. Chun, S. Bajekal, S. Ralph, B. Schmukler, V. Hietala, and J. Laskar, "Realization of multigigabit channel equalization and crosstalk cancellation integrated circuits," *IEEE Journal of Solid-State Circuits*, Vol. 39, NO. 10, Oct 2004, pp. 1659–1670.

[3] Y. Hur, M. Maeng, C. Chun, F. Bien, H. Kim, S. Chandramouli, E. Gebara, and J. Laskar, "Equalization and near-end crosstalk (NEXT) noise cancellation for 20-Gb/s 4-PAM backplane serial I/O interconnections," *IEEE Transactions on Microwave Theory and Techniques*, Vol. 53, NO. 1, Jan 2005, pp. 246–255.

[4] J. G. Proakis, *Digital Communications*, McGraw-Hill Higher Education, 4th Ed., 2001.

[5] F. Krummenacher and N. Joehl, "A 4 MHz CMOS continuous-time filter with on-chip automatic tuning," *IEEE Journal of Solid State Circuits*, Vol. 23, NO. 3, June 1988, pp. 750–758.

[6] H. Wu, J. Tierno, P. Pepeljugoski, J. Schaub, S. Gowda, J. Kash, and A. Hajimiri, "Differential 4-tap and 7-tap transverse filters in SiGe for 10 Gb/s multimode fiber optic equalization," *IEEE International Solid State Circuits Conference*, Vol. 1, Feb 2003, pp. 180–486.

CHAPTER 4

Mixed Building Blocks of Signal Communication Systems

INTRODUCTION

In this chapter, we would like to introduce the readers to the key building blocks of the mixed-signal communication systems. Since the developments of these fundamental circuit operations have been presented in numerous textbooks and technical articles, we provide only the relevant aspects for building integrated mixed-signal communication systems. To cover the wide variety of the circuits used in integrated systems, we focus mostly on the topologies, and provide insight of how they are used in mixed-signal systems.

While studying these topologies, it should be kept in mind that a particular topology can be used in wireless or wireline communication system. Wireless systems would operate using a specific center frequency, as most of the wireless systems are narrowband in nature. Wireline systems, on the contrary, need to be designed for higher bandwidth because of their inherent broadband nature. Because of their immobile nature, wireline systems are usually more relaxed in terms of power, compared with the wireless ones.

Any circuit functionality is a compromise between power consumption and area restraint. The system architect and the circuit designer obtain optimum solutions according to this budget. The choice depends on several aspects, such as (1) operating frequency or data rate, (2) bandwidth (in the case of wireless systems), (3) supply voltage (function of the technology node under consideration), (4) technology platform for implementations, and (5) form factor.

Advanced Integrated Communication Microsystems, edited by Joy Laskar, Sudipto Chakraborty, Manos Tentzeris, Franklin Bien, and Anh-Vu Pham

It is interesting to observe that mixed-signal systems are usually an integration of functionally diverse building blocks. For example, a direct conversion radio front end can be obtained by using the low noise amplifier (LNA), mixer, power amplifier (PA), voltage-controlled oscillator (VCO), synthesizer, and baseband filters. We will address the fundamentals first, to develop insights into developments of integrated communication systems.

4.1 INVERTERS

One of the most popular circuit topology in mixed-signal communication system is an inverter, as shown in Figure 4.1. It is the simplest possible mixed-signal building block, with its origin in the digital circuit domain. Inverters consume power only at the switching instant and no static DC power. In the simple, unclocked static configuration, the voltages at various nodes are obtained by charging and discharging the capacitances.

4.1.1 Key Design Parameters

Key parameters include the aspect ratio of the NMOS and PMOS transistors' (W_N/L_N, W_P/L_P) supply voltage V_{DD} and the threshold voltages of individual devices V_{tn}, and V_{tp}. Like any other static logic combination, the output swings rail to rail without any headroom penalty. This limiting behavior is very suitable for a buffer, which limits the LO signal drive to the mixers. The speed at which such inverters can be used depends on V_{DD}, device geometries, and the output capacitance C_{OUT}, which is a combination of the routing capacitance and the input capacitances of the driven gates. The NMOS transistor passes the "0" (shorting switch to *GND*), whereas the PMOS transistor passes "1" (shorting switch to V_{DD}). Hence, the high-to-low switching operation is handled by NMOS. And the low-to-high switching operation is handled by PMOS; the associated delay timings are denoted by $t_{p,HL}$ and $t_{p,LH}$, respectively. For a symmetric gate, it is desired that $t_{p,HL} = t_{p,LH}$. The maximum output current is obtained at the

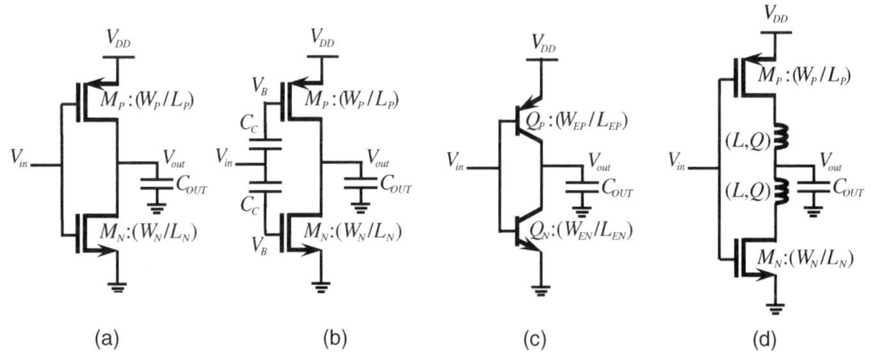

Figure 4.1. Inverter topology/current reuse stages.

midpoint of the DC voltage transfer characteristics (obtained by observing incremental changes in the output voltage with incremental changes in input voltage). For equal strength of the PMOS and NMOS devices, $\beta_N = \beta_P$, and the DC transfer characteristic is obtained symmetric around the point $V_{IN} = V_{OUT} = \frac{V_{DD}}{2}$. The voltage transfer curve is formulated according to the relative drive strengths of the NMOS and PMOS devices. The curve shifts to the right when $\beta_N < \beta_P$ and to the left when $\beta_N > \beta_P$. The static logic family, including inverter, NOR, and NAND gates operates only with the transitions of the input signals, and no other clock-phasing is involved. Inverters are used in almost all IC systems, in logic circuits, to control operation of functional blocks, in LO signal generation paths, as buffers, as drivers, and so on. In high-speed wireline circuits, peaking of the inverter can be obtained to enhance the bandwidth of the circuit. Since inverters need to operate at very high frequencies, a compensation network that is wideband in nature may be used to provide peaking for better high-frequency characteristics. Such a technique is commonly used at high-frequency MUX/DEMUX circuits to boost bandwidth in multi-Ghz range.

4.1.2 Key Electrical Equations

In digital subsystems, inverters are used in a large-signal fashion. Key equations to describe them include:

$$V_{mid} = \frac{V_{DD} - |V_{tp}| + \sqrt{\beta_n/\beta_p}\,V_{tn}}{1 + \sqrt{\beta_n/\beta_p}} \tag{4.1}$$

$$t_{pHL} = \frac{1}{\beta_n(V_{DD} - V_{tn})}\left[\frac{2V_{tn}}{(V_{DD} - V_{tn})} + \ln\left\{\frac{4(V_{DD} - V_{tn})}{V_{DD}} - 1\right\}\right]C_{OUT} \tag{4.2}$$

$$t_{pLH} = \frac{1}{\beta_p(V_{DD} - |V_{tp}|)}\left[\frac{2V_{tp}}{(V_{DD} - V_{tp})} + \ln\left\{\frac{4(V_{DD} - V_{tp})}{V_{DD}} - 1\right\}\right]C_{OUT} \tag{4.3}$$

As can be seen, the symmetry can be maintained, if $\beta_n = \beta_p$ and $V_{tn} = |V_{tp}|$. To obtain the transient responses, the MOSFETs are assumed as simple $R - C$ elements, where $R = \frac{1}{\beta(V_{DD} - V_t)}$, and $C = C_{OX}WL$. For a single transistor, $\beta \sim L/W$, and the intrinsic delay from the gate remain invariant with scaling. However, in a real environment, some load capacitance would always be present, and the gate would have to be sized to meet the specific delay requirements. A larger transistor would consume more power at the switching instants. Circuit designers commonly adjust the aspect ratios of the devices and their sizing to achieve the required delay characteristics through the gate. Figure 4.1(b) shows an optimally biased inverter, where the individual NMOS and PMOS devices can be shifted in DC and can be switched using a smaller input signal. NAND and NOR functionality static circuits are illustrated in Figure 4.2.

Inverters gained popularity because of the complementary nature of the circuit topology. However, in terms of circuit functionality, inverters can very well be realized using bipolar transistors as (PNP/NPN).

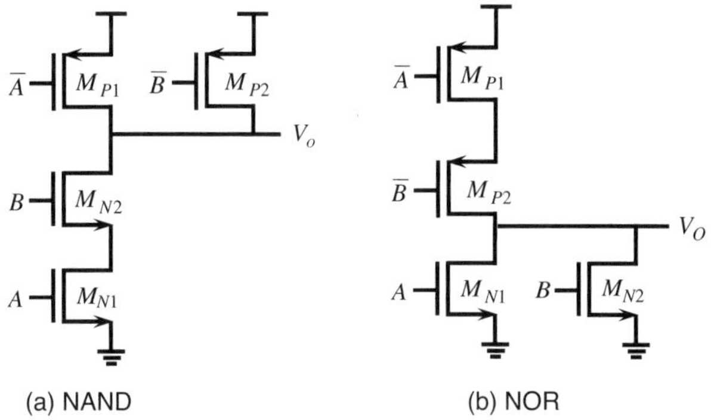

Figure 4.2. Static CMOS circuits.

4.1.3 Current Reuse Amplifier

Inverter topology can also be used as a current reuse amplifier, which provides a phase shift of 180° at the output relative to the input. This is shown in Figure 4.1(c). In this case, both the NMOS and the PMOS devices operate as transconductor ($g_m : g_{mP}, g_{mN}$) and load ($g_L : g_{LN}, g_{LP}$), respectively. The bias voltage at the input of N and P devices can be adjusted such that it switches at lower input signal levels. In this amplifier topology, output currents from the transconductors add together and contribute to the gain. Hence, using the the same current, a higher transconductance can be obtained (the same current is used between NMOS and PMOS devices). The pole is created at the output using the load capacitance C_{OUT}. In this structure, there is no internal pole, and fundamentally it can be used at much higher frequencies. The transconductance of the inverter topology can be varied with V_{DD}. Because of the steep slope of the voltage transfer characteristics, a high gain can be obtained with compromise in the linearity performance. However, because of the presence of only one NMOS or PMOS is in series with the power supply, the power supply rejection performance is also poor for such amplifiers. To obtain superior common-mode rejection, two differential chains are run in parallel. Similar topology is valid for bipolar (only amplifier topology is shown). In the digital domain, MOS gates are advantageous in terms of area and functionality, whereas bipolars are superior in terms of high transconductance and higher output impedance required for analog applications. The class $-$AB type operation makes this topology more efficient in terms of a signal handling capability. Optimally biased inverters can be used to provide current-mode logic to CMOS conversion, which is an often used functionality in mixed-signal communication ICs.

Many short-distance wireless devices operate under low transmit power. Examples of these devices include ultra-wideband applications, implantable medical radios, and so on. In these cases, optimally biased inverters form the power amplifier stages or the predriver stages. Excellent current efficiency can be obtained because of the inherent current reuse in the topology.

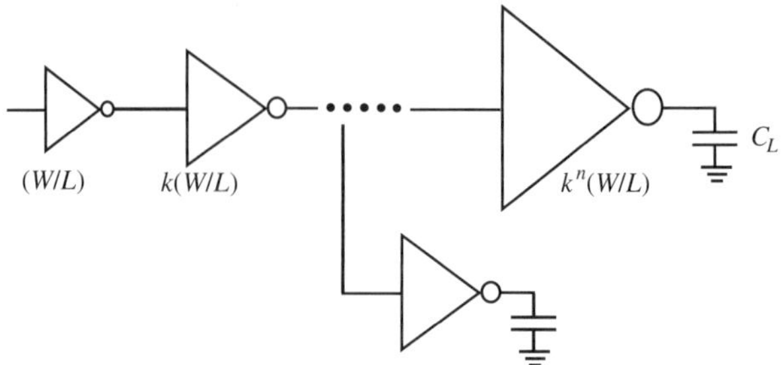

Figure 4.3. Cascades of inverter illustrating potential fanout.

4.1.4 Cascade and Fan-Out

With the basic introduction of inverters, we now would like to proceed to cascades of such stages. This is illustrated in Figure 4.3. In the digital domain, our target is always to minimize propagation delay while being able to drive the desired load impedance (usually input of other logic gates, and purely capacitive in nature). While considering an amplifier application, bandwidth is of utmost importance, as that controls the rise and fall times of the output pulse. In the analog domain, the considerations are the same. Instead of driving a big load by a single stage amplifier, the burden is distributed across several amplifiers, such that the overall response is optimized. To perform this, the aspect ratios of the stages need to be "tapered" as one proceeds to the load capacitance. An optimum ratio of $e = 2.72$ can be used for these tapered stages.

4.2 STATIC D FLIP-FLOP

The scope of the static circuits can be extended to construct D flip-flops, (DFFs) and so on, as illustrated in Figure 4.4. Compared with the various types of flip-flops, DFFs are commonly used because of their area advantage and ease of logic construction over other flip-flop circuit configurations. Static circuits provide no glitch even at low

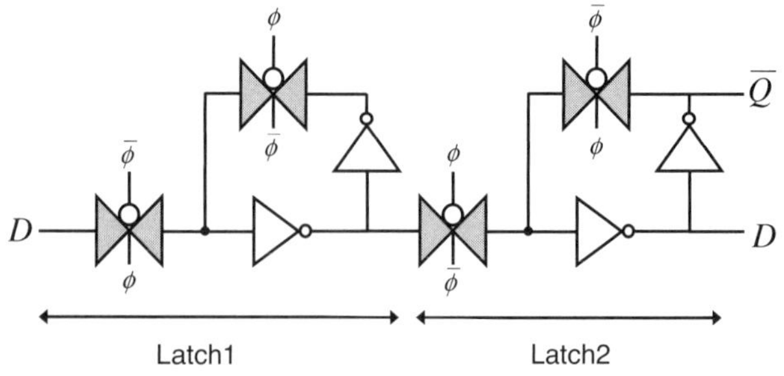

Figure 4.4. Static CMOS-based DFF (transmission gate based).

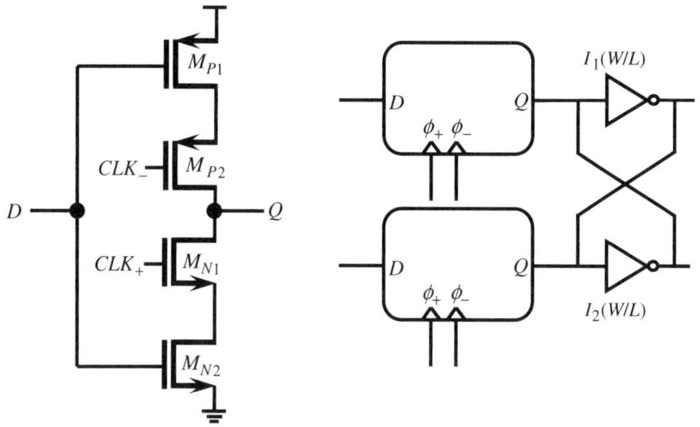

Figure 4.5. CMOS implementation of static D flip-flop.

speeds. However, the difficulty is that they operate on charging and discharging of capacitive nodes, and they use larger input loading (NMOS and PMOS for each input signal). Hence, they are not very power efficient at high-speed operations.

Static DFFs have gained popularity with technology scaling. To many mixed signal designers, this comes as an "easy-to-use" library cell, which operates at increasingly higher speeds with technology scaling. Unless there is a specific need of lowering the supply voltage and delay requirements on the standard cells, standard library cells are easily used in dense designs.

Dynamic implementation of CMOS DFF has gained significant popularity because of their substrate noise immunity, and with technology scaling, they can operate well up to the 4–5-GHz range (see Figure 4.5). Beyond this frequency, they tend to consume higher power currents, and a current mode logic (CML) style is adopted. In integrated systems, CML and CMOS are used together to provide a power-efficient, high-speed circuit solution at reasonable power consumption and area targets. A CML-to-CMOS converter and vice versa are used to transfer the logic levels from one family to the other.

During the late 1970s and early 1980s, some original circuit topologies were designed, mostly directed toward higher integration in microprocessors, to make it power efficient. We shall briefly focus on two circuit techniques, such as dynamic circuits and differential cascode voltage switch logic, which were introduced to improve the power and speed of the digital circuits. The inherent principles, such as charge leakage, and charge sharing, as well as differential signal swing are common in many other mixed-signal circuits. An in-depth illustration of various digital circuit design techniques can be found in [1].

Figure 4.6 illustrates two fundamental methods to implement high-speed digital logic. They can be categorized as voltage-mode (charge based) and current-mode logic families. In a voltage-mode circuit, charging and discharging of various capacitive nodes are involved, and to add to this complexity, the capacitances themselves are a

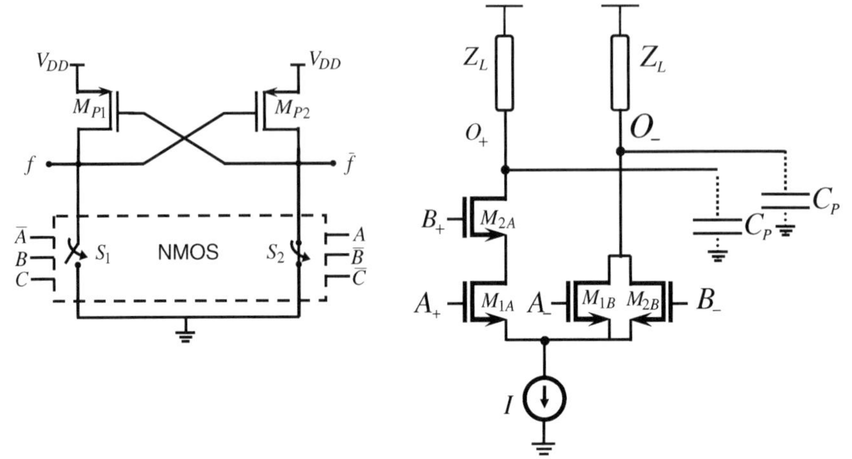

Figure 4.6. CVSL and CML circuit family styles.

function of the voltage. Current-mode logic cells use a bias current, which is switched to each branch of the differential circuit and creates a desired signal swing according to the load impedance. In cascode voltage switch logic, the direct function and its complement is implemented in separate branches, and a latch is used to maintain the complementary nature of the two outputs. Since the "charging and discharging" phenomenon is absent, current-mode logic is suitable and is used in almost all high-speed digital systems, and in the signal generation part of high-speed prescalers. in wireless systems. CML-based latches are most commonly used as part of frequency dividers and so on. Although CMOS is fairly broadband in nature, CML-based latches tend to be sensitive around certain frequency ranges (closer to its self-resonating frequency), and they can be attributed to be somewhat narrowband in nature. The output signal swing is dependent on the load and bias current. The obvious compromise is the required bias current and implementation of the load network. However, most wired applications are not mobile, and although a low-power implementation is desired, it is not critical. Figure 4.7 illustrates CML-based inverter topologies.

Almost all circuits that are encountered in mixed-signal systems are differential, except a few situations, such as an antenna interface, as well as a laser driver interface, PIN diode interface, and so on. Inside the integrated circuit, the signal flow is fully differential, as much as possible. This is followed to provide immunity to substrate noise and to any other common-mode noise in an integrated silicon environment. In most situations, the communication circuits, e.g., the radio front ends and the transceiver front ends, are integrated with high-speed digital circuits such as DSP and ADC. An obvious implication of differential design is that it takes more area than their single-ended counterparts, and in many cases, mismatch between two arms of differential circuitry, along with high Q common-mode impedance, would lead to even order distortions. Hence, while designing in the digital environment, we should pay special attention to the common-mode impedance.

From now on, we will focus heavily on the design of the differential circuits.

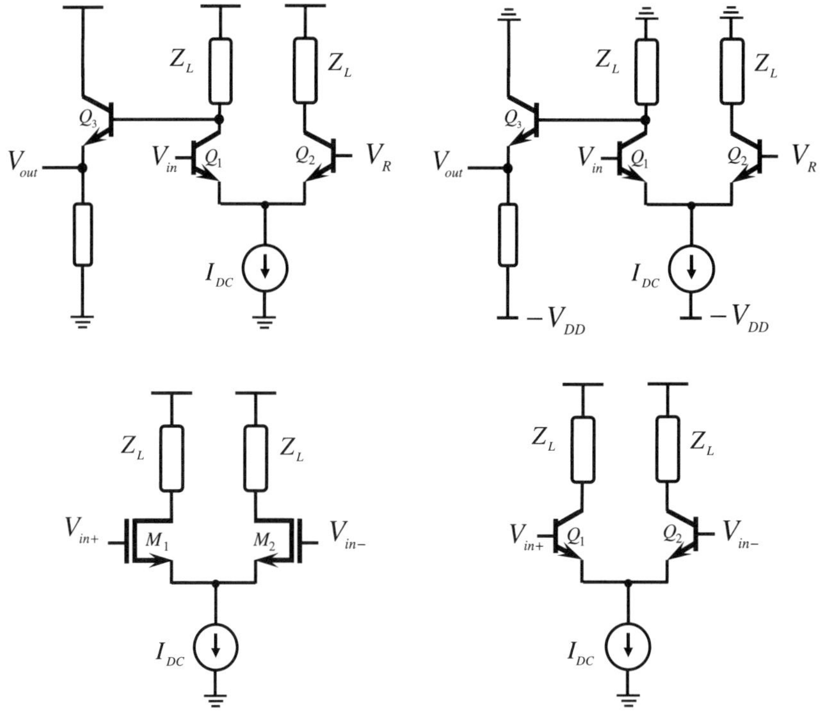

Figure 4.7. Current-mode logic-based inverters.

4.3 BIAS CIRCUITS

Bias circuits play an important role in mixed-signal systems. Key performance criteria of bias circuits include (1) input and output current handling capability, (2) output impedance (high/low), (3) Low headroom, and (4) low noise. Figure 4.8 shows examples of bias circuits.

4.3.1 Current Sources and Sinks

High-output impedance improves the common-mode rejection ratio (CMRR) of the differential transconductors. The first configurations (a) and (b) in Figure 4.8 illustrate current multipliers, which can be generated with MOS, as well as bipolar transistors. In bipolar, emitter degeneration is used to reduce g_m and, thus, to reduce bias source noise, improve current matching, and prevent thermal runaway. In Figure 4.8(c), the bias source would exhibit high output impedance, because of the cascode configuration, which is suitable for a current source, but it consumes more voltage headroom $2V_{D,SAT}$. The configuration in Figure 4.8(d) leads to a lower voltage headroom $V_{D,SAT}$. Figure 4.8(f) illustrates an inductive tail, with some Q associated with it.

As we will illustrate various advanced circuits in detail later, current source performance will be extremely critical. Especially at supply voltages below 1 V,

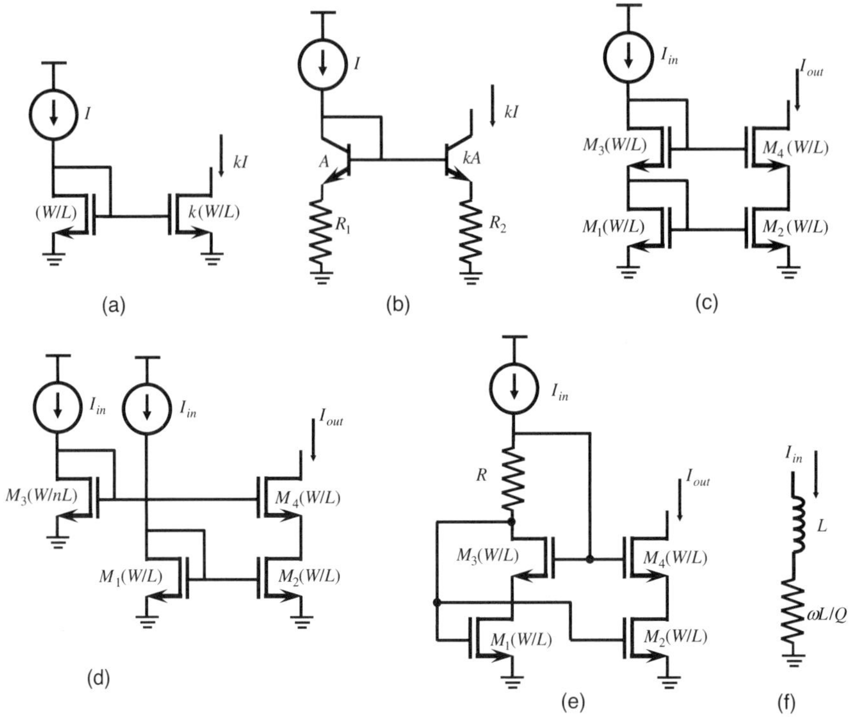

Figure 4.8. Various bias generators used in integrated circuits.

current sources become critical in terms of headroom. The challenges are different for different circuits as well as for the frequency of operation. The output impedance of the current source becomes a critical factor at RF frequencies. Bipolars, simple resistors, and inductive elements provide low-output parasitic capacitances. At low frequencies, this does not limit the performance. However, at high frequencies, the parasitic capacitance becomes a significant performance limiter, especially in the cases of pseudo-differential circuits, such as single-ended to differential amplifiers, and so on. In these considerations, the inductive tail tends to perform superior compared with output impedance ($R(1 + jQ)$), while consuming headroom because of the resistance. The output parasitics is also much lower at RF frequencies. Thus, it is suitable for a low-headroom, high-output impedance bias source. The only tradeoff is w.r.t. on chip area consumption. In an amplifier, the impedance at the operating frequency is of much consideration, whereas for VCOs, impedances at $2f$ become important as well. In general, MOS transistors tend to have the highest output capacitances, on-chip resistors tend to have the next highest, and inductor-based bias tends to have the lowest.

Noise from the bias circuitry is also a concern for integrated circuits. It degrades the performance of the amplifier. In the amplifiers, the noise can be bypassed using a large capacitor (usually 10 pF or so). This would work well for most RF frequencies; however, in VCOs, since noise upconversion degrades the close-in phase noise performances, bypassing capacitors would not help in this situation as the value would be impractical to integrate). Although the illustration has been made for current sinks, similar considerations are valid for current sources as well.

A practical consideration in the construction of current mirrors is the mismatch between various branches. The extent of mismatch is inversely proportional to the overdrive voltage of the transistors, and often, a larger length is used in the current mirrors, with higher $V_{D,SAT}$ values. They are also laid out close to one another to minimize spatial mismatch. In case the mirroring device is very small in geometry, dummy fingers are added to improve matching.

Current sources could be constructed according to need: (1) constant current source, (2) proportional to absolute temperature, (3) constant g_m, and so on.

While constructing bias circuits for RF blocks, one must pay close attention to any loops that exist in the bias circuit. In many cases, the RF signal may leak into one of these paths, and cause stability problems.

4.3.2 Voltage References

Voltage regulators are commonly used in modern ICs to provide supply current to the various building blocks. They need to provide low noise for the sensitive analog circuit blocks. Commonly used in this category is LDOs or "low dropout regulators."

Inherently, a stable voltage reference is required as a comparing point. This is usually obtained by a band-gap reference, in which a base-emitter voltage V_{BE} is added to a scaled proportional to absolute temperature (PTAT) voltage to generate a temperature-stable voltage reference. Figure 4.9 illustrates a conventional band-gap reference circuit technique.

The simplistic scheme illustrated here is prone to second-order effects, such as amplifier offset, finite power supply rejection ratio, and so on. Such effects should be taken into consideration while designing a precise band-gap reference. Band-gap references have become popular, and many regulated supplies are generated from a band-gap reference.

Voltage regulators can be designed using a band-gap reference. Figure 4.10 illustrates a voltage regulator configuration based on resistor ratio and voltage

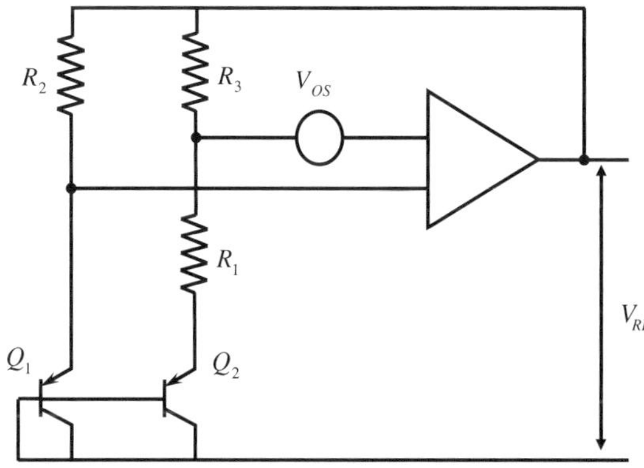

Figure 4.9. Circuit scheme to generate a bandgap reference voltage.

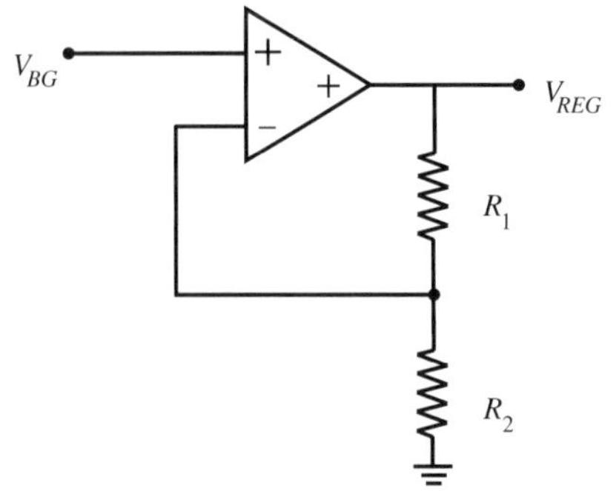

Figure 4.10. Simple low dropout regulator using a bandgap reference.

comparison. The OP-Amp under consideration is usually designed to be high gain (~100–120-dB gain) to provide a very small input-referred offset voltage. It should be also designed to provide adequate phase margin and compensation for the desired load network (the on-chip capacitors). Many other regulators have been used in integrated circuits, but we will not cover them here.

4.4 TRANSCONDUCTOR CORES

Transconductors are used in a wide variety of the circuits and systems in the mixed-signal domain. Their operation can be illustrated under the limits of small-signal inputs, where they provide an output current proportional to the input differential voltage. The key performances of these circuits include (1) small-signal transconductance (g_m), (2) input impedance (Z_{in}) (common-mode and differential), (3) output impedance (Z_{OUT}), (4) input-referred noise ($V_n{}^2$), (5) controllability of g_m, and (6) input-referred linearity. In general, transconductors can be used as open loop signal processing elements, with an input voltage waveform (characterized by the amplitude, frequency, and phase), and the output is current (dictated by the same parameters) proportional to the input voltage. Commonly used transconductor configurations are shown in Figure 4.11. Although emphasis is given on MOS implementations, bipolar and other types of transistors can be used as well.

Figure 4.11(a) illustrates the simplest configuration. The differential transconductance is a function of the bias current I_{DC} and the device geometry. Linear conversion from voltage to current is an important aspect, and a high linearity can be obtained by linearizing the input stage, using an amplifier as shown in Figure 4.11(b) (one needs to be careful about the proper sign within the feedback loop). In this configuration, the input signal appears across resistance R, and the transconductance is independent of the device geometries. Figure 4.11(c) obtains a linearized transconductor, with degeneration resistors, which consume voltage headroom. An inverter-based g_m cell is illustrated in Figure 4.11(d), where the core g_m cell as well as the common-mode feedback is shown. In fact, while designing any of these transconductor cells, a common-mode feedback must be used before analyzing any linearity and noise performances.

Figure 4.12(e) shows a circuit configuration, where the voltage headroom is not consumed by the linearizing resistors. However, it may lead to higher noise

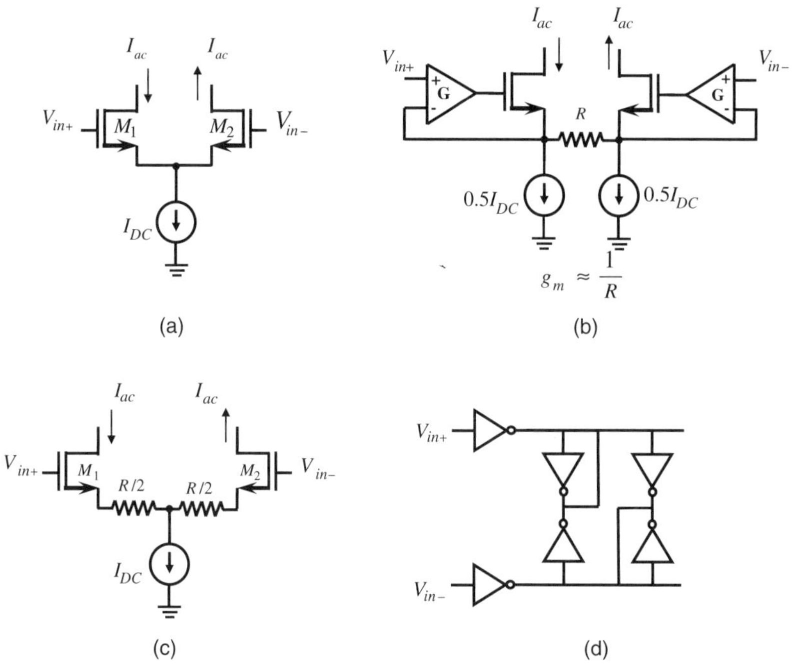

Figure 4.11. Various transconductor cores.

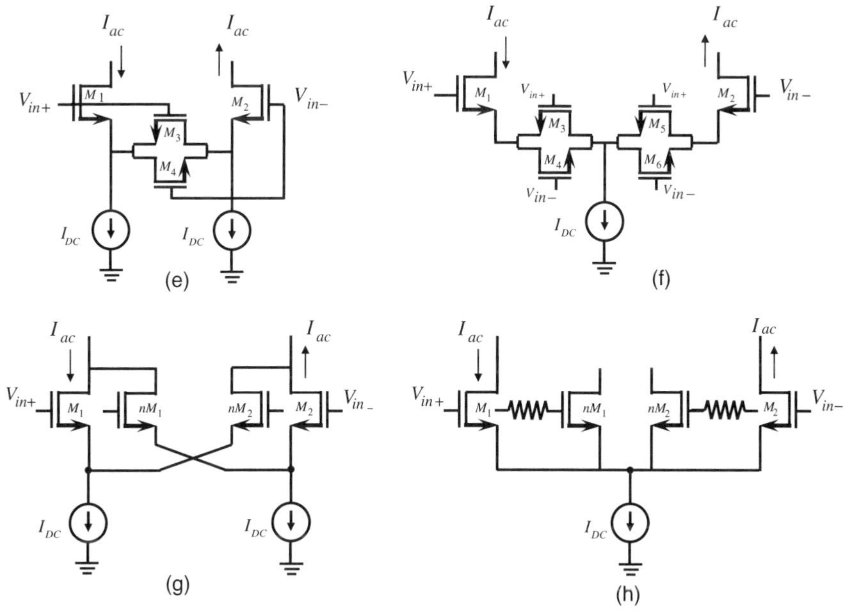

Figure 4.12. Transconductor stages.

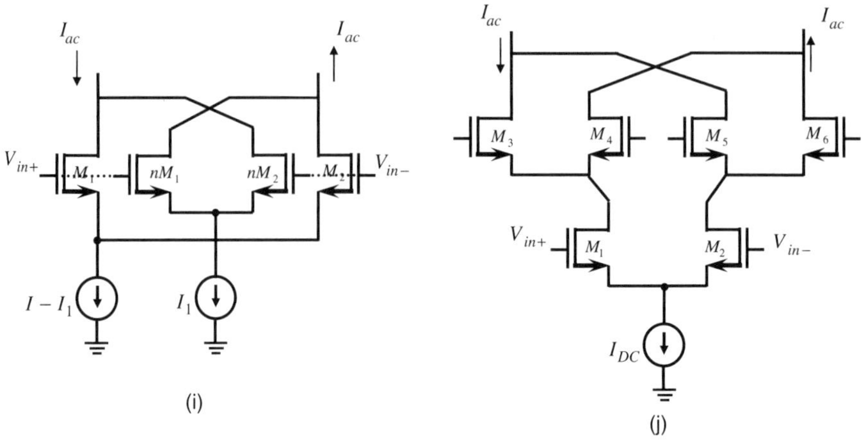

Figure 4.13. Transconductor stages.

performance, as the current source noise is uncorrelated. Figure 4.12(f) illustrates the configuration with headroom consumption. Figures 4.12(g) and (h) illustrate the transconductor configurations with classic methods of providing the offset in the input voltage by sizing input transistors.

Figure 4.13(j) shows a commonly used Gilbertcell configuration, where g_m can be easily controlled by the switching transistors, and the current can be supplied to the output load.

Few other transconductors, which have been recently reported in literature, are shown in Figure 4.14(k), (l), and (m), with a folded cascode stage shown in

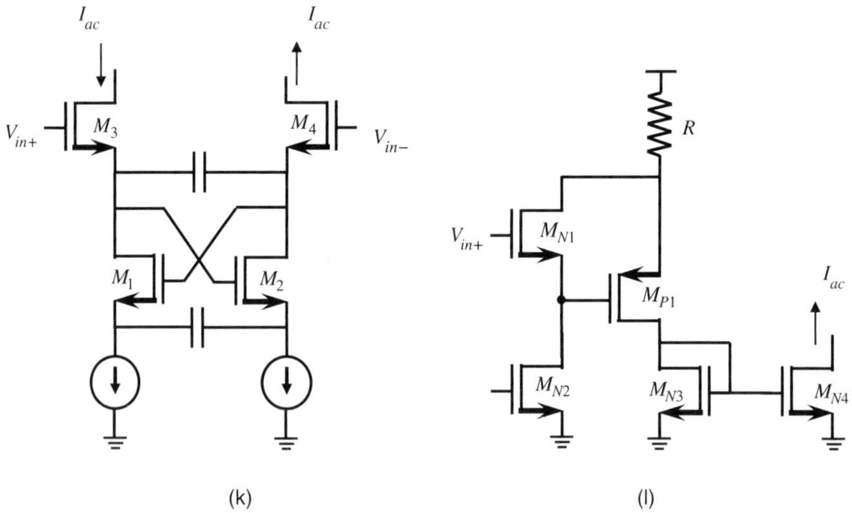

Figure 4.14. Transconductor stages.

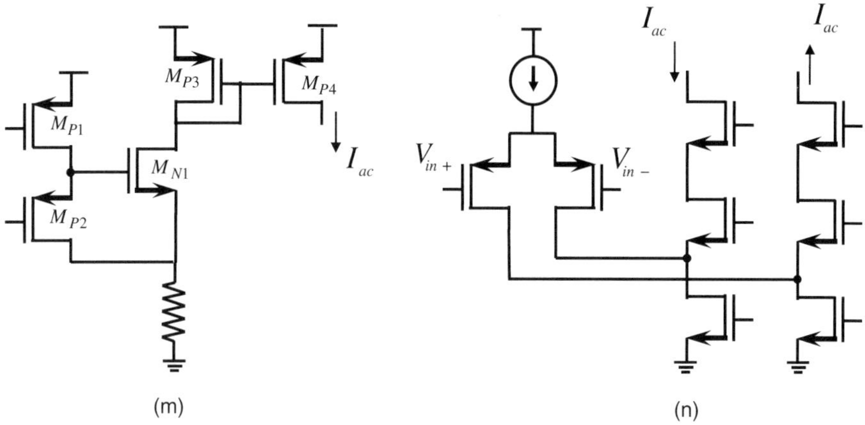

Figure 4.15. Transconductor stages.

Figure 4.15(n). While analyzing transconductor circuits, it is recommended that the designer checks all the branch currents and voltages to understand the sources of nonlinearity.

Transconductors are widely used in (1) operational amplifiers, (2) high-frequency continuous time filters, (3) input stages of downconverters and receivers, (4) equalizer circuits in wireline communication systems, and so on. Although several advanced texts and other literature are available on transconductors, their optimum performance, and their application, we have only covered the basic details to appreciate mixed-signal systems.

4.5 LOAD NETWORKS

Load networks are essential components of active circuits, and several topologies can be used. Figure 4.16 illustrates the various load configurations commonly used in circuits. From a circuit design perspective, the following considerations become important while designing load networks:

1. Voltage headroom consumed in the load
2. Impedance of the load
3. Area consumed by the load
4. Nonlinearity contributed by the load
5. Parasitic capacitance and bandwidth requirements

4.5.1 Passive Load

Passive loads are usually of two types: (1) resistive and (2) inductive. Inductive loads are essential for high-frequency compensation. Inductors do not consume DC

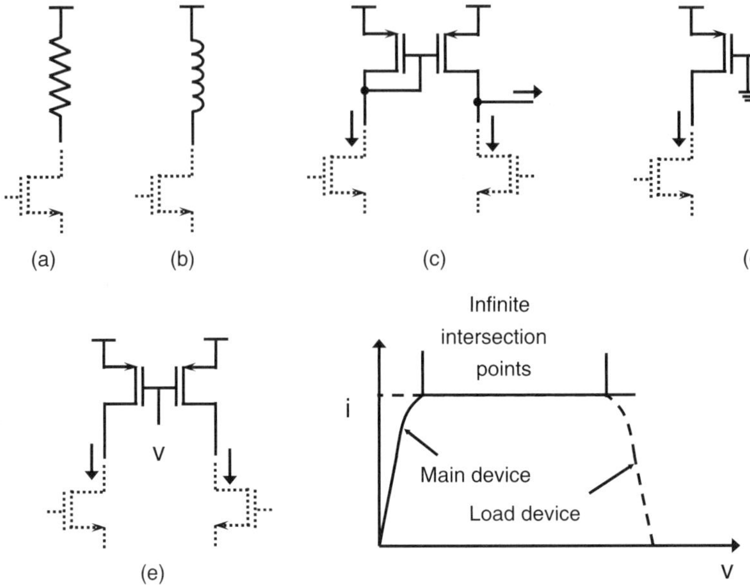

Figure 4.16. Various types of load used in integrated circuits.

headroom, and they can swing on top of the supply voltage, leading to high-dynamic-range circuits. This provides a significant advantage in constructing high-dynamic-range circuits in deep submicron technologies. The core circuit can be NMOS or PMOS type, depending on the specific implementation. Resistive loads consume DC headroom and contribute to the noise.

Resistive load is usually employed in broadband circuits, and at lower frequencies (~100 M–500 M), where use of passive inductors is prohibitive in terms of area and lower Q factor. Early generations of logic circuits used resistive loads to construct logic gates, because of the lack of complementary P-type active element in those technologies. The bandwidth of a resistively loaded circuit is dependent on the RC network formed by the resistance and the output capacitance of the devices. Process and temperature dependencies of resistors must be well modeled prior to their use in the circuit. In many applications, transistors operating at fully ON state (NMOS with the gate tied to VDD or PMOS with gate at 0 V) can be used as resistors to save the on-chip area.

Both resistive and inductive loads are linear with nonzero slope in their DC transfer (I–V) characteristics. When used in series with the active elements, the transfer curve of the load and active elements intersects precisely at a single point; hence, a common-mode feedback is not necessary.

4.5.2 Active Load

Active loads are very often used in circuits to reduce the voltage drop per unit current. This is a critical issue in low-voltage circuits, where the headroom limits the usage of number of transistors and amount of functionality obtained per unit current consumption (true for simple resistor or active device operating in triode). In an active device, $V_{DS} \approx \sqrt{2\beta I}$, and this can be used to save headroom. Figure 4.16(c) illustrates a current subtractor network formed by current mirrors. As the diode-connected end of this network sets the voltage, no common-mode feedback would be necessary. The output impedance is fairly high and given by the parallel combination of the load and main transistors' output conductances. In many modern digital CMOS process technologies, transistors are better controlled, compared with the resistors in the same process. Figure 4.16(e) illustrates a circuit configuration where the load transistors are used in saturation region and their gate voltage is tied to a specific voltage level. Since the DC characteristics of the main device and that of the load device intersects at infinite points, a common-mode feedback would be necessary, as otherwise, when the load and driver transistors operate in a saturation region, they would intersect at infinite points in their DC transfer characteristics. Common-mode feedbacks also employ active or passive loads and current mirroring type loads. The purpose is to correctly define the common-mode voltages, and provide a negative feedback to adjust the operating currents in order to stabilize the common-mode voltage at the output.

Common-mode feedback is very essential, and almost all OP-Amp stages uses them. Common-mode voltage ranges are important in determining the direct interfacing of two circuit blocks, where a capacitive coupling is prohibitive in terms of area. Common-mode voltage ranges also determines the available signal swing at various circuit interfaces, leading to dynamic range optimization.

While designing common-mode feedback networks, one needs to be careful of generating potential instabilities and nonlinearities. Common mode-to-differential conversion may occur, leading to significant degradations to circuit performance.

4.6 A VERSATILE ANALOG SIGNAL PROCESSING CORE

After the discussion of transconductors, we will now concentrate on a multipurpose mixed signal cell as illustrated in Figure 4.17. The input stage is again a transconductor cell, with current switching capability at the output of the transconductor cell.

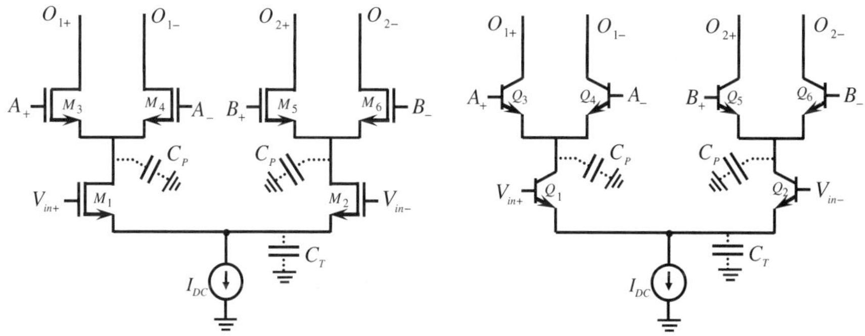

Figure 4.17. A multipurpose analog signal processing cell.

The overall stack is biased using a current source, which can be any of the configurations illustrated before. Depending on the nature of the inputs and the bias conditions, this cell can provide many functions.

1. As a cascode amplifier, eliminate M_4, and M_5 and use A_+ and B_- at DC voltages. This configuration would lead to cascode LNA.
2. As a mixer, when A_+ and B_- are shorted, and A_- and B_+ are shorted, and both are used as input differential switching waveforms. The output current would have components at $\omega_{LO} + \omega_{RF}$ and $\omega_{LO} - \omega_{RF}$, where the input is ω_{RF} and the switching waveform is at ω_{LO}. In a direct conversion receiver, $\omega_{LO} = \omega_{RF}$, and $\omega_{LO} + \omega_{RF}$ tone at the output is filtered by the filtering network. This configuration can also be used as an upconversion mixer, if input is provided at the IF frequency, ω_{IF}, and switched by ω_{LO}. In an upconversion mixer, the tones at the output are $\omega_{LO} + \omega_{IF}$, and $\omega_{LO} - \omega_{IF}$. In a transmit mixer/modulator, a second mixer is also used for the quadrature phase, and at the output, both are combined to provide a single sideband at the output. This will be illustrated in detail later.
3. This multipurpose cell can be used as a variable gain amplifier, where instead of a switching waveform, DC voltages are used and varied to provide a continuous variation in output current. Variable gain amplifiers are characterized by the amount of gain obtained and by ratio of minimum available gain to maximum available gain. In the case of an implementing variable gain amplifier, two drain terminals (in the controlling side) can also be connected to the supply, so that the AC signal is shunted to ground. Usually, RF signal processing is inaccurate (could be a reason to express all our answers in logarithmic scale!). Depending on the standard of interest, a continuous phase may or may not be considered between these gain steps.
4. Single-balanced/double-balanced mixer functionality can be performed if one of the two switching cells, *A* or *B*, is used.
5. Frequency multiplier cell: If *A* and *B* are used in a large signal fashion, at the same frequencies, but 90° out of phase (so as to not disturb the DC levels), the output is a waveform with twice the input frequency, and this can be used as a frequency doubler cell.
6. Sideband combiner/multiplex (selector)/demultiplex cell: In a wireless transmitter, output is contained in a single sideband, and this is obtained by phase rotation and current combination at the outputs of many *V–I* converters. Sometimes many signals need to be multiplexed (in the case of outputs of multiple LNAs to be interfaced with a wideband downconverter mixer in the receiver). Along the same considerations, demultiplexing stages are also important in conjunction with multiplexing stages. These scenarios are illustrated in Figures 4.18–4.19. Multiplexing can also be performed in a switch-like configuration (signal applied to the source/drain of a pass FET switch), where the signal suffers very little nonlinearity.

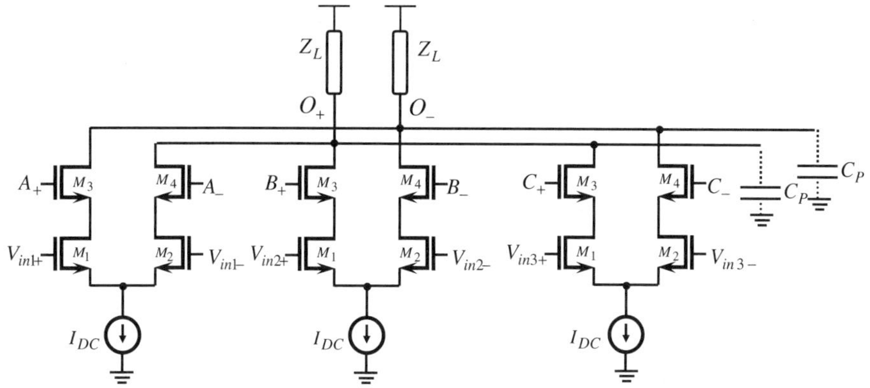

Figure 4.18. N-to-1 MUX circuit.

Design trade-off exists for the cascode transistors in terms of their parasitic capacitance, voltage headroom, and capacitive feedthrough to the input. A large transistor would provide lower voltage headroom but more capacitance. A large capacitance requires lower inductance at resonance, leading to lower overall impedance. At the same time, a larger parasitic (/static) capacitance would reduce the tuning range as well.

In practice, one needs to pay close attention to the parasitic capacitances at each of the nodes at higher frequencies, especially the output node, as it includes contribution of drain to bulk capacitance from all connected transistors. Current source output capacitance is critical at high frequencies, as a little imbalance in the input pair (in practice, nothing is perfectly differential), would lead to signal shunting to ground.

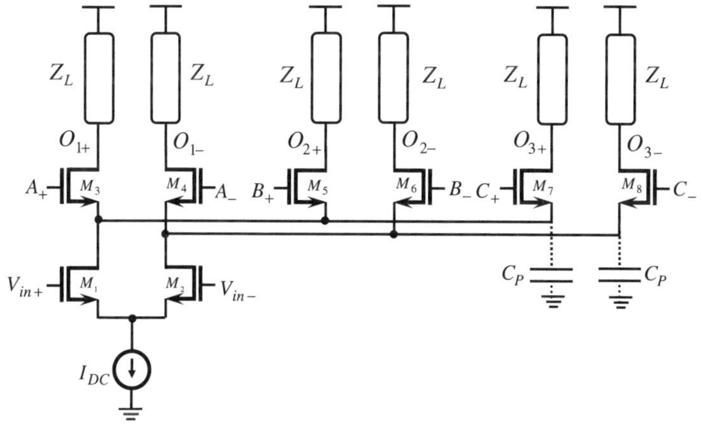

Figure 4.19. 1-to-N Demux circuit.

Bipolar implementation of the g_m cell is beneficial, as the transconductance is higher than a MOS, in the same area, and much lower loading if the input is at RF frequency. Also, bipolar differential pair needs about 78 mV to switch, which motivates a low-power LO chain design.

4.7 LOW NOISE AMPLIFIER

All radio receivers use a low noise amplifier to amplify the signal with lowest possible noise injection. Since the wireless standards are usually of a narrowband nature, these circuits are designed to provide simultaneous power transfer and noise match. The circuit topology is a cascode amplifier stage, and bipolar of MOS can be used as needed, as shown in Figure 4.20.

In the case of bipolar and MOS, the noise formulation is different as illustrated in Chapter 1. Like any other integrated circuits, LNAs are optimized for (1) noise, (2) linearity, (3) gain, and (4) input matching for a specific area and power requirements. At the input, a matching circuitry is used, which transforms the antenna impedance (usually 50 Ω) to an impedance that is optimum for noise and power match. This network amplifies the input voltage by a factor of Q, the quality factor of the matching network. Losses in the matching network provide a noise figure penalty. This can be minimized by using higher Q off-chip components, but too much Q leads to more susceptibility to component variations. The voltage gain obtained from the matching network would improve the noise figure of the LNA. Cascode transistor provides higher gain, reverse isolation. Headroom limitation comes from the current source; otherwise this circuit can use only two transistor in the stack. Choice of on-chip versus off-chip inductors are made w.r.t the area budget, frequency of operation, and electromagmentic cross-talk consideration in a system-on-chip environment.

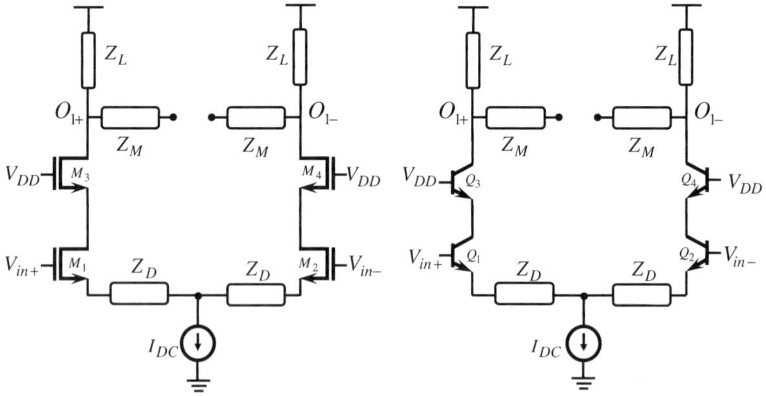

Figure 4.20. Fully differential low noise amplifier topologies.

4.7.1 Single-Ended Interfaces

The choice of single-ended RF input is dictated by the system designer for considerations to antenna interface and so on. An input passive balun implemented on PCB would degrade the noise performance. This causes imbalance in the circuit, which is a function of operating frequency, voltage excursions, and so on. The degeneration inductor, with the single end grounded, would cause an AC voltage swing at the drain node of the current source, and this would cause the output impedance modulation of the current source, leading to a penalty in the linearity performance. To alleviate this issue, the degeneration inductor can be connected in asymmetrical manner (i.e., current source connected at the source of the undriven terminal while the degeneration inductor is connected between source of other transistor and current mirror). This would hold the drain terminal of the current source to AC ground, and no impedance modulation would occur, thereby leading to much superior output impedance of the LNA. In the design of LNA, moderate-to-high Q inductors should be used in order to obtain the desired noise performance. Gain of the LNA should also be adjusted such as it does not saturate the following stages (such as mixers and so on). For this reason, a gain switching is often present in most receivers. In some receivers, more than a gain step is used.

4.7.2 Design Steps

Performance of the cascode LNA is mostly dependent on the input stage: (1) matching network, (2) g_m devices, (3) degeneration inductor (provide negative feedback and linearizes the input stage without any voltage headroom and low noise contribution), and (4) bias current. The design steps can be summarized as follows:

I. Set up a target current consumption.
II. Consider a T-matching network, consisting of off-chip L and C, with package and ESD parasitics considered.
III. Determine the LNA configuration: fully differential or single-ended input.
IV. Optimize the input transistors to provide a desired noise figure. It should be noted that a larger channel length implies that the optimum impedance for input match and optimum impedance to obtain noise match comes closer. Ensure that $V_{DS} \sim 2V_{DSAT}$, with a starting $V_{DSAT} \sim 60$–70 mV for the input transistor.
V. Ensure the input match meets specification.
VI. Use the full circuit and center the output network (cascode device and the output inductor, and the mixer input network) at a desired frequency band. Ensure that the cascode devices are small enough such that the signal feedthrough would not occur.
VII. For a target input P1dB (say -10 dBm), inject a tone at a 40-dB lower power level (-50 dBm, where the circuit would be linear) and observe the level at the output. Next, inject the tone at a desired P1dB power level (-10 dBm) and observe the level at the output. If the difference in the output power level

is 39-dB or less, we meet the P1dB target. If it is more than 39-dB, we then would adjust the input network (input match, degeneration inductor, and the device).

This seemingly simple circuit, however, requires accurate small-signal noise models and high Q on-chip inductors. Many trial-and-error methods have been used by many designers in the past, but a systematic approach leads to finding the optimum device size.

It should be noted at this stage that LNAs are inherently narrowband circuits. Due to the capacitance at the input due to ESD and package pins, the input device dimension should be adjusted to accommodate for the frequency shift while obtaining the desired noise figure. Additional capacitances result from (1) bondwire-to-ground capacitance, (2) bondwire-to-bondwire capacitance, (3) substrate trace-to-the-PCB capacitance, and (4) on-chip parasitic capacitance caused by the trace. Usually the lumped element representation of these capacitances would need to be taken into consideration while performing optimizing of the circuit. In wideband systems, this is difficult to perform, as the tuning range shrinks if the amount of static capacitance is large.

Although the performance of the LNA is often evaluated under the limits of a small signal, the output voltage swing can be significantly large at high signal input levels. In this case, all the transistors, especially the output cascode devices, would need to be optimized for breakdown phenomenon, as well as for any potential linearity degradations caused by the output stages.

An important practical consideration for LNA design is to analyze the stability of the circuit. This can be performed using a microwave "port"-based analysis, as well as analyzing the circuit for input impedance, and observing whether the real part is negative. Port-based analysis calculates the stability factor from the measured S-parameters.

From a device perspective, an LNA design usually uses multiple finger devices. As many fingers are used, g_m increases, and the finger-to-finger capacitance increases. Noise performance becomes slightly better if a large device is segregated in multiple medium-size devices. However, this leads to increased routing capacitance.

Although narrowband LNAs are designed to provide simultaneous noise and power matching characteristics, it should be observed that linearity and noise figures are far more important performance parameters compared to input match level. A 20-dB input match would imply that only 1% power is being reflected. In any practical situation, this can be compromised to obtain a superior noise figure performance. The linearity is mostly set by the negative feedback degeneration inductor, and care should be taken such that the source capacitance of the input stage does not resonate with the degeneration inductor. A larger device with a moderate current consumption seems to perform better in view of this optimization process. With more and more wireless standards and blocker profiles in the license-free bands, LNAs require significantly higher linearity.

Let us concentrate on the effects of the matching network once more. Gain from the matching network reduces the noise by increasing the signal level, but this increased signal level would lead to degraded linearity from the input devices. Also, as we increase the number of sections in the matching network, the matching network performance becomes less susceptible to component tolerances.

4.7.3 Gain Expansion

An interesting observation of pseudo-differential amplifiers and single-ended amplifiers is their characteristics w.r.t. input power. As the input power increases, it can be seen that at specific power levels, an expansion of gain occurs that cannot be captured in small signal analysis of the circuit. A large input signal tends to increase the bias current of the differential amplifier structure, and in the case of a pseudo-differential amplifier, the current is unrestrained. Thus, it would lead to a situation where linearity would depend on input signal level. The linearity limitation of the LNA could occur because of two factors. The input linearity limitation is based on the fact that a large signal can drive the input differential pair into a non-saturation region and can change the operating point of the transistors (or even can drive the current source into linear region). The input signal handling capability is determined by the matching network gain and by the input matching level. The signal handling capability of the differential pair is determined by the degeneration inductance and by the DC current consumption. The output linearity of the LNA is dependent on several factors, such as (1) output ESD, which can limit the signal swing; (2) the cascode device itself; and (3) any other functional blocks connected across the resonating tank load. If ESD is a limiting factor, they can be operated from a higher supply. If the cascode devices are limiting, then the geometry can be adjusted. The testbenches illustrated in Chapter 1 can be used to identify the linearity limiting block in a parallel combination of three or four components forming a resonator tank.

4.7.4 Layout Considerations

Careful layout techniques must be used to minimize common-mode noise (substrate noise, etc.) and parasitic oscillatory behavior. All unwanted capacitors such as gate-drain and drain-source would need to be minimized by using higher layers of metal routing. As the main device, and the cascode device have similar g_m, the voltage gain is unity at the drain node of the input device. Hence, any overlap capacitance created by G–S overlap in metal traces is reflected back to the input as twice the overlap capacitance. In most receivers, LNAs consume a significant amount of current for linearity reasons; hence, any parasitic resistance would be detrimental, both in terms of voltage headroom and, more so, in terms of noise contribution. In fact, parasitic resistance causes problem at all frequencies, and unlike capacitances, they cannot be tuned out. Hence, higher layers of metal levels are most desired for routing purposes. Extensive simulation with package components would need to be considered prior to fabrication, to eliminate any unwanted parasitic oscillation. In a differential layout, any layout asymmetry would give rise to even harmonics.

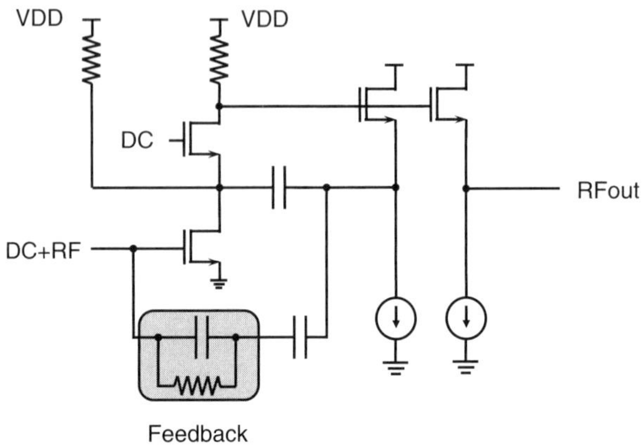

Figure 4.21. Broadband LNA topology using inductorless topology.

4.7.5 Inductorless LNAs

At sub-gigahertz frequencies, or at even lower frequencies (50–100 MHz), use of on-chip inductors becomes prohibitive because of area, lower Q, and cost restrictions. They also tend to generate EM cross-talk with other active devices in the signal path, despite whether an off-chip inductor or inductorless LNAs [7] are used. Resistively loaded LNAs usually lead to lower dynamic range. Resistors are used in the load and feedback networks, leading to a reduced area of these LNAs, at the compromise of higher power consumption. In contrast to the inductors, resistors tend to contribute more noise, leading to degraded sensitivity.

Figure 4.21 illustrates an LNA implemented using resistors.

4.7.6 Gain Variation

Gain variation in circuits allows them to operate in an optimum dynamic range condition as well as to use the supply power in the most efficient manner. In a cellular communication system, signal strength close to a base station is significantly different from that in the cell edge. If gain variation is not implemented, the same LNA operating at the cell edge would be driven into saturation, leading to significant degradation in dynamic range and potential burnout.

Few considerations are important while constructing a gain variation stage:

1. Gain variation range
2. Gain variation accuracy
3. Gain variation mode (continuous vs. discrete step)
4. Phase change during gain variation
5. Noise figure slope

To determine the gain variation range, the maximum and minimum signal strengths (given in RMS power level) are first obtained using channel propagation model and transmitter power levels. Next, the maximum and minimum signal handling capability of the ADC is obtained, and the difference between these two levels provides the gain variation range.

The accuracy of gain variation depends on the accuracy of the components used in the gain variation. Although the accuracy may vary, it is always desired that the gain variation be monotonic w.r.t the gain codeword or the current, as appropriate.

In high throughput and continuous streaming systems, such as video, not much buffering can be allowed in the receiver, and it is desired that there would be no phase change when gain is varied. The noise figure slope determines the extent to which the input-referred noise figure is changed with change in the gain. If the gain variation is implemented after the amplification stage, noise figure degradation is reduced.

Figure 4.22 illustrates the circuit details of gain variation schemes. In (a), the output current from the main g_m transistors is switched using a DC control voltage applied to the cascode transistors. However, it can be observed that this technique does not provide optimum usage of supply current, as a constant bias current is being used no matter whether we operate at higher gain or lower gain. An advantage of this scheme is continuous gain controllability and linear gain variation operation. In (b), three configurations are possible such as (I) changing the load, (II) changing the bias current, and (III) changing the input transistors. In (I), circuit bandwidth is changed, and the matching between components provides a critical design constraint. In (II), headroom changes along the power supply stack, and in (III), an array of transistors can be used, and depending on the input word, a selected set is enabled.

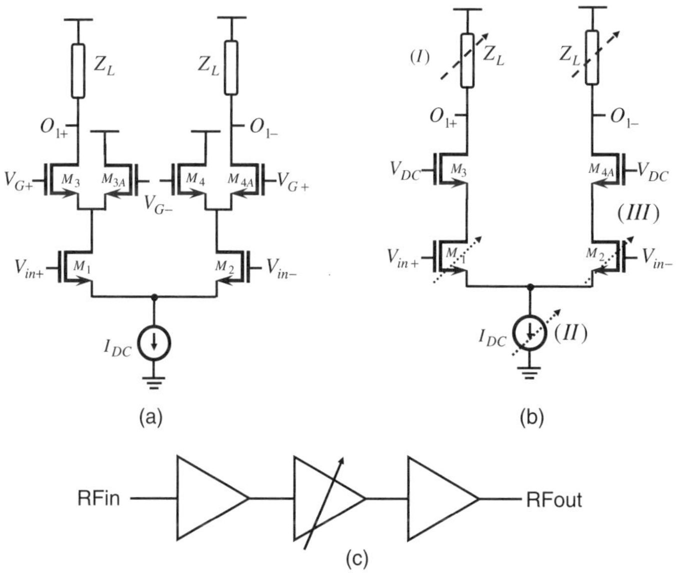

Figure 4.22. Implementing variable gain in amplifiers.

Another attractive arrangement, well suited for RF applications, is shown in (c), where the input and output stages are undisturbed, whereas all the gain variability is provided in the intermediate stage. This scheme does not provide advantage in terms of power consumption, but it keeps the matching and bandwidth requirements unchanged.

4.8 POWER AMPLIFIERS

Power amplifiers have been an area of wide interest because of the cellphone battery life considerations. It is the most power-consuming element in the entire transceiver, and careful optimization must be performed to obtain the required gain, efficiency, and linearity. Apart from cellular communications, a power amplifier is a key building block in radar, jamming, imaging, and RF heating applications. A transmitter may use one or more stages of a power amplifier. The output power from a power amplifier can vary from a moderate −20 dBm until +60–70 dBm. In modern times, almost all PAs employ some signal processing techniques such as predistortion. Similar to the developments of other solid-state circuits, power amplifiers went through spark plugs, vacuum tunes, discrete transistors, and integrated transistors, until modern implementation of DSP-controlled circuits and systems.

4.8.1 Performance Metrics

The usual performance metrics of a power amplifier include (1) linearity, (2) efficiency, and (3) gain controllability. In simple considerations, power amplifiers can simply be thought of as devices that convert DC power to RF.

A linear operation is critical to systems using both amplitude and phase modulation. Modulations such as CPMSK, GFSK, GMSK, and M-ary FSK provide a near-constant envelope. On the other side, many "bandwidth-efficient" modulation schemes such as QPSK, M-ary QAM, and multi-carrier provide a nonconstant envelope, and amplifier linearity is a major consideration in these situations. Data shaping in the baseband is very much required in order to prevent spectral spillover. This functionality is provided by using root-raised cosine filters and so on, and it is critical while operating in an allocated bandwidth of operation.

4.8.1.1 Linearity and its Measures. In most high-data-rate applications, linearity is a critical performance parameter of the power amplifiers. Linearity can be categorized in terms of in-band signal and out-of-band signals. Nonlinearity results in spectral spillover and amplitude-to-phase conversion, leading to overall distortion of the signal. AM–PM effects are caused by the voltage variable capacitances present in semiconductor devices, such as the gate-drain capacitance of a MOS transistor. Various measures of linearity include (1) C/I ratio, (2) NPR, (3) ACPR, and (4) EVM.

When the amplifier is driven with two or more signals of equal amplitude, the intermodulation distortion terms becomes spectrally collocated with the fundamental tones, and the difference between the fundamental and the intermodulation distortion terms is referred to as the C/I ratio.

Noise power ratio (NPR) is another performance metric. In this case, the PA is driven with a Gaussian signal with a notch in a specific frequency location. The amount of energy in that location is measured and that gives an estimate of PA nonlinearity.

In cellular systems, ACPR is very widely used to describe a PA's performance to adjacent channels. It is the ratio of power in a specific band outside the bandwidth of interest to the in-band signal. With an offset frequency of f_{off}, this can be given as

$$\mathrm{ACPR} = \frac{\int_{f_c - f_{off} - BW/2}^{f_c - f_{off} + BW/2} |H(f)|^2 S(f) df}{\int_{f_c - BW/2}^{f_c + BW/2} |H(f)|^2 S(f) df}$$

EVM provides a measure of how linearity impacts the detection process. It is a measure of the distance between ideal constellation and actual constellation points, and RMS average over the entire constellation.

4.8.1.2 Efficiency and its Measures. Efficiency determines the battery life of a wireless device, and it can be classified as (1) drain efficiency, (2) power-added efficiency, and (3) overall efficiency. Drain efficiency is given by the ratio of output power to the DC power

$$\eta = \frac{P_{out}}{P_{in}}$$

Power-added efficiency is determined by

$$\eta = \frac{P_{out} - P_{DR}}{P_{in}}$$

and the overall efficiency is determined by

$$\eta = \frac{P_{out}}{P_{in} + P_{DR}}$$

Often, average efficiency provides more meaningful representation. Instantaneous efficiencies are highest at the peak output power and falls off at lower power levels, and becomes lower at lower power levels. It is given by the ratio of average output power to average DC power, and it is defined as

$$\eta_{AV} = \frac{P_{out,AV}}{P_{in,AV}}$$

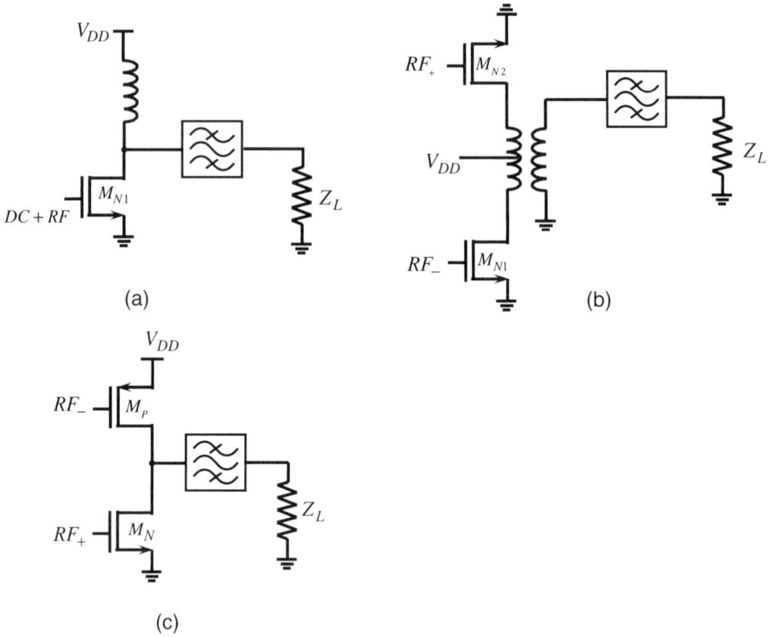

Figure 4.23. Various circuit configurations of the PA: (a) single transistor based, (b) transformer coupled push-pull, and (c) complementary.

The envelope of the modulated signal is a critical factor in determining power amplifier efficiency. The probability density function for a multicarrier profile is given by a Rayleigh profile with a peak-to-average ratio between 8 and 13 dB. To conserve the battery, back-off can be used. These two can be combined together to provide the desired efficiency.

4.8.2 Classes of Amplifiers

Fundamentally, power amplifiers can be categorized into classes A–F, with variations and combinations of two of the fundamental modes to create various other classes of operation. Basic circuit topologies include (1) single ended, (2) transformer coupled, and (3) complementary. Figure 4.23 illustrates the various circuit topologies for a power amplifier. Classes of amplifiers differ in methods of operation, efficiency, and output power capability. Figure 4.24 illustrates the voltage and current waveforms for different classes of PA.

4.8.2.1 Class A. In this mode of operation, the transistor operates in the active region and operates as a current source, controlled by the driving signal. The drain voltage and current are sinusoidal. The output power is given by

$$P_{out} = \frac{V_{out}^2}{2R}$$

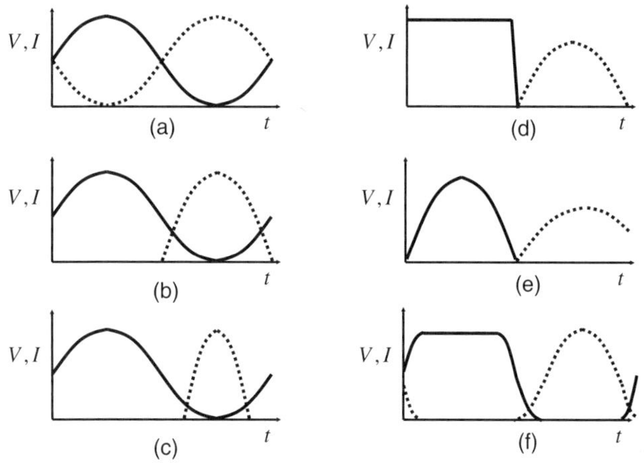

Figure 4.24. Voltage and current waveforms in various classes of amplifier operation: (a) class A, (b) class B, (c) class C, (d) class D, (e) class E, and (f) class F.

with the maximum voltage output as the supply voltage. DC power input is constant with a maximum efficiency of 50%. For amplitude modulated signals, bias current can be varied in accordance with the amplitude to obtain more efficiency. Class A operation is linear in nature, and an increase in bias current or operation at lower signal levels leads to a monotonic increase in intermodulation and harmonic levels. Thus, class A is used in cases where relatively lower power and high linearity. The efficiency of class A amplifiers is affected by the transistor's saturation voltage and load reactance.

4.8.2.2 Class B. In a class B operation, the input active device is biased at its threshold of conduction, the quiescent drain current is almost zero and a large signal input provides current 50% of the time. The device current is determined by the input signal, and thus, class B provides linear amplification. The bias current is proportional to the input signal, and this leads to an efficiency of 78.5% for class B. In practical implementation, two class B amplifiers are used with adjusted bias levels to minimize crossover distortion.

4.8.2.3 Class C. In class C amplifiers, the input stage is biased at slightly less than the threshold to increase the efficiency further toward 100%. Depending on the difference between the device and the actual bias voltage, linearity is determined. In practical implementation, a conduction angle of 150° is used for an efficiency of 85%. At the output, various harmonics of the drain current is filtered out.

4.8.2.4 Class D. This class of amplifiers use two or more switched transistors to provide a square waveshape in the drain voltage waveform. A filter provided at the output selects the fundamental frequency under consideration. This type of topology achieves close to 100% efficiency, and efficiency of such stage is not degraded because of load reactance. Practical considerations such as switching speed, saturation, and drain capacitance play a major role in class D amplifier implementations.

4.8.2.5 Class E. In class E operation, a single transistor operating in switching mode is employed. The drain voltage drops to zero with a zero slope when the transistor turns on. This leads to almost 100% efficiency. Class E eliminates the losses related to the drain capacitance charging and reduces the switching losses. This aspect leads to a wide usage of class E amplifiers at high output power levels.

4.8.2.6 Class F. Classes F and F inverse improves efficiency and output power using harmonic resonator structures. The drain voltage consists of one or more odd harmonics, and the current consists of one or more even order harmonics (approximately a half-wave rectified waveshape). The efficiency of these stages is dependent on the number of harmonics present, and it increases from 50% to 100% as increasingly higher harmonics are added. Harmonic tuning can be easily implemented by stubs, and by series/parallel LC lumped component-based resonators at lower frequencies.

4.8.3 Practical Considerations

Similar to any inductively loaded circuit, the voltage waveforms in a PA exceeds the supply rail, and breakdown of the device under operation must be considered for a safe operation. Various impedance values are determined with maximum voltage and maximum current operation. The impedance states corresponding to maximum achievable power and maximum efficiency are not the same. "Load-pull" measurements usually provide the impedance states on a Smith chart. The optimum impedance to satisfy output power and efficiency should also satisfy the stability criterion of the PA. If the stability condition is not satisfied, then negative feedback or lossy matching component should be used. Usually high gain stages are prone to instability, leading to low-frequency oscillation and in-band instability.

4.8.4 PA Architectures

Several practical considerations come into picture while determining the optimum architecture for PA topolgy. They are as follows.

4.8.4.1 Device Geometry. A larger device often provides higher output power, but because of the large parasitics, unwanted signal coupling leads to oscillation issues. At the same time, heat dissipation becomes a major issue. Multiple lower power PAs are usually combined to obtain higher power. Smaller devices usually achieve higher gain, lower Q factor of matching network, better phase linearity, and lower cost. However, power combiners provide signal loss.

4.8.4.2 Cascades of PAs. In most practical systems, several PA stages are cascaded to obtain higher power. Such stages need to be isolated from one another in order

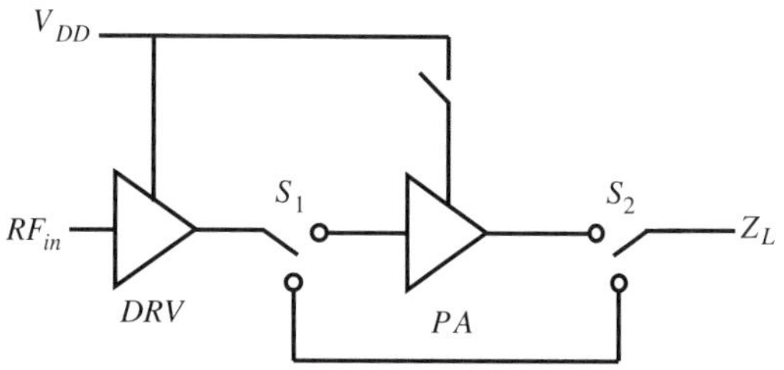

Figure 4.25. One or more stages of PA can be bypassed to increase efficiency.

to minimize interaction from one stage to another. Various coupling mechanisms occur at high frequencies through substrates and so on, aggravating this problem.

4.8.4.3 Bypassing/Switching Stages. In a multistage PA application, some stages can be bypassed by using a switch network around some amplifiers to obtain power efficiency. This process is illustrated in Figure 4.25. In the second case, multiple devices can be connected in parallel to achieve higher efficiency by selectively using one or more stages as necessary.

4.8.4.4 Envelope Elimination and Restoration. Also known as the "Kahn" technique, this type of architecture is very similar in operation to the polar modulator technique. In this technique, the envelope of the modulated signal is eliminated using a limiter, and a phase-modulated carrier is amplified efficiently using high-efficiency topologies. The amplitude modulation is then applied to the last stage to modulate the supply (voltage or current) to restore the original amplifier operation. In modern implementation, this is performed using a DSP-based techniques. This architecture provides a high average efficiency over a wide dynamic range as opposed to the linear power amplifiers. It provides high linearity, as the linearity is not dependent on the RF power transistors. Implementations of this technique, however, need to consider the envelope bandwidth and misalignment between the envelope and the main paths. An implementation is illustrated in Figure 4.26.

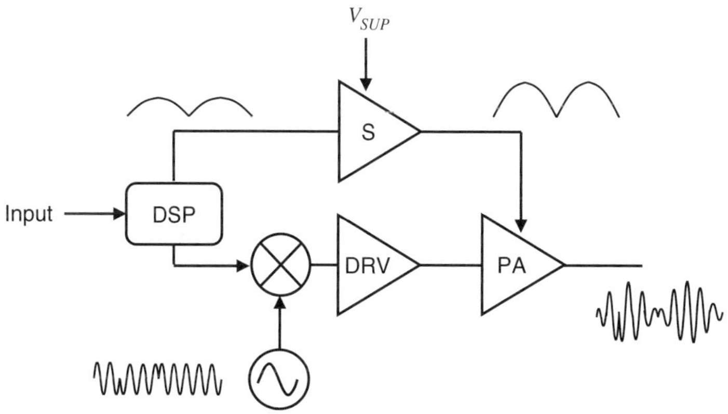

Figure 4.26. Transmitter architecture with separate envelope and carrier paths.

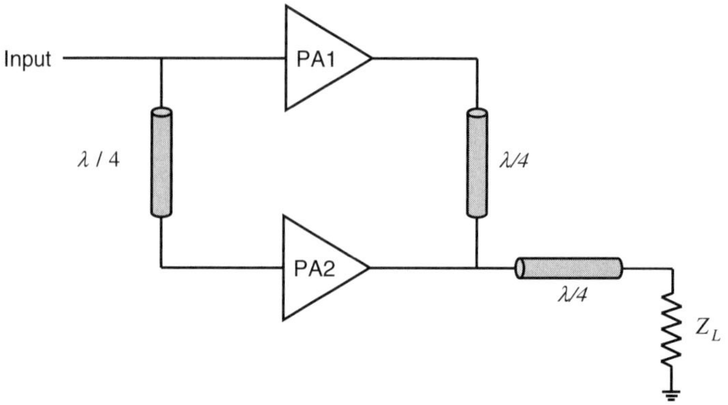

Figure 4.27. A Doherty PA transmitter architecture.

4.8.4.5 Outphasing. In this technique, the outputs from two power amplifiers are vector summed to provide the desired output. The sin-inverse of the input envelope is fed to the two PAs to provide an output proportional to the actual envelope. Both the PAs are ON at all the time, and the vector sum is obtained through the isolation of the two PAs through hybrid coupler as illustrated in Figure 4.27.

4.8.4.6 Doherty Amplifier. In the Doherty technique, two PAs are used called the "main" and the "auxiliary" PA. The main PA is biased in class B, whereas the auxiliary is biased at class C. The main PA is active when the signal is half of the peak envelope power or less. Both of them contribute to the output power when the input is higher than half of the peak envelope power. In the low-power input, the auxiliary PA is OFF, as the input signal strength is lower than the bias level. As the signal amplitude increases, the main PA saturates and the auxiliary PA becomes active. The instantaneous efficiency becomes close to 78.5% (theoretical class B operation). Figure 4.27 illustrates the configuration.

4.8.5 Feedback and Feedforward

Feedback and feedforward techniques are quite essential in order to improve linearity, and operation with reduced back-off, leading to higher efficiency. Several techniques can be employed to provide feedback in the transmitter.

4.8.5.1 Envelope Feedback. Envelope feedback can be employed to improve the in-band distortion. The signal envelope can be used as the feedback parameter to improve in-band distortion. Figure 4.28 illustrates an envelope feedback technique. Both the RF and the input signal envelope are sampled using couplers, and the error signal drives a modulator to compensate for the in-band distortions. However, a drawback of this technique is its inability to correct for AM–PM distortions.

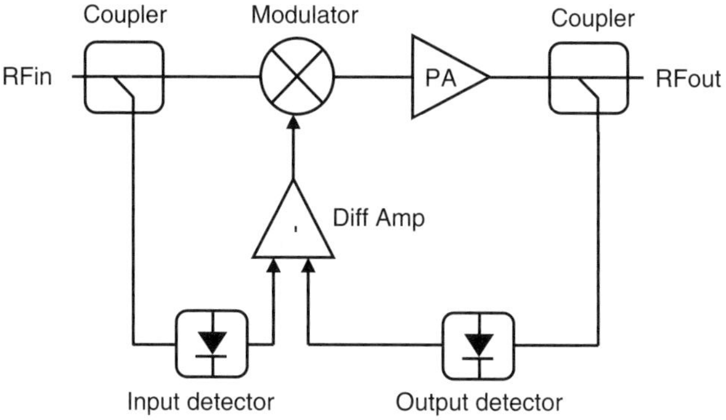

Figure 4.28. An envelope feedback technique for PA linearization.

4.8.5.2 Polar Feedback Technique. Figure 4.29 illustrates the polar feedback-based architecture. Conceptually similar to the polar modulator architecture, as illustrated previously, the PA output is sampled using a coupler, and fed to the PLL and the modulator using the phase and amplitude paths, respectively. A fundamentally common concern of amplitude and phase path synchronization is present here as well.

4.8.5.3 Cartesian Feedback Technique. A Cartesian feedback technique is illustrated in Figure 4.30. For easy accessibility of direct I and Q signals at baseband, this has gained popularity. In this approach, the PA output is sampled using a coupler, attenuated, and downconverted with a shifted phase of the oscillator, which is in synchronization with the main oscillator. The difference between the input and the downconverted signal is fed to the upconverting modulator. The phase shifter is used to

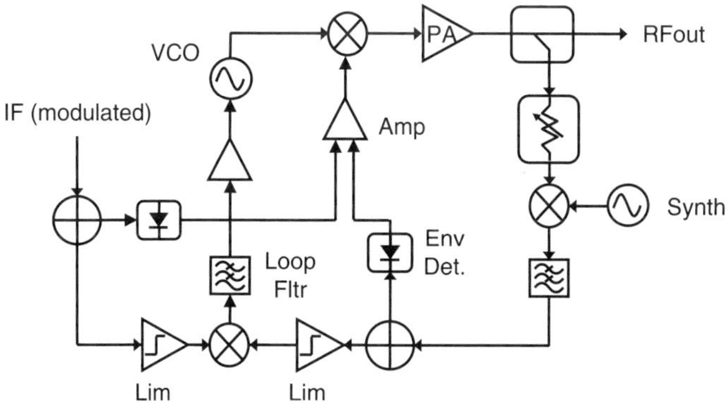

Figure 4.29. A transmitter architecture using a polar loop.

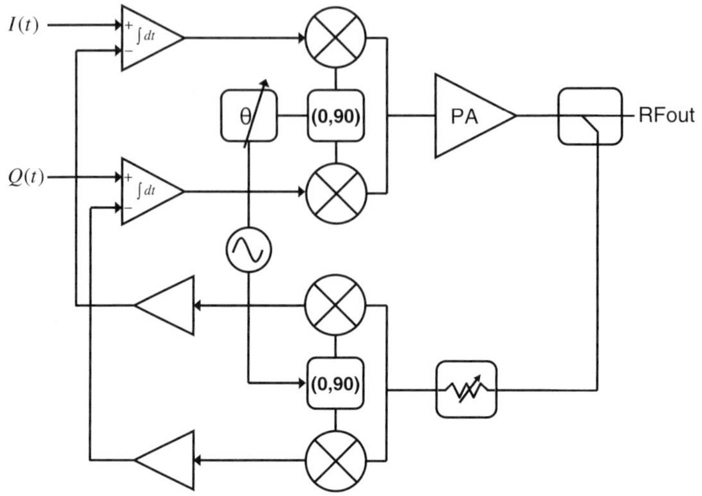

Figure 4.30. A transmitter architecture using a Cartesian loop.

align the phases in order to provide a negative feedback. In general, feedback techniques are prone to settling time and stability concerns.

4.8.5.4 Feedforward Technique. For wide bandwidth systems, feedforward can be used, as illustrated in Figure 4.31. The input signal is split in two paths, one of which goes to the main amplifier and the other to the error amplifier. At the output of the main amplifier, fundamental, as well as distortion tones are present. The subtractor present in the auxiliary path subtracts the fundamental signals, leading to the presence of only distortion terms at the input of the PA. The auxiliary PA then amplifies the distortions to appropriate levels to be canceled by the subtractor at the main output. Both the paths are synchronized w.r.t. the two delay lines.

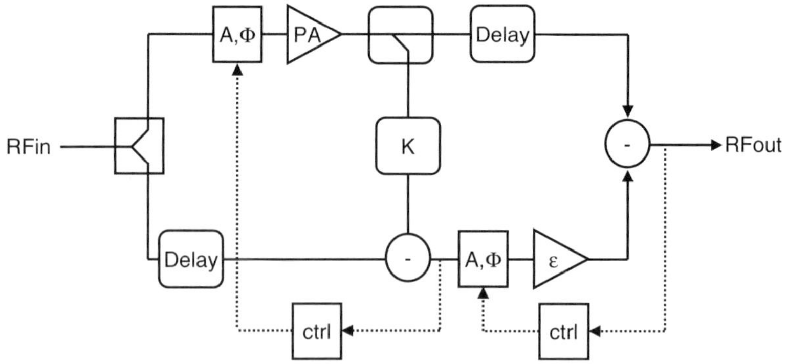

Figure 4.31. A feedforward transmitter architecture.

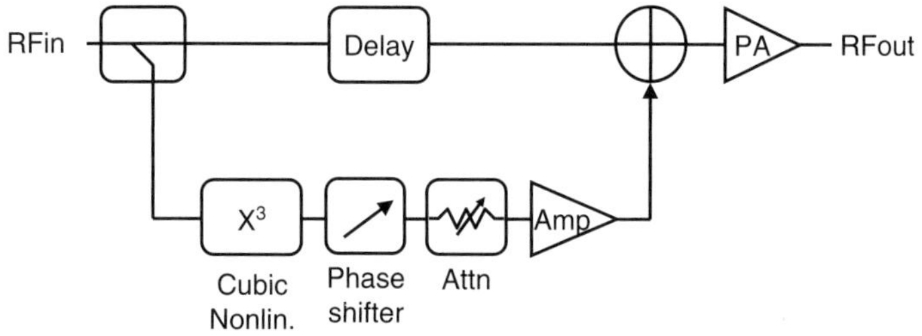

Figure 4.32. A transmitter predistortion implemented in the RF domain.

4.8.6 Predistortion Techniques

Predistortion techniques can be used to compensate for PA nonlinearity. Predistortion can be applied in RF or digital domain. Power amplifiers provide compressive characteristics, and opposite (expansive) input–output characteristics can be devised out of nonlinear elements. The expansive characteristics can be obtained by subtracting a cubic nonlinearity expression from a linear amplification characteristics. Figure 4.32 illustrates a predistortion technique applied in the RF domain. However, component variations are major reasons for performance shift.

Digital predistortion techniques provide flexibility in terms of their programming as well as modeling the expansive characteristics by means of polynomial fit. In this technique, a look-up table can be used to provide the inverse of the PA distortion characteristics. The output of the PA is sampled using a coupler and downconverted to provide I/Q signals to compensate for the PA's compression characteristics. I/Q vector modulation can be employed to provide the upconverted signal. This is illustrated in Figure 4.33.

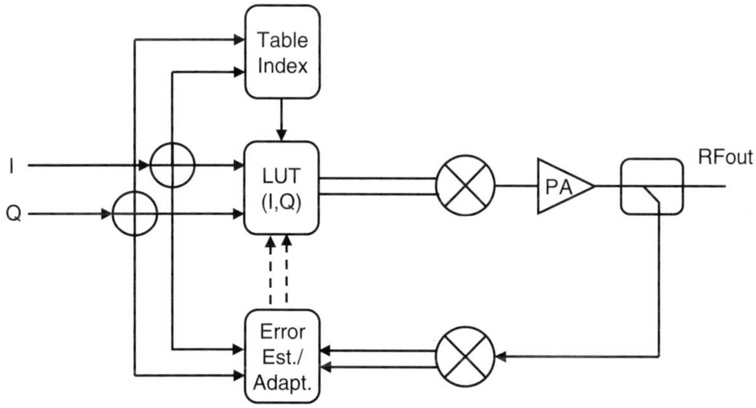

Figure 4.33. A transmitter predistortion implemented in the digital domain.

4.9 BALUN

In mixed-signal systems, often we need circuits blocks, which provide differential signals from single-ended signals and vice versa. These blocks can be implemented in numerous ways, as shown in Figure 4.34.

The key performance parameters include (1) insertion loss, (2) phase imbalance, and (3) bandwidth. Figure 4.34(a) shows a commonly used differential to a single-ended converter, which is commonly used at the low frequency (~10 MHz). The current mirror load provides the difference current to the output, and the gain of the circuit is given by $A_v = g_m(r_{op} \| r_{on})$, where r_{op} and r_{on} denote the output impedances of the PMOS and NMOS, respectively. Figure 4.34(b) illustrates a single-ended input, differential output inductively loaded circuit. At high frequencies, the balance of this circuit is determined by the input stage. An important consideration for this circuit is that the current source is not at virtual ground, which modulates the output impedance of the current source. This configuration is commonly used as a single-ended input LNA to interface with a differential mixer in the front end.

Figure 4.35(c) shows a circuit configuration based on class AB type stage, where the input voltage creates equal amount of currents in M_{N1} and M_{N2}. The input impedance of the individual branches is denoted by $Z_{in} = j\omega L + 1/g_m$. This current is then mirrored by M_{N3}, and the output current is now differential. This circuit is linear in nature, and it can be used as an input stage of mixer as well as in the LO signal generation network without generating harmonic distortions. Figure 4.35(d) illustrates the same concept using a passive configuration, similar to a transformer. The signal can be injected at the primary terminal and can be coupled to the secondary using the coupling coefficient, $K < 1$ to the secondary, where it can be interfaced with a mixer, in a differential LO signal drive. The only disadvantage of this topology is the signal loss associated with it and the area consumption. Otherwise, it performs superior at high frequencies. The last configuration in Figure 4.35(e) is based on

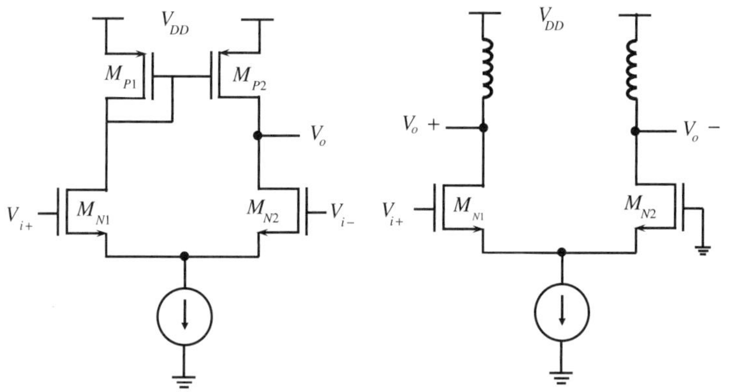

(a) Differential to single end (b) Single end to differential

Figure 4.34. Differential to single ended (analog subtractor).

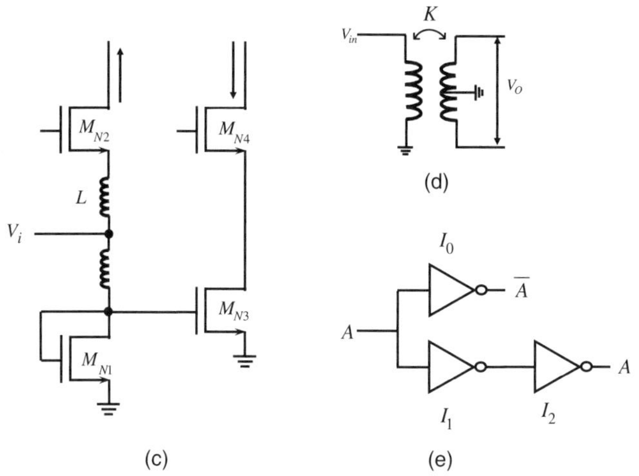

Figure 4.35. Various single-ended to differential configurations.

CMOS-based inverters, and it can be used at any broadband frequency as long as the charging and discharging of the node capacitors take place. The delay between the two paths can be adjusted to provide a 180° phase shift from the input signal. A transmission gate can be used in place of two cascaded inverters. For a specific application, a favorable topology is selected based on the frequency of operation with a certain power and area constraint. Cross-coupled inverters between the two lines (similar to latch-based topology) restores the differential signal levels.

4.10 SIGNAL GENERATION PATH

4.10.1 Oscillator Circuits

Oscillators form an integral part in any transceiver system, as they are directly responsible for on-chip signal generation, on which signal processing is performed. They are responsible for (1) signal down/upconversion, (2) clock/data recovery, and (3) clock generation for sampled data systems such as ADC, DAC, and so on. Key performances of oscillators include (1) tuning range, (2) phase noise, (3) harmonic content, (4) spurious tone, (5) analog Kv, and (6) power supply pulling. All of these are subject to a specific power dissipation and area consumption in integrated systems. Oscillators operate under the limits of large signal levels. We focus mostly on two different oscillator topologies: (1) LC resonator-based VCO, (2) relaxation or ring oscillator-based VCOs. Because of the superior phase noise performance, LC-based oscillators are much preferred over ring-based VCOs.

Fundamentally, oscillators are used in coherent communication systems. The single-ended family of transistor based oscillators includes (1) Colpitts, (2) Hartley, and (3) Pierce. The operating condition of oscillators is given by the Barkhausen condition, which enables the condition of sustainable oscillation with a positive

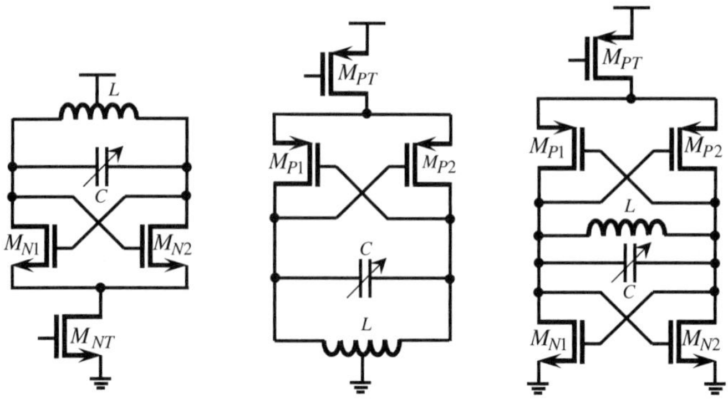

Figure 4.36. Various oscillator topologies.

feedback with a loop gain slightly higher than 1 (usually kept as 1.05). Although in modern times various differential topologies are used, Barkhausen's criterion still holds good for a clear explanation of sustainable oscillations.

4.10.1.1 LC Oscillators. In this section, we will focus on the LC resonator-based oscillators and their performance trade-off. Commonly used LC resonator-based topologies include (1) n-core, (2) np-core, and (3) p-core, as illustrated in Figure 4.36.

These core topologies can then be used in conjunction with various types of current sources (NMOS, PMOS, etc.). n-type transistors offer two to three times lower parasitic than the PMOS counterpart. Hence, the n-type core has higher tuning range, compared with the np-core for the same available g_m. However, in the current limited operating region of VCOs, the np-core performs superior because the tank amplitude is two times larger compared with the standalone NMOS-or PMOS-based cores. At the same time, n-and p-type devices can be optimized to provide symmetry in output waveform. However, np-core topologies have lower voltage headroom, but they consume lower current for the same tank amplitude. A PMOS current source has inherent advantages of tapping the tank common-mode point, which leads to less transient variation. A p-core is usually not suitable for high-tuning-range applications.

The tuning range of the VCOs is determined by the static capacitance in the tank, which results from (1) the bottom plate capacitance of the capacitors used in the tank, (2) parasitic from the devices, (3) parasitic from the inductor, and (4) metal routing for the LC resonator core.

4.10.1.1.1 VCO Startup. A reliable VCO startup is the most important issue, and the negative impedance caused by the cross-coupled devices should be able to withstand the tank losses. This condition is satisfied by the fact that $-G_m > 1/R_T$, where G_m denotes the large-signal transconductance of the cross-coupled devices and R_T denotes the tank impedance ωLQ for the parallel LC network. A differential current injected across the cross-coupled devices can be performed at small and large signals to obtain the negative resistance (a targeted bias current is assumed during this analysis). Tank impedance can be obtained from the parallel combination of L–C as illustrated in Chapter 1. However, it should be observed that at higher frequencies, the tank impedance increases, and a lower core transconductance should be sufficient to start oscillations. The core devices would require maximum voltage across them to start (thus, a low tail voltage is required).

Finally, the loaded tank impedance would need to be evaluated, and the voltage swing can be given by the product of loaded tank impedance with the bias current. In the dynamic behavior during the startup phase, the negative resistance of the core devices would change; however, when the oscillation is sustained and voltage across the tank is settled, the negative resistance is usually a constant value. At higher frequency, inductors enjoy, the advantage of higher Q, and thus, the VCO area can be reduced as well. A cross-coupled VCO structure would usually consist of (1) core devices (n-type or np-type), (2) coarse tuning capacitors (usually binay capacitor arrays), (3) fine-frequency tuning elements (varactors), and (4) inductors (single or differential, depending on the self-resonating frequency). Usually, the inductor is the limiting component in determinng the tank Q. From the above discussion, a tail current of 2 mA, tank impedance of 1000 Ω (differential) (2 nH, $Q = 5$ at 8 GHz), and a tail current of 4 mA, tank impedance of 500 Ω (differential) (0.5 nH, $Q = 10$ at 8 GHz) would lead to the same voltage swing of 2 V diff p-p for an n-core VCO. However, the later option would lead to 4 times more capacitance compared with the first one. Hence, the two optimization variants can be categorized as (1) large transistor area, low L (increased Q and self-resonating frequency), large C; and (2) small transistor area, large L (moderate-to-high Q, lower self-resonating frequency), small C. However, as we will see later, larger devices are sometimes preferred for stringent phase noise performance with the compromise of tuning range. It must be noted that not only the parasitic capacitance is important, but also the Q of the parasitic capacitance is important from a noise perspective. At the same time, too small of a unit capacitance leads to more fringing capacitance. Switches associated with turning ON/OFF these capacitances should also provide lower parasitics when the switch is in the OFF state. It must also be ensured that the capacitances used in the tank circuit should have continuous C–V characteristics without any discontinuity. The center frequency of the VCOs is dependent on the LC product, and it can be changed by scaling these values. Since VCOs operate in large signal limits, their DC operating point can be taken only for benchmarking purposes. While the oscillation sustains, core transistors must operate with a drain-source voltage that is above respective V_{DSAT} values, where V_{DSAT} is determined from the steady-state transient currents. Also, the signal swing across the tank should not lead to a breakdown of device gate oxide for reliability reasons, and should have a minimum transient voltage of $2V_{DSAT}$ across the core devices to avoid device operation in the linear range.

4.10.1.1.2 Tuning Range. Tuning range is another major consideration in VCO design. In modern communication systems, wideband standards demand even higher amount of VCO tuning range (as the frequency is usually divided down). At the same time, it needs to ensure robustness over process and temperature. Tuning range limitation is attributed by the high static capacitance from the VCO core. Cores should be designed in a manner to provide the lowest amount of parasitic capacitance (especially the voltage-dependent ones). Also, reducing component geometries arbitrarily would lead to mismatches, which is detrimental for the flicker noise upconversion from current source when the devices are "hard-switching." Enough coverage should be provided such that the tuning range specification is met over process corners.

A circuit level implementation of the oscillator switch bank is illustrated in Figure 4.37.

4.10.1.1.3 Phase Noise. Phase noise is perhaps the most important consideration for the VCO cores. Phase noise is usually used as a performance metric in wireless transceiver systems, whereas jitter is used in the context of high-speed digital systems. Much analysis have been performed and reported in literature [8–10], based on the

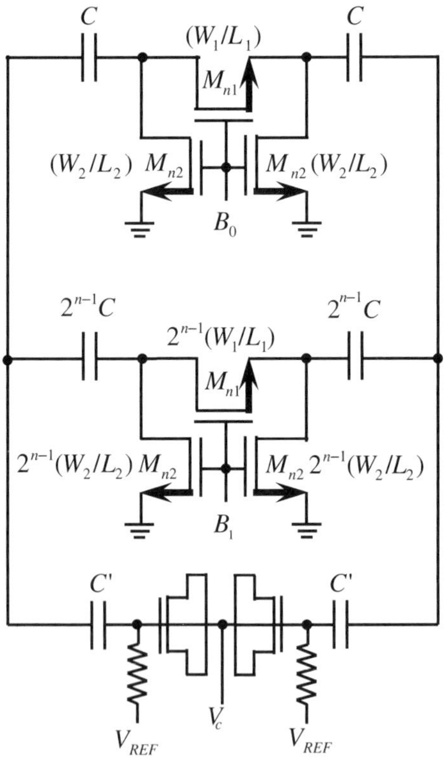

Figure 4.37. An implementation of oscillator switch banks.

linear time-variant nature of oscillators. The phase noise is depicted as follows:

$$L(\Delta\omega) = 10\log\left\{\frac{2FkT}{P_s}\left[1+\left(\frac{\omega_0}{2Q_L\Delta\omega}\right)^2\right]\times\left[1+\left(\frac{\Delta\omega_{1/f^3}}{|\Delta\omega|}\right)^2\right]\right\} \tag{4.4}$$

where the symbols have the illustrations as follows:

F: device excess noise factor
k: Boltzmann's constant
P_s: Power dissipated in resistive part of the tank
ω_0: oscillation frequency
$\Delta\omega$: offset frequency from carrier
$\Delta\omega_{1/f^3}$: frequency of the corner between $1/f^2$ and $1/f^3$

As can be observed from this semi-empirical equation, the phase noise would improve with higher signal amplitude of the core in the operating region, where the oscillator is not saturated. Clearly, the phase noise profile over frequency and consists of three regions: (1) $1/f^3$, (2) $1/f^2$ and (3) constant.

This phenomenon can be viewed intuitively as well. The current source of a VCO provides flicker noise at its output current, and this output noise is upconverted by the chopper action of the switch transistors. Since noise folding occurs from the harmonics of the frequency of oscillation, for a 5-GHz oscillator, current noise at DC would be upconverted and that from 10 GHz would be downconverted (by the switching action). From a system designer's perspective, the phase noise improves with the square of the loaded tank Q factor. At the same time, doubling the power of the VCO core would lead to a 3-dB improvement in the phase noise performance.

In the transmitter, it directly contributes to the out-of-band noise, which is usually attenuated by the PA output stage and the filter following it. In the receiver, phase noise leads to the problem of reciprocal mixing w.r.t. the out-of-band tones, which causes degradations in SNR. In VCOs, both amplitude and phase noise are present, but since amplitude noise can be eliminated by using a buffer, we are interested in optimization of the cores w.r.t. the zero crossing instants, which lead to phase noise considerations (in the time domain, it is illustrated as jitter). Figure 4.38 shows two configurations of VCO core transistors. The base terminals of these transistors are capacitively coupled to the main tank. Biasing the base voltage at a DC level lower than the collector implies that the signal swing at the tank terminals can increase, while not driving the transistors into saturation region (linear region for MOS). When the bipolar transistors are driven into saturation region, they tend to short the tank terminals by providing a low impedance and by degrading the signal amplitudes significantly. This process leads to phase noise degradations. In the current limiting region, phase noise is dependent mostly on the inductor Q, VCO tail current.

To improve phase noise degradations from the current source, frequency traps can be implemented using a combination of L and C components tuned to resonate at $2f$ of

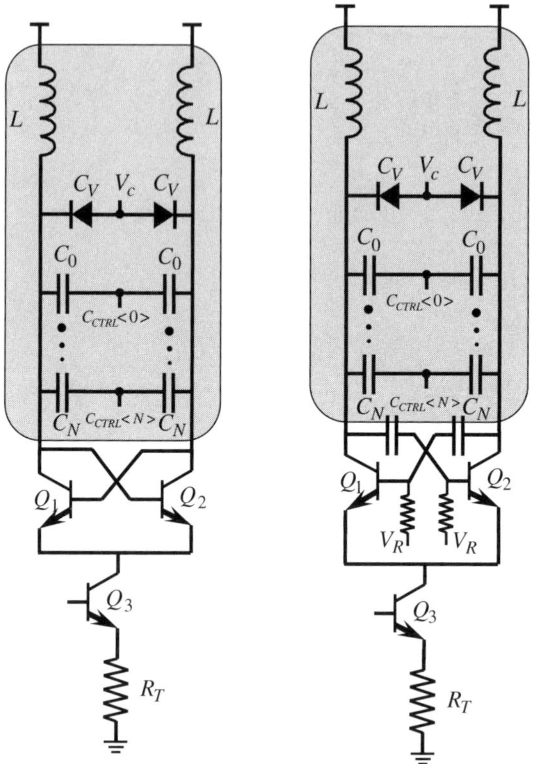

Figure 4.38. Direct-coupled and capacitively coupled oscillators.

the oscillation frequency. This provides low impedance to the noise current from the current source. Additional traps can be placed at $4f$, $6f$, and so on. This is illustrated in Figure 4.39 as described in [9].

4.10.1.1.4 Amplitude Control Loop. Tank amplitude plays a significant role in optimizing the phase noise of the VCOs. At large VCO output amplitudes, oscillation amplitude does not grow with increasing current, and at low current levels, VCOs do not start. At the same time, the tail resistance cannot be reduced significantly, as this would increase noise contribution from the tail when the devices are "hard-switching." Hence, an optimum amplitude level is desired for the best possible phase noise. This is obtained by the amplitude control loop, which can be implemented in a continuous or discrete manner, as described in [12].

An illustration of an amplitude control loop is shown in Figure 4.40, with the circuit illustration in Figure 4.41. Since the tank swing is dependent on the current through the core, the rectifier outputs a voltage and the comparator generates a signal (voltage), which adjusts the current through the VCOs. The low-pass filters using R–C filter the DC component of the rectified voltage. Conceptually an amplitude control loop, digital or analog, cannot "generate" amplitude. It can just

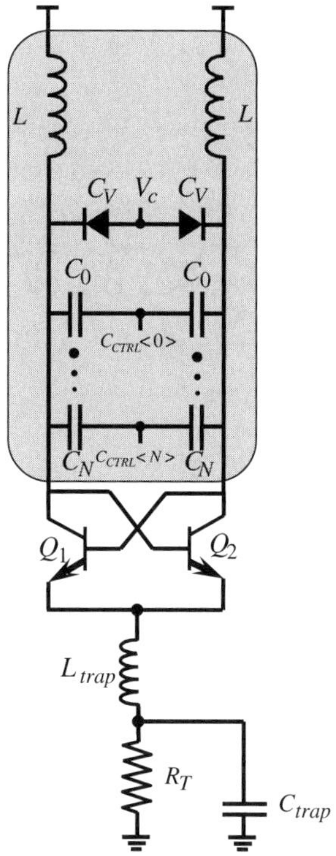

Figure 4.39. Hramonic traps to provide low phase noise.

stabilize the amplitude already generated in the VCO tank. It also contributes additional noise to the VCOs. The amplitude control loop helps a fast and reliable startup, stabilizes amplitude, optimizes oscillator bias for best phase noise, and may reduce the requirement of VCO loop gain [13]. A high loop gain is desired from the amplitude control loop in order to reject any amplitude variations. Although seemingly an attractive circuit for VCO amplitude calibration, two considerations become critical. (1) Any AM noise from the amplitude control loop leads to a bias current variation of the VCO, hence tank swing variation, and (2) modulation of phase by the nonlinear *C–V* characteristics of the capacitors used in the tank. The oscillation condition is also dependent on the phase delay of the active elements,

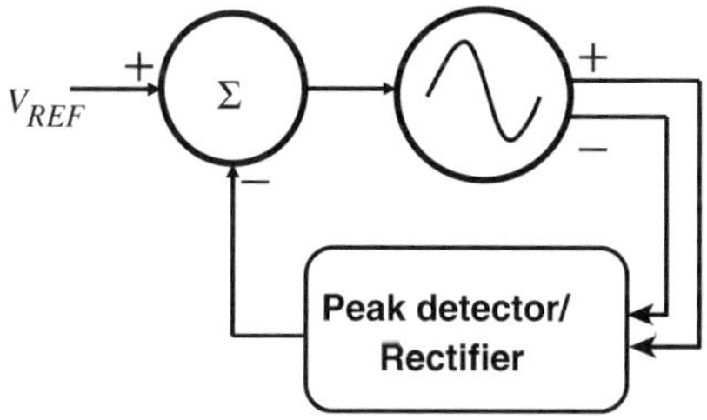

Figure 4.40. An oscillator employing an amplitude detector scheme.

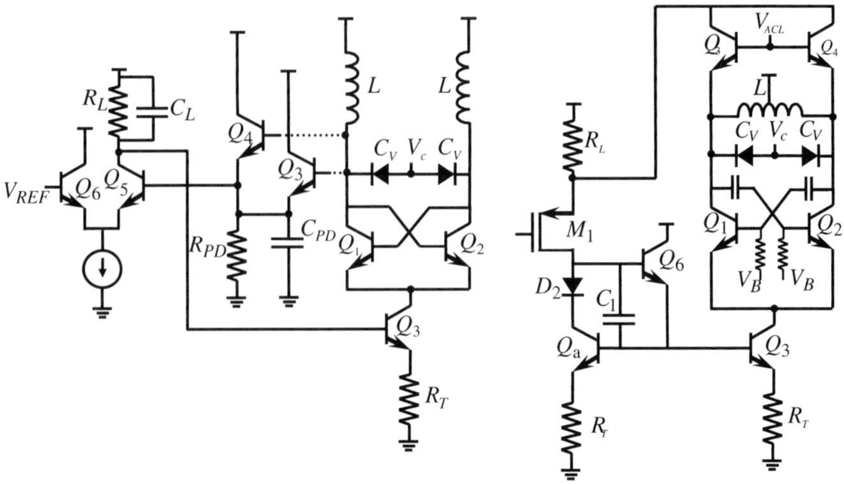

Figure 4.41. Circuit illustration of an amplitude detector.

which is a function of core current consumption; current noise leads to frequency modulation effects as well.

Analysis of the phase noise degradations is also illustrated in [13]. The main difficulty is that the rectifiers turn on only when the signal increases beyond V_t, and moreover, the rectifier operation is nonlinear in nature. Various types of amplitude control loops including that for an np-core [16] have been proposed, and whatever configuration is used, it should provide low noise and prevent formation of any unwanted parastic oscillations. At high frequency, usually the simplest configurations work better. Figure 4.42, and Figure 4.43 illustrates two amplitude detection schemes.

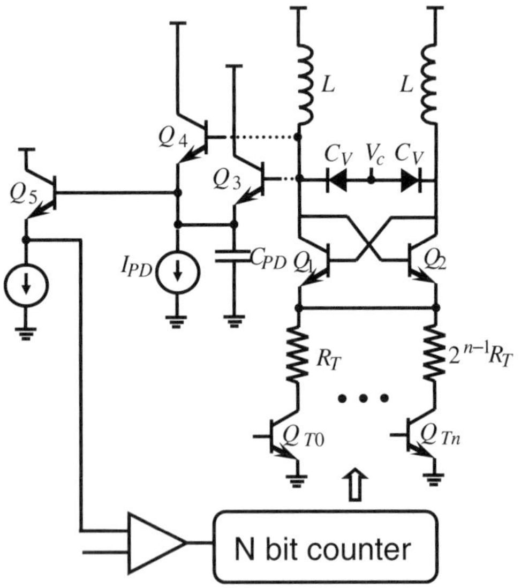

Figure 4.42. An oscillator with a digital amplitude control scheme.

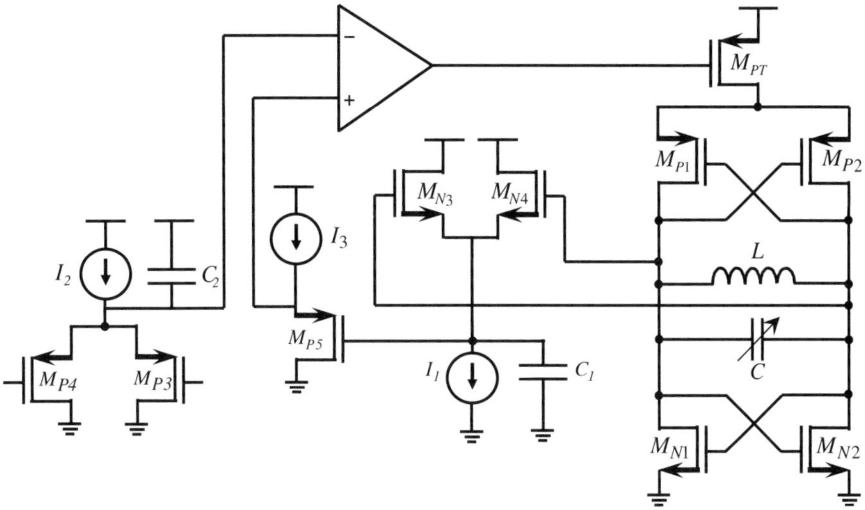

Figure 4.43. An np-oscillator using a dual peak detector scheme.

A variation of *LC* resonator-based oscillators can be obtained by connecting two oscillators to provide quadrature. This configuration is capable of providing quadrature signals directly. However, it must be ensured that the phase noise adds in quadrature, and the phase noise of the individual VCOs should be 3 dB better than if they were operating standalone. Also, the area requirement is increased for such blocks. A spatial mismatch may also lead to I/Q phase imbalance, which results in a system design trade-off for VCO followed by divider versus quadrature VCOs.

4.10.1.2 Ring Oscillators. *LC* tank-based oscillators are extremely popular, and they are routinely used in wireless and wired communication systems. For microprocessor applications, and other digital applications, where low area is a necessity, relaxation or ring-based oscillators tend to perform better. This can be illustrated in Figure 4.44, which consists of several delay cells (inverters or g_m over g_m cells) connected in cascaded and feedback fashion.

Differential topology is used for obvious reasons, and for the possibility of easy phase inversion (just a wire change at the output). Ring oscillators can generate multiple phases with 50% duty cycle, and they can easily be used in wireless transceivers. However, to achieve the desired phase noise performance, they may consume significantly higher power. Also, because of the cascaded delay cells and the parasitics of the active devices (to ensure waveform/delay symmetry PMOS devices are two to three times larger than NMOS), the speed may also be restricted up to 5 GHz, even in advanced technology nodes.

An illustration of this toplogy is provided in [17]. The most attractive feature of this topology is its inherent simplicity. This topology is based on symmetric load configuration, which is formed by a PMOS-connected diode transistor shunted with an equally sized PMOS device. Buffer delay is dependent on the control voltage. In this topology, delay is a ratio of capacitances, which can be precisely controlled. Ring

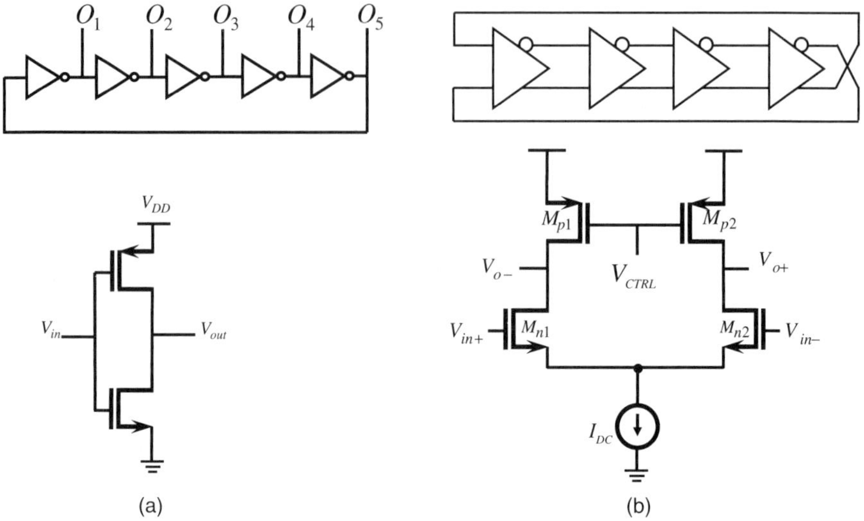

Figure 4.44. Ring oscillator configurations based on a delay cell.

oscillators provide several advantages including: (a) fast startup, (b) wide tuning range, (c) inherent quadrature generation.

4.10.2 Quadrature Generation Networks

Any transceiver system requires quadrature phases for signal processing. In the receiver, the quadrature phase is essential. At the same time, in the transmitter, the quadrature phase is essential for a single sideband combination to save bandwidth. However, not only the quadrature phases are required, but it is also desired that the duty cycle be 50% for such signals. The key performance of such circuits includes (1) frequency of operation, (2) self-oscillation frequency, (3) input signal swing, (4) duty cycle, (5) amplitude imbalance, (6) phase imbalance, and (7) phase noise. Such circuits are usually placed at the output of the VCO buffer.

4.10.2.1 D Latch-Based Divider. The first configuration illustrated in Figure 4.45(a) is a current-mode logic frequency divider with a division ratio of 2 or 4 depending on the clock phasing of the individual D latches.

As this is a current-mode logic circuit, the output swing is dependent on the current consumed and on the load impedance. It is desired that the load impedance provides small parasitic capacitance such that it does not limit the bandwidth of the circuit. For a broadband operation, the D latches can be resistively loaded or a low Q inductively loaded for increasing the bandwidth. These configurations can divide by two or four circuits that provide excellent balanced signals (the amplitude and phase imbalance of the I and Q paths are minimal). A divide-by-2 configuration provides a 6-dB improvement in phase noise, whereas a divide-by-4 configuration provides a 12-dB improvement in phase noise. In a divide-by-2 circuit, the clock skew may be a limiting

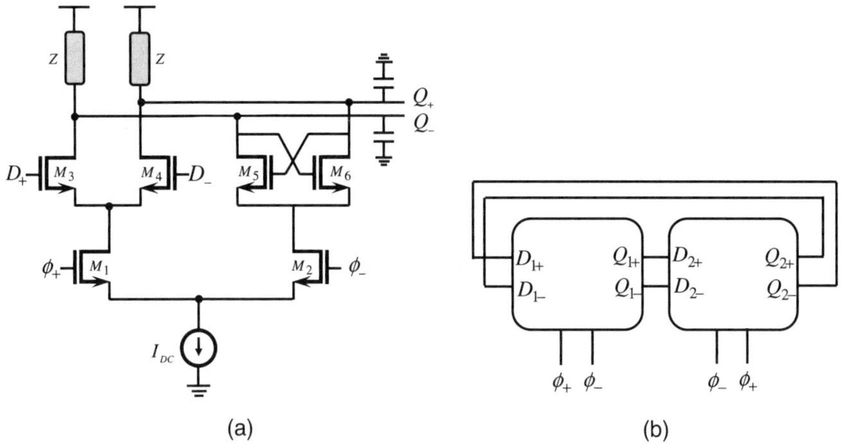

Figure 4.45. Frequency dividers based on D latches.

factor, whereas in a divide-by-4 circuit, since the D latches operate on the same edge, the skew problem is not present. In any communication systems, isolation between the signal generation chain and the transmitter or receiver signal paths is a major consideration w.r.t. feedthrough and unwanted coupling. Dividers alleviate these issues, as the input frequency is at a harmonic above the output frequency. At the same time, they relax the phase noise specification on the VCOs. At low supply voltages, the current mirror tail may be removed, with a compromise of CMRR performance. Headroom is an important configuration of this circuit at low supply voltages. Often, a lower load resistance and higher current provide the desired signal swing. The load used for these circuits is inherently of a broadband nature, as the circuit needs to maintain the same delay at all harmonics of the LO signal. As these frequency divider circuits operate on switching bias current through different branches, these circuits operate with even inaccurate models of transistors. Minimum channel length devices must be used to realize minimum V_{DSAT} and parasitic capacitances at various nodes of interest.

4.10.2.1.1 DFF Delay. Delay in the current-mode logic occurs from multiple *RC* time constants in the circuit nodes. A delay of CML stages has been analyzed extensively in the literature [18,19]. Although the analysis is performed in the case of bipolars, it can be easily extended to MOS. The overall delay is caused by (1) transient response of the input circuit, (2) delay from the junction capacitances, and (3) large signal voltage variation at the output node. The output waveform may have an amplitude variation caused by such a complicated operation (and may look much different from an ideal sinusoidal shape). However, the even order terms can be filtered out, and we are concerned mostly about the zero crossing, which determines the phase noise performance.

Pure CMOS-based differential dynamic logic circuits can also be used at RF frequencies, depending on their speed. At ultra deep sub-micron (UDSM) CMOS nodes, the parasitic capacitances are small, and such circuits would consume currents at the switching instants only. It should be mentioned at this stage that, in a frequency divider circuit, one can also obtain signal phases such as 45°, 135°, 225°, and 315°, respectively, in addition to standard quadrature. These additional phases can be used for multiphase signal processing as well as for frequency doubling. In most cases, a buffer is placed between the frequency divider network and the mixer LO terminals to provide the required signal swing. The buffer provides isolation to the divider and may provide some signal limiting action as well. A CML-to-CMOS logic converter can be used to provide signal swings from rail to rail.

It is imperative that the input signal to the frequency divider network be as "clean" as possible w.r.t. its harmonic contents. Even small levels of harmonic content would lead to an error in zero crossing, which causes unwanted output frequency. This level of clearness can be ensured by using a moderate *Q LC* filter at the driver stage.

4.10.2.1.2 Divider Design Steps. The design steps of a D latch-based CML divide-by-2 can be illustrated as follows:

1. Obtain the transistor stack, with an initial guess of 2 mA per D latch.
2. Bias the transistors in order to maintain transient signal swing $> 2V_{DSAT}$, with an initial guessed load resistance of 200 Ω.
3. Obtain the self-oscillating frequency of the D latches. The self-oscillation frequency should be close to the frequency of input signal to ensure D latch locking at low input signal levels.
4. At the output of a D latch, use a low capacitive load of the buffer and keep the interconnect capacitances low as well.
5. Obtain the optimization point of the circuit by adjusting to a low bias current and high resistance, making sure that the parasitic capacitance of the resistance is lower.
6. Obtain the highest possible signal swing at the output for the desired level of input signal. Keep the margin in the signal swing to include interconnect capacitance, routing parasitics, and so on.
7. Obtain the phase noise to ensure that it is at least 10 dB better than the phase noise of the input signal.

Figure 4.46 illustrates the phase noise improvement and sensitivity curve for a divide-by-2 configuration. To obtain a 6-dB improvement in phase noise, the divider

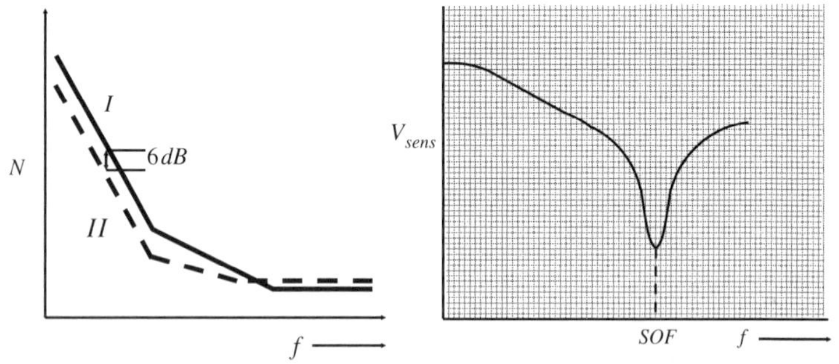

Figure 4.46. Divider phase noise and self-oscillation frequency.

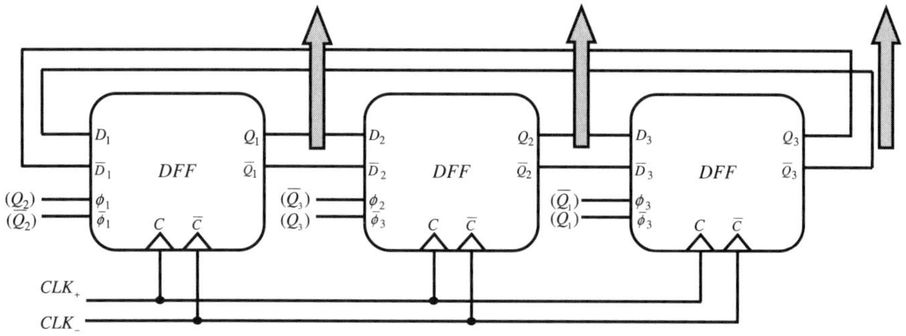

Figure 4.47. Counters—divide-by-3 with a 50% duty cycle.

close in phase noise must be at least 10 dB or more better than the VCO close in phase noise. However, the thermal noise floor of the VCO phase noise would inevitably increase after passing through the divider, and it should be designed to provide minimum thermal noise floor degradation.

4.10.2.1.3 Divide-by-N (Integer) Circuits. An interesting variation can be observed by performing divide-by-N circuits, which operate in both edges of the input clock, and the clock phasing can be selected, depending on some logic combination of the DFF outputs. In this topology, to perform a divide-by-N operation, one requires N DFFs, and some combination logic circuitry, which leads to a less power-efficient cascade, compared to $\lfloor \log_2 N \rfloor + 1$ circuits, but it has the inherent advantage of providing quadrature signals, which is essential for transceivers. Other ways of implementing divide-by-3 would be a: (1) counter-based approach, which would not provide 50% duty cycle, and (2) divide-by-2 followed by divide-by-1.5, which leads to more DFF as well as to logic complexity. This biphase selectable operation is reported in [20] and is illustrated in Figure 4.47.

4.10.2.2 Polyphase Quadrature Generators. Polyphase networks were initially used in the single sideband transmitter/receiver architectures, for a combination of upconverted sidebands, and they were developed using a combination of high-pass and low-pass sections in a ring fashion. These networks have been used extensively in low IF receivers for combining I and Q phases (with a 90° phase shift provided between them).

4.10.2.2.1 Operating Principle. Figure 4.48 shows an $RC-CR$ ladder structure to provide phase shift. At the frequency of interest, $f = \frac{1}{2\pi RC}$, the RC path lags the input signal, whereas the CR path leads the signal. However, since this is a passive network, there is signal attenuation, and the network is inherently linear in nature. Input to this network is a differential signal at f_{LO}, and the output is quadrature signals 0°, 90°, 180°, and 270° at the same frequency. As the input and the output frequncies are the same, signal isolation suffers. To improve the isolation, several stages of cascode buffers (preferably inductively loaded) can be used. In practice, two or more stages of

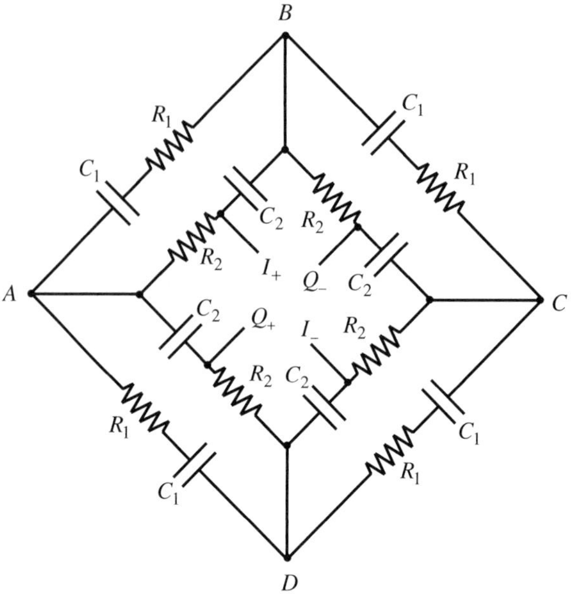

Figure 4.48. A two-stage passive polyphase network.

polyphase networks are used to provide a broad amplitude and phase characteristics. The input impedance of the polyphase network is also determined by RC, and a large R, and small C would lead to a higher impedance, but at RF frequencies, the parasitic capacitances of the resistors would load the signal, as well as C that is too low would lead to component matching issues, causing phase imbalance. Usually the performance of the transceivers is more susceptible to the phase errors; in the signal generation chain, some sort of limiting stage is used before interfacing to the mixer.

4.10.2.2.2 Impedance/Loading Consideration. Polyphase network stages are designed to provide increasing impedance from input to output. Voltage transfer is important for interfacing to the MOS stages. The last stage of the polyphase may have a higher capacitance to minimize the loading effect. The phase noise performance of the cascaded stages is important for the transceiver, as it directly translates to the possibility for reciprocal mixing as well as to transmitter out-of-band phase noise. Usually the RC stages are very quiet in nature. However, the buffer stages tend to contribute thermal noise whenever the input differential swing does not exceed V_{DSAT} by a significant amount. Noise is contributed when both of these devices are "ON" for some amount of time. This can be performed as follows: (a) Reduce V_{DSAT}, which implies reducing the current or increasing the sizes of the devices, and (b) reduce loss through the polyphase network. The input swing to the active stages should be sufficient to hard switch the input transistors. Polyphase networks are also capacitively coupled from the driver and driving stages, and they can be held at 0 V DC so that there is no degradations from the voltage-dependent parasitics.

Loading of the polyphase network is important for LO generation. To minimize loss through polyphase filters, the impedance of successive stages can be made progressively larger. Output of the polyphase network is interfaced to the gate of a differential buffer, and the input capacitance of the buffer may introduce the phase inaccuracy of the quadrature output signal. Although the polyphase networks provide signal loss through them, the output load cannot be increased beyond a certain value; otherwise, the quadrature accuracy would suffer. This loading is a serious concern, as it mandates for more stages of buffering, which is desirable for isolation purposes, with a compromise on the power consumption increase. This is especially important when the mixers require rail-to-rail swing, which is the case for passive mixers. Noise from the polyphase filters is dominated by the output stage, with a broadband noise spectral density of $4kTR$, and it should be sized to provide a lower noise penalty. The input impedance of the polyphase network is $R\|(1/j\omega C)$, and in the frequency of interest, this impedance is $R/(1 + j)$. Depending on the impedance, the driver stage can be a common source amplifier or a source follower. A source follower configuration has lower output impedance, but it leads to additional noise and the signal experiences some loss through it. In bipolars, this option may be attractive, but in MOS, it leads to unwanted signal loss, and loading to the previous stage. A common source buffer may be sufficient in the case of driving the polyphase filter.

4.10.2.2.3 Harmonics in Polyphase Filter. Another interesting property of polyphase networks is the inherent harmonic phase rotation. At the frequency of interest, the low-pass and high-pass sections would provide -45° and 45°, respectively. If the input waveform has a third harmonic content, then the polyphase network would provide -135° and $+135^\circ$ phases at the output. This would impact the zero crossing at the output, which would lead to significant phase imbalance. Hence, the driver stage should preferably use an inductively loaded circuit. After a "clean" signal is generated in quadrature, then various signal limiting stages may be used for lower area considerations.

In the layout of the polyphase filters, a significant amount of "dummy" components is used to improve matching of the components (R and C) in a compact placement. Performance is more susceptible to parasitic capacitances in the layout than parasitic resistances. Careful layout techniques to minimize I and Q phase coupling should be used. Readers may use [21] as a reference to polyphase networks. A multistage polyphase filter can be designed by inserting buffer stages (common source stages) between them.

4.10.2.2.4 Design Steps. Polyphase network-based signal generation chains can be designed as follows:

1. At the operating frequency, we obtain the largest possible resistor, which is a compromise between component matching and parasitic capacitance.

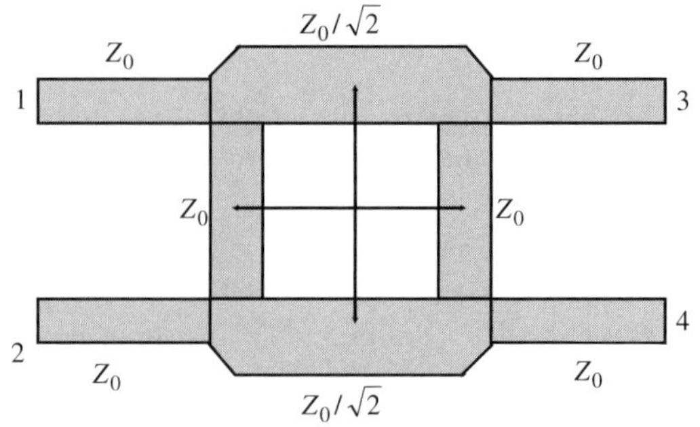

Figure 4.49. A hybrid network to generate quadrature.

2. Ensure that C is not too small to impact matching characteristics but larger than the output buffer load.
3. Start with a two-stage topology, and scale the impedance of the second stage to 30% higher than the first stage.
4. Optimize the resistors in terms of noise requirements.
5. Design the pre-and post-driver of the polyphase network to provide sufficient signal to the mixer circuits.

4.10.3 Passive Hybrid Networks

At frequencies above several tens of gigahertz, and well into the milimeter-wave frequencies, quadrature hybrids are popular, as the wavelength of the signal is comparable with the component dimensions, such that reflections occur. Configuration of a 90° hybrid is illustrated in Figure 4.49. As this is a passive circuit, it cannot generate any harmonics or subharmonics of the input signal. It provides poor signal isolation. Also, the performances of such circuits may be limited by the Q factor of the components on silicon.

4.10.4 Regenerative Frequency Dividers

Another configuration, commonly used at high frequency, is a "dynamic divider" as shown in Figure 4.50. These dividers operate in a mixer-like configuration, where the following equation is valid: $f_{OUT} = F_{IN} - f_{OUT}$. They have been analyzed to be power efficient and are called "dynamic" because of their feedback nature and the generation of frequency on the fly. An observation is that unlike the static dividers, these circuits do not generate quadrature, so some other circuits are used to generate quadrature. However, if not designed carefully, this circuit is prone to generating spurious tones caused by the dynamic mixing operation, which is detrimental for the transceiver performance.

It must be noted at this point that the quadrature signal generating networks must generate signals with a 50% duty cycle. Otherwise the transceiver performance would suffer from the even-order effects (IIP2, for instance, is heavily dependent on the duty cycle and on mismatch effects). Digital counters, which will be illustrated later, also generate a frequency division operation, but they do not generate a 50% duty cycle.

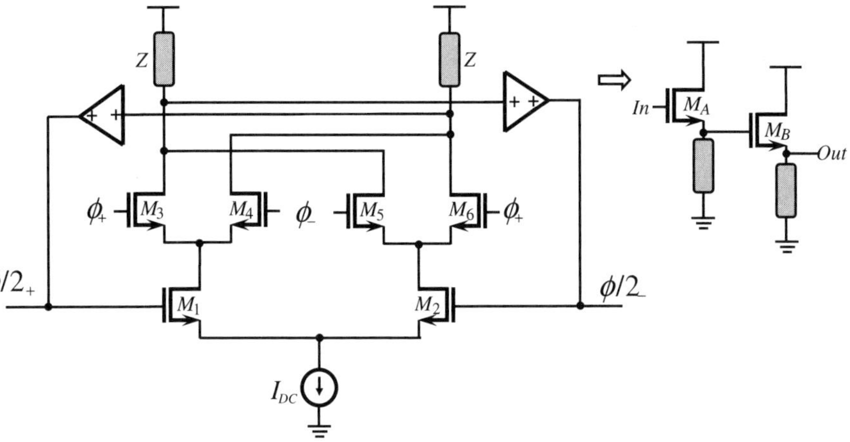

Figure 4.50. A regenerative frequency divider.

4.10.5 Phase Locked Loop

Phase locked loop (PLL) is an integral part of the signal generation chain. It is a feedback loop that converts frequency to voltage and voltage to current ($f-V$, $V-I$, $I-V$). Inherent to the synthesizer, multiple frequency division is used, and comparison with the reference clock (crystal) occurs at low frequency. Since the PLL "locks" the loop and a stable frequency is obtained at the output (dependent on crystal ppm frequency deviation), the output can provide sampling clocks to other building blocks, such as ADC and DAC.

4.10.5.1 Impact of VCO Frequency Resolution. PLLs are always codesigned with VCOs, and the frequency resolution is essential (how much frequency changes with change in the least significant bit (LSB) of the capacitor array). We start the design from a base VCO (some $L-C$ combination in the tank). The center frequency of the VCO can be shifted by changing L or C. The following situations illustrate the frequency resolution w.r.t. a capacitor LSB change.

Case I: Change in center frequency caused by change in capacitance only

$$\Delta f_1 = \frac{1}{2\pi\sqrt{LC_1}} - \frac{1}{2\pi\sqrt{L(C_1+\Delta C)}} = \frac{1}{2\pi\sqrt{LC_1}}\left[1-\sqrt{\frac{C_1}{(C_1+\Delta C)}}\right] \tag{4.5}$$

$$f_{c1} = \frac{1}{2\pi\sqrt{LC_1}}, \quad f_{c2} = \frac{1}{2\pi\sqrt{LC_2}} \tag{4.6}$$

$$\Delta f_1 = f_{c1}\left[1-\left\{1+\frac{\Delta C}{C_1}\right\}^{-1/2}\right], \quad \Delta f_2 = f_{c2}\left[1-\left\{1+\frac{\Delta C}{C_2}\right\}^{-1/2}\right] \tag{4.7}$$

$$\Delta f = f_c[1-\{1+\Delta C(4\pi^2 f_c{}^2 L)\}^{-1/2}] \approx 2\pi^2 f_c{}^3 L(\Delta C) \tag{4.8}$$

With the assumption that $\Delta C(4\pi^2 f_c{}^2 L) \ll 1$, (for 1 LSB). Hence, the ratio of the frequency shift caused by 1 LSB at two center frequencies is given by $\frac{\Delta f_2}{\Delta f_1} = (\frac{f_{c2}}{f_{c1}})^3$.

Case II: Change in center frequency caused by change in inductance only

We assume that the parasitic capacitance of the active devices changes insignificantly

$$\Delta f_1 = f_{c1}\left[1-\left\{1+\frac{\Delta C}{C_1}\right\}^{-1/2}\right] \tag{4.9}$$

$$\Delta f_2 = f_{c2}\left[1-\left\{1+\frac{\Delta C}{C_1}\right\}^{-1/2}\right] \tag{4.10}$$

$$f_{c1} = \frac{1}{2\pi\sqrt{L_1 C_1}}, \quad f_{c2} = \frac{1}{2\pi\sqrt{L_2 C_1}} \tag{4.11}$$

$$\frac{\Delta f_2}{\Delta f_1} \approx \sqrt{\frac{L_1}{L_2}} = \frac{f_{c2}}{f_{c1}} \tag{4.12}$$

The above situations illustrate the impact on the PLL design from a VCO center frequency scaling w.r.t. change in L – C combination. When both components are scaled, a combination of the above formulations should be used.

The feedback loop equation on the PLL is given as follows:

The front end of a PLL (prescaler) operates at very high frequencies and may use CML logic as necessary to be interfaced with VCO. The back-end frequency division is performed in terms of counters implemented using CMOS-based DFFs. Counters are used to translate the input frequency to a lower frequency, and many times they provide a non-50% duty cycle at their output. A quadrature is not needed. A pulse-swallow type divider is commonly used for a complicated division ratio (e.g., to obtain a division ratio of 6.66, the input is divided by 6 once and divided by 7 twice, so that on an average, the division ratio is 6.66). In this section, we would provide several counter-based topologies and division ratios.

4.10.5.2 Complicated Divide Ratios. Divide-by-2 is quite common in nature and used as much as possible because of its low design complexity. Cascaded divisions are possible by connecting the output of one DFF to the clock input of the next DFF. Ring-type counters can be used to provide delays, in which the output frequency is $f_O = \frac{1}{2nt_{pd}}$, where t_{pd} is the propagation delay of each DFF. For division ratios of 3/4 or 4/5, the number of FF required would be given by $\lfloor \log_2 N \rfloor + 1$, where N is the maximum possible division ratio (4 in case of 3/4 and 5 in case of 4/5). Suitable logic combination circuits (NAND/NOR as appropriate) are inserted in the chain to provide the other division ratio. Finally, the desired output is selected. The NAND/NOR functionalities may also be performed using CML techniques to operate at high speeds. Counter design starts by noting the output states (output of each DFF, and the state sequence is maintained to obtain the closest possible duty cycle to 50% to obtain ON/OFF margin in

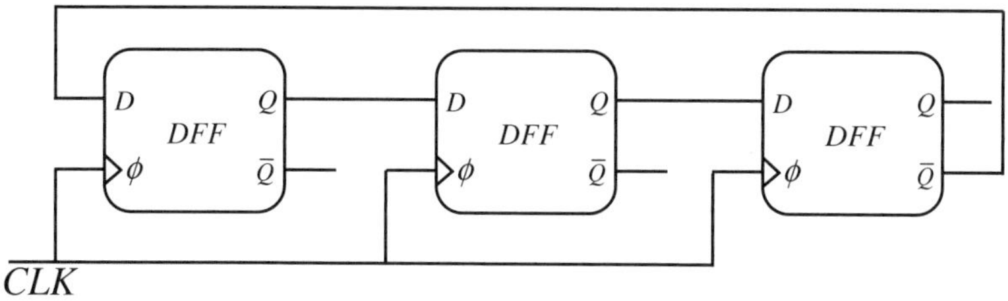

Figure 4.51. A ring counter configuration.

the succeeding logic stages), and optimizing Karnaugh map representation of each state variable by minimizing additional logic circuit to be added (NAND/NOR/INV, etc.). A detailed illustration of counter design is provided in [2]. In these implementations, one must ensure that the DFF can return to one of the desired states even when it goes to an unwanted state. Another simple way to implement counters is to let the DFF cascaded chains count and to reset the input when a specific state is achieved. For a divide-by-11 counter, one can detect that all DFFs are reset to zero input when the output reaches 1011 (binary). Figures 4.51–4.53 illustrate a ring counter, modulo 3 counter, and modulo 5 counter. Modulo 3 and 5 counters provide a non-50% duty cycle.

DFFs in the back end may use static logic or a single-phase clocking DFF to reduce clock routing and skew-related problems. Principles of single-phase circuits are similar to dynamic circuits, and they minimize clock loading as well. For charge sharing and charge leakage reasons, they may provide glitches at very low frequencies, similar to dynamic circuits.

4.10.5.3 PLL Loop and Dynamics. Figure 4.54 illustrates the key building blocks for a PLL system. It consists of (1) VCO, (2) VCO buffer, (3) prescaler, (4) dividers/counters, (5) phase/frequency detector, and (6) charge pump. VCO and prescalers (consists of high-speed dividers) are the most power-hungry blocks in the PLL loop. The loop "locks" the VCO to a specific frequency, determined by the division ratios in the loop. The phase/frequency detector generates a phase output, which essentially causes an output current from the charge pump followed by its integration over the

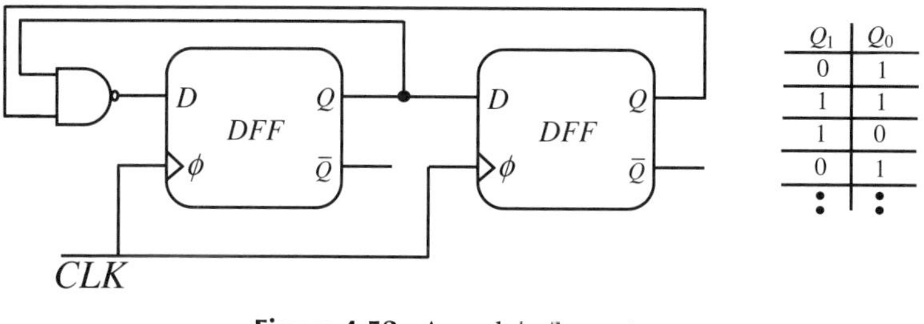

Figure 4.52. A modulo 3 counter.

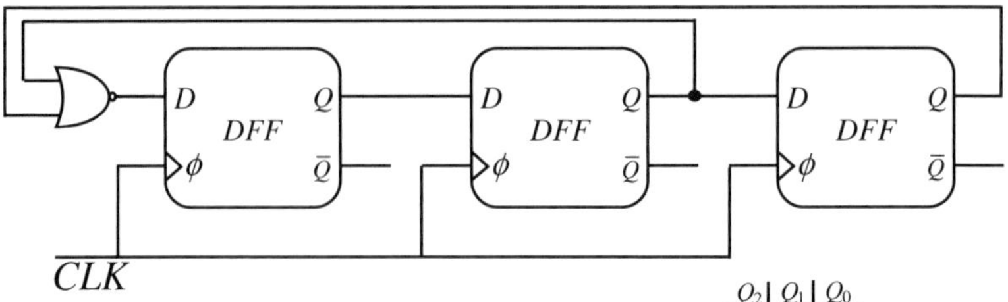

Q_2	Q_1	Q_0
1	1	0
1	0	0
0	0	0
0	0	1
0	1	0
1	1	0
⋮	⋮	⋮

Figure 4.53. A modulo 5 counter.

loop filter, which is essentially a low-pass filter. The loop filter's output is a voltage, which determines the center frequency of the VCO (by speeding up or slowing down). Hence, the VCO converts voltage to frequency inside the loop. Frequency dividers can be thought of as blocks that operate in the frequency domain (frequency to frequency). The phase/frequency detector converts the phase difference to a voltage (phase to phase) and the rest follows.

The challenging design aspects in a PLL loop include

1. Charge pump dynamics, mismatch, and so on.
2. Area of loop filters

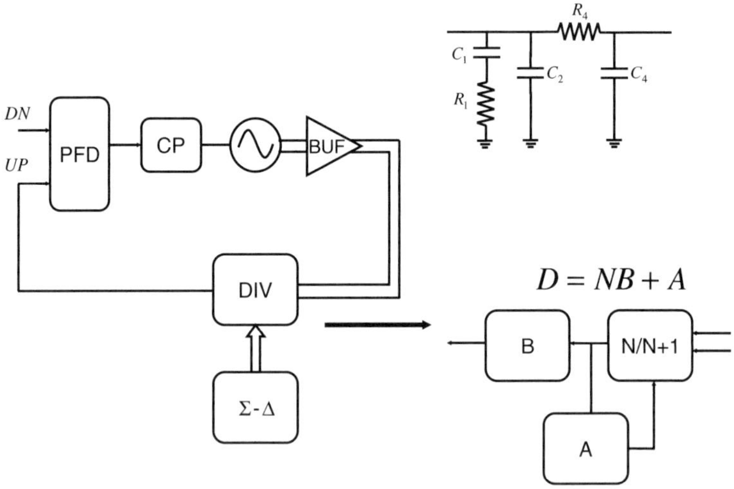

Figure 4.54. Basic blocks involved in a PLL loop.

3. VCO/PLL interface
4. Digital noise coupling from digital logics to VCOs, and so on.
5. Noise transfer functions from different parts of the PLL loop

4.10.5.3.1 Pulse-Swallow Counter. The reference clock for the PLL loop is usually generated by a crystal oscillator. We first start our discussion by illustrating a pulse-swallow counter. The division ratio out of this counter is given by $D = NB + A$. The prescaler is first divided by $N + 1$. When the A counter overflows by counting A cycles of the input signal, the prescaler starts dividing by N. Hence, the overall division ratio is given by, $(B - A)N + (\mathrm{N} + 1)A = NB + A$. In the design phase, this ratio is first calculated by dividing the VCO center frequency by the crystal frequency, with the condition $B > A$. As N increases, more stages are cascaded, leading to higher power consumption. However, a larger N also leads to lower phase noise. Hence, the value of N is chosen to be the minimum number that meets the power consumption and phase noise requirements.

4.10.5.3.2 VCO and Divider. In the VCO block, with a small signal perturbation, the phase changes according to $V_{out}(t) = A\sin(\int K_V(V_0 + V_{ac}\cos(\omega_{in}t)dt) = A\sin[\omega_0 t + \frac{K_V V_{ac}}{\omega_{in}}\sin(\omega_{in}t)]$, where the integrand provides the instantaneous frequency of the oscillator. Hence, the voltage-to-phase gain of the VCO in the frequency domain is represented by $\frac{K_V}{s}$. Because of the presence of the perturbation, the divider triggering instants would experience an error by

$$\Delta t = \left\{\frac{K_V V_{ac}\sin\omega_{in}t}{\omega_{in}}\right\} / \omega_0 T_{ref} = \frac{1}{D}$$

This determines the phase-to-phase transfer function of the divider.

4.10.5.3.3 Charge Pump. Assuming the phase difference of the input signals at the charge pump to be θ, it injects a current of magnitude $\frac{I_P}{2\pi}\theta$, where I_P denotes the pump current. If θ changes, the output current can be represented by rectangular pulse train and may be represented by a delta pulse with the magnitude equal to the area of each of the pulses. In the frequency domain, we obtain the Fourier transform of the time domain waveform of the pulse train, which is given by

$$\theta(f) \times \frac{1}{T_{ref}} \sum \delta(f - \frac{n}{T_{ref}})$$

Obviously, the charge pump provides a low-pass filtering, and the resulting transfer function (phase to phase) is $\frac{I_P}{2\pi}$.

4.10.5.3.4 Loop Filter. The loop filter is illustrated in Figure 4.54, where the dominant poles are given by

$$\omega_{P1} \approx \frac{1}{R_1(C_1||C_2)}, \quad \omega_{P2} \approx \frac{1}{R_4C_4}$$

and the zero is given as $\omega_{z1} \approx \frac{1}{R_1C_1}$. The unity gain frequency is given by

$$\omega_u \approx \frac{I_P R_1}{2\pi} \cdot \frac{K_V}{N}$$

The phase margin is optimized if the unity gain frequency is the geometric mean of the zero frequency and the first dominant pole. The location of the dominant pole is determined by the amount of filtering needed for the reference, spurious, as well as the noise filtering of the sigma-delta modulator.

4.10.5.3.5 Noise in PLL. From a noise analysis perspective, two paths would be necessary to consider. The fundamental noise sources include (1) noise from the reference, (2) noise from the charge pump, (3) noise from the VCO, and (4) noise from the sigma-delta modulator. Extensive literature is available on such analyses, and derivation of transfer functions from individual blocks to the final output determines the nature of the noise profile at the output. From the reference noise source to the output, the transfer function is given by

$$\frac{\theta_{0,ref}(s)}{\phi_{ref}(s)} = \frac{N \times G(s)}{1 + G(s)}$$

where $G(s)$ is the loop gain of the PLL. From the VCO noise source to the output, the transfer function is given by

$$\frac{\theta_{0,ref}(s)}{\phi_{ref}(s)} = \frac{1}{1 + G(s)}$$

The expression for $G(s)$ can be approximated by

$$G(s) = \frac{\omega_u/s}{(1 + s/\omega_{P1})(1 + s/\omega_{P2})}$$

The main contribution for the PLL phase noise results from the white noise of the reference oscillator and the close in phase noise of the VCO. Noise from the charge pump component is thermal in nature, with appropriate filtering transfer function provided to each source, whereas for the Σ–Δ converter, it is usually the quantization noise to be considered.

4.11 MIXERS

4.11.1 Basic Functionality

Mixers are functional blocks, which takes two different signals at its input and produces the sum and difference of the frequencies of the two signals at the output. Inputs and outputs can be taken in voltage or current mode, whereas the switching operation is performed usually in voltage mode. Two types of mixers are reported in literature: (1) switching-type mixers and (2) multiplier-type, large-signal mixers, which are based on the nonlinear characteristics of the active elements. As passive components cannot produce nonlinearity, they alone cannot provide a mixing operation. The terminology of passive versus active mixing is from the conversion loss or gain of the mixer circuits. In the mixer, two operations occur. The transfer characteristics of the input stage (or "port") is small-signal linear operation, whereas the switching operation is inherently large-signal operation. A more conceptual illustration regarding this is provided in [23]. In the switching-type mixers, the input current is switched using LO signals, and the output is filtered by a filtering network (can be $R - C$ or $L - C$ stages in receiver and transmitters, respectively). Output currents can be added or subtracted, depending on the phase relationships of the switching waveform. In a receiver, the switching waveform is a differential input signal (usually a large signal), whereas in transmitters, they are phased in quadrature for a single sideband combination. On the other hand, multiplier-type mixers are based on the nonlinear characteristics of the mixers, which could be square law or exponential, depending on the nature of the transistor's large-signal nonlinear characteristics. During the initial years of radio communications, mixing operation was performed based on the large-signal exponential characteristics of the diode devices, and improvements were made from a single-diode-based structure to a ring architecture. In modern technologies, switching-type mixers are quite popular because of their "hard-switched" behavior w.r.t. large-signal LO drive. Key performance metrics of mixer circuits include

1. Conversion gain or loss
2. Linearity (intermodulation characteristics): IIP3, IIP2
3. Power consumption
4. Noise
5. Signal isolation (LO/RF feedthrough)
6. LO leakage
7. Input impedances

The performance parameters of mixers are heavily dependent on the large-signal LO drive. In practice, a large sinusoidal waveform or a square waveform are used. Zero crossing of the LO waveform determines the noise characteristics of the mixers. In case a sinusoidal shape is used, some sort of "limiter" functionality is used to provide robustness w.r.t. process and temperature variations. In a square waveshape, third and fifth harmonics would be present (even harmonics are eliminated because of differential topology). Although the conversion loss performance of passive mixers become better with square wave drive, they are prone to downconverting the third and fifth harmonics of the desired frequency bands.

4.11.2 Architectures

Various types of mixer configurations are illustrated in Figure 4.55. Figure 4.55(a) illustrates a single diode-based mixer with even harmonic variations shown in (b), (c), and (d). For the finite reverse isolation of the diode, usually some amount of filtering is inserted at each port. Figure 4.55(e) shows a ring mixer based on a single diode, whereas Figure 4.55(f) shows an even harmonic mixer architecture based on ring configuration. Diode-based mixers are usually dynamically biased (sometimes near the cutoff voltage of the diode), and they do not consume any DC current. The signal driving strength of the diode is large, leading to isolation issues. Usually these mixers are linear in nature.

The nonlinearity of the transistors can be used to achieve mixer functionality as well. Figure 4.56 illustrates various mixer topologies based on transistors. Starting

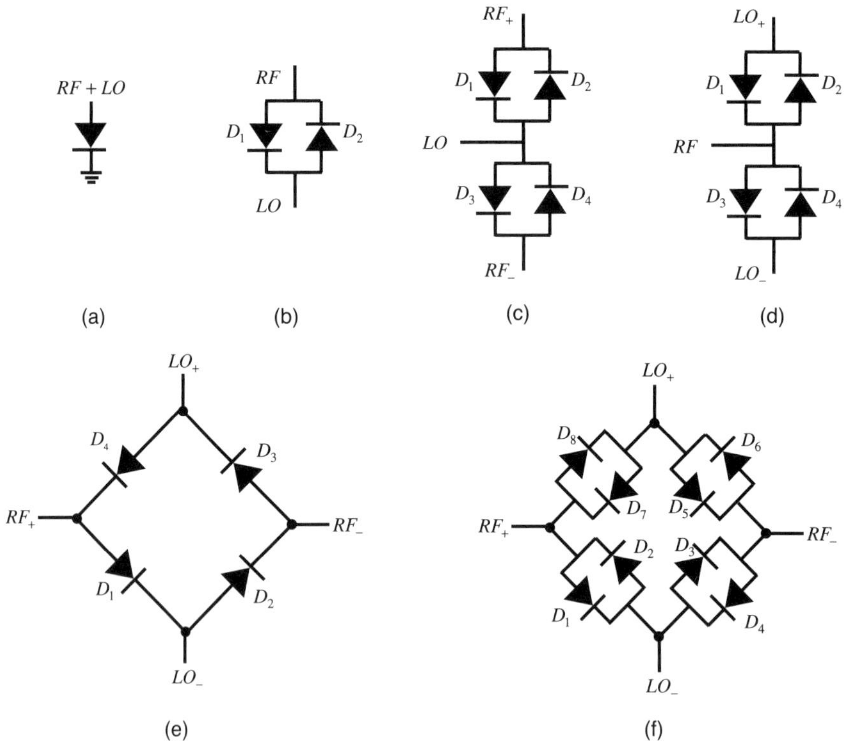

Figure 4.55. Diode-based mixer architectures.

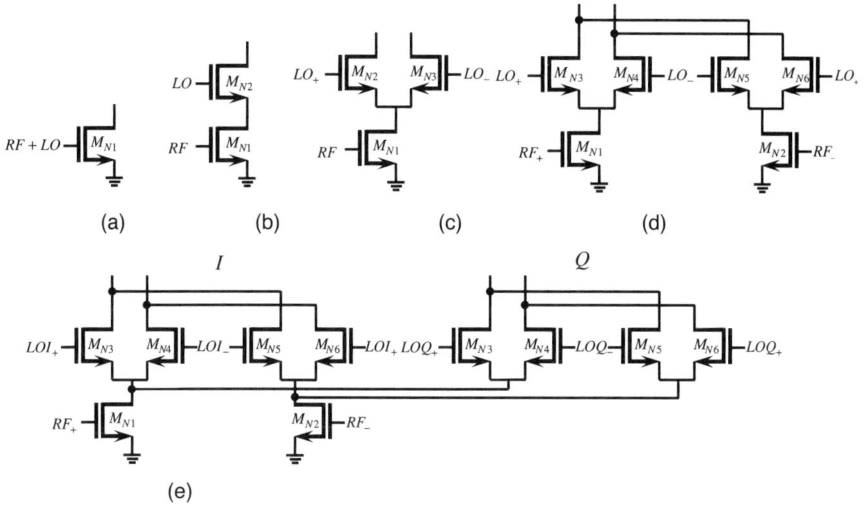

Figure 4.56. Transistor-based mixer architectures.

with single transistor-based mixing illustrated in Figure 4.56(a), single-balanced and double-balanced topologies can be constructed for modern transceiver systems as illustrated in Figure 4.56(c) and (d). Quadrature phase splitting is often a requirement in integrated transceivers, and Figure 4.56(e) illustrates a topology that shares the input transconductor stage between the two phases.

4.11.3 Conversion Gain/Loss

Conversion gain or loss is dependent on the power dissipation and the mixer load. Load network determines the effective voltage headroom across the core devices. The switch network used in mixers is inherently linear, as it operates in current mode (current input and current output), and the small signal linearity is thus determined by the g_m devices. In a passive ring-based mixer, the linearity is based on the intermodulation produced in the previous V–I conversion stage (g_m of the mixer stage or g_m of the LNA as the architecture demands). Usually this sets the power consumption requirement in the mixers. With this power consumption budget, the LO transistors are sized accordingly in order to provide "hard-switched" characteristics with smaller input signal swings. This device geometry then poses the drive and power consumption required from the LO distribution chain. Mismatches in the transistors, load resistors, or any duty cycle mismatches in the LO drive lead to second-order intermodulation products, IM2. Similar to the DC offset effects, IM2, being a second-order effect is strongly dependent on any mismatch, imbalance, and a higher common-mode rejection of the mixers would improve the IM2 performance.

In an upconversion mixer or modulator, two quadrature baseband signals are combined to form a single sideband as illustrated in Figure 4.57. Depending on the architecture, this upconversion can be a two-step process or a direct process. This

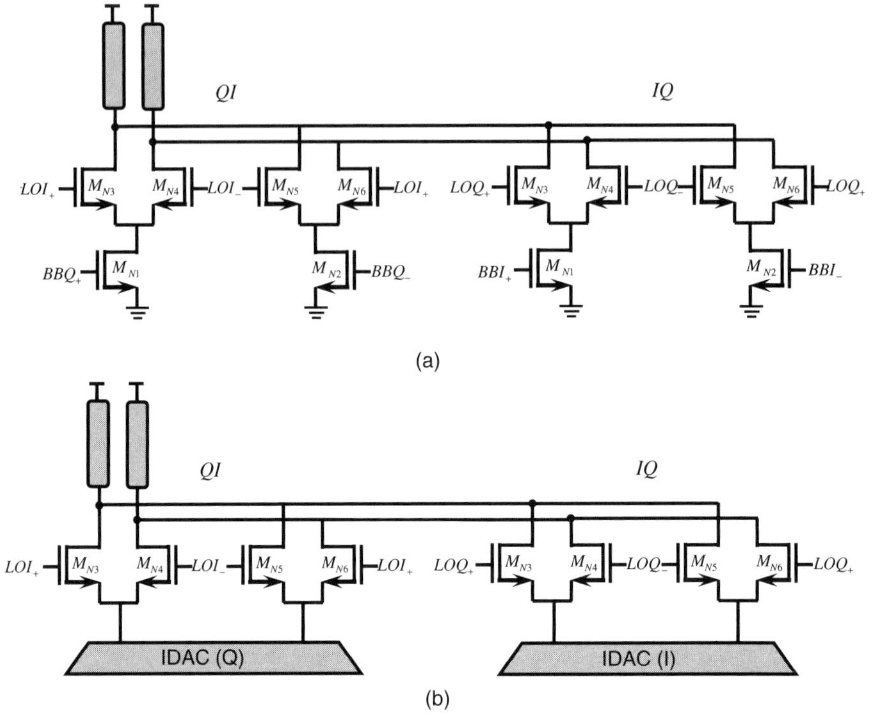

Figure 4.57. I – Q Modulator using (a) voltage mode (b) current mode baseband input.

signal processing is performed using the modulator, and usually modulators are not very efficient because of the signal combination (and rotation of I and Q phases); we lose 3 dB of signal (total signal power $= 2A^2$, where A is the amplitude at baseband I and Q each, whereas after combination with ω_{LO}, signal power is A^2). Thus, depending on the order of combination, and the sign (plus or minus), which can be obtained easily in a differential configuration by flipping wires, the desired sideband (upper or lower) can be obtained. The baseband input can be current or voltage to the modulator as shown in Figure 4.57. In essence, double-balanced topology is adopted to reduce LO feedthrough. To reduce the LO harmonics, sometimes a filter is provided at the output of the modulator, or judicious phase combination is performed at the baseband processing to reduce the third and fifth harmonic (this is called "harmonic reject architecture" as illustrated in Chapter 2). Modulators are usually linearized at their baseband part.

4.11.4 Noise

Noise in mixers originate from (1) noise from the transconductors (thermal noise sampled by the LO waveform for the duration when both transistors are ON (when the transistors are switching, common-mode noise from the current source would be canceled at the output), (2) flicker noise from the LO switch transistors, and (3) thermal noise from the load resistors. An account for detailed noise analysis is presented in [28,29]. The drain-referred flicker noise current, is also dependent on the bias current. In passive ring-based mixers, the mixers do not use a bias current, leading to reduced flicker noise performance. It is also desired that the LO transistors are as much "hard-switched" as possible. A bipolar transistor-based differential-pair requires an input differential signal of 78 mV to hard switch, and provides less loading to the LO generation network, compared with its MOS counterpart, which provides more parasitic capacitances at various nodes. The slope of the LO waveform is also quite important for the switching performance, and a sharp transition around zero crossing is desirable. The drive waveshape is also an important consideration, and they are usually sinusoidal-or square-type waveforms. A limiting functionality is provided to obtain reliable switching of transistors under process and temperature corners. Like any frequency translation device, noise from the transconductor devices is frequency translated at the output.

4.11.5 Port Isolation

Signal isolation is also important for the mixers. Usually, the LO-to-RF leakage performance is important in the downconversion mixers, as this leads to unwanted components at IF frequency, such as DC offset, which needs to be corrected in the baseband. Leakage of individual tones such as LO or RF is not very relevant at the output as they can be filtered out by various filtering elements. Since the mixers are usually 1 × type ($\omega_{IF} = \pm n\omega_{LO} \mp \omega_{RF}$), where $n = 1$, any finite leakage would lead to unwanted signal at DC. In the case of transmitters, they may pull the VCO to a different frequency. For this reason, the VCO center frequency is different from the receive/transmit signal path frequency. Active circuits generate harmonics leading to degradation in the SNR at the output of the mixer by unwanted downconversion. In the Gilbertcell-type mixers, the signal isolation is usually superior, as the two signals experience reverse isolation of transistors. Pasisve mixers, on the other hand, provide high linearity, but they usually provide a worse signal leakage performance.

LO leakage is a major consideration in the transmitter circuits. Similar to the DC offset phenomenon, LO leakage provides an undesirable large signal in the middle of the transmitted frequency band. Usually this is eliminated by using doubly balanced modulators subject to the matching performance of the transistors. Due to port isolation consideration, gain stepping in the transmitter is implemented as a combination of baseband and RF stages. Operating the baseband at lower signals (for low output power) may degrade the output SNR due to the presence of in-band leakage terms.

4.11.6 Receive and Transmit Mixers

Although the mixing functionality is similar in the transmitter and the receiver, there are fundamental differences. In a receiver, the input differential signal is

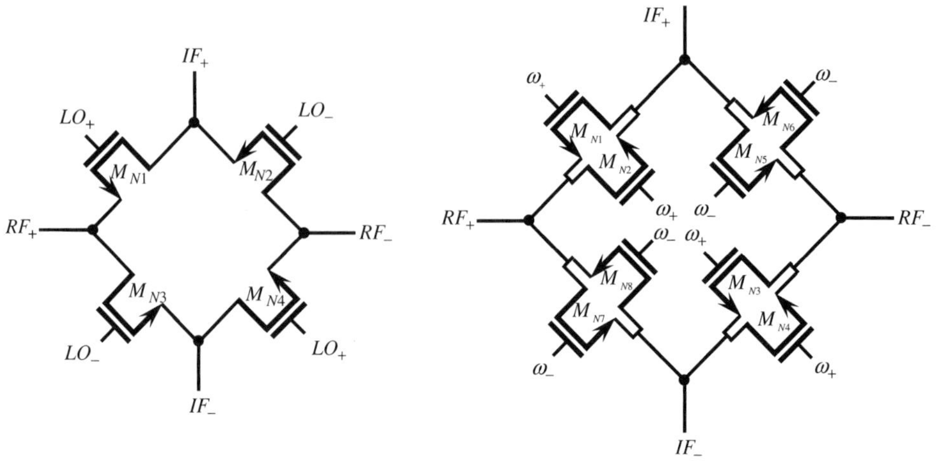

Figure 4.58. Mixer architectures based on transistor rings.

downconverted in I and Q phases with a quadrature differential signal from the LO drive network. Hence, from a single phase, two quadrature phases are obtained. In a transmitter, input signals (baseband) are in quadrature phases, and they are combined with quadrature phases from the LO to produce single sideband at the transmit frequency. I/Q matching of mixers is important and I/Q coupling should be minimized while routing the signals through the LO generation path. In both transmit and receive, the I/Q imbalance is critical mostly in the signal generation path. In the transmitter, I/Q imbalance leads to sideband rejection (also referred to as "phase combination accuracy"). In the receive path, this leads to an I/Q imbalance in terms of amplitude and phase. In the receiver, we are interested in the intermodulation behavior of the RF tones, and not in harmonic distortion, as they fall out of band. In the transmitter mixer or modulator, we are interested in both intermodulation as well as harmonic distortion, as both components are in-band. Figure 4.58 shows high linearity passive mixer realization in regular and sub-harmonic LO injection.

4.11.7 Impedances

Input and output impedance of mixers are important considerations in terms of the drive capability from the RF or IF stages. Input impedance determines whether the voltage or current-mode interface would be optimum, and what the signal swings at various interfaces would be. Similarly, the output impedance would imply how much output signal swing could be obtained. In a fully differential circuit, performance is dependent on differential impedance and its Q factor, whereas the common-mode impedance needs to be large in order to provide necessary rejection to unwanted common-mode signals. This is particularly an important consideration while interfacing pseudo-differential circuits to fully differential circuits.

4.12 BASEBAND FILTERS

Baseband filters play a major role in radio systems in filtering out-of-band signals. In transmitters, they filter out the unwanted aliases and spurious components from the DAC outputs. In receivers, they filter the out-of-band components in order to improve the signal-to-noise ratio. Usually, in integrated systems, both filters may use the same topology, and the transmit filter is usually designed to be of much lower gain compared with the receiver filter. The key functionality of the receive baseband filter is to provide enough channel select filtering and strong attenuation to adjacent blockers. Transmit filters are used to reject the aliases obtained from the DAC to provide a "clean" baseband signal to the modulator/upconverter, and essentially, they are viewed as "smoothing filters." The cutoff frequency of baseband filters extends until the channel under consideration. In modern communication systems, baseband filters may operate as high as 250 MHz (in UWB wireless systems). All of the functionality of baseband filters should be achieved with the lowest possible input referred noise. The key performance of baseband filters include

1. Voltage gain
2. Input and output common-mode ranges
3. Out-of-band rejection at adjacent blockers
4. IIP3 referred to in terms of voltages
5. Input-referred noise (current or voltage)
6. Input impedance
7. Group delay

All of the above provide the system considerations for the baseband filter block, and it must be achieved at a predefined area and power consumption targets. In modern standards, multiple bandwidth baseband filters are frequently used in order to operate in various data rates.

4.12.1 Classification of Integrated Filters

Integrated filter topologies are well understood, and the filter design is quite a mature technical domain. In this section, we will illustrate the aspects relevant to integrated communication systems. The most popular topologies include (1) switched capacitor filters, (2) Gm-C filters, (3) multifeedback structure, (4) OP-Amp RC filters, and (5) passive filters. For modern wireless standards, often the adjacent channel blocker requirement is stringent, and the filters use multistage configurations. A higher order filter is usually obtained as cascades of biquadratic stages, and in the following sections, we introduce details of various design issues related to biquad stages.

4.12.2 Biquadratic Stages

Figure 4.59 illustrates a generic biquad configuration. Any higher order filter can be realized using cascades of biquad stages.

Only the methodology to realize this transfer function changes. Setting up nodal equation leads to

$$\frac{V_i(s)}{R_5} + \frac{V_2(s)}{Z_1} - \frac{V_o(s)}{R_4} = 0 \tag{4.13}$$

$$\frac{V_o(s)}{Z_2} + \frac{V_2(s)}{R_3} = 0 \tag{4.14}$$

$$\frac{V_i(s)}{R_5} = \left(\frac{R_3}{Z_1 Z_2} + \frac{1}{R_4}\right) V_o(s) \Rightarrow \frac{V_o(s)}{V_i(s)} = \frac{Z_1 Z_2 R_4}{R_5(Z_1 Z_2 + R_3 R_4)} \tag{4.15}$$

$$\frac{V_o(s)}{V_i(s)} = \frac{\left(\frac{R_1 R_2}{R_3 R_4}\right) \cdot \left(\frac{R_4}{R_5}\right)\left(\frac{1}{R_1 R_2 C_1 C_2}\right)}{s^2 + s\left(\frac{1}{R_1 C_1} + \frac{1}{R_2 C_2}\right) + \left(\frac{1}{R_1 R_2 C_1 C_2} + \frac{1}{R_3 R_4 C_1 C_2}\right)} \tag{4.16}$$

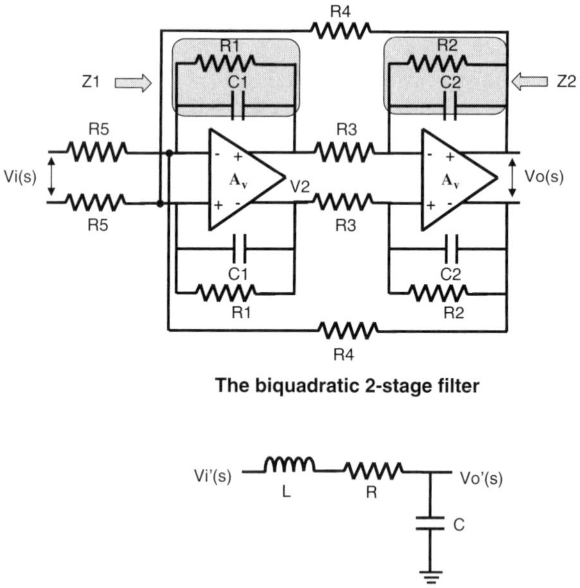

The biquadratic 2-stage filter

LRC pole transfer function

Figure 4.59. A two integrator-based biquad stage.

This is analogous to the passive LRC transfer function

$$\frac{Vo'(s)}{Vi'(s)} = \frac{\frac{1}{LC}}{s^2 + s\frac{R}{L} + \frac{1}{LC}} \tag{4.17}$$

In general, the transfer function of a biquad stage is given as follows:

$$H(j\omega) = \frac{A}{(\omega_0{}^2 - \omega^2) + jB\omega}$$

where ω_0 is the angular frequency and the phase response crosses 90°. Gain at the center frequency $\omega = \omega_0$ is given by $|\frac{A}{B\omega_0}|$, whereas the DC gain is given by $|\frac{A}{\omega_0{}^2}|$. Thus, the peaking of biquad is given by $|\frac{\omega_0}{B}|$. This transfer function can also be represented using a Z transform using bilinear transformation from analog-to-digital frequency. It must be kept in mind that at $\omega = \omega_0$, the small-signal gain peaks, and the output phase transitions through 90°. As the gain is frequency dependent, when mentioning gain of biquad stages, we should also indicate the frequency.

4.12.3 Switched Capacitor Filters

The fundamental principles of switched capacitor filter are quite old. However, unlike digital filters, the input and output samples do not coincide in time for switched capacitor filters. Such filters consume only switching power, which is dependent on clock frequency. They are quite suitable for use in systems with large-signal swing at the input. Switched capacitor-type filters are extremely precise in terms of the filter coefficients (and center frequency). Switched capacitor filters do not provide any gain, and there are issues related to leakage of clock harmonics that are used to design the filter stages. However, for higher frequency applications, the power consumption of such filters, and the oversampling ratio for realizing resistances, become quite prohibitive. Also, to achieve the SNR required for wireless systems (or to reduce noise), they tend to use a large capacitor (lower KT/C noise), leading to an increase in area.

As the frequency response of switched capacitor filters depends on the capacitor ratios and clock frequency, it can be easily changed by adjusting the clock frequency. Switched capacitor filters do not require any tuning, thereby reducing any area overhead in terms of calibration. A differential configuration is used, in order to reject common-mode noise, as well as DC effects because of clock feedthrough.

Design and implementation of switch is of prime consideration. These are illustrated in Figure 4.60(a). A switch can be modeled by its ON state resistance, and to reach 1% of the signal level, it takes six to seven time constants. At the same time, the overlap capacitance of the MOS devices C_{GD} provides a clock feedthrough, and it creates a pedestal in the output voltage characteristics. Instead of charging to V_{in}, the output voltage now charges to $V_{in}\frac{C_{GD}}{C_{GD}+C}$, where C is the capacitance to be charged. Figure 4.60(c) illustrates a charge injection mechanism from the channel, which is dependent on the input signal level, and the body bias, leading to a nonlinear operation

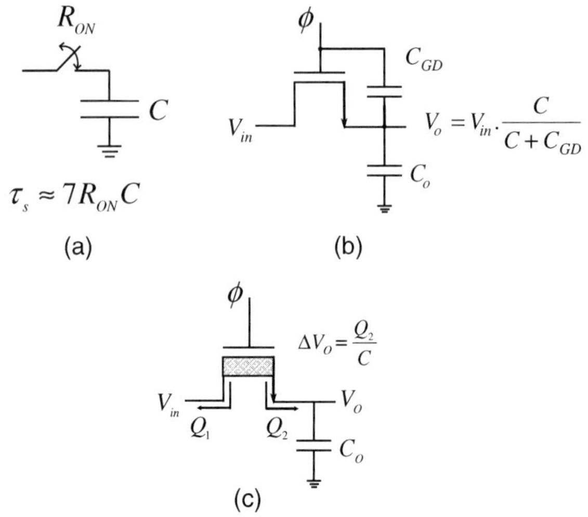

Figure 4.60. Switched capacitor nonidealities.

from the switch itself. This can be canceled in a differential configuration. As switched capacitor circuits operate on different phases of the input clock waveform, a nonoverlapping clock generator is also needed, which is illustrated in Figure 4.61. In addition to all of these errors caused by the switches, finite DC gain from the OP-Amp results in amplitude and phase errors as well [32].

A practical implementation of the differential switched capacitor integrator stage is shown in Figure 4.62. The gain-bandwidth of the OP-Amp is usually four to five times larger compared with the clock frequency. An interesting aspect of this configuration is that it is insensitive to the parasitic capacitors, C_{P1}, and $C_{P2} \cdot C_{P1}$ appears at the virtual ground of the OP-Amp, where the signal swing is minimal, whereas C_{P2} appears in parallel to a voltage source. In terms of noise characteristics, OP-Amps often tend to be the most significant contributor. In the switched capacitor configuration, flicker noise is usually not aliased, as the $1/f$ corner is less than $f_c/2$. However, wideband thermal noise is usually aliased. If the circuit can inherently limit the noise bandwidth less than $f_c/2$, the noise aliasing would not be present. However, usually these circuits have to

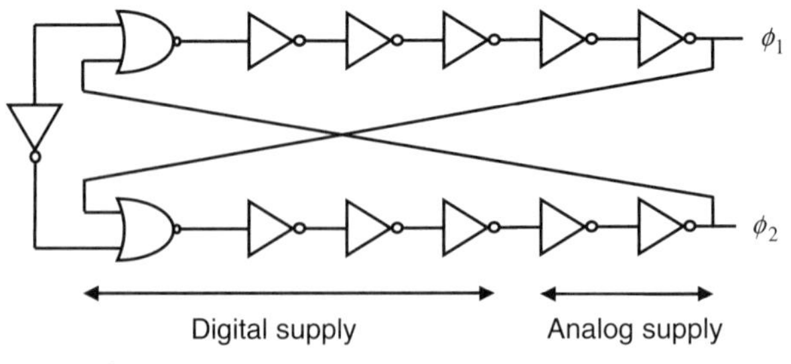

Figure 4.61. Nonoverlapping clock generation.

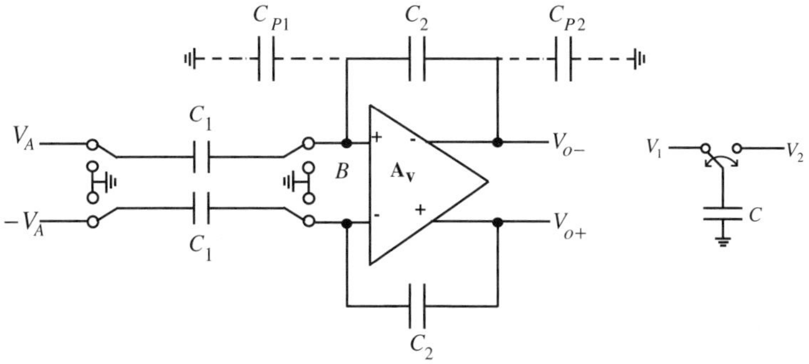

Figure 4.62. A switched capacitor integrator stage.

meet the speed requirements, and the noise power is folded to the band of interest by aliasing [30].

4.12.4 Gm-C Filters

A wide variety of systems use this configuration. Contrary to the OP-Amp RC filters that operate in a closed-loop fashion (with feedback), Gm-C based filters operate in open loop and not very linear, but highly stable. The common-mode stability and linearity degradations associated with the Gm-C filters are important considerations. Stability must be noted in a multistage configuration, where there are multiple feedbacks, and the overall common-mode stability needs to be observed.

The basic G_m core cell is illustrated as before. Several configurations of G_m are illustrated in Figure 4.63: (1) lossy integrator, (2) transconductor, (3) integrator, (4) resistor, and (5) gyrator. An inherent advantage of the Gm-C structure is that the

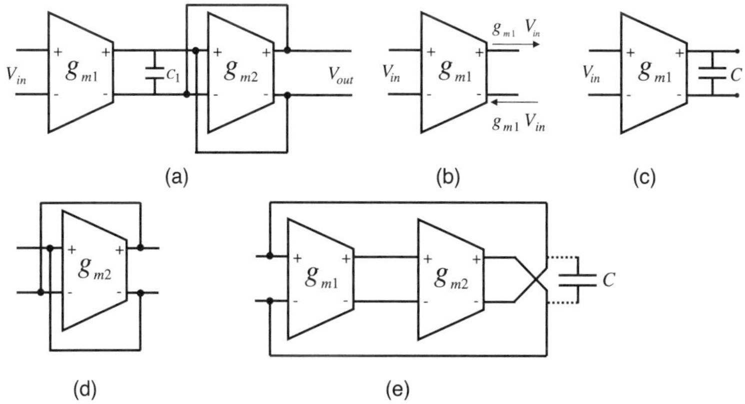

Figure 4.63. Gm stages to achieve various functionalities.

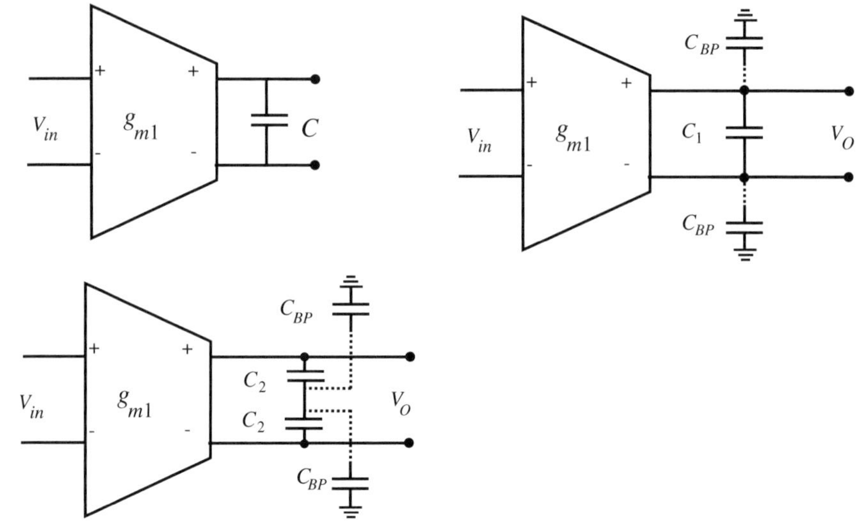

Figure 4.64. Gm-C symmetric capacitor arrangements.

output current can be easily used for sum or difference. The output capacitance can be arranged in a symmetric manner in order to compensate for the bottom plate parasitic capacitor.

Two such configurations are shown in Figure 4.64, with the bottom plate capacitance denoted by C_{BP}. A fundamental difference with the switched capacitor configuration is that the frequency response of the Gm-C filters is given by G_m/C, which changes significantly with process and temperature. Thus, some on-chip calibration is required, which requires additional area (in order to accommodate the replica of the original circuit) and/or various clock phases to perform calibration. A Gm-C biquadratic stage is shown in Figure 4.65, which is capable of providing low-pass and band-pass outputs.

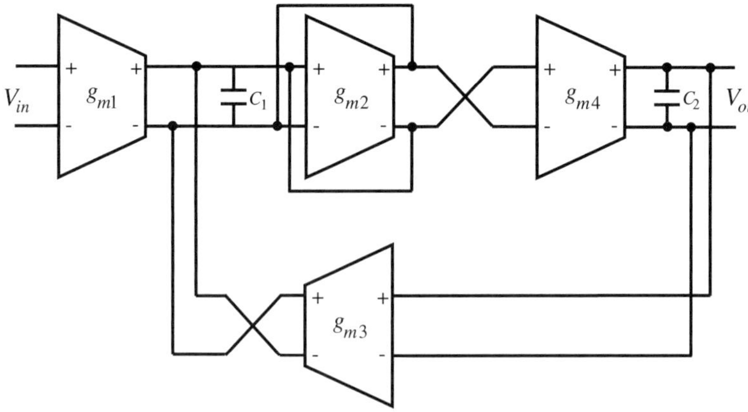

Figure 4.65. A Gm-C-based biquadratic stage.

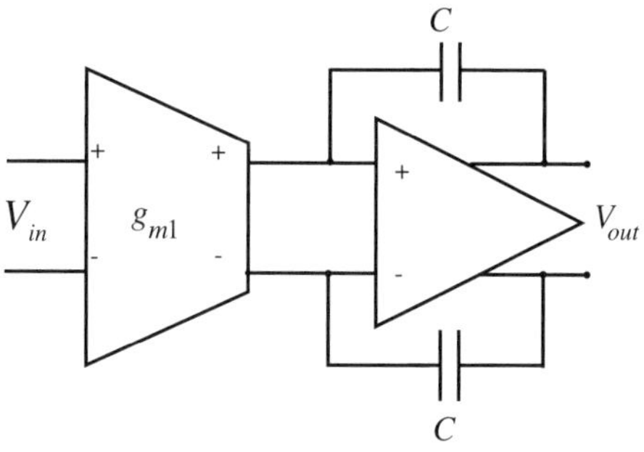

Figure 4.66. Gm-C followed by an OP-Amp configuration.

Often, a Gm-C stage followed by an OP-Amp-C integrator leads to several advantages with some possible compromise of speed consideration. This is shown in the Figure 4.66 configuration.

As the output of the Gm-C filter is held at virtual ground, the signal swing is much reduced, and linearity considerations are relaxed. The output parasitic capacitance is connected across a voltage source, which could, in principle, source any amount of current. Thus, this structure is insensitive to parasitic capacitance as well. At the same time, this structure is not very sensitive to OP-Amp gain. Because of the linearity limitations, Gm-C filters tend to perform poorly in the presence of large out-of-band blockers. In a Gm-C structure, however, the input impedance is not low, and it can be lowered by connecting the Gm stage in a feedback mode from input to output. The lower input impedance helps to reduce the signal excursion at the input. Gm-C filters are well suited for high-frequency applications because in Gm-C filters, parasitic limitation from the resistance is absent. As for other filters, common-mode feedback is a very important aspect in Gm-C filters, and it should be kept at some fixed level while observing the large-signal Gm of the filter stages.

4.12.5 OP-Amp-RC Filters

Possibly the oldest of all active filters are the OP-Amp RC-based filters. There are two possible configurations: (1) multifeedback-type architecture and (2) standard OP-Amp-based RC integrators. Figure 4.67 illustrates a multifeedback filter architecture along with the interface with front-end receive mixers.

The multistage feedback architecture uses a single OP-Amp in each biquad, thus reducing the power consumption of the overall filter solution. However, the usual problems related to these filters include (1) presence of high value of input impedance (usually 500–600 Ω); (2) poor noise performance, as the first stage is usually a passive RC integrator; (3) variation of noise performance with the filter's corner frequency (input-referred noise of the filter would vary when the corner frequency needs to be tuned); (4) capacitors in a multifeedback structure also tend to consume much more area on chip, and are difficult to optimize toward a low-area solution; and (5) moderate-to-high input impedance (500–600 Ω) of the filter tends to degrade the

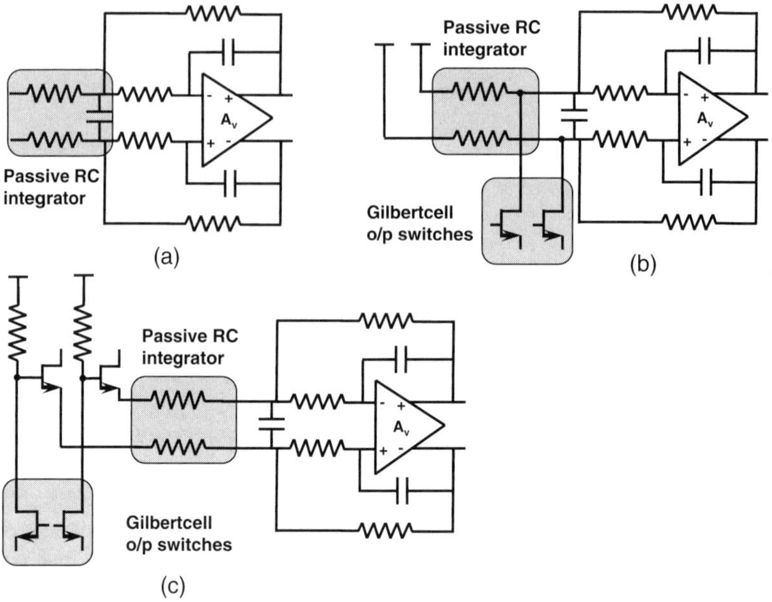

Figure 4.67. Multifeedback structures and mixer/BBF interface.

mixer's performance if the mixer is of a switching type. This structure usually requires a large capacitor to reduce noise. Also, the resistances cannot be increased significantly, and various circuit nodes would need to source more transient currents.

One can also use a two OP-Amp-based architecture for the biquads as shown in Figure 4.68. Biquads are usually designed to provide a fixed frequency response and a fixed gain. No gain change can be performed in the biquad stage itself. In this

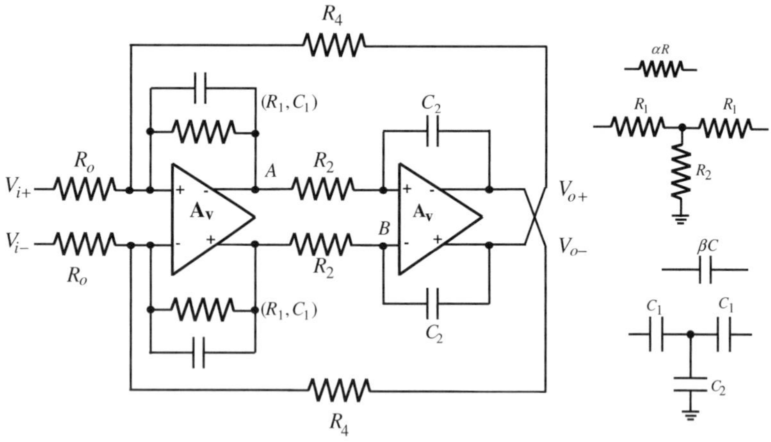

Figure 4.68. A two Op-Amp-based biquadratic stage.

approach, the noise contributions from the resistors appear at the output of the OP-Amp (as the input is held at virtual ground). The input-referred noise is divided by the voltage gain of the integrator. As the noise is governed by $\sqrt{\frac{KT}{C}}$, the capacitance can be reduced by a factor of A_v^2, where A_v is the voltage gain of the OP-Amp stage in the feedback configuration. In this topology, the gain control is also independent of the noise, and one can optimize both separately. Usually in a two-stage OP-Amp-based integrator, sufficient gain is placed in the first stage, and the signal is amplified to use the OP-Amps maximum voltage headroom. The output current from the second stage should be close to the output OP-Amp's maximum current delivery capability. Any additional programmable gain steps can be placed after the fixed gain of the first biquad stage to reduce noise degradations resulting from later stages of the filter. Both considerations set the power consumption of the biquad stages. Driving higher impedance is beneficial, as it reduces the transient current swings, essentially reducing the DC power consumption of the OP-Amp At the same time, capacitances can be reduced significantly to a value for KT/C noise limits, while not exceeding the component-matching limits. Smaller capacitance values also provide higher impedance, favoring the low current consumption. Noise floor is set by the desired signal sensitivity, and the intermodulation performance sets the gain of each of the stages. It is desired that the OP-Amp outputs should swing to their maximum limits in the presence of large adjacent channel blockers, such that no clipping should take place. This provides maximum current efficiency from such stages. The degradations are observed more at the band edge of the channel under consideration compared with the middle of the band. Clipping of the blocker usually leads to spectral expansion of the blockers (in case they are modulated, or of multicarrier-type themselves). Successive filter stages provide blocker rejection as the signal progresses in the chain. Usually, the first adjacent blocker is the most dominating one to be considered. Modern cellular standards operate in the presence of high adjacent channel blockers, leading to the use of higher order filter.

OP-Amps used to implement the core amplifier can use either the simple two-stage implementation, (Figure 4.69) or the folded cascode-type topology (Figure 4.70). Both must use a common-mode feedback for stable operation. Folded cascode is preferable in submicron technologies because of their higher gain. A high gain is desired form these OP-Amps, whereas the bandwidth can be four to five times the filter bandwidth. The overall cascaded filter structure must be stable with common-mode and differential perturbations. Two types of OP-Amps are considered: (1) low noise and (2) low power. The first stage uses a low-noise, low-input offset OP-Amp, whereas the subsequent stages use a low-power OP-Amp. DC offset cancellation is implemented in the chain, and it is usually combined with the common-mode level circuit.

4.12.5.1 *Voltage-Limiting Behavior.* In the presence of large blockers, the filter stages are designed such that the output swings to the maximum possible extent while not clipping the blocker signals. The output transistors in the OP-Amp would swing well into their V_{DSAT} limit but not lead to signal clipping. Depending on the intermodulation distortion performance, some back off can also be used between

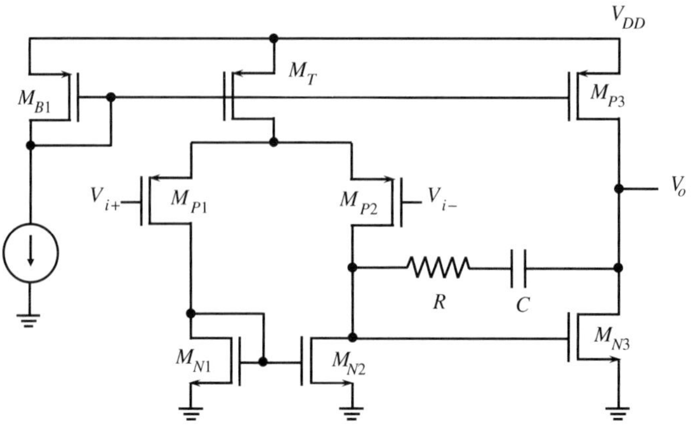

Figure 4.69. A simple two-stage OP AMP.

the stages. Thus, the OP-Amp can behave as a voltage-limiting stage. As the output stage would swing into the V_{DSAT} of the transistors, the output currents may assume a highly nonlinear shape. The individual integrator gains would be given by ratios of resistances, and to maintain the shape of transfer function characteristics, it is desired that, $R_1R_2 = R_3R_4$. The gain of the first integrator is determined by R_1/R_5, and the gain of the second integrator is determined by R_2/R_3. These ratios should be optimized in order to maximize the voltage swing at the outputs of each integrator at the maximum gain of the receiver. Starting with the mixer's output impedance, the first stage is optimized w.r.t $R-C$ values for maximum swing. Then the second stage is optimized to provide rail-to-rail swing at its output. Both are performed in conjunction with one another to optimize for the Q factor for individual stages.

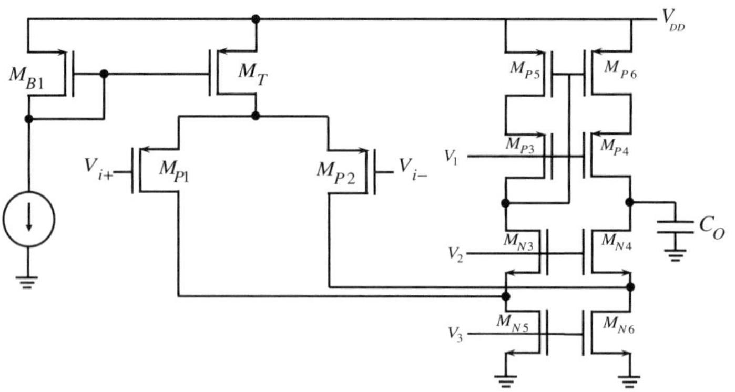

Figure 4.70. A folded cascode OP AMP.

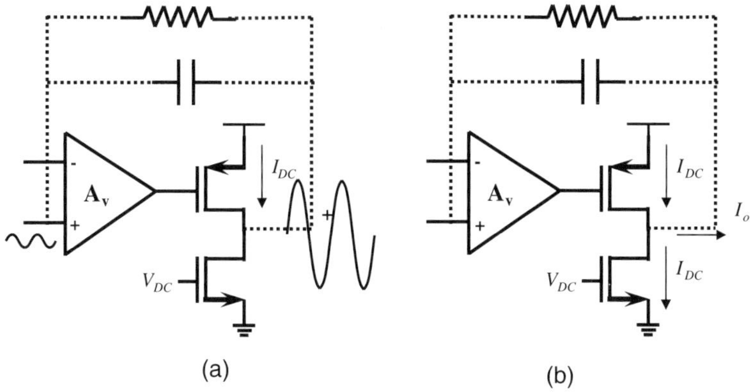

Figure 4.71. Output driving stage and power considerations: (a) voltage swing limited and (b) Current limited behavior.

4.12.5.2 Current-Limiting Behavior. Let us now consider the current-limiting behavior of the filter output stages. The OP-Amp would need to drive (1) capacitance and resistance at its own outputs, (2) the output feedback resistance, (3) output feedback capacitances, and (4) resistance between the output and the virtual ground of the next stage. At high frequencies, however, the driving impedance presented to the OP-Amp is determined mostly by the capacitances connected to the output, with the output transient current peak being set by the output voltage divided by the capacitive impedances. The OP-Amp output stage must be able to source this current. A lower capacitance implies higher impedance at blocker frequency (hence, lower current to be sourced), which results in power savings in the OP-Amp. Slewing of the OP-Amp may also happen depending on the input current level and the frequency. Figure 4.71 shows the voltage and current limited behavior of baseband filter stages.

4.12.5.3 Phase Rotation. As the blocker moves through various cascaded integrator stages, it experiences phase rotation through *RC* stages, and essentially, it does not add up in phase at the output of the integrator stages. When the blockers are rotated by a perfect 90°, the intermodulation floor rises by 3 dB, as opposed to the in-phase addition impact of 6dB (without any rotation). Thus, phase response in the filter stages is important to consider, in addition to the amplitude response.

In addition to the *KT*/*C* noise contribution from the capacitors, the OP-Amp also contributes noise. Usually a low input-referred noise, low input-referred offset voltage OP-Amp is used in the first integrator, and the subsequent integrators are designed to be low-power topologies. Providing sufficient gain in the first stage allows insensitivity to the input-referred noise from the blocks beyond the first integrator stage. High Q, high-density capacitances should be used to minimize area, and resistances should be increased upto a value at which they are limited by the parasitic capacitances. Usually the voltage gain; A_v is proportional to R, and the noise is proportional to $\sqrt{R}$; hence, a large value of resistance would lead to lower input-referred noise. The programmable gain steps can be performed by using resistive ladder networks. Increased gain steps in the RF front end would need lower gain steps in the baseband stages. Instead of OP-Amps, the OTA stage can also be used. OTAs are fundamentally different from Gm stages in the fact that their output is a voltage, and the input signal is small. Details of analysis can be found in [35].

4.12.5.4 Architectural Considerations. There is a fundamental advantage in using a continuous-time filter prior to the ADCs in the receive chain. If the ADC is directly placed at the output of the mixer, then the input signal swings are much lower, and the ADCs would need more capacitances in the circuit in order to reduce *KT/C* noise. This would increase the overall area of the circuit. The dynamic range would depend on the number of bits, and the noise would be governed by *KT/C*.

4.12.5.5 Multiorder Continuous-Time Active Filters. As a specific example, we can consider designing any type of classic filter topologies. Multiorder Chebyshev, Butterworth, elliptical, and Cauer filters are commonly used in most practical systems. A Chebyshev filter has its poles on an ellipse, whereas a Butterworth filter has its poles on a circle. The poles of the ellipse can be adjusted to result in a circular pole constellation. This is the basis of reprogramming one type of filter to another with no area penalty. Although theoretically any filter would continue to provide monotonic attenuation at out-of-band frequencies, in practice, the rejection would be limited by the component mismatches. Chebyshev provides ripple in both passband and stop-band, whereas Butterworth is monotonic in the passband, providing excellent group delay characteristics in the case of wideband systems. However, Chebyshev filters tend to be superior compared to Butterworth filters in achieving out-of-band rejection performance for the same amount of current consumption. This result is because Chebyshev out-of-band rejection is much higher at the same number of poles. The ripple in the passband of Chebyshev would be dependent on the mismatches in the location of the poles, as well as on the Q of the biquad stages. Elliptic filters are well behaved in terms of flat in-band response, but out-of band, they cause ripples, which is detrimental to large blockers. A seventh-order Chebyshev filter can be designed by using a cascode of (1) low Q biquad, (2) medium Q biquad (b) real pole, and (3) high Q biquad. A biquad transfer function can be easily mimicked by using an L–R–C-based configuration. In a seventh-order response, there would be four crests and three troughs in the passband ripple characteristics.

Many times ripple from the Baseband filter can be tightened by defining a performance metric such as EVM, which is defined as follows:

$$U = 10\log_{10}[G^2 + \tan^2\phi]^{1/2} < -35\text{ dB}$$

where

$$G = \left(\int_{f_L}^{f_H} \left[\frac{A - A_{RMS}}{A_{RMS}} \right]^2 df \right)^{1/2}$$

and

$$\phi = 2\pi\left(\int_{f_L}^{f_H} [\phi(f)-\phi_{lin}]^2 df\right)^{1/2}$$

are the amplitude and phase responses, respectively, with ϕ_{lin} being the best possible linear fit to the phase response curve. Thus, it can be seen that a smaller amplitude ripple would lead to superior EVM performance.

4.12.5.6 Common-Mode Levels. In a filter chain, it is important to provide the right common-mode levels at the input and outputs of each integrator. The input common mode must be compatible with the output common mode of the mixer, as at low frequency (~MHz), no coupling capacitors can be placed on-chip because of the area requirements. Similarly, the output should be compatible with the ADC input common mode. Sometimes level shifter networks can be placed in the input and output networks to provide the various voltage levels, with the requirement that they will not produce any unwanted linearity degradations, as well as to produce a minimum amount of noise.

4.12.5.7 OP-Amp Design. The OP-Amps to be used in these systems can usually be implemented with the standard two-stage topology or a variation of that. To eliminate issues related to the voltage headroom, several other topologies can also be used. Common-mode feedback is a definite requirement in these OP-Amps, and one must ensure their low noise and distortion performances. The filter capacitors connected in a feedback fashion provide phase margin to the OP-Amp; hence, an internal compensation capacitor may be eliminated, which leads to area savings. The gain of the OP-Amp is dependent on the g_m of the input devices and on the g_{ds} associated with the output nodes. g_m is dependent on the bias current and on the geometry of the input devices, and the g_{ds} is dependent on the bias current, as well as on the channel length under consideration. In deep submicron technologies, g_{ds} is usually much worse, and larger channel length helps in the OP-Amp performance. Power dissipation is dependent on the gain bandwidth product of the OP-Amp, which in turn is dependent on the signal bandwidth under consideration. The parasitic capacitance of the output devices can be a limiting factor to the OP-Amp bandwidth.

4.12.5.7.1 Design Considerations for the First Stage. The first stage in the filter chain should be designed for (1) low offset voltage, (2) low noise, (3) input common-mode voltage compatible with mixers, (4) high gain, (5) low supply voltage, (6) desired signal swing at the output to handle the large blockers, and (6) low input differential impedance at the blocker frequencies. The subsequent stages would be designed to provide; (1) moderate noise, (2) low power, and (3) high swing capability, moderate gain. Partitioning into low-noise and low-power categories lead to optimized power and area solutions.

The input stage of the first OP-Amp in the chain should be optimized for low input-referred noise and low static DC offset values. This static offset (as well as the dynamic offset) is multiplied by the filter gain, and it appears as a large signal at the output of the filter. To meet both of these requirements, the input stage usually requires large devices, as spectral density of drain flicker noise current and the input-referred static DC offset are inversely proportional to the square root of the area. Device dimensions are increased until the stage becomes parasitic limited. Also, the overdrive voltages are kept low, in order to reduce noise from these devices. The input-referred static offset (and the receiver's DC offset) is dominated by the first stage in the IF filter, and DC offset compensation needs to be used in several stages in the filter in order to nullify the unwanted DC offset effects.

4.12.5.7.2 DC Offset. Many implementations of DC offset cancellation are reported in literature [40], and we will discuss two categories of these. The first in this category uses a feedback loop to low-pass filter component around zero frequency, and subtract the average signal of the envelope as shown in Figure 4.72.

In the second option, a trimmable current is injected at the input nodes using binary-weighted IDAC as illustrated in Figure 4.73, which leads to DC null at the input (subject to the resolution of the DAC). In the first option, however, since it is a feedback loop, it requires additional active elements such as OP-Amps, and consume more current, and the filter characteristics provides a DC notch, which degrades the phase characteristics of the filter response around zero frequency. At the same time, the input offset of the feedback amplifier may pose a limitation to this technique as well. In the offset DAC option, the output impedance of the offset DAC should be high and the current elements should match precisely with one another. This implementation requires much lower current to be injected and consumed in the circuit, as the common-mode impedance of the OP-Amp stages is significantly higher. The noise contribution from the DC offset cancellation sources must be kept at a minimum level in order to reduce the noise figure degradation of the receiver. Usually DC offset cancellation blocks are distributed along the filter stages in order to best nullify the DC at the output. If this is not followed, then the baseband stages may get saturated because of even some residual DC offset in the previous stages. The cancellation is limited by the resolution of the IDACs in the discrete mode, and in the continous calibration

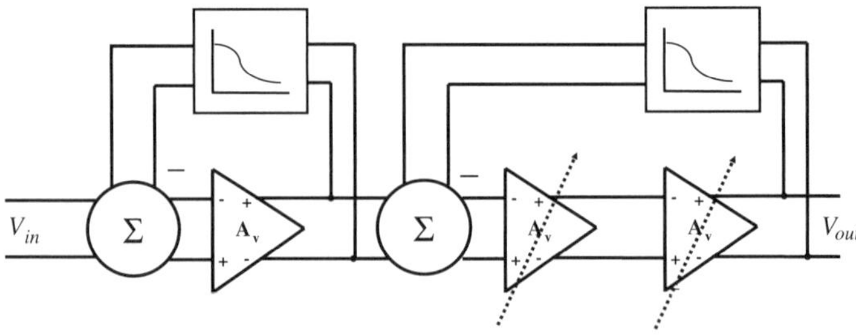

Figure 4.72. DC offset compensation using low-pass filtering.

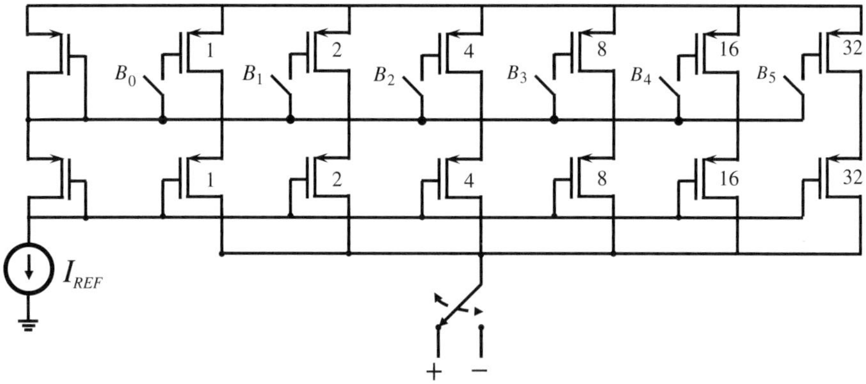

Figure 4.73. DC offset compensation using IDAC.

mode, one may require multiple such loops, leading to stability considerations. DC offset through multiple chains is obtained through linear superposition of the various DC offset cancellation stages.

4.12.5.7.3 Input Impedance. We will now pay close attention to the input impedance of the OP-Amp, which is given by

$$Z_{in} = {}^{Z_f}\!/_{A_v}$$

where Z_f is the feedback resistance and A_v is the open-loop gain of the OP-Amp. Hence, with a 10k resistance in the feedback loop of 60-dB voltage gain, the OP-Amp would lead to an input impedance of 10 Ω. It should be noted at this stage that the open-loop gain of the OP-Amp has a low-pass response itself

$$A_v = \frac{A_{v0}}{(1 + j^{\omega}/_{\omega_0})}$$

This leads to an inductive input impedance at high frequencies, and care should be taken in order to prevent any resonance occurring from the mixer's output impedance. This may even boost up the blocker level and should be alleviated by using careful design. It should also be noted that although a low input impedance (hence, a high gain of the OP-Amp) is desirable, it leads to low bandwidth (gain-bandwidth trade-off), so a compromise should be obtained. Much of the basic materials on various OP-Amp topologies are covered in [4].

4.12.5.8 R–C Switching Banks. In an OP-Amp-based R–C configuration, the R–C banks are usually placed between two terminals, the output of an integrator and the virtual ground of the succeeding stage. Thus, the values of the R–C components should be evaluated with a similar configuration of connecting voltage source at one

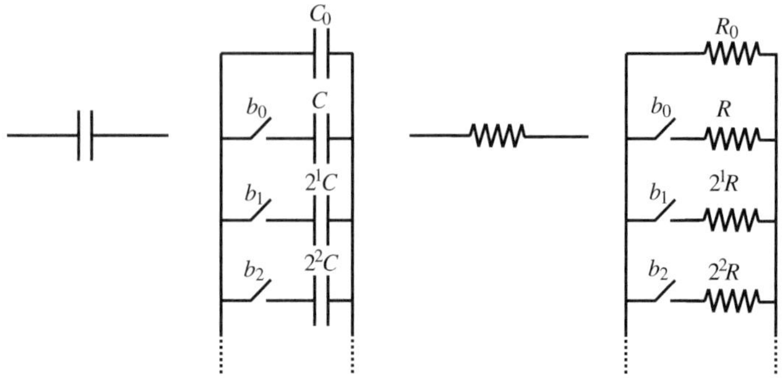

Figure 4.74. RC switching banks used in analog filters.

terminal, and observing the current flowing into the virtual ground of the OP-Amp. Figure 4.74 illustrates the switched RC banks that are used for filter design. The switches are selected to provide a high Q factor, and they are weighted in a binary fashion to provide constant Q irrespective of number of ON switches.

The selection of capacitance is dependent on their density, voltage handling capability, and process variation. Voltage swing across capacitors increases as the signal propagates from the input to the output. Hence, the capacitors should be selected in order to provide (1) high density, (2) high breakdown limits (lower defect numbers), (3) low-voltage-dependent linearity, and (4) low bottom plate parasitic capacitance. Usually, in IF filters, the first stage uses more capacitor values than the following stages, and it consumes more area. The Q factor of the capacitors is not much of a concern as the operating frequencies are well below gigahertz ranges. These capacitors are then placed in parallel branches switched by MOSFETs (NMOS or PMOS switches are used, depending on whether they are referred to zero potential or supply). Parasitic capacitances from the switches may lead to shrinkage of the tuning range. The resistances are increased until the performance degradation caused by the associated parasitic capacitance does not degrade the performance of the biquad stage. The values of the resistances must be adjusted in conjunction with the ON resistance of the switch transistors. As the voltage swing increases, there would be modulation of the ON resistance leading to linearity degradations. When a capacitance is ON, voltage across the switch would be given by the ratio of the switch impedance to the capacitive impedance of the branch. In the OFF state of the switch, the voltage across the switch transistor is governed by the ratio of branch capacitance to the OFF state switch capacitance. This is prevented by ensuring that the ON resistance of the switches is about 5–10% of the actual branch resistance. A high Q biquad stage would need lower component mismatch levels and degradations because of the parasitic capacitances in the individual components. Hence, these aspects must be evaluated around the highest Q biquad in the filter chain. It must also need to be ensured that the capacitances do not "leak"; otherwise, the DC voltages would shift unnecessarily.

4.12.5.9 Stability Analysis of Filters. Stability is a major consideration in IF filters, as they may contain multiple feedback loops. Common-mode feedback loops are also key to this analysis, as they lead may lead to stability issues. Stability usually refers to the phase margin around the loop, and it can be understood by using a small signal in frequency domain analysis or a large signal in time domain analysis. Both may be used to correlate one another. Intuitively, as the operating frequency increases, capacitive impedances become dominating and may lead to phase margin degradations. At further higher frequencies, they may short out part of the circuits, which leads to unwanted oscillatory behavior. The stability analysis leads to the understanding of the circuit behavior in the presence of an unwanted voltage or current spike at the input. This must be performed in both common-mode and differential-mode configurations to analyze individual stability and any possibility of cross-mode conversion (common mode to differential mode and vice versa).

Small signal analysis is based on a Middle-Brook "loop cutter" technique. This technique disconnects the feedback loop and analyzes both the voltage and the current gains, and one needs to combine the two results to obtain the exact gain expression in the feedback loop configuration. Figure 4.75 illustrates the three situations. The first is our original configuration of the circuit. Let us assume that the OP-Amp has a voltage gain of A, and as

$$V_{out} = \frac{R_L}{R_L + r_O}[A(V_{in} - V_{out})]$$

implying

$$G = \frac{V_{out}}{V_{in}} = \frac{AR_L}{AR_L + R_L + r_O}$$

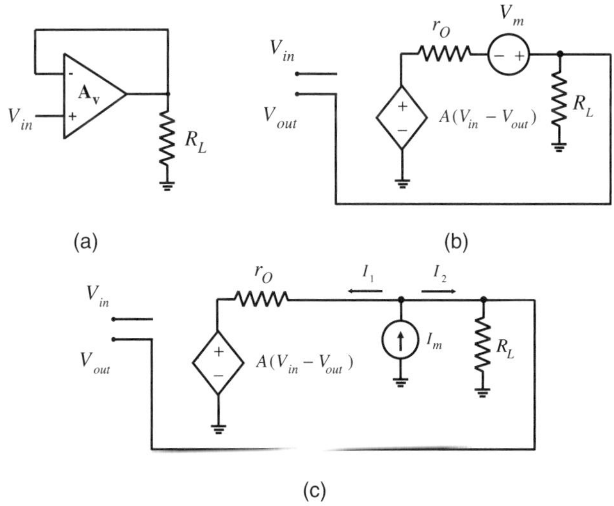

Figure 4.75. Stability analysis of filters using Middle Brook analysis.

For a voltage-only configuration, as shown in Figure 4.75(b), we obtain

$$\frac{A(V_{in}-V_{out})-(V_{out}-V_m)}{r_O}=\frac{V_{out}}{R_L}$$

with $V_{in}=0$, simplification leads to

$$G_v=\frac{V_{out}-V_m}{V_{out}}=A+\frac{r_O}{R_L} \tag{4.18}$$

Similarly for current we obtain,

$$\frac{A(V_{in}-V_{out})}{r_O}=-I_1$$

with $I_2=\frac{V_{out}}{R_L}$, with $V_{in}=0$, current gain is given as follows:

$$G_i=\frac{I_1}{I_2}=(A+1)\frac{R_L}{r_O} \tag{4.19}$$

leading to

$$G=1-\frac{1}{G_i+1}-\frac{1}{G_v+1} \tag{4.20}$$

The above analysis is performed in the limit of small signals. However, actual dynamics of the system is captured better in terms of large signal spikes. In this case, voltage pulses ($\sim$10–20 mV) can be placed at any of the branches in the input of the circuit (for a duration of $\sim$10 nS), and the output would provide a decaying sinusoidal waveform. The decay rate would imply the phase margin of the system. In the stability analysis, all the capacitors and resistors are switched in the filter circuit in order to provide the maximum possible phase shift. Sum and difference of the voltages at the output terminals provide common-mode and differential-mode stabilities, respectively.

4.12.6 Calibration of On-Chip Filters

As discussed in the previous sections, the filter coefficients, and hence the corner frequency, can be ratios or products of two dissimilar quantities (g_m/C or RC). With process variation, these quantities will change, which leads to the change in frequency response, gain, and linearity characteristics. To retain the desired filter characteristics, calibration is needed. Many filter calibration procedures have been reported in literature, and they are based on amplitude and phase characteristics of the filter. Calibration can be performed using two broad categories: (1) using a replica of the main filter and (2) using the actual filter.

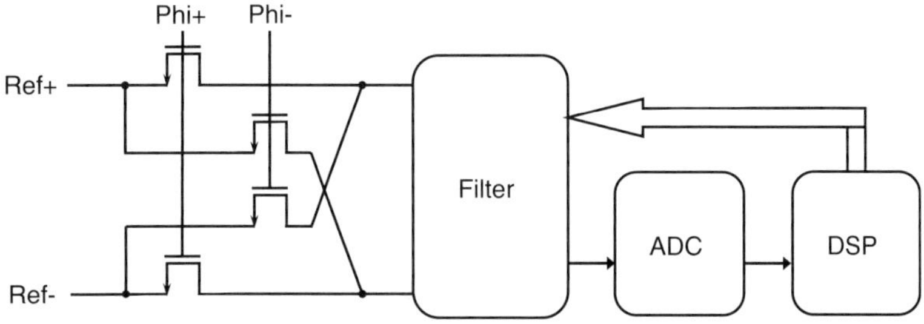

Figure 4.76. Calibration of a continuous-time filter using an amplitude ratio.

Figure 4.76 illustrates a technique to detect corner frequency based on amplitude ratios at different parts of filter characteristics. In this technique, the filter characteristics is obtained with Phi = 1, and average power is measured, and represented as in Figure 4.76.

Figure 4.76 illustrates a calibration procedure based on magnitude comparison. In this case, output of the filter at DC and a sinusoidal input are compared w.r.t. their power, and the code is adjusted as long as their ratio has the desired value. First, a DC reference is provided in the filter to obtain average power at DC, $\overline{V_{out},_{DC}^2}$. A square wave is provided next, and average power is measured to provide AC magnitude, $\overline{V_{out},_{AC}^2}$. Selection of the resistance and capacitance code is performed by a digital code, and the search continues until $\overline{V_{out},_{AC}^2}/\overline{V_{out},_{DC}^2}$ has the desired value. While performing this computation, $\frac{4V_{REF}}{\pi}$ comes into picture as the fundamental component of the square waveform. Reference [34] illustrates more details of this technique.

A tuning mechanism based on a replica filter is illustrated in Figure 4.77. It is based on the phase of the filter output. The phase of the replica filter output is compared with the phase of the reference signal and is provided to a phase detector. The filter's

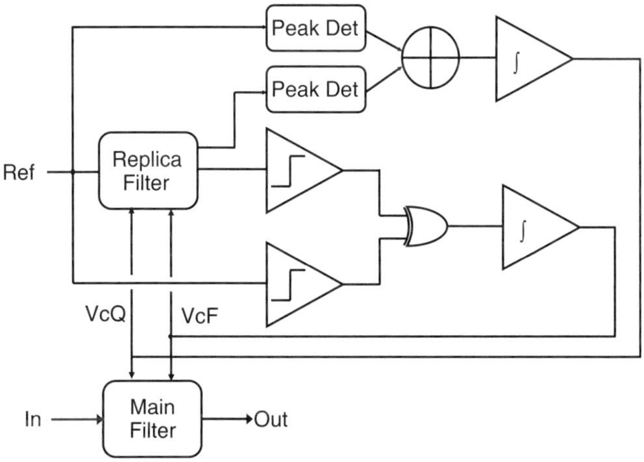

Figure 4.77. Calibration of a continuous-time filter using Q and center frequency tuning loops.

codeword to select resistance and capacitance values is adjusted until the phase reaches the desired value (for a high Q biquad, the output phase is at 90° at the center frequency). The reference frequency is selected in order to provide maximum immunity to clock feedthrough.

In the replica method, a smaller version of the main components is placed on-chip, and techniques similar to a built-in self-test, using analog signals, are used. However, the bias current and geometry of the circuit components should be adjusted such that the transistors represent the actual circuit under consideration. In conjunction to the replica testing, the passive components (resistors/capacitors) can be characterized on-chip by comparing them w.r.t. off-chip accurate components with tight tolerance (off-chip precision resistor/capacitors). Calibration can be performed using a main filter also, in which the test signals are provided to the filter in the "calibration phase" where processing can be interrupted. When processing cannot be interrupted, a replica filter is useful.

The above concept can be used to compensate for process and temperature variations of components in a circuit. In a replica testing, the signal path is completely separated from the main signal path, which leads to elimination of spurious signals, resulting from the clock and other analog stimuli. However, a finite amount of spatial mismatches would be present between the main and the replica circuits. Calibration circuitry in the main signal path may degrade linearity, and noise figure and may create unwanted feedback loops. At the same time, replica testing consumes more area. Depending on the performance criticality, each can be used as appropriate.

4.12.7 Passive Filter Configuration

At high channel frequencies, passive filters can be used in the transceiver paths. These filters can be designed using an actual L–C combination in the signal path. With increasing frequency, the inductive impedance would increase, and the L-C stage may provide some voltage gain (but not power gain!), depending on the available Q from the on-chip inductor (which poses a significant difficulty below GHz frequency ranges). Such filters are usually very quiet in terms of noise. However, the driver stage prior to this filter may consume significant current as the input impedance may not be significantly higher at IF frequencies.

4.13 SIGNAL STRENGTH INDICATOR (SSI)

In any communication system, we need to detect the signal strength. This can be performed in both the receiver and the transmitter. Depending on the signal strength, and amplification, the power dissipation of various blocks can be adjusted. Signal strength indication can be performed in analog and digital domains. In practical situations, the implementation usually consists of both domains. Providing full-range SSI is impractical in digital, as that may overload the ADCs and so on. in the back end. Output of the SSI block is usually taken in digital, which decides on the power consumption and on the gain partitioning of the building blocks.

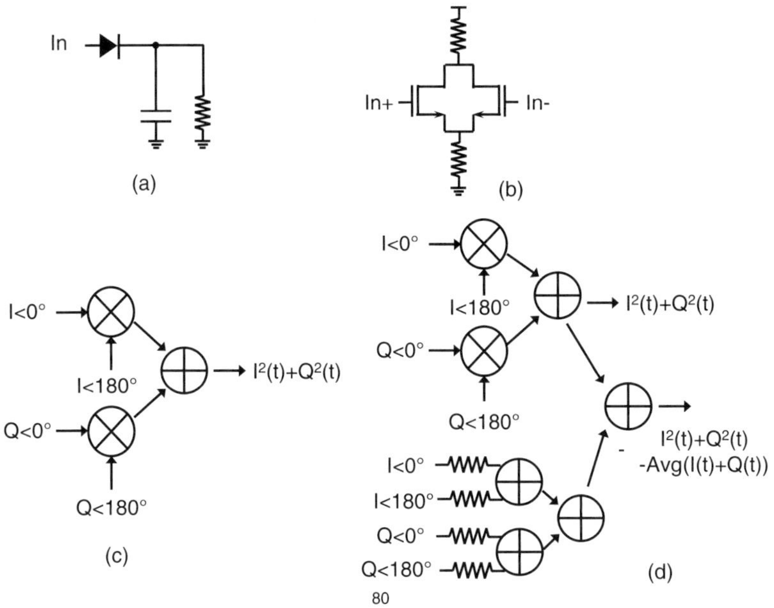

Figure 4.78. Circuits used to perform signal strength indicator functionality.

Figure 4.78 illustrates various configurations of signal strength indicators. Fundamentally it is an envelope detection mechanism, which can be obtained by:

1. Diode-based envelope detection
2. Transistor-based rectifier configuration (similar to frequency doublers)
3. Mixer-based envelope formulation from two quadrature phases.

$$E = I(t) \times I(t) + Q(t) \times Q(t)$$

4. Envelope minus the average value (the average value is subtracted from the envelope energy) $E' = [I(t) \times I(t) + Q(t) \times Q(t)] - [Avg[I(t)] + Avg[Q(t)]]$. SSI is a very important block in high-dynamic-range communication systems, such as cellular devices, which would operate at very low signal levels near the cell edge and at high signal levels close to the base station.

4.14 ADC/DAC

ADC/DAC has reached reasonable technical maturity over the past 20 years. Fundamental performance parameters of ADCs include:

1. Architecture: Nyquist or oversampling
2. Resolution (determines dynamic range)
3. Speed

Noise shaping in oversampling ADCs is a very important concept that is used in modern communication systems. In a simple oversampling ADC, noise power is reduced because of the high sampling frequency. Noise shaping is performed by placing the quantizer in a feedback loop, and a filter is used in the forward loop. In this case, only the quantization noise is shaped, whereas the signal spectrum is preserved.

Delta-sigma is most widely used because of its immunity to the circuit imperfections. Fundamentally, it is based on oversampling, noise shaping, and trades off resolution with speed. Starting from moderate sampling speeds, they are now being widely used in wireless communication systems. The advantages of delta-sigma converters include the potential of high resolution and easy integration in CMOS technology with a zero mask adder, the relaxed requirement of external filters, and the easy digital filtering of out-of-band noise.

Figure 4.79 illustrates the first-and second-order delta-sigma architectures. The input–output relationship for a first order modulator is given by

$$Y[nT_s] = X[(n-1)T_s] + Q[nT_s] - Q[(n-1)T_s] \tag{4.21}$$

A z transformation leads to $Y(z) = z^{-1}X(z) + (1 - z^{-1})Q(z)$, with the signal transfer characteristics as z^{-1} and the noise transfer characteristics of $(1 - z^{-1})$. Hence, the signal path provides low-pass characteristics, whereas the quantization noise is shaped by a high-pass transfer function. In the case of a second-order loop, the equations are given as

$$Y(z) = z^{-1}X(z) + (1-z^{-1})^2 Q(z)$$

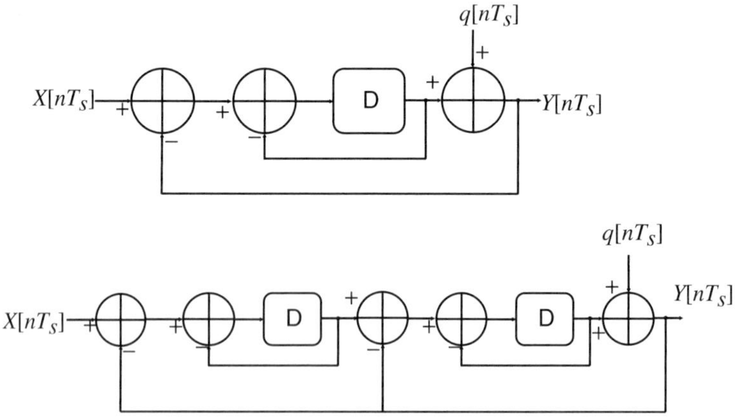

Figure 4.79. A first- and second-order Σ–Δ modulator.

similarly for an L-level quantizer, the noise transfer function is given as $(1 - z^{-1})^L$. The dynamic range is given by the ratio of signal power to the integrated noise power, which leads to

$$DR^2 = \frac{\left(\frac{\Delta}{2\sqrt{2}}\right)^2}{\frac{1}{f_s}\int_{-BW}^{BW} |1-e^{-j2\pi f T_s}|^L \cdot \frac{\Delta^2}{12} df} \approx \frac{3}{2}\frac{(2L+1)}{\pi^{2L}} M^{2L+1} \tag{4.22}$$

where M is the oversampling ratio.

For a B-bit Nyquist rate converter, the dynamic range is given by $DR^2 = 3.2^{2B-1}$ Hence, it is observed that for a first-order sigma-delta converter, the dynamic range increases by 9 dB (number of bits by 1.5) for every doubling of the oversampling ratio. For a Nyquist ADC, an increase of one bit provides a 6-dB improvement in dynamic range.

The oversampling ratio and dynamic range are the fundamental considerations while constructing an appropriate architecture for sigma-delta modulators. For a high bandwidth signal, the sampling frequency may be limited by technology. It should be noted that the sigma-delta configuration simply "shapes" the quantization noise power; it cannot "reduce" noise power. The total noise power is the same without using the noise shaping functionality. Increasing the number of integrator stages further improves the in-band noise shaping. However, this leads to potential instability as multiple loops are included in the structure.

Instead of providing multiple integrators, the same dynamic range can be achieved by using a cascade of multiple lower order stages. Such structures are commonly known as MASH, where outputs of each stage are provided to a digital cancellation stage. This is illustrated with a two-stage cascade in Figure 4.80. If N

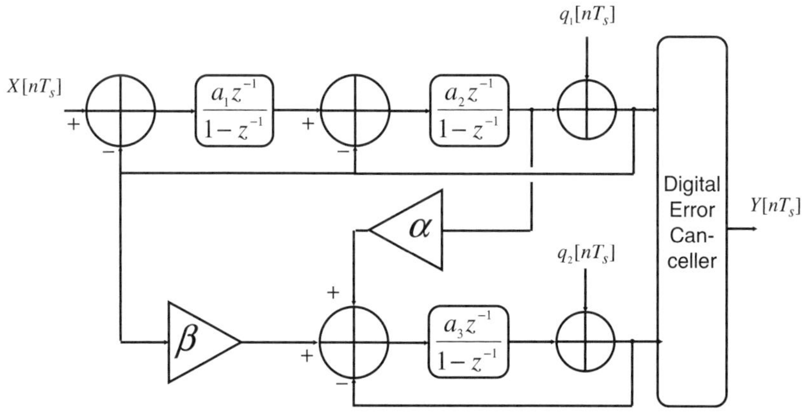

Figure 4.80. A Multistage S–Δ modulator.

cascades are used, then the digital cancellation logic cancels the quantization noise from the previous ($N-1$) stages. The scaling factors can be adjusted accordingly, to provide quantization noise cancellation. Cascaded implementation of low-order stages are used extensively in modern architectures. Successful implementation of digital correction of quantization noise depends heavily on the matching of components and accuracy.

In practical implementations, thermal and flicker noise from various sources also needs to be taken into consideration. In-band thermal noise is determined by $kT/(C.M)$, and large capacitors would be needed to improve the SNR. The input signal can also be increased to improve SNR. Flicker noise in deep submicron CMOS provides a significant design challenge.

The core circuits used in a sigma-delta converter include a (1) a Gm cell and (2) a comparator. Power consumption is mostly dependent on these blocks.

4.15 LATCH

Latches are extensively used in mixed signal circuits and are shown in Figure 4.81. It illustrates two configurations using inverters as well as MOS transistors (NMOS shown for simplicity; PMOS can also be used). Circuits using latch are given as follows: (1) comparators, (2) D flip-flop, and (3) oscillators. Fundamentally, latches use positive feedback to establish the operating voltages.

A latch has two modes of operation. In the first phase, the signal is applied to the terminals. In the second phase, positive feedback is enabled and the latch operation

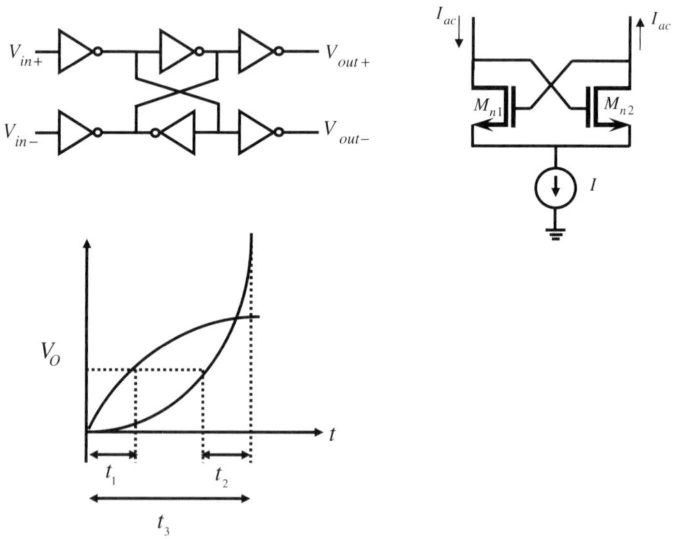

Figure 4.81. Implementation of cross-coupled latches: (a) gate level and (b) transistor level.

starts. The time taken to establish stable outputs determines the circuit characteristics employing a latch.

In the time domain, latches provide positive time constant exponential characteristics with a time constant of $\tau = \frac{C}{g_m}$, where C is the gate-source capacitance and g_m is the transconductance of the transistor. This is different from the time domain response of an amplifier stage, which provides a negatively decaying exponential. To reach a certain output voltage level, negative time-constant characteristics provide faster transient response near the origin, whereas positive time constant characteristics provide faster transient response at large time instants. These two characteristics can be used judiciously to construct high-speed comparators, where the signal levels are sufficiently amplified prior to providing to the latch, and this provides a faster response compared with operating in the latch curve alone $(t_1 + t_2) < t_3$. Preamplifier stages help keep the input-referred offset to a minimum value for the combined latch-preamplifier stage.

CONCLUSION

In this chapter, we have introduced the readers to various circuit blocks required to build mixed-signal communication systems. Research and developments in the area of circuit design have been quite slow, and many circuit techniques used in the vacuum tubes are still commonplace using transistors in advanced semiconductor technologies. In essence, we attempt to provide the desired functionality and specification targets within a power and area budget. Technology scaling reduces power and area consumption, and many circuit blocks are being designed in advanced technology nodes to exploit this advantage.

Another aspect of the material presented in this chapter is observing the similarity among various circuit design styles and techniques. We have used an intuitive approach to describe key functionalities, as much literature is already present on specific topics. This intuitive approach helps the readers to focus on fundamental cores without having to understand the same circuit topology, depending on different circuit design domains. This way, we could cover much material within a limited scope while focusing on salient points and design trade-offs.

REFERENCES

Digital Circuit Design

[1] J.P. Uyemura, *CMOS Logic Circuit Design*, Kluwer Academic Publishers, 1999.

[2] H. Taub and D. Schilling, *Digital Integrated Electronics*, McGraw-Hill International Edition, 1977.

Basic Circuit Concepts

[3] P.R. Gray, P.J. Hurst, S.H. Lewis, and R.G. Meyer, *Analysis and Design of Analog Integrated Circuits*, 4th edition, John Wiley, 2001.

[4] P.E. Allen and D.R. Holdberg, *CMOS Analog Circuit Design*, 2nd edition, Cambridge University Press, 1999.

[5] D. Johns and K. Martin, *Analog Integrated Circuit Design*, John Wiley, 1997.

Amplifiers

[6] J.-H.C. Zhan and S.S. Taylor, "An inductor-less broadband LNA with gain step," *IEEE VLSI Symposium*, Sept 2006, pp. 344–347.

[7] F.H. Raab et al., "Power amplifiers and transmitters for RF and microwave," *IEEE Transactions on Microwave Theory and Techniques*, Vol. 50, No. 3, Mar 2002, pp. 814–826.

VCOs

[8] D.B. Leeson, "A simple model of feedback oscillator noises spectrum," *Proceedings of IEEE*, Vol. 54, Feb 1966, pp. 329–330.

[9] A.A. Abidi and R.G. Meyer, "Noise in relaxation oscillators," *IEEE Journal of Solid State Circuits*, Vol. SC-18, Dec 1983, pp. 794–802.

[10] A. Hajimiri and T.H. Lee, "A general theory of phase noise in electrical oscillators," *IEEE Journal of Solid State Circuits*, Vol. 33, No. 2, Feb 1998, pp. 179–194.

[11] T.K. Johansen and L.E. Larson, "Optimization of SiGe HBT VCOs for wireless applications," *IEEE RFIC Symposium Digest*, Jun 2003, pp. 273–276.

[12] M.A. Margarit, J.L. Tham, R.G. Meyer, and M.J. Deen, "A low-noise, low-power VCO with automatic amplitude control for wireless applications," *IEEE Journal of Solid State Circuits*, Vol. 34, No. 6, Jun 1999, pp. 761–771.

[13] A. Zanchi, C. Samori, S. Levantino, and A. Lacaita, "A 2-V 2.5Ghz -104dBc/Hz at 100 kHz fully integrated VCO with wide-band low-noise automatic amplitude control loop," *IEEE Journal of Solid State Circuits*, Vol. 36, No. 4, Apr 2001, pp. 611–618.

[14] J.W.M. Rogers, D. Rahn, and C. Plett, "A study of digital and analog automatic-amplitude control circuitry for voltage controlled oscillators," *IEEE Journal of Solid State Circuits*, Vol. 38, No. 2, Feb 2003, pp. 352–356.

[15] A. Berny, A.M. Niknejad, and R.G. Meyer, "A 1.8-Ghz LC VCO with 1.3GHz tuning range and digital amplitude calibration," *IEEE Journal of Solid State Circuits*, Vol. 40, No. 4, April 2005, pp. 909–916.

[16] D. Miyashita, H. Ishikuro, S. Kousai, H. Kobayashi, H. Majima, K. Agawa, and M. Hamada, "A phase noise minimization of CMOS VCOs over wide tuning range and large PVT variations," *IEEE Custom Integrated Circuits Conference*, 2005, pp. 583–586.

[17] J.G. Maneatis, "Low-jitter process-independent DLL and PLL based on self-biased techniques," *IEEE Journal of Solid State Circuits*, Vol. 31, No. 11, Nov 2006, pp. 1723–1732.

Quadrature Generation

[18] M.Y. Ghannam, R.P. Mertens, and R.J. Van Overstraeten, "An analytical model for the determination of the transient response of CML and ECL gates," *IEEE Transactions on Electron Devices*, Vol. 37, No. 1, Jan 1990, pp. 191–201.

[19] W. Fang, A. Brunnschweiler, and P. Ashburn, "An analytical maximum toggle frequency expression and its application to optimizing high-speed ECL frequency dividers," *IEEE Journal of Solid State Circuits*, Vol. 25, No. 4, pp. 920–931.

[20] R. Magoon and A. Molnar, "RF local oscillator path for GSM direct conversion transceiver with true 50% duty cycle divide by three and active third harmonic cancellation," *IEEE Radio Frequency Integrated Circuits Symposium* Jun 2002, pp. 23–26.

[21] F. Behbahani, Y. Kishigami, J. Leete, and A.A. Abidi, "CMOS Mixers and Polyphase Filters for Large Image Rejection," *IEEE Journal of Solid State Circuits*, Vol. 36, No. 6, June 2003, pp. 873–887.

Mixer

[22] D.M. Pozar, *Microwave Engineering*, 2nd edition, John Wiley, pp. 379-381.

[23] S.A. Maas, *Microwave Mixers*, Artech House, 1993.

[24] M. Cohn, J.E. Degenford, and B. Newman, "Harmonic mixing with an antiparallel diode-pair," *IEEE Transactions on Microwave Theory and Techniques*, MTT-23, Aug 1975, pp. 667–673.

[25] H.C. Torrey and C.A. Whitmer, *Crystal Rectifiers, MIT Radiation Lab series*, Volume 15, McGraw-Hill, 1948.

[26] R. Feinaugle, H.-W. Hubers, H.P. Roser, and J. Hesler, "On the effect of IF power nulls in Schottky diode harmonic mixers," *IEEE Transactions on Microwave Theory and Techniques*, Vol. 50, No. 1, Jan 2002, pp. 134–142.

[27] L. Sheng, J.C. Jensen, and L. Larson,A wide-bandwidth Si/SiGe HBT direct conversion sub-harmonic mixer/downconverter, *IEEE Journal of Solid State Circuits*, Vol. 35, No. 9, Sept 2000, pp. 1329–1337.

[28] C.D. Hull and R.G. Meyer, "A systematic approach to the analysis of noise in mixers," *IEEE Transactions on Circuits and Systems I: Fundamental Theory and Applications*, Vol. 40, No. 12, Dec 1993, pp. 909–919.

[29] H. Darabi and A.A. Abidi, "Noise in RF-CMOS mixers: A simple physical model," *IEEE Journal of Solid State Circuits*, Vol. 35, No. 1, Jan 2000, pp. 15–25.

IF Filters

[30] A.B. Williams, *Electronic Filter Design Handbook*, McGraw-Hill, 1981.

[31] A.I. Zverev, *Handbook of Filter Synthesis*, John Wiley, 1967.

[32] R. Gregorian and G.C. Temes, *Analog MOS Integrated Circuits for Signal Processing*, Wiley-Interscience, 1986.

[33] Y. Tsividis, "Integrated Continuous-time Filter Design- an overview," *IEEE Journal of Solid State Circuits*, Vol. 29, Mar 1994, pp. 166–176.

[34] Y. Tsividis, "Principles of operation and analysis of switched-capacitor circuits," *Proceedings of IEEE*, Vol. 71, No. 8, Aug 1983, pp. 926–940.

[35] M. Banu and Y. Tsividis, "An Elliptic continuous-time CMOS filter with on-chip automatic tuning," *IEEE Journal of Solid State Circuits*, Vol. SC-20, Dec 1985, pp. 1114–1121.

[36] H. Khorrambadi, M.J. Tarsia, and N.S. Woo, "Baseband filters for IS-95 CDMA receiver applications featuring digital automatic frequency tuning," *IEEE International Solid State Circuits Conference*, Sep 1996, pp. 172–173.

[37] F. Krummenacher and N. Joehl, "A 4-Mhz CMOS continuous-time filter with on-chip automatic tuning," *IEEE Journal of Solid State Circuits*, Vol. 23, June 1988, pp. 750–758.

[38] H. Voorman and H. Veenstra, "Tunable high-frequency Gm-C filters," *IEEE Journal of Solid State Circuits*, Vol. 35, No. 8, Aug 2000.

[39] M. Koyama, T. Arai, and Y. Yoshida, "A 2.5V active Low-pass Filter Using All-n-p-n Gilbert Cells with a 1Vpp linear inout range," *IEEE Journal of Solid State Circuits*, Vol. 28, No. 12, Dec 1993.

[40] G. Brenna, D. Tschopp, J. Rogin, I. Kouchev, and Q. Huang, "A 2-Ghz carrier leakage calibrated direct-conversion WCDMA transmitter in 0.13-μm CMOS," *IEEE Journal of Solid State Circuits*, Vol. 39, No. 8, Aug 2004.

CHAPTER 5

Examples of Integrated Communication Microsystems

INTRODUCTION

In this chapter, we illustrate two examples of wireless and wired communication front ends. The wireless example is a direct conversion receiver front end at 5–6 GHz using a SiGe BiCMOS process, and wired example is a 10-Gbps front end using 180-nm CMOS technology. To achieve higher data rates, higher center frequency is used. Both of these examples are directed toward the high-frequency front-end circuits. In the back end, digital gates can be suitably integrated to realize a system-on-chip solution. In terms of integration, system-on-chip is often preferred for cellular and consumer electronics applications because of compaction and ease of finding a single technology to fabricate. Such considerations, however, are heavily dominated by the high digital content of these chips, which improve with technology scaling. Applications demanding very high performance often use hybrid solutions consisting of multiple semiconductor technologies resulting from superior isolation.

5.1 DIRECT CONVERSION RECEIVER FRONT END

Two approaches can be taken in developing the front end. A 50-Ω-based system would lead to deriving separate system level specifications, of low noise amplifiers (LNAs) and mixers, whereas an integrated system can use a non-50-Ω-based approach for power reduction. A 50-Ω based approach simplifies the system-level calculations to

Advanced Integrated Communication Microsystems, edited by Joy Laskar, Sudipto Chakraborty, Manos Tentzeris, Franklin Bien, and Anh-Vu Pham

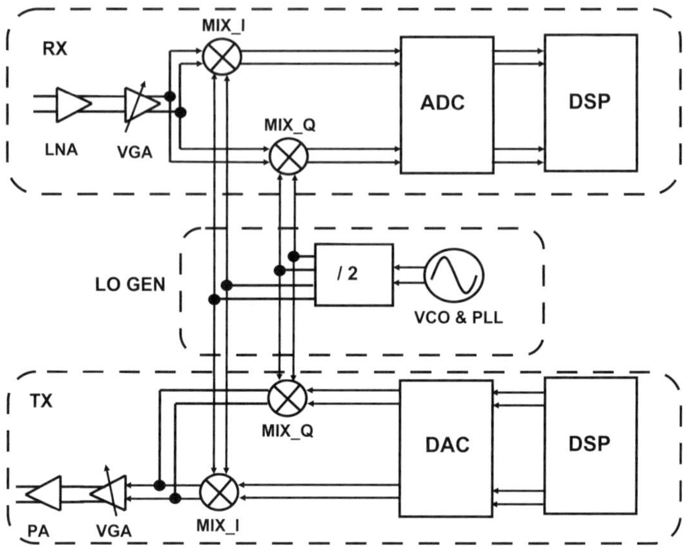

Figure 5.1. A transceiver block diagram.

derive the specifications of building blocks. A non-50-Ω-based system approach, progressively higher impedance is driven to reduce current consumption. Elimination of impedance matching network at individual functional block interfaces lead to area compaction as well. Placement of the filter before the LNA (non-50-Ω architecture) provides a better system design for rejecting out-of-band interferers and facilitates comparison of both system architectures. However, the placement of the filter in-between the LNA and mixer (50-Ω-based system) helps to lower the noise figure of the entire receiver and to reject out-of-band blockers.

A differential configuration in the front-end mixers minimizes the effect of common-mode noise and provides high IIP2 needed for direct conversion receivers. A 2× frequency scheme was adopted in this design for the ease of design for the frequency divider and for the relatively high accuracy of the quadrature signals in the signal generation path. Thus, the frequency divider accepts differential signals at 11.6 GHz, while providing quadrature phases at 5.8 GHz at its output, to be interfaced with the downconverting mixers. Figure 5.1 illustrates a direct conversion front-end architecture based on a 2× frequency scheme.

5.1.1 Circuit Design

The design platform under consideration is a second-generation SiGe BiCMOS technology consisting of five metal layers with an additional thick metal layer for high Q inductor realization. The substrate resistivity is 10–20 Ω-cm, and bipolar transistors with emitter widths of 0.32 μm, 0.44 μm, and 0.8 μm are available. The process uses a high-performance graded band-gap base NPN with f_T and $f_{\max}$ of 47 and 60 GHz, respectively, for standard NPN bipolar junction transistors (BJTs), and with f_T of 27 GHz for high-breakdown NPN BJTs. It uses breakdown voltages of 3 V and 5 V, respectively, for standard and high-breakdown voltage devices. The high f_T transistors use a pedestal structure, which prevents spreading of the currents in different directions as opposed to the conventional BJT structure, thus improving the f_T. The NMOS and PMOS transistors have gate lengths of 0.24 μm. The top copper metal (called the analog metal, AM), is 4 μm thick, and it is used for inductors and low-loss interconnects. Isolation mechanisms include deep trench (DT) and shallow trench (ST).

When selecting the technology modules for circuit design, several trade-offs exist. The minimum width devices exhibit lowest noise contribution because of small base spreading resistance and poor device matching properties resulting from susceptibility to process variations. However, lowest width devices exhibit higher peak f_T compared with the other geometry variants. The choice of passive components (such as resistors and capacitors) is dependent on their parasitic component, temperature coefficient, and tolerance limits. These considerations result in an appropriate choice of components from a semiconductor platform, while building circuits.

5.1.1.1 LNA Design. Cascode topology is used in the design of LNA. At the input, the *L–C* network with high *Q* provides voltage gain and reduces noise. A degeneration inductor is used to provide simultaneous noise and power match at the input. At the output, the *L–C* network at resonance provides the desired output matching. Isolation is a critical aspect of the direct conversion receiver, and the cascode proves adequate for such purposes. The input and output matching is implemented using fully monolithic inductor implementation.

As the receiver demands to operate in the limits of small and large signals, gain switching is essential. Gain switching functionality has been realized by an NMOS transistor pair, M1 and M2. When the control voltage is "LOW" or 0 V, both M1 and M2 are turned OFF, and the LNA functionality is governed by the bipolar transistors Q1 and Q2. This is referred to as the "high-gain" mode operation of the LNA. The low-gain mode occurs when the control voltage is at a "HIGH" state, typically raised to the supply voltage of the circuit. In this case, both transistors M1 and M2 are turned ON. M2 provides a bypassing path to the input RF signal, and M1 provides a short between input and output terminals. This ensures that very little current flow to the base of the transistor, Q2, reduces the gain significantly. Figure 5.2 illustrates the circuit design.

This particular gain switching approach disturbs the input match in the low-gain mode. The gain step is about 20–25 dB. However, if it is desirable that the input match and the radio frequency (RF) characteristics (such as broadband noise and linearity) remain constant, the approaches shown in Figure 5.3 prove to be better alternatives than that is shown in Figure 5.2. Figure 5.3(a) provides a scheme where a current steering approach has been adopted, whereas Figure 5.3(b) provides a scheme where input and output match is kept constant while using the middle stage as a gain control stage (or bypassing stage).

In practical receivers, the control voltage is determined by a digital control from the baseband digital signal processing (DSP), after the signal strength has been detected by the receiver signal strength indicator (RSSI). A proportional-to-absolute temperature (PTAT) current source is desirable to obtain a constant transconductance ($g_m = \frac{I_c}{V_t}$) over the temperature range.

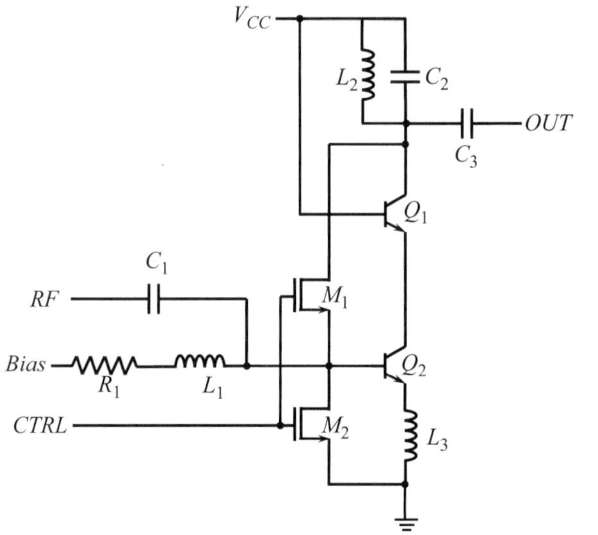

Figure 5.2. A single-ended bipolar LNA structure.

5.1.1.2 Mixer Design. Mixer designs are based on Gilbert-cell topology and its modification. Essentially the topology consists of a g_m core, switch, current source, and the load networks. Switch transistors use a bipolar device because of their low parasitic loading to the local oscillator (LO) network, smaller differential LO swing to "hard-switch" the transistor pairs, and lower 1/f noise contribution. As the LNA provides a single-ended output, the mixer input in the 50-Ω-based system should be single ended. The input g_m core should be a single-ended voltage input to differential output current, to be switched by the switching networks. The current source should

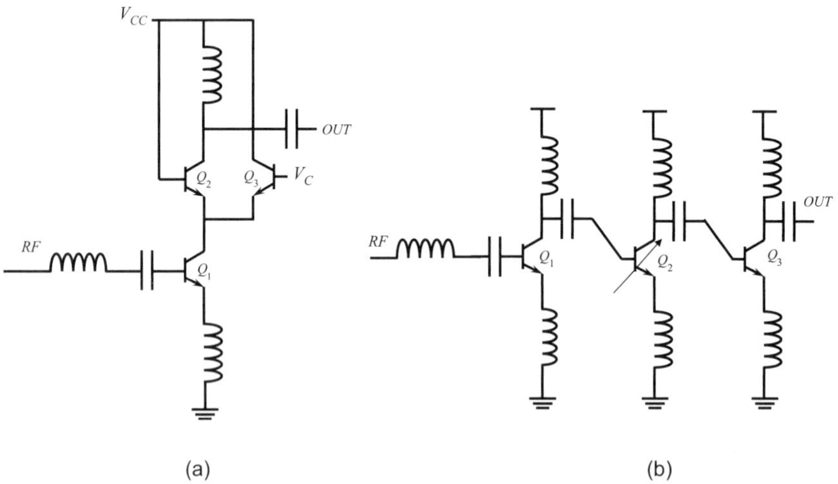

Figure 5.3. An interstage gain step bipolar LNA structure.

provide high output impedance at DC and even harmonics of the input frequency. Low output parasitic capacitance is also an essential requirement of these sources. Bipolar transistors consume lower headroom compared with MOS devices and are suitable in this regard.

The bipolar transconductor provides lower linearity (because of the "tanh" transfer characteristics, which is limited up to 78 mV of differential swing), compared with a MOS differential pair (square law characteristics, limits set by the overdrive voltage). However, use of a MOS differential pair as a high-impedance stage is frustrated at high frequency by the presence of large input device capacitance. To achieve similar transconductance performance with a given bias current, MOS usually consumes a larger area compared with bipolar devices, which leads to higher parasitics. Thus, use of bipolars at current densities slightly lower than that to achieve peak (to accommodate variations) f_T is usually preferred for higher transconductance, and a degeneration inductor can be used to provide higher linearity, at the expense of reduced transconductance caused by the presence of negative feedback. Inductors do not consume voltage headroom and do not generate noise. In a downconversion mixer, the thermal noise of the transconductor stage becomes an important design criterion. It must be observed that the degeneration inductor does not resonate with the base-emitter capacitance of input devices (gate-source in the case of MOS). The switching stage could be considered using hard-switched current steering logic, driven by the switching waveform (local oscillator), and high output impedance of the transconductor and low input impedance of the switches lead to an efficient current transfer stage. Double balanced mixer topology leads to LO feedthrough cancellation and to other even-order terms.

Let us now focus on the practical implementations of two different variants of Gilbert-cell mixers. The first one is a micromixer as shown in Figure 5.4 [1] using a class AB input stage using quasi-symmetric bipolar transistors (Q5 and Q10). The transistors Q5, Q6, Q9, and Q10 form a translinear loop, and the product of collector currents of Q5 and Q10 is equal to that from Q6 and Q9. The switching core is similar to a traditional Gilbert-cell mixer. The input impedance of the circuit is formed by the parallel combination of the impedances observed in the two paths formed by (1) the combined series impedance of L3, diode connected Q10, and R3; and (2) the combined series impedance of R1, L2 and emitter resistance of Q5. At high frequencies, it is important to maintain the balance between the two paths.

5.1.1.2.1 Micromixer. Micromixers exhibit several interesting properties. The first one is the large signal operation of the class—AB stage. Each of the transistors Q5 and Q10 exhibits exponential nonlinearity, but the difference of the currents (which is relevant for the switching operation) remains linear [1]. Thus, even if the individual branch currents exhibit nonlinearity, the differential operation remains linear, which leads to very high linearity of such mixers at low bias currents. Another interesting property is their low input impedance caused by high transconductance of bipolar transistors. The level of input impedance also varies with the input signal, which leads to gain expansion at higher input signal swing as illustrated in [1]. The imbalance in the

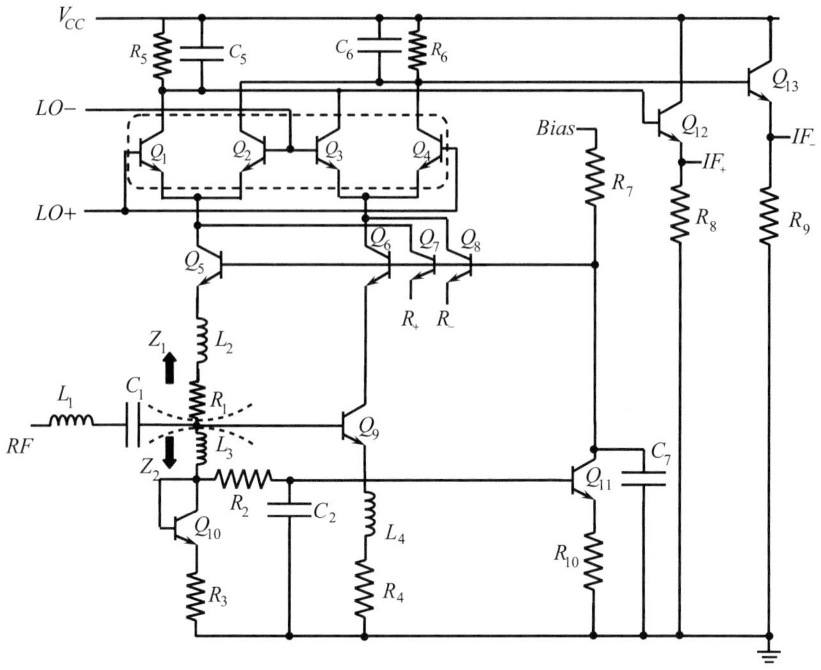

Figure 5.4. A single-ended bipolar micromixer structure.

quasi-symmetric stages is adjusted by a series combination of the passive elements. However, the diode-connected transistor at the input contributes significant noise, and thus, micromixers remain attractive for a 50-Ω impedance standard, and high linearity, while contributing higher noise.

Transistors Q12 and Q13 form emitter-coupled buffers to isolate the mixer from the testing equipment. Capacitors C2 and C7 have been used as noise bypassing capacitors to minimize noise contributions from the bias circuit. Transistors Q7 and Q8, along with pull-down resistors, are used as offset nulling elements to cancel static DC-offset effects.

5.1.1.2.2 Single-Ended Gilbert-Cell. The second variant is a high impedance inductively degenerated Gilbert-cell mixer (shown in Figure 5.5) used for a non-50-Ω-based architecture, and the similar descriptions would apply. The operating frequency of the degeneration inductors should be less than (1) the self-resonating frequency (SRF) of the inductors (L1 and L2) and (2) the resonating frequency of the network formed by the degeneration inductor (L1) and the base-to-emitter capacitance of the bipolar transistor (Q5) at the input stage. The resonance is also made of fairly low Q, such that it does not cause stability problems.

5.1.1.2.3 Load Network. The load network should use highly linear passive elements with a low-voltage coefficient. The key aspects of these load networks include (1) headroom consumed, (2) matching between the components to minimize

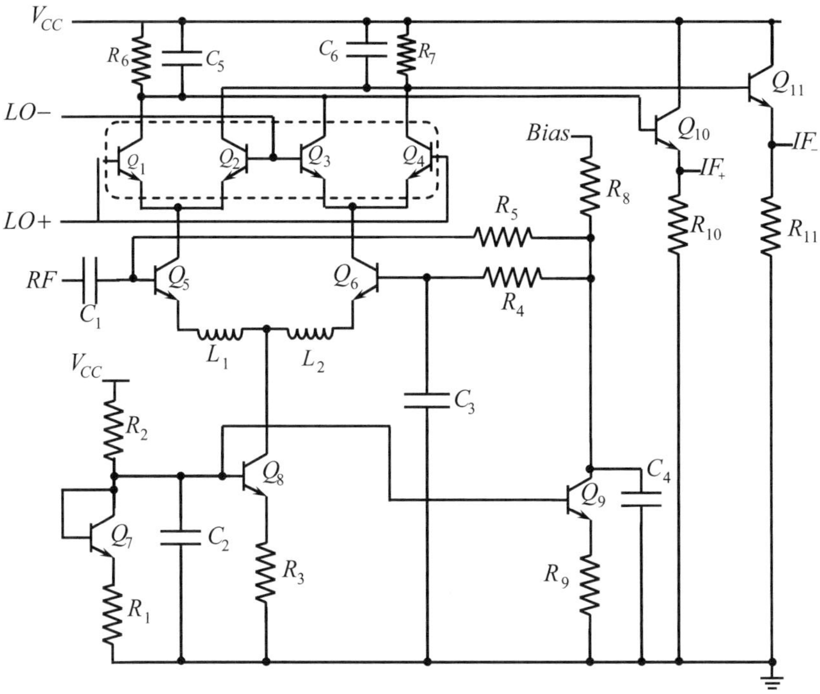

Figure 5.5. A single-ended bipolar Gilbert-cell mixer structure.

DC offset (static), (3) noise contribution, and (4) bandwidth. The signal swing can get higher at the transmitter and the receiver, which leads to significant nonlinearity and shift in the filter poles when the voltage dependence of such components is considerably higher. The parasitic component of the load network would lead to bandwidth shrinkage and gain degradation.

5.1.1.3 Signal Generation Path. In the signal generation path, a 2× frequency scheme has been adopted because of the considerations illustrated earlier. Two emitter-coupled D flip-flop circuits have been used to realize the master–slave configuration needed for a static frequency divider, which operates at positive and negative clock edges.

The scheme is shown in Figure 5.6. It starts with the generation of differential signals from a single-ended input at 11.6 GHz, followed by the conversion to suitable

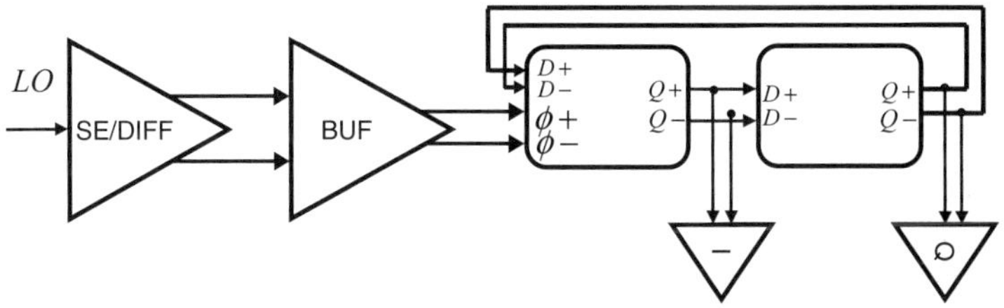

Figure 5.6. A signal path of the direct conversion receiver.

voltage levels by appropriate buffers. The D flip-flop core generates quadrature signals, which in turn is amplified by the buffers to provide a necessary LO swing to the mixer switching core. The accuracy of such an implementation is strongly dependent on the balance in the differential signal path, and any imbalance at the input of the divider network leads to the imbalance of the outputs in quadrature. The quasi-symmetric micromixer stage can be used here to achieve the desired matching for a 50-Ω interface. The load of these dividers should be optimized to provide lower parasitic capacitances, thereby enhancing signal bandwidth.

5.1.2 The Integration: Interfaces and Layout

Having illustrated the details about the operation of the building blocks, we now focus on the integration of these circuits. From a circuit interface point of view, the frequency divider, along with the buffers, should be interfaced with the LO terminal of the mixers and the chain should be optimized for the desired signal swing at the LO terminals needed for switching. A resistive load in the LO buffer might suffice in this regard because of the lower parasitic capacitances of the bipolar transistors. In the signal path, the voltage gain peak must be observed at the LNA–mixer interface, and the output impedance of the LNA should be combined with the input impedance of the mixer for area optimization. The integrated blocks should be designed for gain, noise, time domain behavior, frequency response, linearity, and robustness to process and temperature variations and so on.

Although the circuit design considerations are somewhat straightforward and intuitive to explain, tremendous challenges exist in the layout of the integrated blocks, which usually come from experience, and mistakes performed in the early test chips. For any high-frequency silicon design, the layout plays a very important role in the circuit performance because of the lossy silicon substrate. There are many generic design guidelines toward mixed-signal layout, but in this section, we will consider only those, which are pertinent toward the chosen application. Since the circuit integrates both sensitive analog and digital building blocks, care should be taken to separate the two power supplies, thus minimizing the effect of any common-mode noise from the power supplies. Also, the digital circuitry (a frequency divider in this case) should be surrounded by deep trench and substrate contacts to minimize the noise generated by digital switching on the other analog circuit blocks of the receiver. Substrate contacts (also referred to as "taps") also help in isolating one circuit block from another and fully absorb the ground current in the substrate. An input trace to the LNA should be kept small to minimize any noise figure degradation. Usually the top metal is used for interconnection and routing because of lower parasitic capacitances. Separation and orientation of inductors are important factors to minimize the flux coupling effect. Shielding of the input signals from the substrate is also very important. This is performed by having the input signals in the topmost metal layer with the ground as stacked up M1 and M2 running together surrounding the chip and underneath the pads.

Layout symmetry is extremely critical in receiver front ends, and any asymmetry in this regard destroys the fully differential nature of the front end, which leads to unwanted components. I/Q imbalance associated with the mixers is very sensitive to

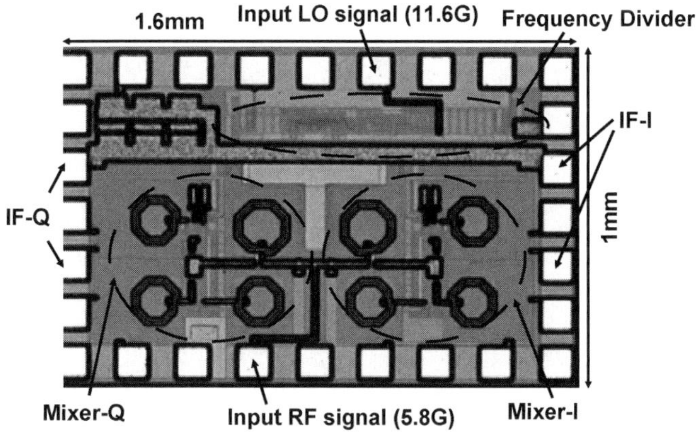

Figure 5.7. Die photograph of a micromixer implementation.

the layout symmetry. Hence, in the layout, equal trace length must be respected from the RF input to the mixers. This is even more critical in the D flip-flop layout in terms of the clock distribution network. If the transistors are not identical and the layout traces are not equal, it results in deviation of the duty cycle from 50%, directly increasing the even order intermodulation distortion terms.

Sometimes shielding of the layout traces becomes very important. If the metal line connecting the base goes underneath the metal line connecting collector and vice versa, it gives rise to additional Miller capacitance, which will degrade the high-frequency performance. Hence, another metal layer can be used as shielding the two traces. DC bias lines may lead to unwanted inductive effect, which may affect the RF performance.

Large DC blocking capacitors should be provided between the power supply and the ground to minimize the effect of noise from the supply voltages (supply voltage variation) as well as to prevent any kind of ringing effect from DC supplies. The building blocks for both the receiver architectures were laid out for on-wafer probing, and all the inputs at high frequency were assumed to be single ended to minimize the number of pads, which leads to area compaction. Figures 5.7 and 5.8 show the microphotographs of the fabricated micromixer and integrated receivers.

5.1.3 Compensation and Corrections

The above description of direct conversion radios addresses only its open-loop operation mostly in the signal path. Although such description is simplistic to appreciate the fundamental operations from the circuits and device perspectives, it is far from being complete from a system viewpoint. A major part of any transceiver system (and more so in the case of the direct conversion radio) is embedded in its baseband controls and signal processing functionalities, offset cancellation, and tuning circuitry for robustness of operation. Any transceiver continuously communicates with its baseband controller during its operation, sometimes in an adaptive way. This is essential for DC offset cancellation and for any offset corrections.

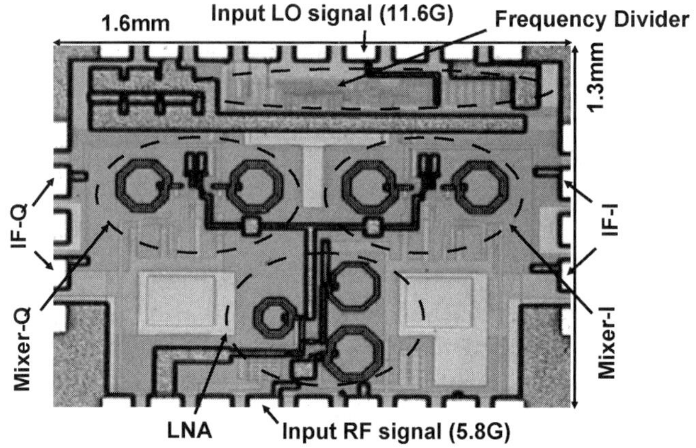

Figure 5.8. Die photograph of an integrated receiver implementation.

Fundamentally, the resultant effect of all mismatches leads to unwanted offset voltage at the baseband, which is stored on a capacitor, and by controlling the polarity of the capacitors, the offset is subtracted from the incoming signal. There could be other examples, such as comparison of circuit performances with a reference design to detect the process corner and to compensate the circuit elements according to the process variations. Even if the principles are simple, such methods enhance the robustness of any successful prototypes. Although compensation schemes could be fundamentally feasible at RF, robust implementations and parasitic considerations at high frequency hinder their applicability in successful prototypes.

Although static offset corrections are easier to perform (at the expense of more area on chip), and usually require power up calibration, dynamic corrections are extremely difficult to perform. For example, to compensate for dynamic DC offsets (whose levels can vary over several orders of magnitude depending on the surrounding environment), one can employ averaging techniques, or other second-order techniques, depending on the allowable time window in the communication protocol. It is inefficient to provide a filtering function, as it would degrade the signal-to-noise ratio (SNR) by considerable amount, consume area (large capacitors would be needed), and result in a long settling time.

5.2 DEBUGGING: A PRACTICAL SCENARIO

The measurement result on I/Q imbalance in the integrated receiver was higher than the simulation, and it was found to be consistent among different chips that were tested. Since the layout had been done symmetrically, there could be two possible explanations to this effect.

One reason for I/Q imbalance is the imbalance in the quadrature signals generated from the frequency divider. As explained, the input stage for conversion from single-ended to differential signals is amenable to deviate from its optimization point, while fabricated. This contributes to some extent, as some amount of I/Q imbalance exists in

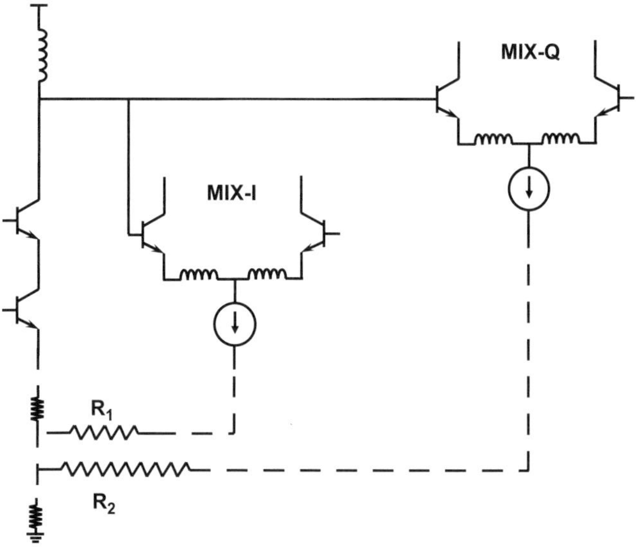

Figure 5.9. Distributed substrate resistance and unequal feedback caused in a global grounding scheme (only AC-equivalent path shown).

the micromixer measurement results (0.25 dB amplitude imbalance, 2.4° phase imbalance). As the same circuit was used for the frequency divider in both cases, it would give rise to some I/Q imbalance.

In the integrated receiver layout, the LNA ground and in-phase mixer (Mixer-I) grounds were connected together and then connected to the overall chip ground. However, this scenario was not present in the case of the quadrature mixer (Mixer-Q), and the LNA ground was not connected to the Mixer-Q ground. Hence, Mixer-Q ground is farther away from the LNA ground, compared with Mixer-I ground. This causes an unequal feedback mechanism between the LNA and two mixers in quadrature.

The feedback mechanism takes place through the finite resistance of the ground connection, as shown in Figure 5.9. The resistors R1 and R2 come from the on-chip ground connection. They cause unequal positive feedbacks to the input, and since R1 < R2, the I channel amplitude is higher than the Q channel. To solve this problem, LNA and mixer grounds should be separately connected to the common circuit ground, and the interconnections of LNA and mixer ground in any other form should be avoided.

5.3 HIGH-SPEED WIRED COMMUNICATION EXAMPLE

5.3.1 Bandlimited Channel

The backplane channel loss characteristics are frequency dependent. Specifically, high-frequency components of the input signal experience larger loss than the

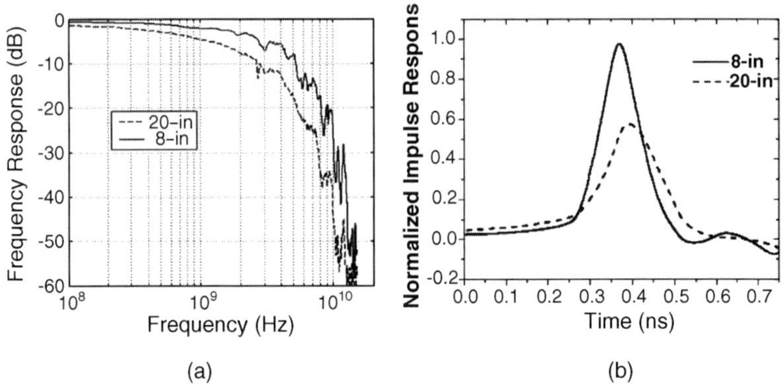

Figure 5.10. Forward transmission of 4, 20-in FR4 backplane traces: (a) frequency responses and (b) impulse responses.

low-frequency components around DC. This high-frequency loss becomes worse in the longer backplane channel environment, as shown in Figure 5.10.

Figure 5.10(a) shows that a 20-in FR-4 backplane has much larger attenuation or loss compared with an 4-in FR-4 backplane. The resulting impulse response of the 16-in FR4 backplane has more DC signal power loss and more widened pulse shape compared with a 4-in channel, as shown in Figure 5.10(b).

Meanwhile, Figure 5.11(a) shows the forward transmission characteristics of the multimode fiber (MMF) with different lengths (100 m and 500 m). As expected, the 500-m MMF has more channel loss compared with the 100-m MMF. Figure 5.11(b) shows the impulse response of the 500-m MMF channel. The second major mode is followed by the first major mode by around 100 ps, which is the symbol space for 10-Gb/s data rate.

When the signal propagates through the dispersive bandlimited channel with the impulse responses shown for both backplane and MMF in Figure 5.10(b) and Figure 5.11(b), respectively, its output signal power spreads in time. This spreading

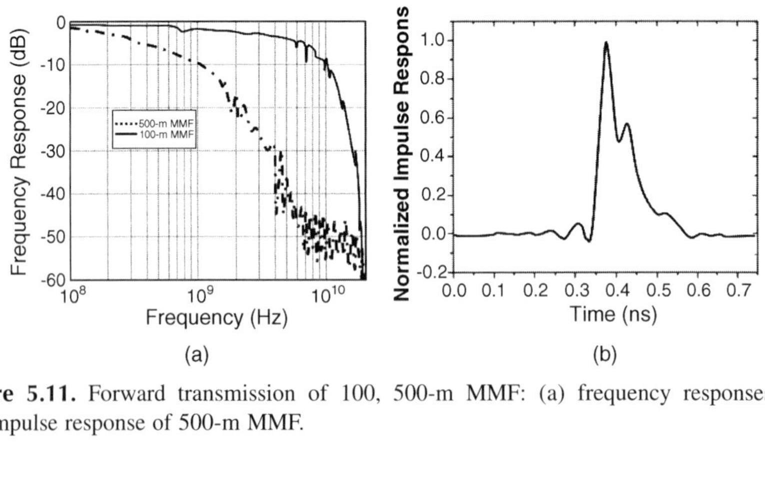

Figure 5.11. Forward transmission of 100, 500-m MMF: (a) frequency responses and (b) impulse response of 500-m MMF.

of signal power causes the inter symbol interference (ISI). In other words, transmitting a square pulse through such a dispersive channel results in a widening and flattening of the pulse at the far end. This implies that each data bit of information overlaps with its adjacent bits. This overlap can cause major distortions of the signal. At high data rates and in long channels, the ISI can be so severe that it becomes impossible to recover the original transmitted data. This is a major phenomenon limiting data transmission, and it must be addressed in bandlimited channels for a 10-Gb/s data transmission.

Furthermore, the difference in width of the dispersive channel from the impulse response shown in Figure 5.10(b) and Figure 5.11(b) suggest that different tap-spacing is needed for optimal equalization. These impulse responses of the dispersive channel also support the need for reconfigurable equalization for effective channel compensation.

5.3.2 Design Example

5.3.2.1 Feed-Forward Equalizer (FFE). Receiver equalization can be implemented with analog finite impulse response (FIR) filter structures, which can be classified as zero forcing-linear equalizer (ZF-LE) or minimum mean-squared error-linear equalizer (MMSE-LE) depending on the tap coefficient calculation principles. As the transfer function of ZF-LE is the reciprocal of the channel transfer function as is illustrated in Figure 5.12, it can remove ISI completely but does so neglecting the impact of high-frequency cross-talk (Xtalk) noise, which is also amplified. In contrast, MMSE-LE can ameliorate this noise enhancement problem, since its tap coefficients are calculated to minimize overall signal degradation from both ISI and Xtalk noise. A frequency response relationship among bandlimited channel, ZF-LE, and MMSE-LE is illustrated in Figure 5.12. Optimum tap coefficients for various bandlimited

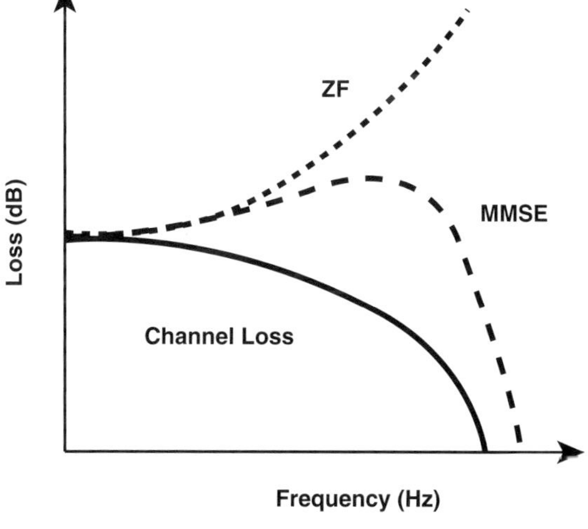

Figure 5.12. Frequency response of bandlimited channel, ZF-LE, and MMSE-LE.

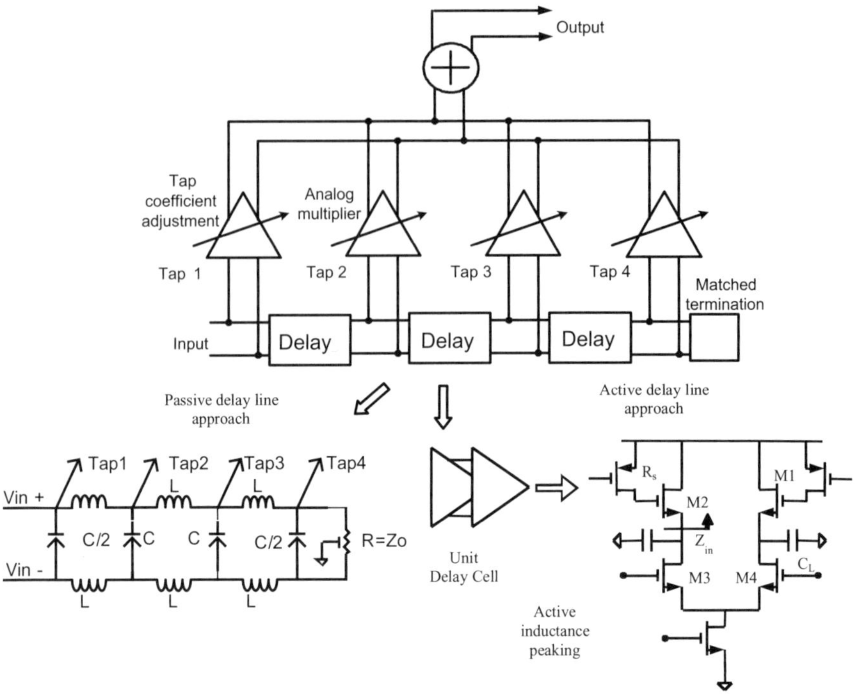

Figure 5.13. The FFE configuration with passive LC ladder delay line and active inductance peaking delay line approaches.

channels are extracted with the use of impulse response information illustrated in Figure 5.10(b) and Figure 5.11(b).

The overall circuit is implemented using the continuous-time analog signal processing technique. The discrete-time equalization requires clock phase information from the incoming data and the sampling circuitry, resulting in limited bandwidth characteristics. Thus, the continuous-time analog signal processing technique makes it easier to operate in high data rate compared with the discrete-time-based equalizer. The overall block diagram for the equalizer is shown in Figure 5.13.

The continuous-time delay line in the FFE is implemented via two different implementation approaches. The passive delay line-based equalizer and the active delay line-based equalizer are shown in Figure 5.13 where the artificial transmission line is used as the delay element for the passive delay line and the active peaking load is used for the active approach. Preliminary results achieved from each approach are discussed in the following sections.

5.3.2.2 FFE with the Passive Delay Line Approach. In this equalizer design, the distributed passive network is used for the continuous-time analog signal delay as illustrated in Figure 5.13. For the matched termination to the characteristic impedance of the artificial transmission line, a half-section for the termination is used as shown in

Figure 5.13. By using the half-section termination, the bandwidth for the artificial transmission line is increased further.

The artificial transmission line constructed by LC ladder is designed with 50-Ω differential characteristic impedance. This is achieved by appropriate choice of inductance L_t and capacitance C_t as shown in the following equation, where L_t and C_t are the inductance and capacitance of each LC section. The characteristic impedance Z_o and the time delay of each segment T_{delay} is defined as follows:

$$Z_0 = \sqrt{\frac{L_t}{C_t}} \quad T_{delay} = \sqrt{L_t \cdot C_t} \tag{5.1}$$

The lumped element analog LC delay line can work as a transmission line below the cutoff frequency; however, over this frequency, the input impedance of the lumped LC line will be eventually purely reactive. The input impedance of the lumped LC delay line can be presented as Eq. (5.2):

$$Z_{in} = \frac{jwL}{2}\left[1 \pm \sqrt{1 - \frac{4}{w^2 LC}}\right] \tag{5.2}$$

So the cutoff frequency, where the input impedance is purely reactive, is

$$w_{cutoff} = \frac{2}{\sqrt{LC}} \tag{5.3}$$

From the simple calculation, we can verify that the lumped LC line has enough bandwidth to work as a delay line for a given equalizer structure.

A 1.5-nH inductor is designed using an analysis and simulation of inductors and transformers in integrated circuits, also known as the ASITIC simulator, and is implemented by optimizing the line space and number of turns to increase the self-resonance frequency of the inductor via enhancing the Q factor.

Both the simulation and the measured result for the LC ladder structure is shown in Figure 5.14(a) and Figure 5.14(b), respectively. From the figure, the zero crossing point for each adjacent signal is delayed by a fixed 33-ps tap-spacing as designed.

The performance of the FFE with a passive delay line is demonstrated with an equalization measurement over a 500-m MMF channel. Figure 5.15(a) shows the received 10-Gb/s NRZ signal eye diagram without equalization. The signal is severely impaired by ISI of the 500-m MMF, and it cannot be mitigated with additional optical power. Figure 5.15(b) shows the same eye diagram after the 4-tap FFE. Most of the ISI is removed, where a 10-Gb/s data can be retrieved.

Figure 5.16 illustrates the chip microphotograph of the FFE with a passive delay line. As it is clear from the figure, CMOS on-chip inductors occupy a large die area. The overall chip size is 1.09 mm by 1 mm including pad area. Total power dissipation for the passive delay line-based FFE is about 9 mW.

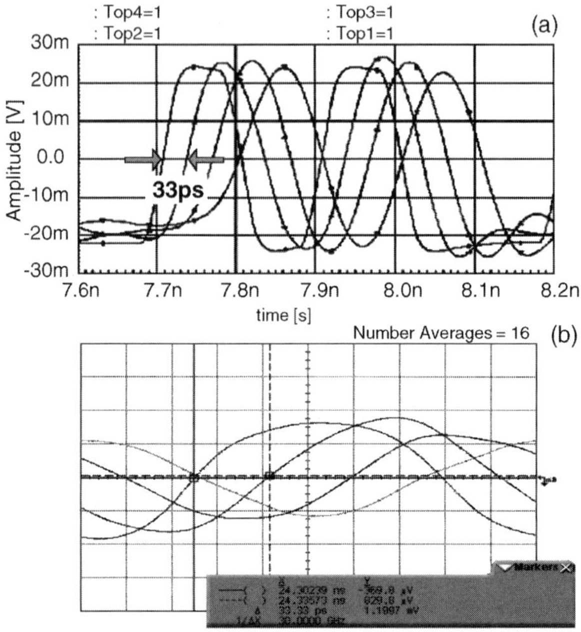

Figure 5.14. Passive delay line performance shows good agreement between (a) simulation and (b) measured performance for the targeted fixed-Ts/3 tap-spacing.

5.3.2.3 Reconfigurable Equalizer System Overview. An equalization technique compensates the frequency-dependent channel loss characteristics. The bandlimited channel has a low-pass frequency response, as shown in Figure 5.17. The larger loss in high-frequency range obstructs fast data transitions of high-speed data. This slow data transition causes the signal power to smear into the neighboring symbols. The equalization technique restores the high-frequency component of the original transmitted signal. Figure 5.18 shows the conceptual illustration of three backplane channels and the corresponding optimal equalizer responses. With more loss across the channel, more frequency boost is required in the equalizer.

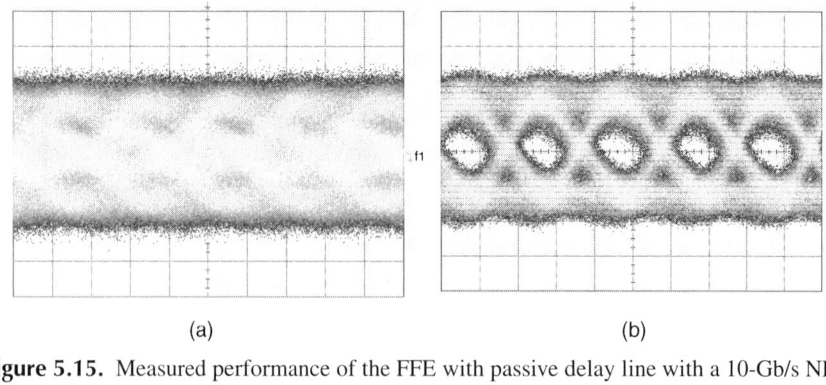

Figure 5.15. Measured performance of the FFE with passive delay line with a 10-Gb/s NRZ data input; (a) before and (b) after the equalization. A 50-ps/div is displayed.

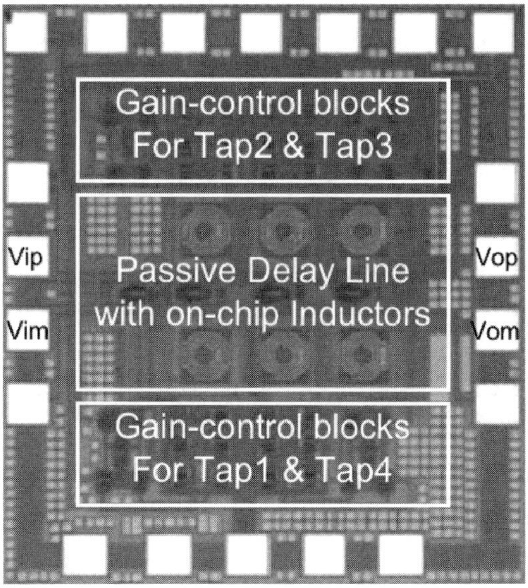

Figure 5.16. Microphotograph of the FFE with the passive delay line approach.

To make the system adjustable to various channel environments, a reconfigurable FFE with FIR structure is integrated at the receiver side as illustrated in Figure 5.19. The FIR filter consists of variable gain amplifiers (VGAs) for tap gains and tunable tap delay lines. Based on the measured channel response, system simulation is performed to derive optimal equalization requirements, including the optimal number of taps, tap delay value, and tap coefficients. To ameliorate the noise enhancement problem, the minimum-mean-squared error algorithm is used for calculation of the tap coefficients.

Figure 5.20 illustrates the simulated eye-diagram of a 10 Gb/s signal before and after equalization over 8-in and 20-in backplanes, respectively. The optimal equalizer system requirement varies for different channel configurations. An optimal requirement for different backplane configurations is summarized in Figure 5.21, demonstrating the need for a reconfigurable equalization technique to adjust over various backplane signaling environments.

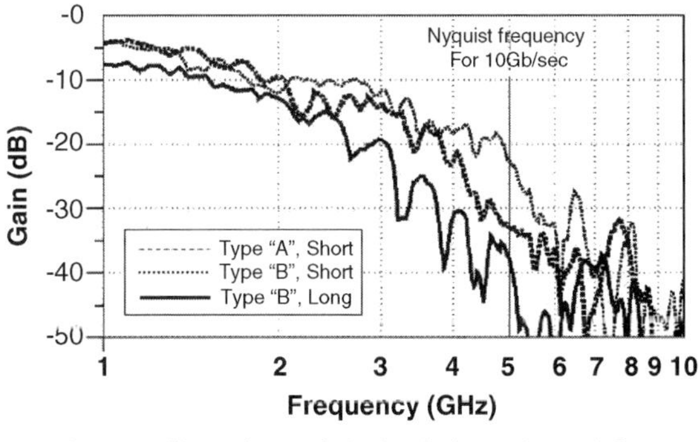

Figure 5.17. Various configurations of the backplane channel frequency response.

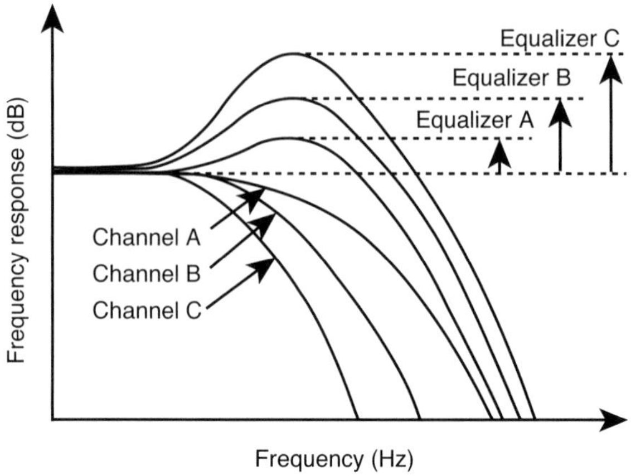

Figure 5.18. Frequency responses of three example channels and the corresponding optimal equalizer responses.

5.3.2.4 FFE with Active Delay Line. The equalizer introduced in Section (5.2.2) uses a passive LC ladder emulating a transmission line to ensure the broad bandwidth for the delay elements. However, the delay value of the passive delay line cannot be tuned once fabricated and would not be able to reconfigure for optimum equalization for various bandlimited channel configurations.

To verify the feasibility of reconfigurable equalization, the first step is to demonstrate the possibility of an active delay line at a 10-Gb/s data rate as it is more challenging to satisfy the bandwidth requirements for 10-Gb/s signal compared with its passive counterpart. To meet the broad bandwidth requirements for the active delay line, active inductance peaking is considered with a simple differential pair as illustrated in Figure 5.13. The overall voltage gain for differential pair in unit delay cell is

$$A_v = g_{m3} \cdot (Z_{in} // C_L) \tag{5.4}$$

$$= \frac{sC_{gs2}R_s g_{m3} + g_{m3}}{C_{gs2}C_L R_s s^2 + (C_{gs2} + C_L)s + g_{m2}} \tag{5.5}$$

where $C_L = C_{gd3}(1 + A_v) + C_{db3}$ and $Z_{in} = \frac{sC_{gs2} \cdot R_s + 1}{sC_{gs2} + g_{m2}}$

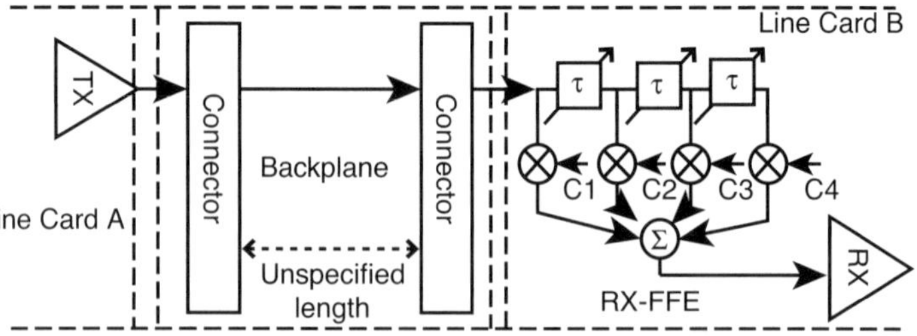

Figure 5.19. Functional block diagram of the proposed receiver-side reconfigurable equalizer system architecture.

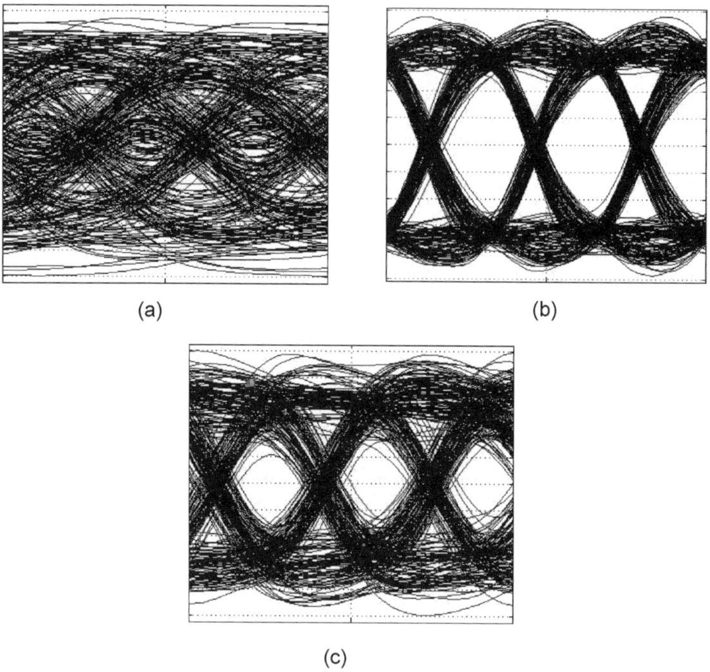

Figure 5.20. System simulation results (a) before equalization, (b) equalization over 8-in backplane, and (c) 20-in backplane.

The two poles are at

$$-\frac{1}{2}\left(\frac{C_{gs2}+C_L}{C_{gs2}C_LR_s}\right)\pm\frac{1}{2}\sqrt{\left(\frac{1}{C_LR_s}+\frac{1}{C_{gs2}R_s}\right)^2-\frac{4g_{m2}}{C_{gs2}C_LR_s}} \tag{5.6}$$

and the zero is at $\frac{1}{R_sC_{gs2}}$.

By varying the R_s (R_s is the turn-on resistance of M1 in Figure 5.13), the zero location can be controlled. In other words, the bandwidth can also be controlled. The peaking adjacent to the dominant pole enhances the 3-dB bandwidth of the delay line by approximately 3.9 GHz compared with the delay line with passive resistor load.

Backplane type	# of taps	Tap delay
"A", short (8-in)	2	50ps
"A", long (20-in)	4	33ps
"B", short (8-in)	3	25ps
"B", long (20-in)	4	33ps

Figure 5.21. Optimal equalizer system requirements for various channel configurations.

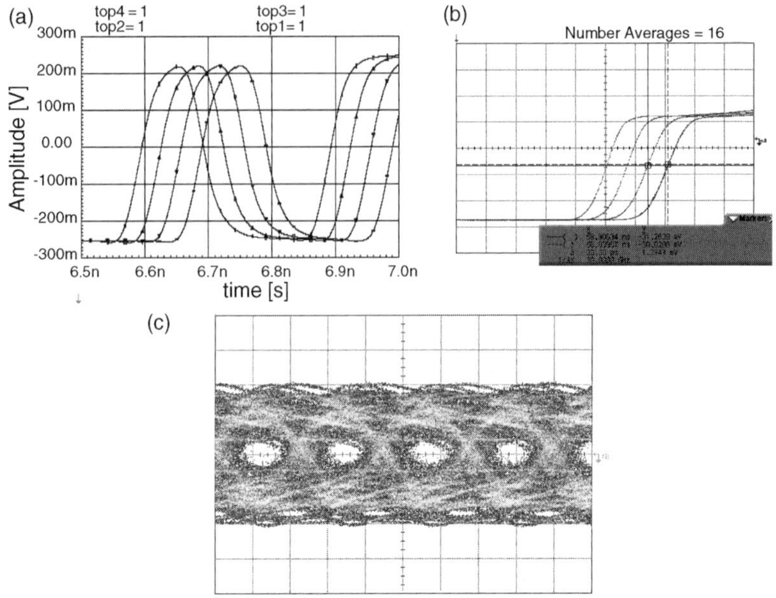

Figure 5.22. Active delay line performance shows good agreement between (a) simulation and (b) measured performance for the targeted Ts/3 tap-spacing.

The simulation includes all the parasitic capacitance from the multiplier cell and the adjacent delay cell.

For the initial active delay line implementation, which will be the foundation for a tunable delay line, two cascaded NMOS differential pairs are used generating a 33-ps ($T_s/3$) delay per unit delay cell. The effective inductance value, which is proportional to $R_s/(1/g_{m2})$, is optimized for 10-Gb/s NRZ signal transmissions.

Figure 5.22(a) and (b) shows the simulation and the measured result for the designed active delay line. The measurement result shows accurate 33-ps tap delay spacing through the active delay line cells. No apparent slewing occurred on the measurement result demonstrating enough bandwidth for high-speed data throughput.

Figure 5.23 shows the eye diagram of a 10-Gb/s signal after the equalization with active delay line FFE. Again, most ISI is removed, providing an eye opening at 10 Gb/s. This initial result clearly demonstrates the possibility for active delay FFE that can lead to a tunable delay FFE.

Finally, Figure 5.24 shows the microphotograph of the fabricated FFE with active delay line. The overall chip area is 1.15 mm by 0.89 mm including pad area. The total power dissipation of the active delay approach FFE is about 27 mW.

5.3.2.5 CMOS Building Blocks for Reconfigurable Equalizer. The suggested reconfigurable equalizer integrated circuit (IC) requires two essential features: tunable tap delay and variable tap gain. Figure 5.25 shows the architecture of the proposed reconfigurable FFE IC. The backplane channel output signal goes through the tap delay line. The delay amount determined by system simulation is controlled by the

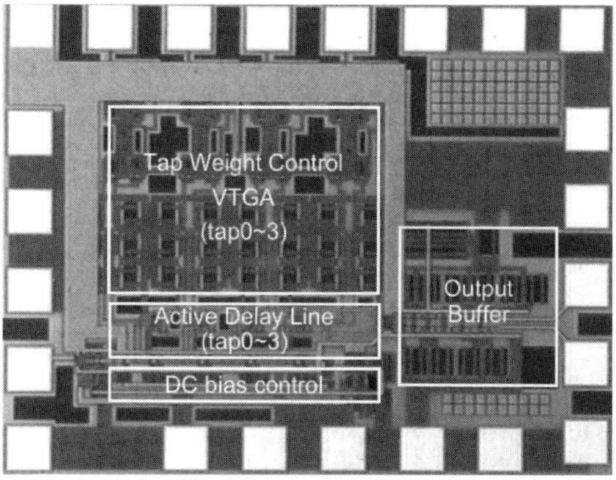

Figure 5.23. Microphotograph of the FFE with an active delay line approach.

DAC 1–3. Then these delayed signals are multiplied by the tap weights given by system simulation. The DAC 4–7 generate the corresponding tap weights being offered to the VGAs. The resulting amplified signals are combined in the current domain. Finally, the output voltage signal swing is obtained with the total current applied to the common load as illustrated in Figure 5.25.

Since this signal processing is performed in the analog domain, the suggested analog equalizer consumes less power and provides broad bandwidth compared with the digital equalization approaches.

5.3.2.5.1 Wide-Range Tunable Active Delay Line. From the system simulation, the tap delay line of the proposed FFE requires ~25–50-ps delay tuning range. An active delay line approach is chosen to provide a tunable delay feature.

The active delay line consists of cascaded differential amplifiers, as shown in Figure 5.25(a). The delay amount is determined by the RC response of the parasitic capacitance and the drain load. These R, C values are highly sensitive to process variation.

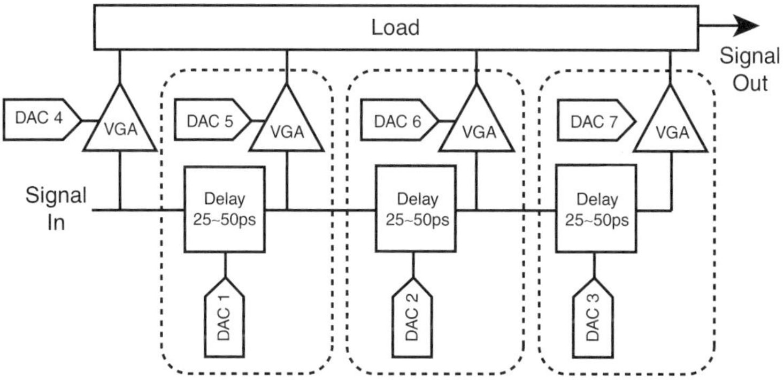

Figure 5.24. Proposed reconfigurable FFE block diagram.

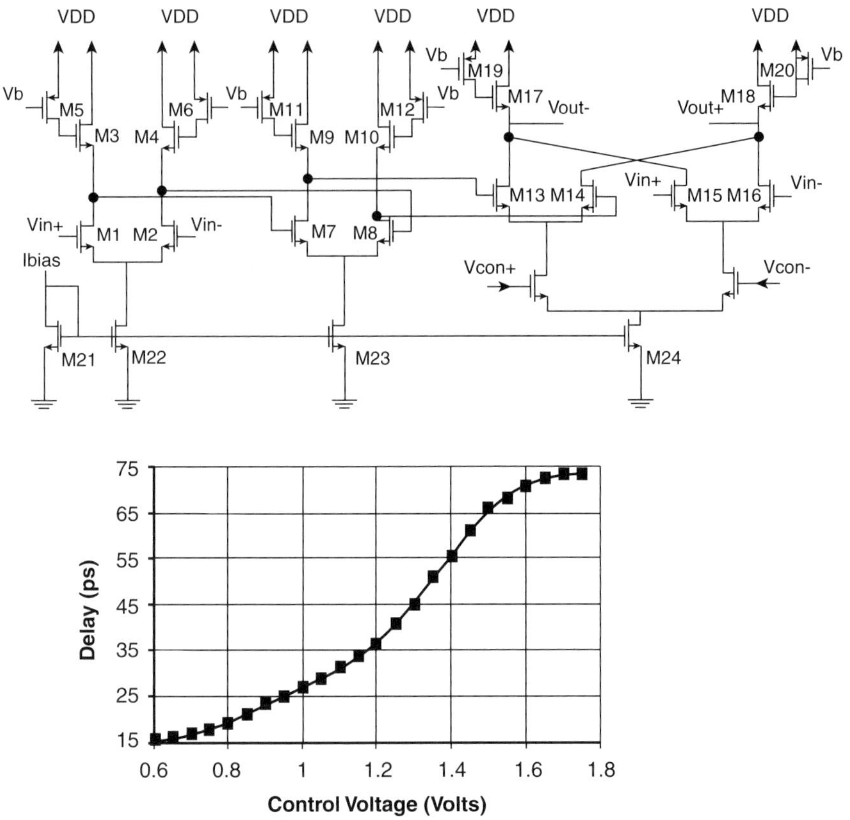

Figure 5.25. (a) Wide-range tunable active delay line topology and (b) illustration of the tunable delay performance.

To make the delay line immune to the process variation effect, the three stages of differential amplifiers are cascaded to cover a ~25–50-ps tuning range per unit tap delay. The differential input voltage is connected to M1, M2 and M15, M16.

With control voltages Vcon + and Vcon-, the delay amount can be varied. In other words, with maximum Vcon + applied, Vin + and Vin− follows M1, M2, M7, M8, M13, and M14, resulting in a slow delay path. With maximum Vcon− applied, a Vin + and Vin− signal follows through only M15 and M16, resulting in a fast delay path. With analog control voltage applied, a delay value can vary continuously.

Meanwhile, this active delay line approach faces a bandwidth challenge compared with the passive delay line approach. Active peaking inductors M5, M6, M11, M12, M19, and M20 were adopted to enhance the bandwidth characteristics enough to process 10 Gb/s.

Figure 5.25(b) shows the result of the tunable delay performance for the designed active delay line. A wide tuning range is achieved from 15 ps (fast delay path) up to 74 ps (slow delay path) for the control voltage from 0.6 V to 1.75 V.

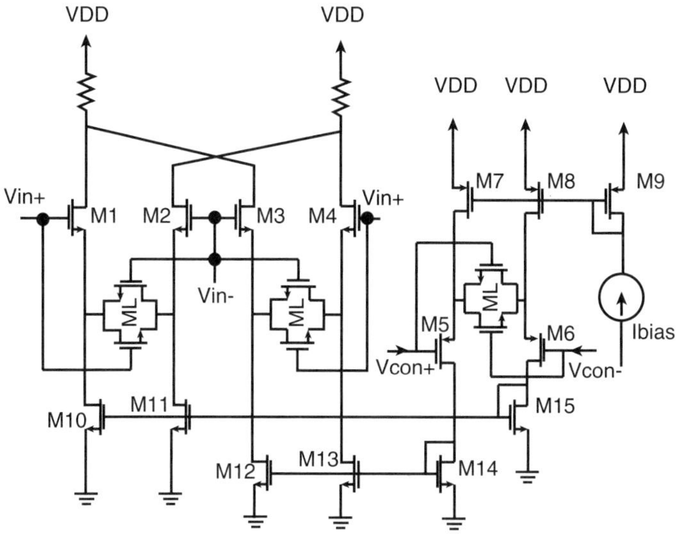

Figure 5.26. Schematic of the linear variable tap gain amplifier.

5.3.2.5.2 Variable Tap-Gain Amplifier. From the system simulation results, the tap gain amplifier needs linear bipolar gain values between −1 and + 1. The Gilbert-cell architecture is adopted to meet this requirement as shown in Figure 5.26.

The proposed FFE has four VGAs connected to the common load. Each amplifier flows DC current, resulting in four times the DC current flow through the 50-Ω resistor. This phenomenon induces the voltage headroom issue. To ensure proper voltage headroom for each VGA, a conventional Gilbert-cell is modified with a current steering bias scheme. Instead of applying control voltage directly to the differential pair below the common source, control voltages, Vcon + and Vcon−, are applied to M5 and M6 to provide the bias currents proportional to the control input. This modified architecture reduces the total number of stacked devices resulting in alleviated headroom condition.

In addition, both linearity and voltage headroom are enhanced by applying an active degeneration scheme between divided common source branches. The ML transistor pairs represent such active degeneration with M7, M8 and M10~M13 being the divided current sources.

5.3.2.5.3 8-bit Digital-to-Analog Converter (DAC). 8-bit DAC is integrated to provide the control voltage to the tunable active delay line and the VGAs. The DAC architecture is based on the R-2R ladder network as illustrated in Figure 5.27. This R-2R-based DAC uses the ratio of single value poly resistors, and these ratios can be highly tolerant over process variation especially with inter-digitized layout and added dummy resistors.

Meanwhile, the error performance of the DAC is critical for accurate equalizer-control. The NMOS switches with large (W/L) ratio were adopted to minimize the offset error caused by a V_{DS} voltage drop across the M1 and M2 switches. Moreover, to

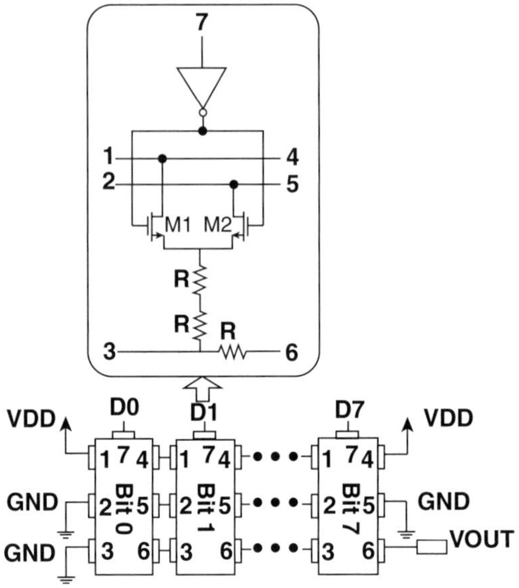

Figure 5.27. 8-bit DAC with R-2R topology.

improve matching and minimize offset, adjacent bit input unit cells were placed further apart from each other in the layout.

As a performance measure, offset error defined as the deviation from ideal value is monitored. Over all process corners, supply voltage, and temperature variations, the DAC performed within less than 10% of the least significant bit offset across all cases.

CONCLUSION

In this chapter, we have illustrated practical design and integration aspects of wireless and wired communication systems. Depending on the specific applications, the design considerations and integration methodologies will vary significantly. In addition to the system design and circuit developments, many additional factors come into account while building solutions for commercial applications such as die area, packaging and so on. Many reports of practical implementations are available in open literature in the past 30 years, and interested readers are encouraged to explore more, depending on the specific systems under development.

REFERENCES

[1] B. Gilbert, "The MICROMIXER: A highly linear variant of Gilbert mixer using a bisymmetric class-AB input stage," *IEEE Journal of Solid State Circuits*, Vol. 32, NO. 9, Sept 1997, pp. 1412–1423.

[2] B. Gilbert, "The multi-tanh principle: A tutprial overview," *IEEE Journal of Solid State Circuits*, Vol. 33,NO. 1, Jan 1998, pp. 2–16.

[3] F. Bien, A. Raghavan, Z. Nami, C.-H. Lee, A. Kim, M. Vrazel, E. Gebara, S. Bajekal, B. Schmukler, and J. Laskar, "A 0.18-μm CMOS fully integrated 6.25Gbps single aggressor multi-rate crosstalk cancellation IC for legacy backplane and interconnect applications," *8th IEEE Workshop on Signal Propagation on Interconnects*, 2004, pp. 73–76.

[4] F. Bien, H. Kim, Y. Hur, M. Maeng, S. Chandramouli, J. Cha, E. Gebara, and J. Laskar, "A 10-Gb/s reconfigurable CMOS equalizer employing a transition detector-based output monitoring technique for band-limited serial links," *IEEE Transactions of Microwave Theory Techniques*, Vol. 54,NO. 12, Dec 2006.

[5] F. Bien, Y. Hur, M. Maeng, H. Kim, E. Gebara, and J. Laskar, "A reconfigurable fully-integrated 0.18-μm CMOS feed-forward equalizer IC for 10-Gb/s backplane links, " *IEEE International Symposium on Circuits and System* (ISCAS), May 2006.

[6] J.L. Zerbe, C.W. Werner, V. Stojanovic, F. Chen, J. Wei, G. Tsang, D. Kim, W.F. Stonecypher, A. Ho, T.P. Thrush, R.T. Kollipara, M.A. Horowitz, and K.S. Donnelly, "Equalization and clock recovery for a 2.5–10 Gb/s 2-PAM/4-PAM backplane transceiver cell," *IEEE Journal of Solid-State Circuits*, Vol. 38, Dec 2003, pp. 2121–2130.

[7] M. Maeng, F. Bien, Y. Hur, H. Kim, S. Chandramouli, E. Gebara, and J. Laskar, "0.18-μm CMOS equalization techniques for 10-Gb/s fiber optical communication links," *IEEE Transactions of Microwave Theory and Techniques*, Vol. 53, NO. 11, Nov 2005, pp. 3509–3519.

[8] J. Laskar, B. Matinpour, S. Chakraborty, *Modern Receiver Front-Ends*, John Wiley, 2004.

CHAPTER 6

Low-Voltage, Low-Power, and Low-Area Designs

INTRODUCTION

In the previous chapters, we have developed our understanding of how the basic blocks would work in a communication system from a functionality perspective with regard to operating frequency, bandwidth, modulation scheme, and impedance. This chapter would illustrate the techniques for low-power, low-voltage, and low-area implementations and design considerations. The essential impact is on the battery life, longevity, and ease of portability (thereby integration of several heterogeneous systems), as well as on the smaller form factor. Various wireless devices benefit from these developments; however, it has become a necessity for biomedical applications. Specifically, we would provide insights from device technologies, circuit techniques, and architectural decisions.

Early reports of low-power microsystems have been in the area of wristwatches for extended battery longevity as reported in [2,3]. Because of the current technology scaling, the supply voltage has gradually reduced, and the unity gain cutoff frequencies (f_T, f_{MAX}) have enhanced significantly. This improvement has led the designers to optimize circuits for power, resulting in utilization of circuits and systems based on several techniques including the subthreshold behavior of MOSFETs. To accommodate for lower supply voltage, several circuit design techniques can also be adopted. Although all of these aspects are attractive, one can also think about architecture-level issues such as super-regenerative receivers, subharmonic receivers, injection locked circuits, and low duty cycle radios. A low

Advanced Integrated Communication Microsystems, edited by Joy Laskar, Sudipto Chakraborty, Manos Tentzeris, Franklin Bien, and Anh-Vu Pham

area is also facilitated by the downscaling of the minimum feature sizes of semiconductor technologies. Digital processing blocks tend to work well according to the scaled technology direction. However, analog circuits do not tend to scale as well as digital ones. To combat these issues, one can think about eliminating the most area-intensive blocks such as passive inductors and use their active counterparts.

6.1 POWER CONSUMPTION CONSIDERATIONS

In this section, we would like to focus on power consumption considerations. In Chapter 2, we have introduced the impacts of system-level parameters on the power consumption. Power consumption in any signal processing operation is dependent on the center frequency, signal bandwidth, modulation scheme, and driving point impedance. Usually, it is proportional to the center frequency, and the square of the bandwidth, and it is inversely proportional to the driving point impedance. Intuitively, narrowband circuits should usually provide lower power than their broadband counterparts, as they need to process signal over a smaller, frequency selective bandwidth. Thus, we process lower bandwidth and obtain lower power in a narrowband implementation. In the presence of blockers, the power consumption is also dependent on the received power levels of the blockers. High-performance radios for cellular communications demand the highest possible dynamic range because of the presence of high out-of-band blockers. Wireline communication systems tend to operate over fairly broader bandwidth, and secure communication links, and the power consumption is a strong function of bandwidth (hence, data rate). On the contrary, wireless communication systems have to withstand and coexist with many other standards, and often, weak signals at various frequency bands are present along with significantly higher levels of signals at other frequency bands.

In the digital processing sub-blocks, the current consumption is related to the capacitive load, the supply voltage, and the operating clock frequency, and the power consumption is mostly dynamic in nature. Several techniques can be adopted to optimize digital circuit designs in terms of power consumption. Digital circuits do not consume any static current, and they dissipate power only at the switching instant. A fundamental difference thus exists in terms of the power consumptions in digital and analog circuit design domains; analog circuits need static power consumption, and the dynamic power consumptions are dependent on the signal amplitude (a typical receiver operates with signal strengths of millivolts to microvolts). It should also be noticed that in scaled technologies, the leakage current is an important factor as that tends to dominate the design of complicated integrated systems-on-chip.

6.1.1 Active Inductors

We have clearly seen the trends in the inductor design and their physical construction. At high frequency, the choice is clear; it needs to be an octagonal structure with the major considerations being the L, the Q, and the area consumed. Figure 6.1 shows an

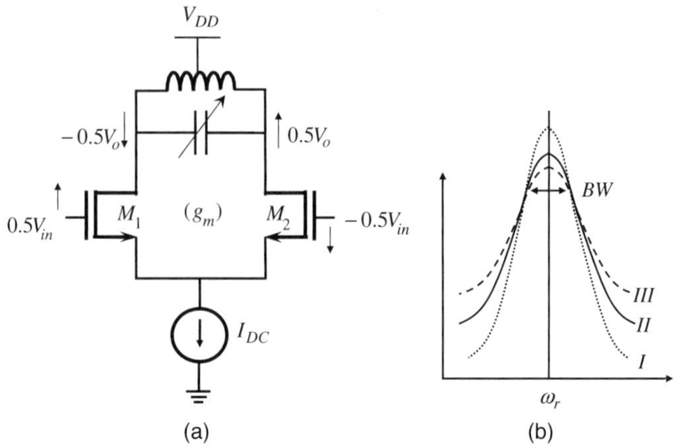

Figure 6.1. Inductively loaded amplifiers: center frequency and Q.

inductively loaded amplifier. At a milimeter-wave frequency regime, microstrip line inductors are used because of their smaller physical aspects and fundamentally higher inductance per unit length. Since the very beginning of commercial radio deployment, the targeted frequencies have been several hundreds of Megahertz, and passive inductors have not been prohibitive. However, with a judicious partitioning, active inductors can provide significant advantages over the passive counterpart, and they are suitable for scaled CMOS technologies (Figure 6.2). As shown in Figure 6.2(b), the common mode can set the signal level at the output of the amplifier node. These techniques are equally applicable for bipolars, and the lower parasitic capacitance, higher output impedance, and higher transconductance realizes a higher Q inductor.

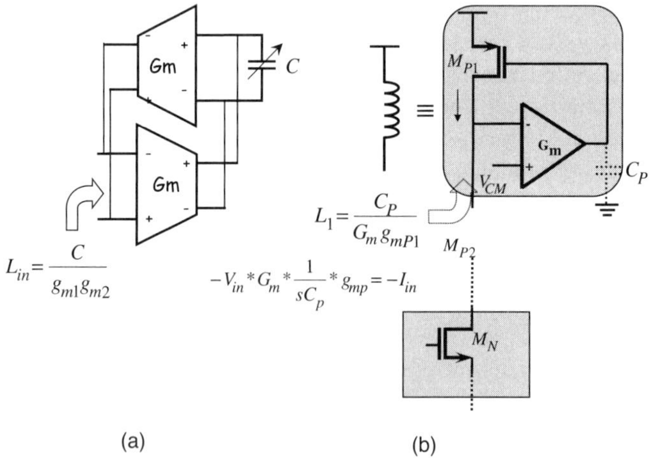

Figure 6.2. (a) Gyrator-C configuration and (b) active inductor implementation.

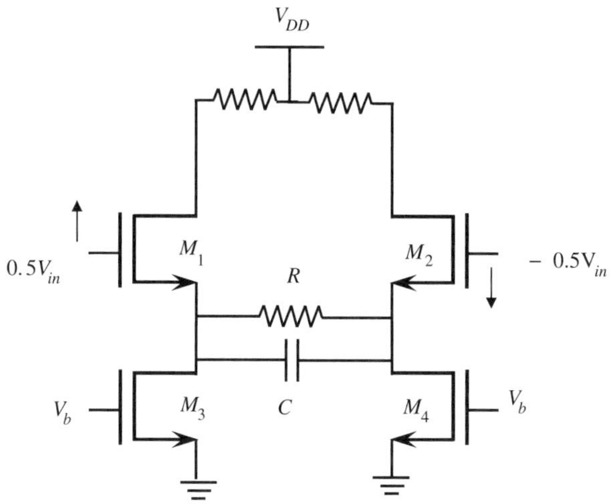

Figure 6.3. Creating transfer function zero.

Although active inductors use a reduced area, they consume voltage headroom and power dissipation as well as introduce noise. Another drawback in using active inductors is the nonlinearity in the inductor structure itself, forcing the signal swing to a reduced level, essentially reducing the dynamic range.

Table 6.1 Shows the various inductor architectures and performance trade-offs.

6.1.2 Adding Transfer Function Zero

In many cases, a resistance and capacitance connected in parallel at the source terminals of a differential amplifier configuration leads to the formation of a transfer function zero. This can provide effects similar to an inductor without much additional area consumption. The input impedance should be observed for stability. In many practical situations, one desires the inductor value, and in various large signal operations, noise may not be a concern, and an active inductor type configuration can be successfully used. However, such a creation of zeros in the transfer function is almost always associated with formation of a pole, and their relative position should be observed in the case of process variation and so on. An active inductor implementation using transfer function zero is shown in Figure 6.3.

6.1.3 Driving Point Impedance

Impedance plays a major role in determining system power. In a resonating tank, the impedance is proportional to the LQ product at resonance, which leads to the formulation of a "real" impedance. The higher the LQ product, the lower the power consumption, at the expense of area. The consideration of power versus impedance is also present at lower frequency systems, as in baseband filters and so on. Let us consider a class-AB output stage of an amplifier as shown in Figure 6.4.

Table 6.1. Comparison Among Various Inductor Implementation Methodologies

	Passive spiral inductor	Microstripline inductor	Active inductor
Area	Moderate/Large (spiraling used for compaction)	Large (No spiraling, used at GHz range)	Low (uses analog technique)
Power/Headroom	No additional current for implementation	No additional current for implementation	May need additional current
Noise	Low	Low	Moderate/High
Linearity	High	High	Moderate/Low
Crosstalk	Electric and magnetic	Electric and magnetic	No EM crosstalk
Areas of application	1–15 GHz circuits -high Q (LNA, VCO, PA) -Medium Q (Mixer, buffer) -Low Q (digital circuits, inverters, high speed MUX, frequency dividers) - DC bias - lossless feedback	15–100 GHz -high Q (LNA, VCO, PA) -Medium Q (Mixer, buffer) -Low Q (digital circuits, inverters, high speed MUX, frequency dividers) - DC bias - lossless feedback -Harmonic match (stub based design)	1 M–3 GHz -modrate Q (LNA, mixer) -Medium Q (Mixer, buffer) -Low Q (digital circuits, inverters, high speed MUX, delay cell frequency dividers)

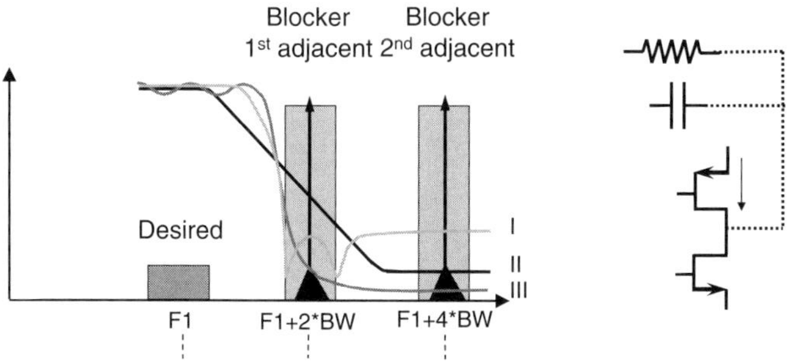

Figure 6.4. Filtering stages: (I) elliptic, (II) Butterworth, and (III) Chebyshev transfer curves and the output stage.

The driving stage may consist of a two-stage amplifier where only the output driving stage is shown for illustration. This output stage can be a part of a multistage elliptic, Butterworth, and Chebyshev filter topology, as appropriate for the specific wireless standard. The rejection of an integrator to an out-of-band blocker is dependent on the *RC* product. Since the capacitors consume more area, to realize a specific 3-dB corner of an integrator, we should target for smaller capacitance values. However, the voltage swing at various output stages of the filter in the presence of the worst-case out-of-band blocker should be kept as close to rail-to-rail swing in order to ensure maximum utilization of bias currents. The output driver stage is either a voltage- or current-limited one. Hence, the output stage needs to be biased such that the blocker can swing inside the V_{DSAT} of both transistors. The input of the amplifier stage would be held at virtual ground, and the signal swings are much smaller, which leads to much lowered distortions and reduced sensitivity to parasitic elements. The current in this driving stage needs to be larger than the small signal current swing that needs to be sourced from that stage. It is very easily observed that for a given voltage swing, if one drives higher impedance on chip, the small-signal current swing would be reduced, which implies a reduction of DC current in this stage. As this block could potentially be used in the output stages of an operational amplifier, which is commonly used and replicated in a multistage baseband filter, it helps optimize the power. However, to realize higher impedances, one can use a higher value of on-chip resistances, or a lower value of on-chip capacitances. Both of these are useful approaches, with the restraint that (1) large resistances would lead to larger parasitic capacitances, (2) a lower value of nominal capacitance would lead to degraded component matching (proportional to component area). Although the resistance parasitics can be reduced using a lower component width, it leads to poor voltage variation as well as to process variation.

6.1.4 Stacking Functional Blocks

A popular technique of low-power circuits is stacking multiple functional blocks within the supply rails. The essence of this technique is current reuse between various functional blocks, and any additional current can be supplied or taken away using current bleeders. Many examples of stacked circuits and systems has been widely reported in the literature. Although the current is shared, these systems compromise voltage headroom with stacked functionalities, and the amount of stacked functionalities needs to be properly chosen. In essence, this configuration uses current mode processing.

6.2 DEVICE TECHNOLOGY AND SCALING

Technology scaling impacts the design of integrated circuits and systems. Recently, the effective gate length of CMOS technologies has gone well below the deep submicron regime toward a more nano-technology domain, and circuits at 90-nm and 65-nm nodes are becoming a common practice. Bipolar technologies have also experienced similar scaling in terms of their emitter lengths, pushing the effective cutoff frequency f_t toward the 500-GHz range in the fourth generation of bipolar devices. This technology scaling has multiple effects on integrated systems, as follows.

6.2.1 Digital and Analog Circuits

Technology scaling for digital circuits would reduce the chip size progressively, with increasing gate leakage currents. The supply voltage and threshold voltage reduces progressively; as the technology nodes continue to scale down, the threshold voltage scales at a slower rate compared with the supply voltage. Any switching CMOS gate experiences a quadratic reduction in its dynamic (switching) power corresponding to a linear reduction in supply voltage. However, because of the various leakage mechanisms, such as gate leakage, drain-induced barrier lowering, and gate-induced drain leakage, digital MOS circuits tend to behave as imperfect switches in deep submicron process technologies. Hence, although dynamic power and area reduction proves to be attractive, the leakage current requires careful attention in terms of circuit design techniques, and technology considerations, such as multi-threshold circuits and so on. At the same time, the parasitic capacitances from the transistors reduce significantly, which leads to faster circuits, and more functionality per "dynamic miliwatts" or per "unit area." Bipolar transistors are usually not used as digital signal processing blocks in transceiver architectures because of their relatively larger area requirements and lack of high-speed complementary PNP transistor support at high speeds. Hence, even though the leakage increases, CMOS proves to be a power and or area-efficient approach. However, several methods and techniques such as "threshold voltage adjustment," "clock gating," and "signal enabling" are commonly used to reduce the leakage of the functionally OFF blocks as needed by the system. Leakage current plays a significant role in determining the standby performance of a mobile device, as it directly impacts battery longevity.

6.2.2 Supply Voltage, Speed, and Breakdown

Keeping the same topology, the power consumption for an analog signal processing reduces proportional to the supply voltage with technology scaling. Hence, as technology scaling leads to progressively reduced supply voltages, new circuit topologies need to be investigated. The breakdown voltage also reduces for the high-speed devices because the product of breakdown voltage and f_T remains constant. Area scaling of the circuits (and, hence, the aspect ratio of an analog front end) is a complicated function of the minimum feature size. Contrary to the long channel generations, the transconductance of a MOS transistor does not hold the same proportionality function in scaled technologies, although bipolar technologies maintain the same exponential law characteristics as predicted by physics-based formulations.

Technology scaling enhances the operating frequencies and reduces breakdown voltage as understood well by Johnson's limit ($f_T V_m = EV_s/2\pi$, where V_m is the maximum allowable applied voltage and V_s is the saturated drift velocity). f_T, for the specific transistor under consideration, is dependent on the biasing point and parasitic loading. Thus, with the circuit under consideration, the transistor f_T can be different from what is reported as a peak f_T.

6.2.3 Circuit Impacts of Increased f_T

f_T has twofold meaning in terms of analog radio Frequency (RF) circuits. Increased f_T enhances the possibility of higher frequency circuits in the analog front end. For example, enhancing the f_T by a factor of 2 implies that one can design the front end at higher frequencies. On the other hand, it implies a reduction of static power by a factor of 2, while operating at the same frequency. Hence, increasing f_T has a significant improvement in terms of overall system power and area, while they are considered differently from digital circuits. MOS transistors perform twofold functionalities, as an amplifier (g_m is the key parameter) and as a switch (R_{ON} and C_{OFF} are key parameters). Technology scaling enhances g_m and reduces R_{ON}. Thus, the MOS switch can be close to an ideal one because of its low parasitics. At the same time, the output impedance of the MOS transistor also reduces, which implies a strong necessity for cascode topologies. The front-end circuit performance is heavily dependent on the maximum power gain frequency f_{MAX}, which takes into consideration the parasitics associated with the transistors and provides a meaningful benchmarking parameter as it impacts the efficiency and noise of the circuit performance.

6.2.4 MOSFETs in Weak Inversion

Technology scaling results in another interesting possibility of using subthreshold MOS circuits as part of analog/RF systems. The gate bias of the MOSFET is adjusted such that the device operates inside the threshold region. In the subthreshold operation regime, the transistor uses the highest value of the transconductance-to-bias-current ratio (g_m/I_d). However, keeping current to a much reduced level, the device sizes need to be increased significantly to achieve a specific g_m. Under a subthreshold operation, MOS transistors exhibit an exponential I–V relationship and, from a functionality perspective, comes close to a bipolar transistor. Since it also provides the gate leakage current, many bipolar circuit design techniques can be incorporated toward low-power

MOS circuits. Thus, technology scaling also enables the possible operation of low-power strongly inverted and low-power weakly inverted circuits at the same technology node. More on this will be illustrated later in this chapter.

6.2.5 Millimeter-Wave Applications

Since scaled technologies provide the promise of increasingly higher operating frequencies, many components that used to exist in the microwave circuit domain tend to appear attractive for integration in planar silicon CMOS substrates. Successful realization of low-loss microstrip lines, stubs, and so on, are commonly used in high-frequency circuits. At higher frequencies such as tens of Gigahertz, the component dimensions are smaller and do not pose any significant difficulty in terms of an integration. However, care should be taken in order to design such components in a high Q fashion. An illustration of an integrated mmW circuit is illustrated in [5]. Microstripline inductors are preferred at these frequencies as they provide higher inductance per unit length and eliminate many interwinding capacitances without much area penalty. Stub sections are very useful in providing frequency isolation. If a stub section is designed at a length of $\lambda/4$ to provide a high impedance (open) at frequency f, then it provides a low impedance (short) at frequency $2f$. This is quite attractive for a harmonic receiver, where the Local oscillator (LO) and RF frequencies can be provided at the two ends of the diode pair without any interference between them.

6.2.6 Practical Considerations

In summary, technology scaling impacts integrated system developments in several ways as mentioned before: (1) improved circuit density (more gates per unit area, and more die per wafer), (2) lower power operation in analog circuits, (3) increased capability of mmW frequency circuits and systems, and (4) increased capability of subthreshold MOS-type circuits. With all of these advantages, leakage current increases, component mismatch increases, and breakdown voltage reduces. Since the wireless devices operating in a specific frequency band need to deliver a specific amount of power, designing large signal circuits at transmitter, such as a PA driver, poses a significant challenge. This becomes even more challenging in MOS devices because of the presence of hot carriers, which can rupture the gate oxide, leading to reliability issues. Component mismatch is a major limitation in submicron CMOS nodes, and sometimes multiple calibration circuits are used to alleviate the performance degradations caused by such effects, and to ensure higher yield. While migrating between technology nodes, it may be possible to retain the same geometry of the devices in analog/RF circuits, to provide a faster time to market, as essentially it retains the resonance frequency to be the same as before. However, two transistors with the same geometry parameters perform very differently in two process technology nodes as the gate oxide changes. To alleviate the component mismatches, additional mask steps and precise fabrication steps can be taken, which leads to cost ineffectivity of deep submicron technology. Hence, the selection of a semiconductor platform is a significantly important decision, and many times, a single-chip may not be the most optimum in terms of performance. Single chip implementations bring their fundamental challenges of isolating digital and analog blocks, which often Leads to major architecture and floorplanning issues.

6.3 LOW-VOLTAGE DESIGN TECHNIQUES

Low-voltage systems reduce the number of batteries, thereby leading to smaller form factors of the solutions. The key approaches usually include circuits as well as device operation. Several approaches have been reported in the literature, which include:

1. Reduction of voltage headroom across individual AC-coupled functional blocks. The functional blocks are separated in their DC biasing path.
2. Using MOSFET devices in a subthreshold regime.
3. Using passive components to allow maximum voltage across the transistor stack used for functionality.
4. Using circuits to operate in current mode fashion.
5. Using multiple transconductor paths from input to output.

6.3.1 Separate DC Paths per Circuit Functionality

Let us illustrate these techniques in detail. Figure 6.5 illustrates a circuit topology of two functional blocks stacked between the supply rails. As the supply voltage reduces, the transistors tend to operate in nonsaturated regions, lowering their effective transconductance.

However, the same topology can be transformed to provide the desired DC operating point for each of the transistors, while coupling the output signal of the first functional block to the input of the second functional block. The L/C tank can be designed to resonate at the desired RF frequency. This technique is generic, and it can

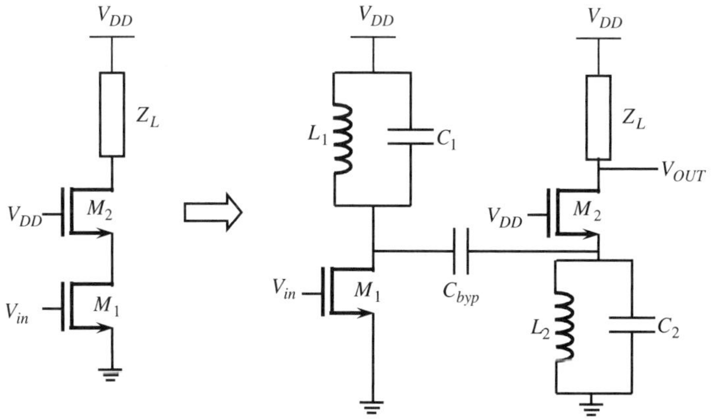

Figure 6.5. Low-voltage design by AC coupling.

be applied to any RF functional block or a combination of blocks. The independent biasing provides a degree of freedom to the designer such that the optimum bias point can be chosen for each functional sub-block. The technique can be extended to amplifiers, mixers, and so on. It is analogous to the well-known "folded-cascode" approach of the analog design domain, where the high output impedance of the current source MOSFETs allow the signal-dependent current to flow to the desired load impedances. In this case, the high impedance is realized by the L–C resonating tank, which consumes no headroom at DC. Although this approach optimizes voltage, it does not help in power consumption and leads to a significant increase in area caused by the presence of additional inductors in the biasing paths.

Passive components help in generating maximum voltage headroom in the functional transistor stack. In addition to using inductor-based approaches, transformer, autotransformer (tapped inductor), and coupled line approaches are fairly common in RF circuits and systems.

6.3.2 Transformer Coupled Feedback

From the previous section, it is found that in the low-voltage design space, it is necessary to provide a single transistor across the supply rail. However, the gate-drain overlap capacitance C_{GD} in submicron CMOS significantly reduces the achievable f_T of the transistors, $f_T = \frac{g_m}{2\pi(C_{GS}+C_{GD})}$, and the equivalent input referred capacitance is also reduced by the amplifier gain. To avoid this bandwidth-limiting phenomenon, a neutralization capacitor is provided across the amplifier stage, which provides a positive feedback. Hence, while employing a neutralization capacitor, amplifier stability is a major consideration.

An alternative to this approach of neutralization can be manifested by transformer coupled negative feedback as illustrated in Figure 6.6. As the input voltage increases, the drain current increases, and the voltage drop across the drain part of the transformer (primary/secondary) increases, which leads to an increase in the source part of the transformer (secondary/primary). This feedback phenomenon reduces the gate-source voltage V_{gs}.

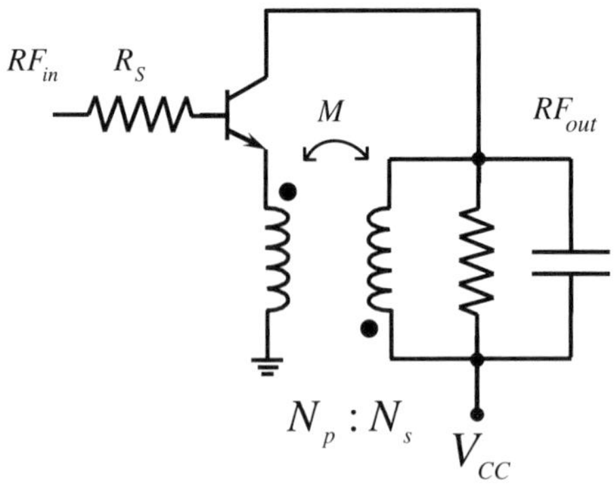

Figure 6.6. Transformer coupled feedback.

The primary and secondary windings can be connected in a symmetrical spiral fashion, providing maximum area efficiency, while maintaining the self-resonating frequency much higher compared with the frequency of operation. A single-ended illustration is illustrated here for simplicity, whereas in practical implementation, a differential configuration would be necessary. However, an area penalty may be inevitable while realizing a transformer.

6.3.3 Positive Feedback

Feedback can be used in many ways, and a side effect of using feedback can be independent control of various performance parameters. [15] provides an illustration, where the positive feedback provides independent control of the input impedance and noise performance of a common-gate Low noise amplifier (LNA). Positive feedback can be employed to improve the gain of the circuit as well. Figure 6.7 illustrates a common-gate LNA topology employing positive feedback. The output voltage is sensed by the PMOS transistor and fed back to the input as a current signal, which leads to a shunt–shunt feedback topology. The current feedback path is added in parallel to the main signal path branch, which causes an increase in overall amplifier current gain.

The shunt–shunt positive feedback increases the input impedance of a pure common-gate amplifier stage. From the open-loop impedance, $Z_{OL} = R_S + 1/g_m$, the closed-loop impedance is given by, $Z_{CL} = \frac{Z_{OL}}{(1-G_{loop}(\omega))}$, where the loop gain is given by, $G_{loop}(\omega) = \frac{R_s \cdot g_{m1}}{1+R_s \cdot g_{m1}} \cdot g_{m2} \cdot Z_l(\omega)$. This leads to the input impedance of $Z_{in}(\omega) = \frac{1}{g_{m1}(1-g_{m2}Z_{load}(\omega))}$. Hence, at resonance of the load, the input impedance would assume a real, resistive value.

The transconductance of the positive feedback LNA is given by $G_{fb} = \frac{g_{m1}}{1+g_{m1}R_s(1-g_{m2}Z_l(\omega))}$, which equals $\frac{g_{m1}}{2}$ at the matching condition, which is significantly higher than the regular common-gate LNAs $\left(\frac{1}{2R_s}\right)$.

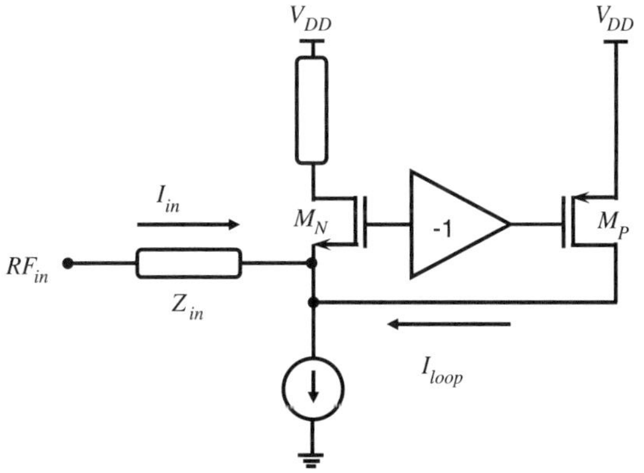

Figure 6.7. Common-gate LNA with positive feedback.

6.3.4 Current-Mode Interface

Many times, the voltage headroom is helped by using current-mode circuits. We can start with the classic example of simple-looking passive mixer topology. It can be observed that, in passive mixer-based topologies (based on NMOS ring or transmission-gate-type architectures used as a passFET), the current is switched using the MOS transistors being used as switches. Although use of passive structures (such as passive mixers and passive RC filters) seem attractive in terms of their zero DC power consumption (and, hence, absence of flicker noise), and voltage headroom, the input impedance of a structure poses significant considerations in power consumption. Also, passive structures provide conversion loss, and thus require additional gain stages to boost up the signal level. If the passive building blocks provides a lower input impedance, the block that drives it needs to consume sufficient DC current in order to drive it. We can consider several examples such as follows:

1. A passive polyphase filter using multisection RC/LC topology in the signal generation path of a quadrature low IF architecture provides attenuation, and the LO buffers following it need to consume current in order to provide a desired signal level to the mixer switches.
2. A passive LC filter at the baseband may provide a low input impedance, thereby increasing the power consumption of the preceeding stage (usually the mixer) that drives it.

Current-mode circuit construction uses an inherent advantage of faster signal processing at RF and improves linearity, and it eliminates parasitics of long traces (as the various building blocks communicate in terms of current, and not voltage); if constructed carefully, it does not alter the impedance levels at the output of the building blocks. This is shown in Figure 6.8, where the input RF signal is switched using an LO drive, and the output current is injected into the virtual ground of the baseband amplifier. The signal swing at the baseband input terminals is significantly reduced,

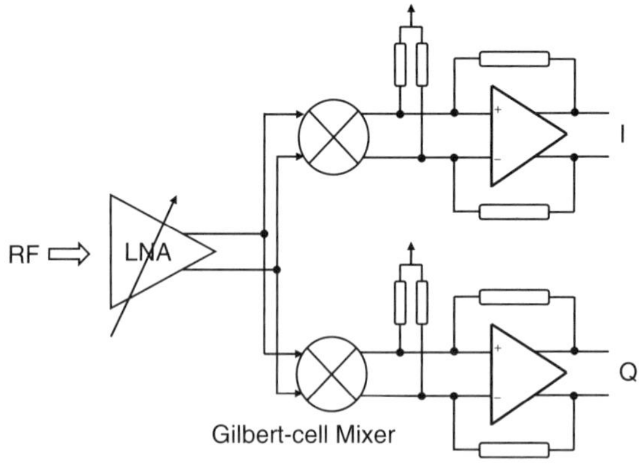

Figure 6.8. System architecture based on current-mode output Gilbert-cell mixer.

which leads to higher linearity at that interface. Current-mode processing is inherently linear, leading to high dynamic range for the building blocks.

6.3.5 Circuits Based on Weak Inversion

A MOS device operates in weak inversion when the gate-source voltage is lower than the threshold voltage V_t such that a depletion region is formed at the surface of the MOS device. In this region of operation, the drain current is mostly caused by diffusion. The current-voltage relationship is given by

$$I_D = K\left(\frac{W}{L}\right)e^{V_{GS}/nV_T}(1-e^{-V_{DS}/V_T})(1+\lambda V_{DS}), \text{ and with, } V_{DS} > 0,$$
$$I_D = K\left(\frac{W}{L}\right)e^{V_{GS}/nV_T}(1+\lambda V_{DS}) \qquad (6.1)$$

MOS devices at weak inversion are extensively used in cases of low-power applications, usually at about Megahertz frequency ranges. The transconductance-to-current ratio is highest at weak inversion. The exponential nature of transfer characteristics suggests that MOS behaves as bipolar transistors (without the base current consideration) at weak inversion, and it can be used in many circuits and systems, similar to the bipolar transistors. However, with scaled geometries, these devices tend to operate at a cutoff frequency well into the gegahertz range. However, they tend to perform poorly in terms of thermal noise, and hence, they should be used carefully while designing circuits such as LNAs. The feasibility of bipolar-looking characteristics provides circuit designers with low-voltage, low-power variants of circuits in standard digital MOS technologies. The possibility of realizing exponential characteristics, combined with square law MOS devices, provides a powerful combination to the circuit designer.

6.3.6 Voltage Boosting

Technology scaling reduces the supply voltage progressively, but the reduction in threshold voltage is slower than the supply voltage. This impacts the design of switched capacitor circuits, amplifiers, and so on. Sometimes native devices are found in the technology, where no threshold voltage adjustment has been performed. While designing circuits with these transistors, leakage current is a concern for integrated systems. These devices may require an additional mask step, and they are usually less well characterized in terms of their models.

Clock boosting techniques are employed in the circuits in order to increase the supply voltages. Fundamentally, the increased level of voltage is obtained by providing current to isolated capacitive nodes. Usually these techniques are used to boost the clock signals locally. Figure 6.9 illustrates various clock boosting techniques. In the first variation, the input voltage is boosted to $2V_{DD}$, whereas in the second implementation, the input voltage is boosted to $V_{DD} + V_{sig}$. In the second approach, the gate overdrive voltage V_{OV} remains constant (except for body bias effects), which improves the

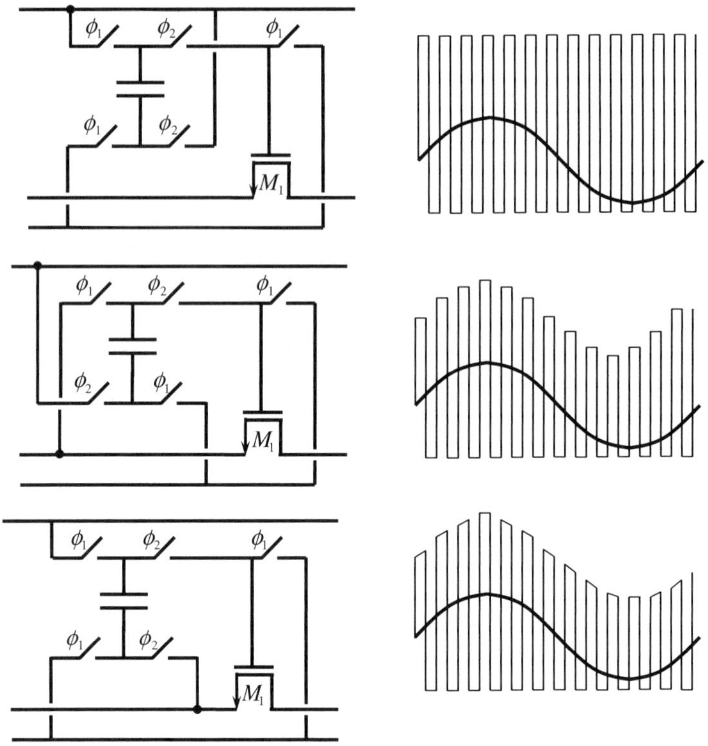

Figure 6.9. Clock boosting techniques.

linearity of track-hold circuits. In the second configuration, a signal is sampled before a clock is boosted to a high value, whereas in the last option, V_{DD} is sampled on the capacitor, which is connected between the G–S terminals, and the gate signal tracks the input signal, leading to even lower distortion. During the clock boosting phase, all voltages of the MOS transistors V_{GS}, V_{GD}, and V_{DS} must be kept below V_{DD} for reliability reasons, whereas in some cases, the drain-bulk voltage V_{DB} may go up to $2V_{DD}$.

6.3.7 Bulk-Driven Circuits

Another interesting aspect of low-voltage design techniques is to exploit the four-terminal MOSFET device itself. The back-gate of the MOS transistor can be used as a second transconductor, and it can be combined with the main transconductor path to provide the desired small- or large-signal characteristics. Thus, with the same voltage headroom, one can consider two parallel transconductors, or one terminal enhancing the transconductance of the other. Large-signal illustrations are also possible. Mixer configurations using LO injected at the bulk terminal and the RF injected at the gate (and vice versa) have been reported, as well as variable gain amplifiers. An illustration of a four-quadrant multiplier has been shown in Figure 6.10.

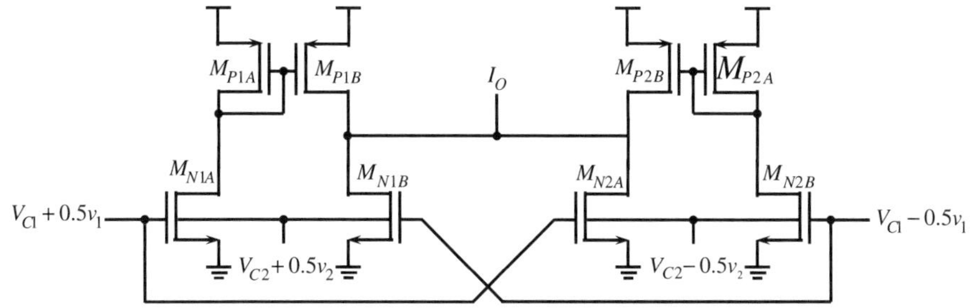

Figure 6.10. Bulk-driven circuits: four-quadrant multiplier.

Although these approaches are fundamentally feasible, they are sometimes not practicable because of the need for additional steps such as deep NWELL and the risk of contaminating the substrate.

The threshold voltage of a MOS device determines its speed as well as the voltage headroom to be consumed by a stack of transistors in analog circuits. The drive requirements of analog signals also depend on the threshold voltage. The threshold voltage can be adjusted by adjusting the source-bulk reverse bias V_{SB}, and it is given by

$$V_t = V_{t0} + \gamma\left(\sqrt{2|\phi_F| + V_{SB}} - \sqrt{2|\phi_F|}\right) \tag{6.2}$$

The bulk terminal can be driven by a small signal (similar to the operation of a small-signal amplifier) or a large signal (within the limits of not forward biasing the junction). A corollary of the second phenomenon lies in the implementation of a harmonic mixer when the source-bulk voltage is varied using a large-signal LO tone in order to obtain a threshold voltage modulation, which leads to a harmonic mixing using the square-law MOS transfer characteristics. With an RF signal applied at the gate, the output current is given by

$$i_D = K\left(\frac{W}{L}\right)\left[V_{RF}\cos\omega_{RF}t - \left\{V_{t0} + \gamma\left(\sqrt{2\phi_F + V_{LO}\cos\omega_{LO}t} - \sqrt{2\phi_F}\right)\right\}\right]^2 \tag{6.3}$$

This provides the frequency terms of $(\omega_{\mathrm{RF}} \pm n\omega_{LO})$ at the output. This scheme is illustrated in Figure 6.11.

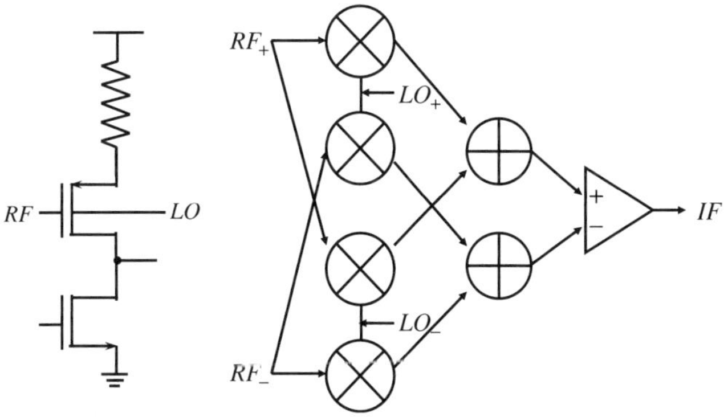

Figure 6.11. Bulk-driven circuits: harmonic mixer.

6.3.8 Flipped Voltage Follower

Figure 6.12 illustrates different variations of a flipped voltage follower topology. This topology has been used in various circuits over many years. We start the discussion with a source follower stage, or unity gain buffer.

It can be observed that the current through transistor M_{P1} depends on the output current, and V_{GS1} is not constant with various load levels. Hence, the voltage gain deviates from unity with a resistive load, and with capacitive loads at high frequencies. In configuration (b), the circuit operates as a source follower where the current through M_{P1} is held constant, independent of the output current. This is similar to voltage follower with shunt feedback, as the transistor M_{P2} senses the output voltage and feeds a current proportional to that to the input. Ignoring the body bias effect, V_{GS1} is held constant, and the voltage gain approaches unity. This circuit is able to source a large amount of current, but the sinking ability is limited by the bias current. The output impedance is given by $r_o = \frac{1}{g_{m1}g_{m2}r_{01}}$, and it is of the order of 20–100 Ω.

It has been shown in [16] that biasing the lower transistor M_{P1} in the saturation region becomes difficult with a low value of input voltage V_i. This problem can be solved by using a level shifter transistor between the drain of M_{P2} and the gate of M_{P1}.

Because of its low-output impedance, the topology can be used as a current sensor. At the circuit node connecting M_{P2} and M_{P1}, a large amount of current can flow,

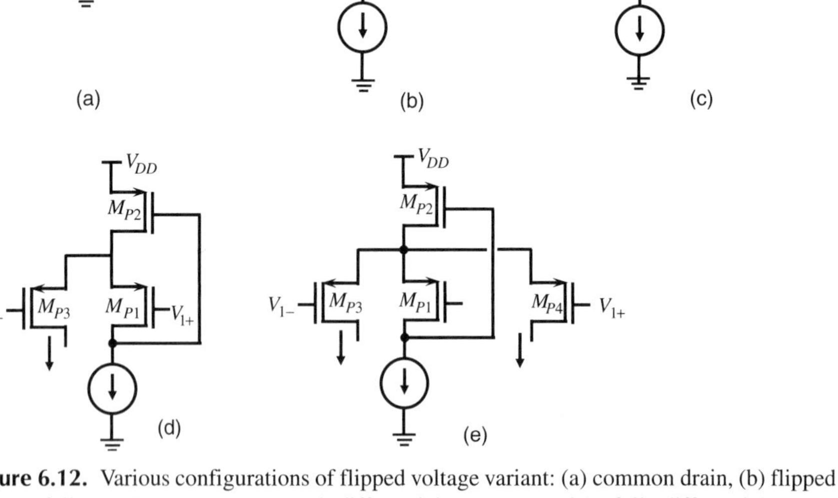

Figure 6.12. Various configurations of flipped voltage variant: (a) common drain, (b) flipped voltage follower, (c) current sensor, (d) differential structure, and (e) fully differential output.

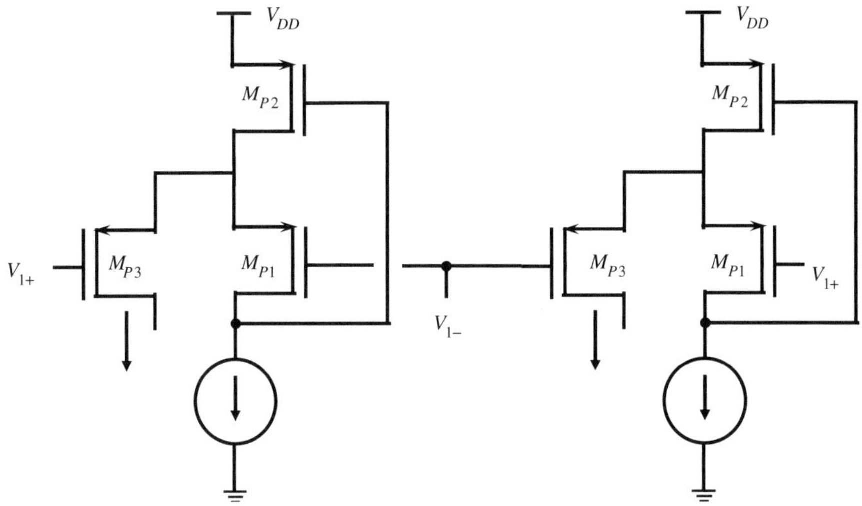

Figure 6.13. Flipped voltage OTA circuit.

without changing the voltage levels. This voltage is then translated at the drain node of the transistor (top of the current source), and the current is mirrored through M_{P3}.

Configuration (d) illustrates a current- and a voltage-mode output circuit. The input differential signal can be assumed to be a difference of $V_{1+}-V_{1-}$, and the output can be a current (taken through transistor M_{P3} or mirroring the current through M_{P2}). In terms of voltage, the output can be taken at the top of the current source. Because of the low-output impedance, the circuit can source a large output current, much higher than the quiescent current.

Configuration (e) provides a modification on the single-ended structure, and this is a true differential configuration, and the differential output current is obtained from transistors M_{P3} and M_{P4}. The gate of transistor M_{P1} can be held at the common-mode voltage, $0.5(V_{1+}-V_{1-})$

Figure 6.13 illustrates an OTA core cell. OTAs based on this core can be used in switched capacitor filters, gm-C filters, and so on. Similar to the earlier fully differential configuration, it provides a quiescent current that is much smaller than the maximum available current. The large-signal behavior can be nonlinear.

6.4 INJECTION-LOCKED TECHNIQUES

In the signal generation path, the oscillator and the buffers consume a significant part of the overall power, which is present in both transmit and receive modes. In this part, we would discuss injection-locked techniques in order to reduce the power of the oscillators.

Traditional prescalers in a phase-locked loop (PLL) structure use DFF-based dividers that use positive feedback. They provide a wideband operation in order to accommodate various harmonics of the input signal, and the power consumption

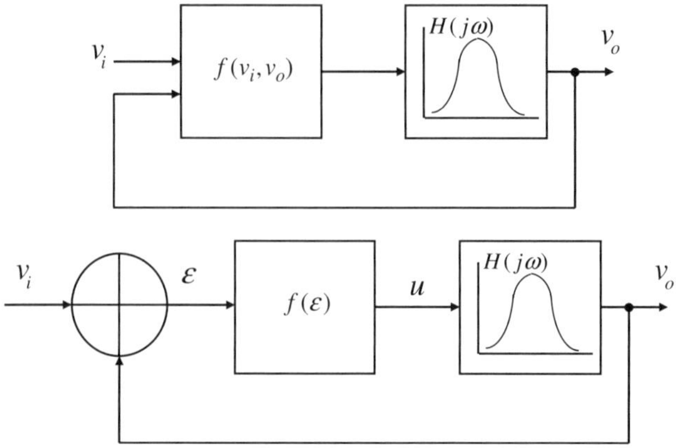

Figure 6.14. Injection-locked frequency divider model.

increases with operating frequency. In contrast, injection-locked frequency dividers operate on the basis of synchronizing an oscillator with the input signal. Hence, fundamentally, such dividers trade off power consumption w.r.t. the bandwidth. In general, depending on the input signal frequency, such dividers are categorized as follows: (1) first-harmonic (input signal is the same as the oscillation frequency), (2) superharmonic (input signal is an integral multiple of the oscillation frequency), and (3) subharmonic (input signal is an integral submultiple of the oscillation frequency). A generic, as well as a simplified, model for an injection-locked divider is illustrated in Figure 6.14.

It can be noted that

$$\begin{aligned} v_i(t) &= V_i\cos(\omega_i t + \theta) \\ v_o(t) &= V_o\cos(\omega_o t) \\ u(t) &= f[v_o(t) + v_i(t)] \\ H(j\omega) &= \frac{H_0}{1 + j2Q\left(\frac{\omega - \omega_r}{\omega_r}\right)} \end{aligned} \tag{6.4}$$

Because of the resonant tank, all frequency components far away from the resonating frequency ω_r are filtered out, and the output frequency is given by $\omega_o = \omega_r + \Delta\omega$. Hence, we are interested in terms where the output frequency is obtained because of the intermodulation products caused by the nonlinear function. In other words, we obtain an output, where $|m\omega_i - n\omega_o| = \omega_o \sim \omega_r$. The locking range of such dividers can be increased by lowering $\frac{H_0}{Q}$, or by increasing the level of input amplitude V_i.

Two circuit configurations for an injection-locked divider are illustrated in Figure 6.15. In the single-ended configuration using a Colpitts oscillator, transistor M_{N1} is used to implement the summing function of the loop. The cascode transistor M_{N2} is smaller in size compared with the main transistor to reduce parasitic loading on the tank network, thereby increasing the locking range. In the differential structure

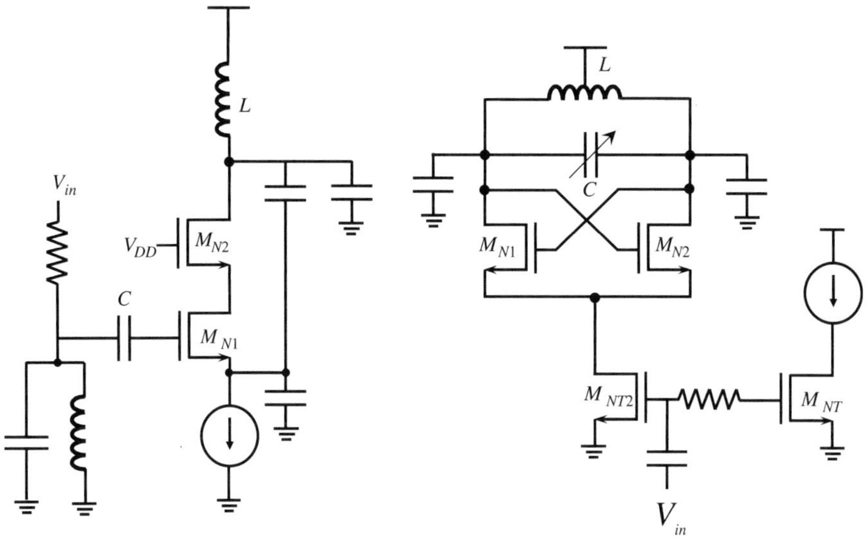

Figure 6.15. Injection-locked frequency divider topology.

with a resonating tank, the signal is in current mode and is injected at the common source point. Summation is obtained at the gates of M_{N1} and M_{N2}. The choice of the injection node also coincides with the fundamental mode of operation; as in the absence of the injected signal, the source node would oscillate at the second harmonic.

6.5 SUBHARMONIC ARCHITECTURES

Subharmonic architectures originated from the microwave domain, and they are mature in their state-of-the-art. These architectures use a fraction of the transmitted/received RF signal frequency to up/downconvert signals in a radio front end. In the history of subharmonic receiver development, passive-mixer-based techniques preceded their active counterparts. To optimize for power, the frequency planning of a communication system can be performed in such a manner that the lowest number of circuit nodes could operate at the highest possible frequency. From this perspective, subharmonic transceivers can be considered as a multirate RF signal processing front end.

6.5.1 Formalism

Figure 6.16 illustrates the subharmonic architecture in detail. The local oscillator frequency ω_{LO} is an integral submultiple of the RF frequency, ω_{RF} and the intermediate frequency is related to both by the equation $\omega_{IF} = \omega_{RF} \pm n\omega_{LO}$.

As the frequency in the LO chain gets multiplied, their individual phase waveforms also do the same, thus requiring more phase granularity in the LO waveform to create a differential signal for down/upconversion. From a system and a circuit design

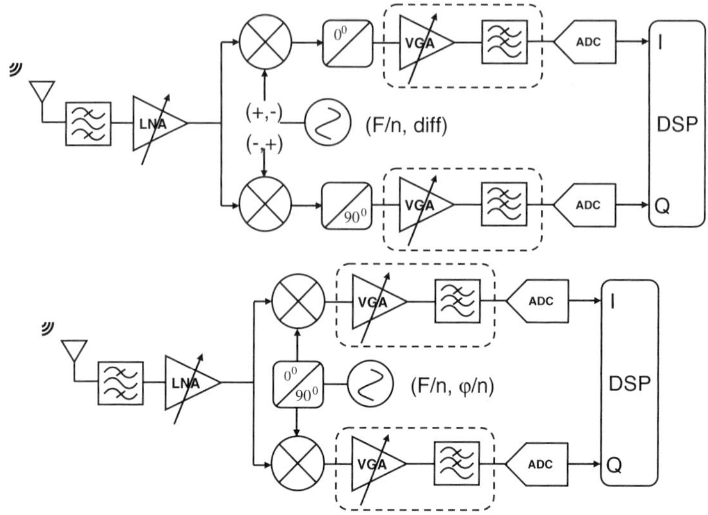

Figure 6.16. A subharmonic receiver using single-stage direct conversion topology: (a) I/Q in the RF path (top), and (b) I/Q in the LO path (bottom).

perspective, we expect the value of n to be large, implying lower frequency generation using on-chip voltage-controlled oscillators (VCOs), and power optimization considerations. As shown in Figure 6.16, the two other aspects of such transceivers are as follows: (1) phase shift in the RF signal path, which can be a compromise for the noise figure; and (2) higher phase division granularity for higher n, and the phase error associated with individual-phased waveforms, would accumulate due to the frequency multiplication in various stages. Hence, in the first architecture, we pay a penalty in terms of increased levels of thermal noise, and in the second, we pay a penalty in terms of worse-phase noise from the oscillators. In the second architecture, we also pay a penalty in terms of area, as multiphase oscillators would require two inductors per phase of signal generation. However, the large value of n might allow one to use more digital and mixed-signal schemes to generate those phases accurately.

6.5.2 System Considerations

To date, there have been many reports of subharmonic mixing approaches for integrated RF systems [26,27]. These mixing techniques can be broadly categorized between passive and active. There are two approaches in developing a subharmonic mixing core: (1) use of an antisymmetric transfer characteristic of nonlinear devices such as diodes and (2) use of a frequency multiplier as part of a mixer's switching core. The principles of diode-based mixing are about a century old, and diode-based detectors were popular during the world war. The basic configuration of antiparallel diode pair (APDP) mixers based on Schottky barrier diodes were originally proposed by M. Cohn, and it was successfully used for CDMA cellular applications by Mitsubishi Inc.

In any nonlinear circuits and systems, it is difficult to generate a subharmonic frequency from an input signal. Usually nonlinear circuits and systems generate harmonics and frequency multiplicative operations. Hence, when one uses subharmonic architecture, cross-talk between PA and VCO, and LNA and VCO, reduces (lower frequency pulling), which leads to superior isolation performance.

6.5.3 Antiparallel Diode Pair

Let us illustrate the fundamental operating principles of subharmonic mixers as illustrated by M. Cohn. Exponential characteristics can be realized easily with the help of diodes. Figure 6.17 illustrates the various core topologies of diode-based mixers.

In this case, we can replace the nonlinear devices as shown in Figure 6.17(b) by Schottky barrier diodes, leading to the following equations:

$$i_1 = -i_s(e^{-\alpha V} - 1) \tag{6.5}$$

$$i_2 = i_s(e^{\alpha V} - 1) \tag{6.6}$$

where α is the diode slope parameter ($\alpha \approx 38V^{-1}$ for GaAs Schottky diodes). Similar analysis as before results in

$$g_T = g_1 + g_2 = \frac{di_1}{dV} + \frac{di_2}{dV} = \alpha i_s(e^{\alpha V} + e^{-\alpha V}) = 2\alpha i_s \cosh(\alpha V) \tag{6.7}$$

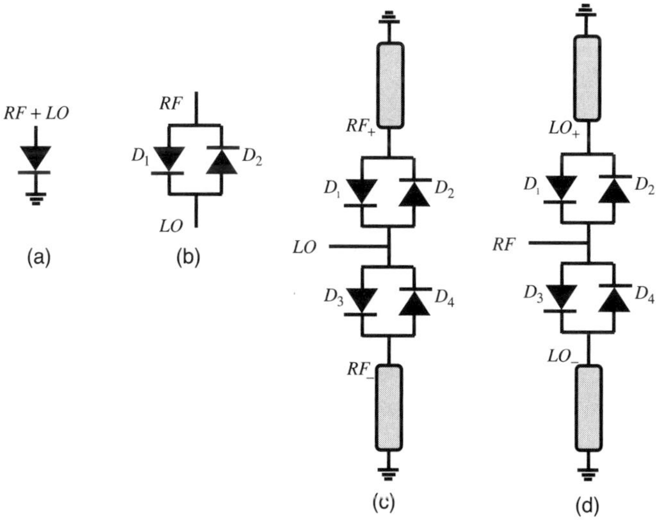

Figure 6.17. Diode-based mixers: (a) single diode mixer, (b) APDP, (c) APDP with differential RF, and (d) APDP with differential LO.

With the applied voltage of $V = V_{LO}\cos(\omega_{LO}t) + V_{RF}\cos(\omega_{RF}t)$, as before, and including only the LO waveform in the transconductance term under the assumption $V_{LO} >> V_{RF}$, the transconductance expression in Eq. (6.7) reduces to

$$g_T = g_1 + g_2 = 2\alpha i_s[I_0(\alpha V_{LO}) + 2I_2(\alpha V_{LO})\cos(2\omega_{LO}t) + 2I_4(\alpha V_{LO})\cos(4\omega_{LO}t) + \cdots] \tag{6.8}$$

where $I_n(\alpha V_{LO})$ are modified Bessel functions of the second kind. For the applied voltage $V = V_{LO}\cos(\omega_{LO}t) + V_{RF}\cos(\omega_{RF}t)$, the output current expression is

$$i = g_T V = g_T[V_{LO}\cos(\omega_{LO}t) + V_{RF}\cos(\omega_{RF}t)] \tag{6.9}$$

Equation (6.9) results in the output current as follows:

$$\begin{aligned} i = {} & A\cos(\omega_{LO}t) + B\cos(\omega_{RF}t) + C\cos(3\omega_{LO}t) + D\cos(5\omega_{LO}t) \\ & + E\cos(\omega_{RF} + 2\omega_{LO})t + F\cos(\omega_{RF} - 2\omega_{LO})t + G\cos(\omega_{RF} + 4\omega_{LO})t \\ & + H\cos(\omega_{RF} - 4\omega_{LO})t + \ldots\ldots \end{aligned} \tag{6.10}$$

As observed in Eq. (6.10), the output current contains frequency terms $(m\omega_{LO} \pm n\omega_{RF})$, where $(m+n)$is an odd integer; i.e., $(m+n) = 1, 3, 5, \ldots$ As the output current is formed by the interactions with even harmonics of LO waveform, this type of mixer is often referred to as even harmonic mixers (EHMs). Terms incorporating an odd order of the LO waveform are confined within the diode pair loop, and they cannot be taken outside because of the two-terminal nature of the diode devices. Hence, diode-based APDP structures can only be used in their even LO harmonic mixing variant. Transistors can be used to provide outputs at flexible harmonic frequencies.

Although APDP-based mixers have been used in GaAs technologies in the past, they can be implemented in silicon as well. Because of the lower parasitic capacitance of Schottky barrier diodes, these mixers are the highest operating frequency devices in the technology. In a silicon technology, one can easily obtain a Schottky barrier diode; however, these components may not be well modeled in a particular technology node.

At this stage, let us consider two fundamental aspects associated with passive harmonic mixers such as (a) conversion loss and (b) large local oscillator swing. Even if one can design these mixers with the highest possible design elegancy, they provide conversion loss, which must be compensated by higher gain of the front-end amplifier for a high sensitive receiver, which leads to higher power consumption. At the same time, the LO drive needs to be large to obtain an optimum conversion loss performance. At higher frequencies, this also leads to a significant amount of static power dissipation in the front end. However, the mixers themselves do not consume any DC current and are self-protective (when one diode is ON, the other can provide an ESD protection).

An inherent advantage associated with using a passive APDP-based architecture is the simplicity of the topology and lower voltage headroom as well as elimination of DC bias voltage. This removes the necessity of DC biasing, as the large-signal LO drive drives the individual diodes into the cutoff and ON regions. When the diode is OFF, only the reverse-biased capacitance is important, whereas for the forward-biased diode, the voltage variation of the ON resistance, which varies from infinity (OFF state) to a very small value (ON state). Such a large variation in the input impedance complicates the task of mixer design using antiparallel diode pair topology. As the topology is compact, the mismatch performance of the diode pair topology is inherently superior. As discussed, mismatch leads to the even-order terms including DC offset and second-order intermodulation values. Being a passive topology, the dynamic range is significantly higher compared with its active counterpart, while providing a much lower level of LO-to-RF isolation. This isolation is at 1X frequency, but in practice, it is the second harmonic of the LO frequency where this isolation needs to be considered.

Antiparallel diode pair topologies can also be constructed using transistors, and the two diodes to be used could be realized using gate-source or base-emitter terminals as illustrated in Figure 6.18, whereas the drain terminals can be used as the current output terminal. If the transistors are biased near their threshold voltage such that the excursion in the "dead-zone" of a transistor's input characteristics is mimimized, which leads to more efficient utilization of the LO waveform. This can lead to a significant power reduction, while providing conversion gain at the output. Because of the exponential nature of the nonlinearity, the harmonic strengths of the LO waveform decrease slowly, which leads to its use at fairly higher integral multiples of the LO frequency.

Beyond the normal operation of the antiparallel diode pair as a mixing core, interesting aspects with respect to their harmonic and amplitude performance can be observed. A diode-based frequency detector would lead to the formation of $N-1$ nulls in the downconverted waveform, whereas the LO drive amplitude is varied

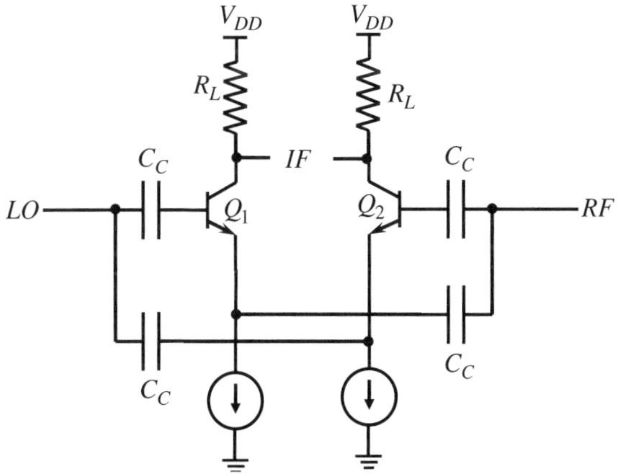

Figure 6.18. A transistor-based active APDP structure.

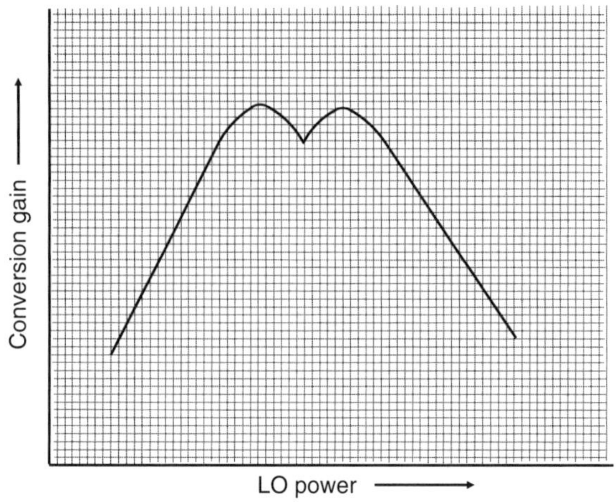

Figure 6.19. Conversion null formation in APDP mixer topologies.

[28]. Following the same analysis, it can be shown that the antiparallel topology would provide $\frac{N}{2}-1$ nulls in the conversion characteristics. This is illustrated in Figure 6.19.

The even-order terms resulting out of mismatches can also be mitigated by injection of an even-order LO harmonic, and they have been successfully implemented in [29].

A variant of passive subharmonic mixing can also be realized by ring field effect transist (FET) mixers using CMOS technology as shown in Figure 6.20. However, this topology uses an explicit frequency doubler in its architecture. A doubler-based approach can also be used in active mixers.

6.5.4 Active Subharmonic Mixers

An active subharmonic mixer-based transceiver is attractive from a system perspective, in terms of system budget, as it requires a low signal drive on the LO terminal, and it provides conversion gain. At the same time, transistors provide superior isolation between the LO and RF ports. Compared with what has been described before, active harmonic mixers tend to use explicit frequency doublers. A commonly used topology is shown in Figure 6.21.

In the first set, the frequency multiplication action is implemented by a simple digital logic function using transistors. These waveforms are arranged in a 90° out-of phase version. As can be easily observed, this topology increases transistor stacking (leading to difficulty in lower voltage operations) with increased subharmonic order (higher N). An alternative approach is to use the parallel transistor approach, which saves headroom, while increasing more current consumption as the stack width (and not the height) increases with higher subharmonic order. This is essentially the rectification of the incoming waveforms performed using two parallely connected transistors as illustrated earlier. All of these architectures are generic and can be used for bipolar and CMOS technologies.

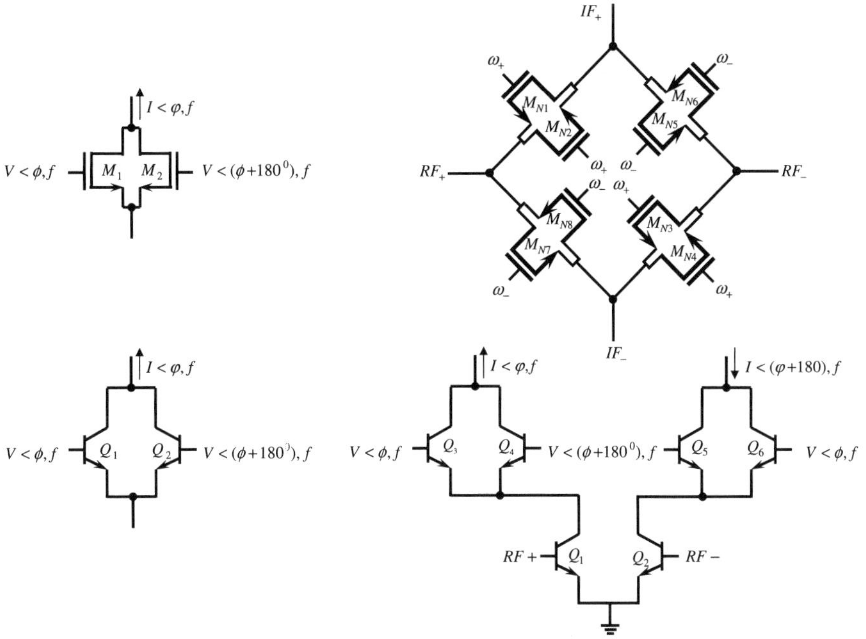

Figure 6.20. Subharmonic circuits based on explicit doublers: (a) passive ring FET, and (b) bipolar.

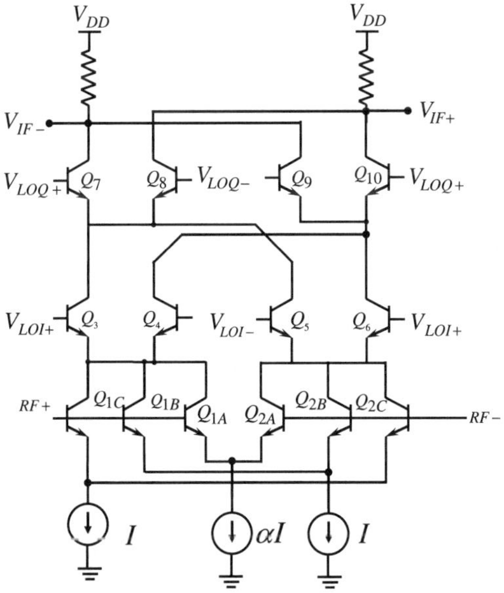

Figure 6.21. An active subharmonic mixer based on LO I-Q multiplication.

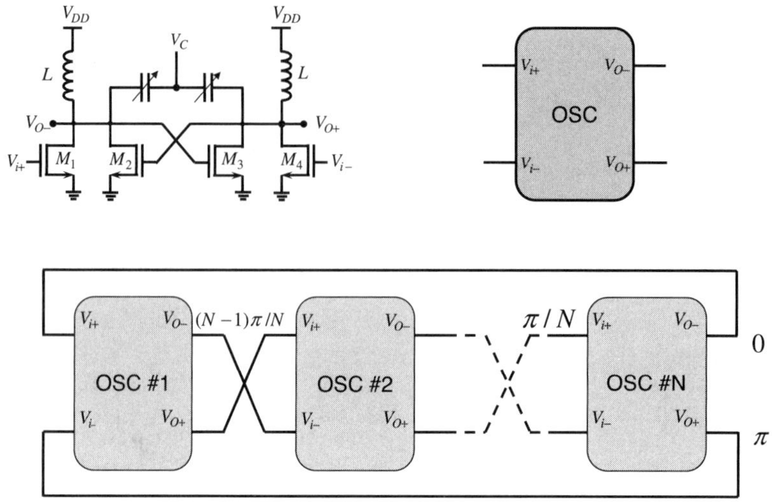

Figure 6.22. Multiphase oscillator architecture.

6.5.5 Subharmonic Architecture Building Blocks

Subharmonic transceiver approach is a judicious way of frequency planning, and is quite generic to be used in the case of direct conversion, low IF and super heterodyne architectures. In general, the salient features of subharmonic receivers include:

1. Inclusion of explicit/implicit frequency multipliers
2. Minimum number of circuit nodes operating at maximum frequency
3. Multiphase LO drive waveforms, either generated as part of VCO or a frequency divider, or polyphase network

Figure 6.22 shows an oscillator architecture to provide multiphase LO dignal. Subharmonic architectures are especially useful for mmWave frequency band, where lower power becomes quite attractive, and it's difficult to generate high frequency signal sources.

6.6 SUPER-REGENERATIVE ARCHITECTURES

The super-regenerative architecture was invented by Armstrong in 1922. A block diagram of such an architecture is illustrated in Figure 6.23.

It consists of a low noise amplifier (also called "isolation amplifier") followed by an oscillator and then by an envelope detector. The oscillator is represented by an amplifier with gain A, followed by the frequency selective network β. The quench

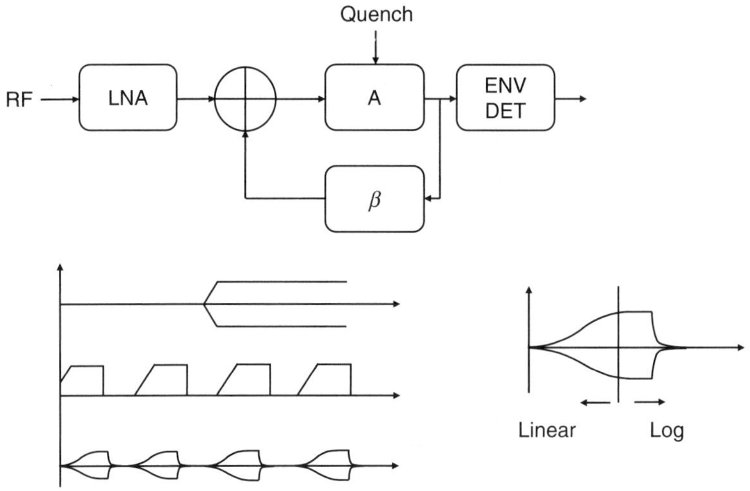

Figure 6.23. Super-regenerative architecture.

signal is used to modify the gain of the amplifier in a periodic manner, thereby setting conditions of oscillations. If the RF signal is not injected, the oscillation condition is governed by the quench signal, which sets the loop gain and the oscillator will have a certain start-up time. In presence of the RF signal, the start-up time would vary, depending on the input signal level.

Figure 6.23 also illustrates the key signals in the receiver system. The RF signal can be assumed to be OOK modulated, and the quench signal is simply a periodic signal with a period smaller than the data period. In the presence of the quench signal, oscillations build up, and depending on whether the RF signal is applied, the start-up condition would differ. Hence, the output is a train of start-decay envelopes, which can be detected using an envelope detector.

6.6.1 Formalism

Fundamentally, the receiver can operate in two fundamental modes: (1) linear and (2) logarithmic. In the linear mode, the relationship between the demodulated amplitude and the input RF signal is linear in nature. The oscillator does not reach its steady state because of high quench frequency or low quench current. In the logarithmic mode, however, the output from the envelope detector provides a logarithmic relationship w.r.t the input RF amplitude. The oscillator reaches its steady-state amplitude at each quench cycle (low quench frequency or high current). A typical operation of the receiver consists of a combination of these two modes of operation. Transitions between two modes are determined by change in amplitude, frequency, and duty cycle of the quench signal.

The other mode of this receiver is "regenerative mode," during which the bias current of the receiver is not sufficient to cancel all the losses, which leads to a

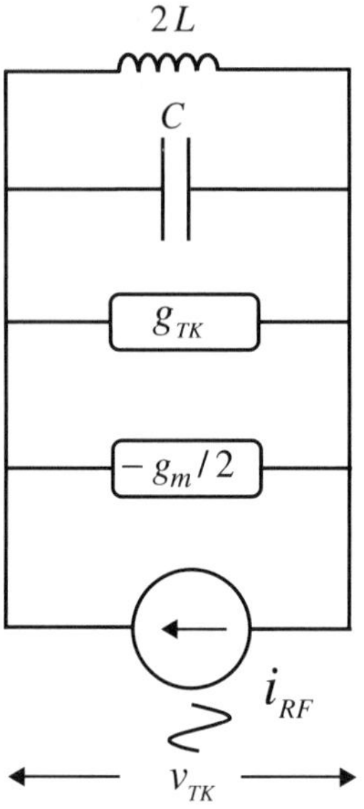

Figure 6.24. The tank circuit.

Q-enhanced selective amplifier. The gain is essentially higher, but it is difficult to control the Q of the amplifier in a robust manner.

The basic principle of a super-regenerative receiver is illustrated in Figure 6.24. This is similar to the resonating tank of a VCO circuit, where the components can be placed on/off-chip. The negative active conductance, denoted by $-g_m/2$, is varied periodically w.r.t. the quench signal. Tank loss is given by the positive conductance g_{LC}, and the overall conductance is given by

$$g(t) = g_{LC} - \frac{g_m(I,t)}{2}$$

and $g(t)$ can assume positive and negative values, and varies periodically below and above zero at the quench frequency f_q. For stability, $\bar{g}(t) = f_q \int_{t_0}^{t_0+1/f_q} g(t)dt > 0$ should be satisfied.

The general equation describing the time domain behavior of the tank voltage can be represented as follows:

$$\frac{d^2v}{dt^2} + 2\frac{dv}{dt} \cdot \frac{g(t)}{2C} + \left(\frac{1}{2LC} + 2\frac{d\omega(t)}{dt}\right)v = -\frac{1}{C} \cdot \frac{di_{RF}}{dt} \tag{6.11}$$

where $i_{RF} = i_{RF0}\sin(\omega_{RF}t)$. If $g(t)$ is positive, the equation provides a stable solution, and the free oscillation is damped with a time constant $\tau = \frac{2C}{g(t)}$. In this mode, the circuit works as a filter. If $g(t)$ is negative, the time domain response exhibits an oscillatory behavior with increasing time constant.

Solution of the above equation leads to the time domain representation of the tank signal given by

$$v(t) = e^{-\int_0^t \omega(t)dt} \cdot \left[\alpha \cos(\omega_0 t) + \beta \sin(\omega_0 t) + \frac{1}{\omega_0} \int_0^t \varphi(z) \sin\{\omega_0 (t-z)\} dz \right. \quad (6.12)$$

where $\omega_0 = \frac{1}{\sqrt{2LC}}$.

Solutions of the time domain waveform should be found out by providing different quench waveforms such as square, sinusoidal, and triangular, which are feasible to be generated in an IC. Such analysis has been provided in [35], and the best selectivity of such receivers is obtained at a Sawtooth waveform shape. Traditionally such receivers are operated at lower data rates, and an amplitude modulation (ON–OFF keying) is commonly employed.

6.6.2 Architecture and Circuit Illustration

From a system perspective, low power is achieved by elimination of several blocks such as switching mixers and frequency dividers. Low area can be achieved by sharing of inductive load between the LNA and the VCO core.

Let us consider the circuit level architecture as shown in Figure 6.25. After the LNA and the VCO circuits, an NMOS-based envelope detector provides a voltage

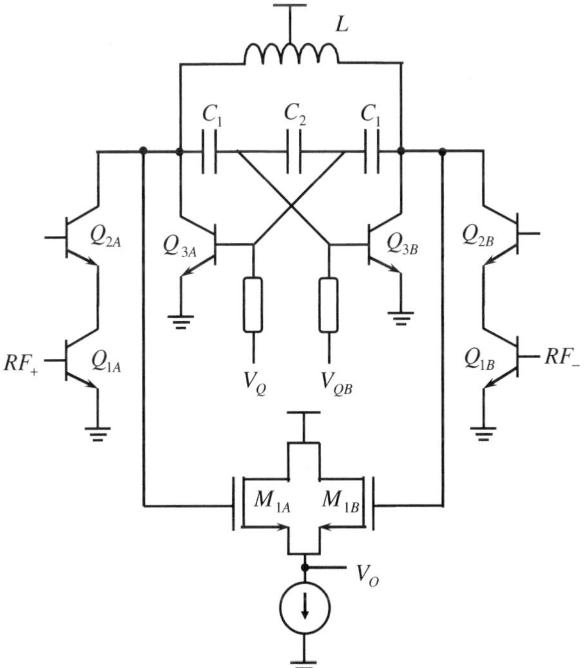

Figure 6.25. Transistor-level implementation of super-regenerative receiver.

proportional to the rectified envelope of the input RF signal. A cascode configuration provides a superior reverse isolation performance, and it provides the RF output current to the VCO core. Bipolar and MOS devices can be used as appropriate. In the analog back end, the filters can be designed using weak inversion MOS transistors to obtain further power reduction. Although they are attractive from a low-power implementation, super-regenerative receivers provide degraded sensitivity levels, and they operate under the limits of fairly low data-rate systems.

6.7 HEARING AID APPLICATIONS

Low-power, low-voltage applications are critical toward the development of hearing aid systems. In general, hearing aid systems can be (1) digitally programmable analog processors and (2) embedded DSP platforms. The first type usually provides low area, low power, whereas the second approach provides more flexibility. To combat the low supply voltage, voltage boosting can be extensively used. These systems operate from 100 Hz to 80–10-kHz frequency ranges. Figure 6.26 illustrates an architecture for a hearing aid system. These systems use a single-cell voltage supply as low as 1 V and operate at the kilohertz frequency range.

6.7.1 Architecture Based on Digital/Mixed-Signal Circuits

In a digital hearing aid system, the flexibility and programmability is obtained through the DSP. The accuracy and dynamic range of the analog front end degrade because of the low supply voltage, and the task of power management unit becomes stringent. The entire system consists of (1) preamplifier, (2) Σ-Δ ADC, (3) DSP, (4) Σ-Δ DAC, and (5) receiver driver. The analog front end consisting of preamplifier and Σ-Δ ADC consume most of the system power. Various approaches have addressed adaptive control of the SNR, by changing the loop structure and the clock frequency of the Σ-Δ ADC as illustrated in Figure 6.27.

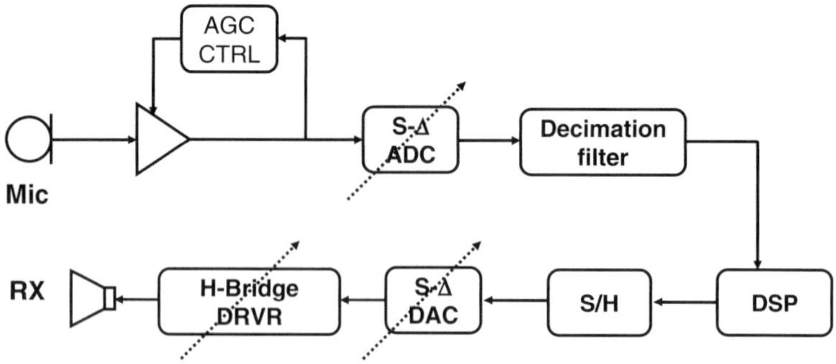

Figure 6.26. The signal path in a mixed-signal hearing aid system.

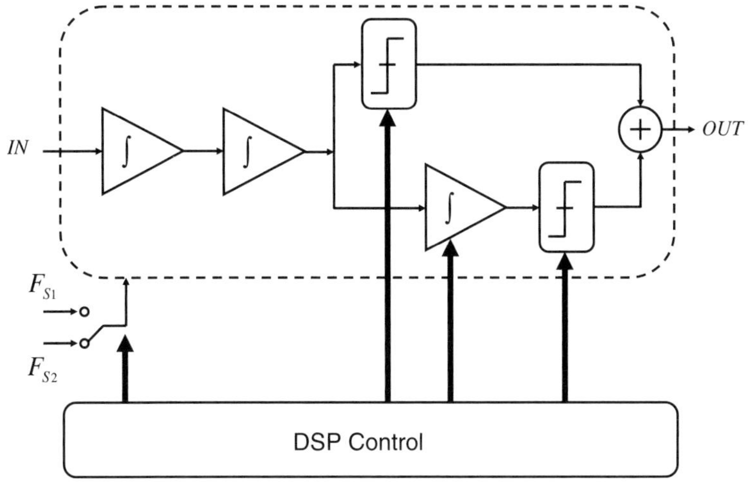

Figure 6.27. An adaptive-SNR Σ-Δ modulator.

High-order Σ-Δ converters are usually avoided because of their stability issues. Gain control of the preamplifier can be performed by combining the exponential gain control and an automatic gain control.

These functionalities can be combined using the combined loop as illustrated in Figure 6.28. The threshold gain and threshold knee point of the preamplifier can be set by varying V_{VC} and V_{TH}. The peak detector circuit is used to sense the outputs of a gain amplifier, and it generates the scaled envelope. When the value of V_{VC} is low, the

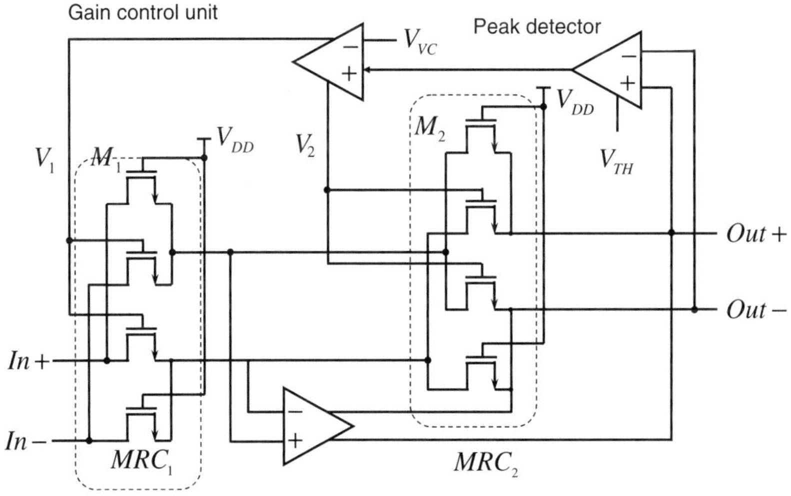

Figure 6.28. Combined automatic gain control and exponential gain control for hearing aid application.

threshold gain decreases, and the power consumption of the preamplifier is reduced. With a high V_{VC}, the threshold gain increases and the required resolution of S-Δ reduces. A logarithmic function can be expressed in terms of a pseudo-polynomial expression as given by

$$G = \frac{1+x}{1-x}$$

Using the input MOS transistors in the linear region, this leads to

$$A_v = \frac{(L_2/L_1)}{(W_2/W_1)} \cdot \frac{(V_{DD}-V_1)}{(V_{DD}-V_2)}, \quad V_1 = V_{VC}-V_x, \; V_2 = V_{VC}+V_x$$

Because of the low-frequency operation, low flicker noise devices are used in the OTA design employed in the preamplifier block. In the DAC, the noise contribution originates from the phase noise of the clock (jitter) and from the order of the S-Δ modulator. The edge symmetry (rise and fall) of the oscillator can be obtained by the symmetry of the P and N devices to reduce clock jitter performance.

6.7.2 Architecture Based on Subthreshold Current-Mode Circuits

Analog implementation of hearing aid systems programmed by digital word is illustrated in [41]. This architecture operates on a 1-V supply and uses a subthreshold circuit operation of CMOS transistors. The front end can adapt to various impedances of different input sources, and it can mix them using programmable weights. A voltage regulator is used to bias the input microphone and other transducers. The AGC block is responsible for all dynamic range corrections. The optimum point of all analog processing blocks in the chain can be controlled by digital word. The feedback loop computes an effective signal envelope through combined fast and slow filters to ensure protection against overshoots and speech intelligibility. Other blocks of the loop provide a steady compression ratio and open loop gain. In the back end, low impedance receivers are driven by a class-D output amplifier stage using pulse-duration modulation. The SNR of the signal path can be optimized by adjusting AGC gain and volume control through digital word, and these controls can bypass a few stages to optimize for power consumption. The analog signal processing can exploit the nonlinear characteristics of MOS devices, operating in the subthreshold regime. Incoming current signals are compressed in voltage using compression functionality. After the conversion in the voltage domain, a nonlinear function (expansion) maps the signal from the voltage domain to the output current. In this manner, the external signal processing appears linear in nature. An exponential function is used to realize the compression and expansion characteristics, which can be realized using bipolar or MOS transistors in weak inversion. From the EKV model, the subthreshold MOS can be modeled as follows:

$$I_D = I_S e^{\left[\frac{V_{GB}-V_{T0}}{nV_T}\right]} \left(e^{\left(-\frac{V_{SB}}{V_T}\right)} - e^{\left(-\frac{V_{DB}}{V_T}\right)} \right) \tag{6.13}$$

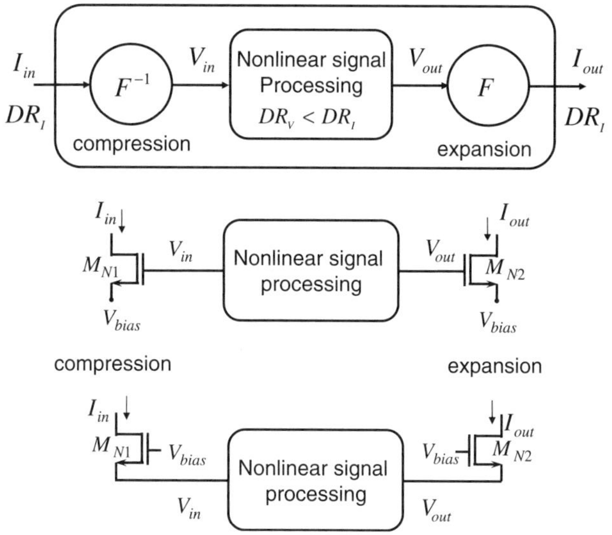

Figure 6.29. Compression and expansion using subthreshold MOS devices: (a) gate driven and (b) source driven.

Log companding functions can be used according to gate driven or source driven, resulting in

$$F(V) = I_s e^{\left[-\frac{V_{T0}+nV_{bias}}{nV_T}\right]} e^{\left[\frac{V}{nV_T}\right]} \text{ (gate driven)} \tag{6.14}$$

$$F(V) = I_s e^{\left[\frac{V_{bias}-V_{T0}}{nV_T}\right]} e^{\left[-\frac{V}{V_T}\right]} \text{ (source driven)} \tag{6.15}$$

Various functionalities of the hearing aid block can be realized using MOS subthreshold logic. The basic functionalities include (1) amplification, (2) automatic gain control, (3) filtering, (4) bias generation, and (5) pulse-duration modulator. Gate- and source-driven configurations are illustrated in Figure 6.29.

In the current domain, an amplification can be illustrated as $I_{out} = GI_{in}$, which is reflected in the voltage domain by $V_{out} = \Delta V_{gain} + V_{in}$. The gain of the logarithmic amplifier is given by

$$G = 20\log(e) * \frac{\Delta V_{gain}}{V_T} \tag{6.16}$$

An automatic gain control loop is illustrated in Figure 6.30, followed by the logarithmic amplifier circuit in Figure 6.31. The effective output envelope is computed by applying a full-wave rectified and filtered version of amplifier output. The error signal $V_{\log}$ is obtained from the comparison of this signal and from the output of auxiliary amplifier using I_{tk}. The closed-loop compression ratio is obtained

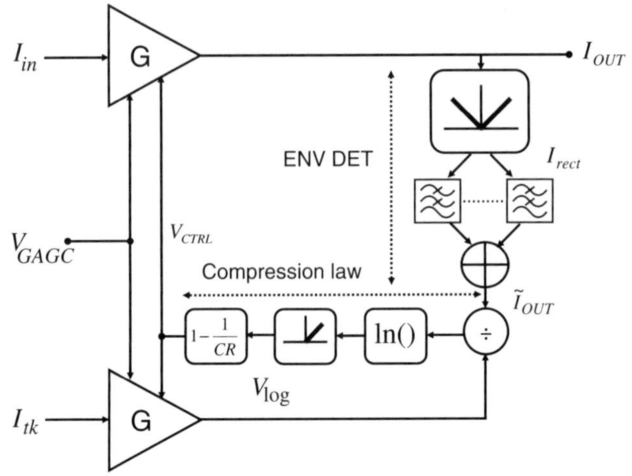

Figure 6.30. Automatic gain control using a logarithmic amplifier.

by properly scaling $V_{\log}$ and by injecting a control signal V_{ctrl} to both amplifiers. The overall input–output characteristics are given by

$$\begin{aligned} I_{out} &= I_{in}, I_{in} \leq I_{tk} \\ I_{out} &= I_{tk}^{1-\frac{1}{CR}} I_{in}^{\frac{1}{CR}}, I_{in} \geq I_{tk}, 1 < CR < \infty \end{aligned} \tag{6.17}$$

Figure 6.32 illustrates the various circuits under consideration in a hearing aid system. The logarithmic output voltage is given by

$$\begin{aligned} V_{\log} &= 0, I_{out} \leq GI_{tk} \\ V_{\log} &= V_T \ln\left(\frac{I_{out}}{GI_{tk}}\right), I_{out} > GI_{tk} \end{aligned} \tag{6.18}$$

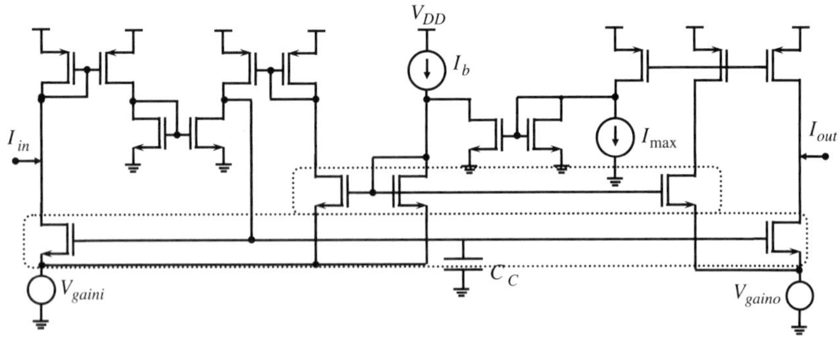

Figure 6.31. Circuit schematic of a logarithmic amplifier.

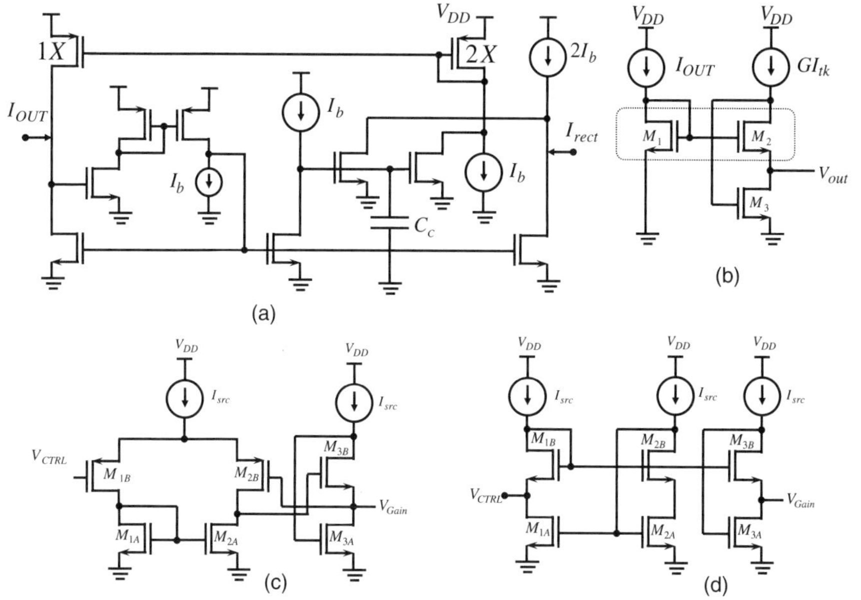

Figure 6.32. Circuit schematic of (a) low-voltage full-wave rectifier, (b) first quadrant log divider, (c) gain control block for worse matching, and (d) gain control at low noise.

The filtering circuit can be characterized by the current domain equation

$$\frac{dI_{out}}{dt} = 2\pi f_0 (I_{in} - I_{out})$$

processing in the voltage domain to keep the filter externally linear, the source driven companding function leads to

$$\frac{dV_{out}}{dt} = 2\pi f_0 V_T (1 - e^{\frac{V_{out} - V_{in}}{V_T}}) \tag{6.19}$$

A PTAT circuit can be developed using exponential characteristics of subthreshold MOS devices in the main transistors, whereas in the loop, the ratio is adjusted by a factor K, which leads to the reference voltage as follows:

$$V_{ref} = V_T \ln K \tag{6.20}$$

Apart from the main transistors, others would be in strong inversion, which leads to the output current as (assuming switch is OFF and no R_{ref} is present) follows:

$$I_{ref} = Q\left[\frac{\ln K}{2(M+1)}\left\{\sqrt{\frac{M}{N}} + \sqrt{\frac{M}{M+N+1}}\right\}\right]^2 \tag{6.21}$$

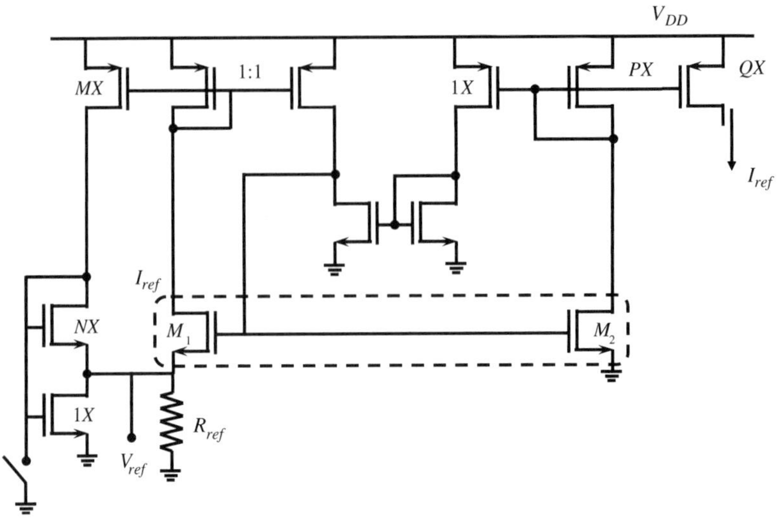

Figure 6.33. Low-voltage PTAT generation.

With the switch in circuit, with R_{ref} in the circuit,

$$I_{ref} = Q\frac{V_T \ln K}{R_{ref}} \tag{6.22}$$

The PTAT circuit is illustrated in Figure 6.33.

The pulse duration modulator can be obtained from comparing the baseband input (I_{in}) and high-frequency signal (I_{tri}). Figure 6.34 illustrates a block diagram of a pulse duration modulator. A triangular waveform can be obtained by integrating a constant reference I_{ref}, which is periodically inverted according to output window

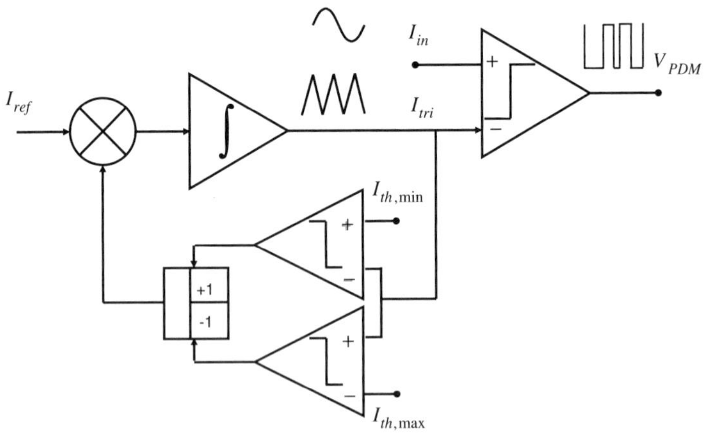

Figure 6.34. Pulse duration modulator.

$I_{th,\min}$ and $I_{th,\max}$. In the linear region, the time domain slope can be given by $\frac{dI_{tri}}{dt} = \pm\frac{I_{ref}}{\tau}$, $I_{th,\min} < I_{tri} < I_{th,\max}$; with gate driven companding, this leads to

$$\frac{dV_{tri}}{dt} = \pm\frac{nV_T}{\tau}e^{\left(\frac{V_{ref}-V_{tri}}{nV_T}\right)} \tag{6.23}$$

6.8 RADIO FREQUENCY IDENTIFICATION TAGS

RFID technology is being widely used to identify, locate, and exchange information with remote devices. The concept of RFID systems is quite old, but has attracted a lot of attention recently. These types of systems provide uniqueness in terms of their low-power operation, and the application is much more different from a conventional cellular communication system.

6.8.1 System Considerations

In comparison with higher frequency bands, the UHF band (400 M) provides superior diffraction around and penetration through obstacles where a line of sight may not be feasible. Compared with the lower frequencies (13.56 M), this band can support higher data rates, making it suitable to communicate with multiple devices at the same time. Various bands that are in consideration are as follows: 315 M, 433 M, 868 M, and 915 M.

RFID tags use two major topologies: (1) active RFID tags and (2) passive RFID tags. Active tags use a battery to power the circuits. Passive tags use rectifiers and generate DC voltage from incoming RF signals. These systems are usually of very small form factors and are targeted toward high-volume applications. The RFID tags communicate with a central base unit (reader) by pulsed RF communication. In the forward link (reader to tag), the presence of RF energy denotes bit "1" and a lower level or absence of RF energy denotes "0." In the reverse link of communication (tag to reader), a constant RF signal level is input from the reader, and the tag provides "1" by changing the input impedance of the system by a shorting switch to ground (change in impedance of the antenna). This provides a large RF power to the reader, and it is commonly known as "backscattering." Bit "0" is obtained by retaining the desired impedance level and by reflecting minimum power to the tag.

In the two cases of forward and reverse links, the chip needs to operate at very low RF power levels, and to operate from a large on-chip capacitor. This capacitor is precharged at the instant when RF power is available. Because of the area requirement, the on-chip capacitor cannot be arbitrarily large. Usually the chip consumes about 5μA, and hence, the regulator must be designed at the 100nA range of power dissipation. Hence, a significant design challenge exists in the design of the regulator.

6.8.2 System Architecture

Figure 6.35 illustrates an integrated RFID system. The received RF signal is used in the rectifier, converting RF energy to DC. The rectifier also allows current flow in one

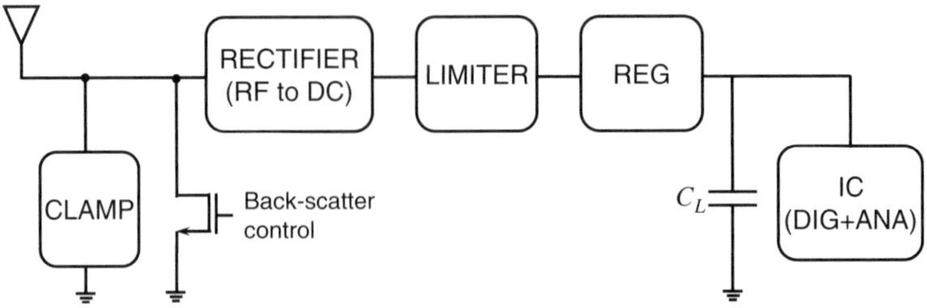

Figure 6.35. RFID voltage regulation block diagram.

direction, which prevents any leakage from the on-chip capacitor. Depending on the reader-to-tag distance, this DC level can vary significantly. To provide almost constant voltage to the chip, this DC voltage is processed through a limiter circuit, which uses clamping of the voltage. Afterward, a regulator is used to regulate the voltage at the desired level.

Figure 6.36 illustrates a multistage rectifier consisting of N diodes and N capacitors. The transistors and capacitors are equal in all branches. The RF lines can be connected directly through an antenna or through a matching network. The input RF signal is assumed to be sinusoidal with $RF_{in} = V_{RF} \cos\omega_{RF}t$. The load capacitor is chosen to be sufficiently large to store enough charge for signal processing tasks.

6.8.3 Rectifier, Limiter, and Regulator

Let us consider the steady-state voltage distribution in the rectifier network. Starting from the initial condition, in the negative half cycle of the input RF waveform, capacitor C_1 is charged to its peak value V_{RF}. C_1 retains the charge even in the decreasing part of the charging waveform as long as the MOS connected diode M_1 is

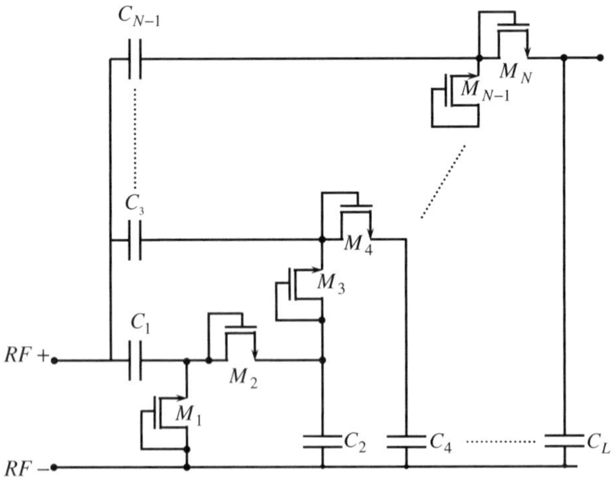

Figure 6.36. Circuit illustration of an N-stage rectifier.

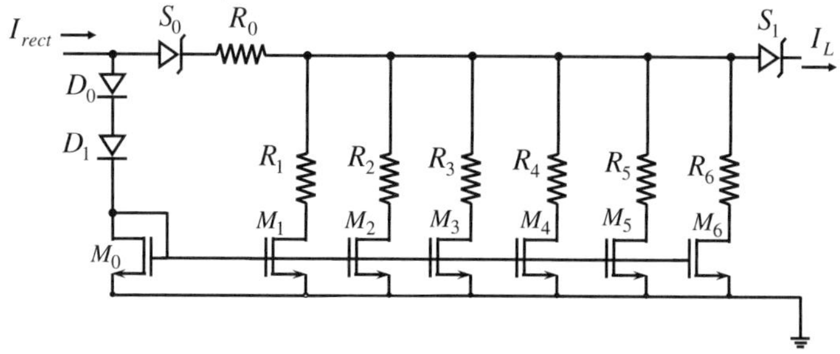

Figure 6.37. Limiter circuit schematic.

reverse biased. In the positive half cycle, the top plate of the capacitor C_1 is charged to $2V_{RF}$, as M_2 is ON. Hence, C_2 is also charged to $2V_{RF}$. In the next positive half cycle, C_3 is charged to $3V_{RF}$, and so on, which leads to a final voltage of NV_{RF} across the load capacitor, in the steady state. The number of stages can be even or odd.

Figure 6.37 illustrates an RFID limiter circuit. At high RF input levels, there needs to be a bypassing path to ground; otherwise, the voltage would increase significantly, resulting in the breakdown of transistors. In this limiter, the current shunting is obtained through D_0, D_1, and the MOS diode, M_0. The device sizes should be optimized such that at low RF levels, the limiter does not load the rectifier, and the entire current from the rectifier flows to the regulator. At high levels of output signal from the rectifier, the shunt path provides a bypassing path for flow of excess current to ground. This excess current can be sensed through some mirroring circuit using M_0, and this can be used to bias the regulator dynamically for current savings. Current through the shunt branch is mirrored inside the limiter to control the slope of the limiter DC I–V characteristics at the output. The limiter output voltage V_{LIM} is given by

$$V_{LIM} = V_{D0} + V_{D1} + V_{GS,M0} - V_{S0} - V_{S1} - \frac{N}{N+1}(I_{rect} - I_L) - I_L R_0 \tag{6.24}$$

with N being the number of branches in the mirror. The slope of this output characteristics is given by

$$\frac{dV_{LIM}}{dI_{rect}} = \frac{1}{\sqrt{2(N+1)K(W/L)}} \cdot \frac{1}{\sqrt{I_{rect} - I_L}} - \frac{N}{N+1} R_0 \tag{6.25}$$

At low values of input current I_{rect}, the first term is dominant in determining the slope. Using more stages smoothens out the output characteristics, thus avoiding any breakdown issues. In terms of breakdown, the drain/source breakdown voltage is higher compared with the gate-source breakdown voltage, and in submicron nodes, the gate-source voltage is the key determinant of the breakdown voltage.

The regulator implementation is illustrated in Figure 6.38. The load current varies from about 100 nA to 25 μA, and the rectifier current varies from 2 μA to 3 mA; under these conditions, the regulator stability is a challenging design consideration. The

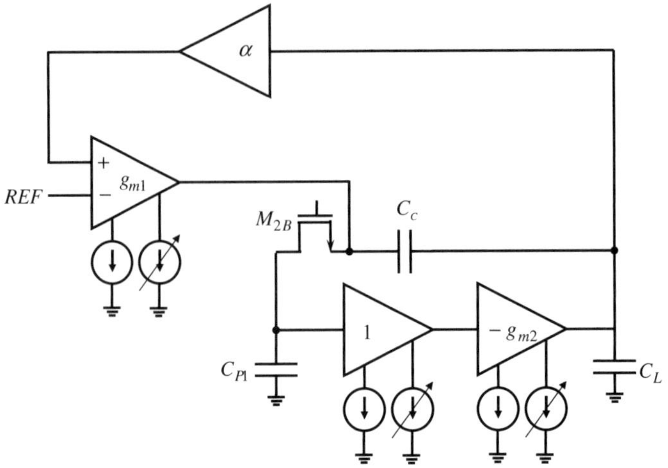

Figure 6.38. Regulator feedback loop.

currents of the g_m blocks are dynamically adjusted to solve phase margin and stability issues at various load limits. This process is also known as "dynamic bandwidth boosting," and it increases the bandwidth of the circuit and reacts fast to the supply variations caused by a sudden change in RF energy. Implementation of the error amplifier is based on a pair of MOS devices operating in subthreshold. The input devices are deliberately mismatched, and because of the exponential characteristics at subthreshold, this creates an offset voltage that is a PTAT voltage. This PTAT voltage is scaled using a resistor string, and it is added to the V_{BE} of the transistor Q_1, to provide a band-gap voltage. Figure 6.39 illustrates the error amplifier.

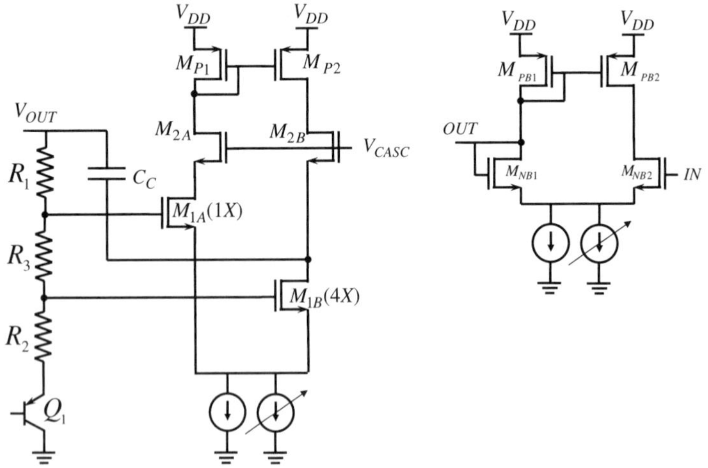

Figure 6.39. (a) First transconductor (gm1) loop and (b) unity gain buffer.

6.8.4 Antenna Design

The design of antennas provides a significant challenge to the success of UHF frequencies (300 M to 1 G). It is desired that the RFID antennas be quite small, usually 0.1λ, and physically thin. The most popular types include spirals, loops, and meander lines to achieve trade-off between antenna performance and size. Active RFID implementations also include a battery, and their size needs to be optimized for the best possible system trade-off.

Sometimes it may be desired that the antenna would be self-resonant without the necessity of any matching network, and that the geometry would scalable. [43] illustrate the meander-line-based antenna topology. It uses a wire or metal trace that is continuously folded in a two dimensional (2-D) plane. This folding adds inductive loading and reduces an antenna's size in comparison with a resonant dipole. Figure 6.40 illustrates the antenna implementation. It consists of two sections, a capacitive strip and a meander line. The resonance of the antenna is determined by these two sections. The design parameters include antenna width w, length of capacitive strip d, number of turns N, meander line width w_l, and meander line spacing w_s. The ground plane should be comparable with the size of the antenna to obtain performance similar to monopoles.

Instead of providing a mathematical illustration, let us obtain intuitive insights into this design. Keeping everything the same, as w_l increases, the effective inductance of the meander line decreases, which leads to a large capacitive component to keep the resonance frequency unchanged. At the same time, an increase in w_s would lead to a reduction in inductance, and the capacitive strip dimension d would need to be increased. An increase in w_s would also lead to a decrease in the number of turns N as the resonant length of the meander section is dependent more on the spacing between

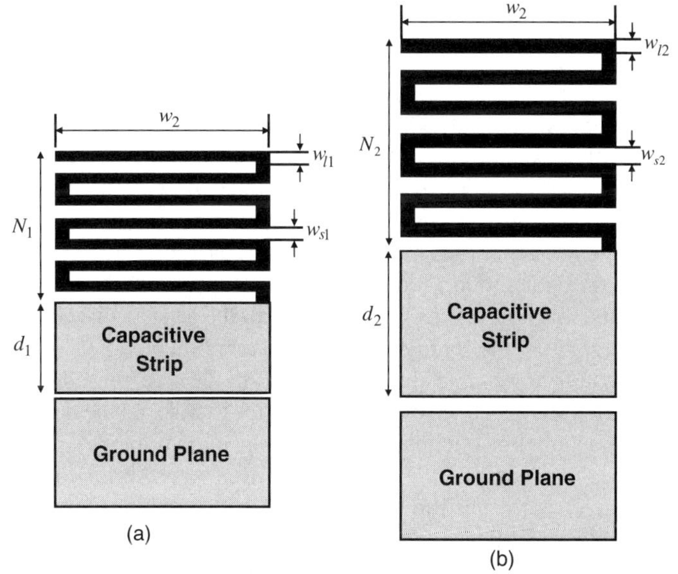

Figure 6.40. Two illustrations of RFID antennas: geometry shrinkage and expansion.

the turns. Antenna gain is proportional to the meander length. Hence, if the width and spacing of these antennas are altered, while keeping the same length of the antenna, the gain would be similar. However, as the dimensions become smaller, antenna structures become more susceptible to variations w.r.t. material parameters. In the design phase, EM simulators must be used to optimize w.r.t. all considerations, and integration with the other parts of the transceiver system, as necessary.

6.9 ULTRA-LOW-POWER RADIOS

In recent times, ultra-low-power, short-distance radios have attracted a lot of attention because of the increasing interest in health care and implantable telemetry. Various commercial solutions have already been reported [51–56]. These radios are designed for a range of several meters, starting from as low as 1–3 m and consume less than 3 mA in their peak operation. They usually target lower data rates (although some target higher data rates, in a duty cycled fashion), and reduced sensitivity. Low duty cycle radios can achieve significantly lower average power. The duty cycle of operation can be as low as 0.1% for many medical applications. This consideration is especially critical to the implantable devices, which need to last for several tens of years inside the human body. The performance metric for these radios can be stated in terms of nJ/bit, or mW/Mhz.

Ultra-low power radios can be categorized into two broad domains: (1) body area network radios, and (2) implantable radios. In a body area network application, the radio is outside the body, and its function is to establish a gateway between the human body to a wireless hub at home or hospital. Implantable radios, however, are inside the body, and once implanted, they should last for many years, ideally the lifetime of a person. Thus, they need to be designed and sequenced toward this application.

Body area network radios can be narrowband as well as wideband in nature. Power consumption considerations in the case of narrowband radios have been well illustrated in terms of a compromise between cost and form factor. These radios can operate in the ISM bands, such as the 315/433/868/915-MHz bands. Wideband radios usually use the power spectral density profile in the 3–10-GHz bands. These radios wake up and transmit a large burst of data. Because of the wide bandwidth of the system, the phase noise of the oscillator is not a stringent consideration, and a ring oscillator can be used to implement the pulse generation mechanism. Thus, oscillator start-up and the power consumption in their operation become less important in such radios.

Fundamentally, this implies considerations in all the aspects that we have already covered: (1) power reduction for high current consumption blocks, such as LNA, PA, and VCO, (2) form factor; and (3) low-power sniff mode receiver radios that operate on polling or interrupt signal from the external device to wake up the main transceiver. In terms of architecture, this implies minimum usage of continuous signal processing blocks at the highest possible frequency. However, apart from the peak power requirements, these radios need to exhibit ultra-low standby current and ultra-low sniff mode current. Implantable radios also enjoy the low-temperature variation of the human body environment, which simplifies many considerations for integrated circuits. However, they need to be very robust, reliable, and immune to any effects of MRIs, X-rays, and so on. For these reasons, specific semiconductor technologies are preferred over others, and dedicated frequency bands (402 M–405 Mhz) are allocated to the implantable devices.

It is clear to the readers at this point that ultra-low power cannot be achieved by simply powering down the radio part of the signal processing. Radios draw attention because they operate at the highest possible frequencies. However, a significant amount of power reduction needs to take place at the DSP, sensors, and software protocols that are used in many of the low-power standards. As the duty cycle heavily impacts the average power consumption, many times these radios are viewed as "energy efficient" rather than as "ultra-low power."

When the data rates are low/moderate, a loop modulator can be used to provide the transmitter solution, with the modulation scheme being FSK or a variant of frequency modulation to obtain a low peak-to-average ratio (leading to power-efficient amplifiers). In the case of moderate/high data rates, I/Q-based architectures can be used for transmitters as well as for receivers. In implantable devices, the transmit versus receive operation is heavily imbalanced, and an ultra-low-power transmitter would help improve the solution. The power levels are significantly lower compared with the existing cellular specifications.

Readers are recommended to study the various references that are listed at the end of this chapter. Ultra-low-power radios, body area, and implantables tend to solve the last mile problems of wireless communication. In many cases, they can enhance the quality of life significantly, especially for the hearing- and vision-impaired population.

CONCLUSION

To conclude this chapter, we would like to highlight fundamental considerations of low-power, low-voltage, and low-area systems. The power consumption of a radio is dependent on center frequency, bandwidth, modulation scheme, and driving point impedance. Given a technology node, designers often come up with several options based on the required performance. In advanced deep sub micron technologies, one can obtain higher speed, lower supply voltage, higher mask cost, higher process variation with reduced area and power. In older technologies, processing is quite well controlled and cheap, but the power and area increases. It is also clear that the packaging technologies also need to move in the direction of silicon feature size; otherwise, very soon the complicated million transistor systems would be bond-pad limited for functionality and automated testing purposes. Also, one has to look for judicious technology for radio design when the digital complexity is not much. Another fundamental question to address this issue is as follows: What is the application under consideration? To deploy wireless solutions to meet the needs of medical applications such as hearing aids, one need not have to consider highly sophisticated processes. However, for a high-data-rate communication system, one does not have the alternative. The integrability of MEMS-based resonators above a standard CMOS process may prove to be a viable device option for ultra-low-power transceivers at small form factors, as off-chip, high Q resonators tend to consume a significant area. In every implementation of low-power design, one needs to think about (1) source of power (battery/inductively coupled/environmentally scavenged), (2) duty cycle of the radio (always ON vs. rarely ON), (3) performance (output power/phase noise/sensitivity/linearity), (4) the form factor requirements, and (5) development cost. In consumer applications, small form factor, low cost, and low power may be the key aspects, whereas in medical applications, form factor and cost can be compromised in favor of ultra-low energy. In the future, many systems may even use scavenged energy to power up the radios and signal processors.

REFERENCES

Low-Power Designs

[1] A. Abidi, "Low-power radio-frequency ICs for portable communications," *Proceedings of the IEEE*, Vol. 83, No. 4, 1995.

[2] M. Degrauwe, et al., "A micropower CMOS instrumentation amplifier," *IEEE Journal of Solid State Circuits*, Vol. SC-20, 3, Jun 1985.

[3] A.J. Leeuwenburgh, et al., "A 1.9Ghz fully integrated CMOS DECT transceiver," *IEEE International Solid-State Circuits Conference*, Vol. 1, 2003. pp. 457–507.

Device Technologies/Scaling

[4] M. Namba, et al., "An analytical and experimental investigation of the cutoff frequency f_T of high-speed bipolar transistors," *IEEE Transactions on Electron Devices*, Vol. 35, No. 7, July 1988, pp. 1021–1027.

[5] J.D. Cressler, "SiGe HBT technology: A new contender for Si based RF and microwave applications," *IEEE Transactions on Microwave Theory and Techniques*, Vol. 46, No. 5, May 1998, pp. 572–589.

[6] S.K. Reynolds, et al., "A Silicon 60-Ghz receiver and transmitter chipset for broadband communications," *IEEE Journal of Solid State Circuits*, Vol. 41, No. 12, Dec 2006.

Low-Voltage Designs

[7] T. Manku, G. Beck, and E.J. Shin, "A low-voltage design technique for RF integrated circuits," *IEEE Transactions on Circuit And Systems-II*, Vol. 45, No. 10, 1998.

[8] A.L. Coban, P.E. Allen, and X. Shi, "Low voltage analog IC design in CMOS technology," *IEEE Transactions on Circuit And Systems-I*, Vol. 42, No. 11, 1995.

[9] P.E. Allen and D.R. Holdberg, *CMOS Analog Circuit Design*, 2nd edition, Cambridge University Press, 2002.

[10] S. Pennisi, "A low-voltage design approach for class AB current mode circuits," *IEEE Transactions on Circuits and Systems II*, Vol. 49, No. 4, April 2002.

[11] R. Castello, F. Montecchi, and A. Baschitotto, "Low-voltage analog filters," *IEEE Transactions on Circuit And Systems-I*, Vol. 42, 1995.

[12] M. Harada, T. Tsukahara, et al., "2-GHz RF front-end circuits in CMOS/SIMOX operating at an extremely low voltage of 0.5V," *IEEE Journal of Solid State Circuits*, Vol. 35, No. 12, Dec 2000.

[13] M. Steyaert, et al., "Low-voltage, low-power, CMOS-RF transceiver design," *IEEE Transactions on Microwave Theory and Techniques*, Vol. 50, No. 1, Jan 2002.

[14] D. Cassan, and J.R. Long, "A 1-V transformer-feedback low-noise amplifier for 5-Ghz wireless LAN in 0.18um CMOS," *IEEE Journal of Solid State Circuits*, Vol. 38, No. 3, 2001, pp. 427–435.

[15] A. Liscidini, et al., "A 0.13um CMOS front-end, for DCS 1800/UMTS/802.11b-g with multiband positive feedback low noise amplifier," *IEEE Journal of Solid State Circuits*, Vol. 41, No. 4, April 2006, pp. 981–989.

[16] R.G. Carvajal, et al., "The flipped voltage follower: A useful cell for low-voltage low-power circuit design," *IEEE Transactions on Circuits and Systems-I*, Vol. 52, No. 7, July 2005, pp. 1276–1291.

Injection Locking Techniques

[17] R. Adler, "A study of locking phenomena in oscillators," *Proceedings of the IEEE*, Vol. 61, No. 10, Oct 1973.

[18] A.S. Daryoush, T. Berceli, R. Saedi, P.R. Herczfield, and A. Rosen, "Theory of subharmonic synchronization of nonlinear oscillators," *IEEE Microwave Theory and Technique Symposium*, June 1989, pp. 735–738.

[19] R.G. Harrison, "Theory of regenerative frequency dividers using double-balanced mixers," *IEEE Microwave Theory and Technique Symposium*, June 1989, pp. 459–462.

[20] G.R. Sloan, "The modeling, analysis, and design of filter-based parametric frequency dividers," *IEEE Transactions on Microwave Theory and Techniques*, Vol. 41, No. 2, Feb 1993.

[21] H.R. Rategh and T.H. Lee, "Superharmonic injection-locked frequency dividers," *IEEE Journal of Solid State Circuits*, Vol. 34, No. 6, June 1999.

[22] H.R. Rategh, H. Samavati, and T.H. Lee, "A CMOS frequency synthesizer with an injection-locked frequency divider for a 5 ghz wireless LAN receiver," *IEEE Journal of Solid State Circuits*, Vol. 35, No. 5, May 2000.

[23] S. Verma, H.R. Rategh, and T.H. Lee, "A unified model for injection-locked frequency dividers," *IEEE Journal of Solid State Circuits*, Vol. 38, No. 6, June 2003.

Subharmonic/Multirate Signal Processing

[24] M. Cohn, J.E. Degenford, and B. Newman, "Harmonic mixing with an antiparallel diode pair," *IEEE Transactions on Microwave Theory and Techniques*, Vol. MTT-23, Aug 1975, pp. 667–673.

[25] D.N. Held and A. Kerr, "Conversion loss and noise of microwave and milimeter-wave mixers: Part 1- theory," *IEEE Transactions on Microwave Theory and Techniques*, Vol. MTT-26, No. 2, Feb 1978, pp. 49–55.

[26] K. Itoh, M. Shimozawa, N. Suematsu, and O. Ishida, "Even harmonic type direct conversion receiver ICs for mobile handsets: Design challenges and solutions," *IEEE Radio Frequency Integrated Circuits Symposium*, June 1998, pp. 53–56.

[27] M. Shimozawa, K. Kawakami, H. Ikematsu, K. Itoh, N. Kasai, Y. Isota, and O. Ishida, "A monolithic even harmonic quadrature mixer using a balanced type 90 degree phase shifter for direct conversion receivers," *IEEE Microwave Theory and Technique Symposium*, June 1998, pp. 175–178.

[28] R. Feinaugle, H.-W. Hubers, H.P. Roser, and J. Hesler, "On the effect of IF power nulls in schottky diode harmonic mixers," *IEEE Transactions on Microwave Theory and Techniques*, Vol. 50, No. 1, Jan 2002, pp. 134–142.

[29] B. Matinpour, S. Chakraborty, and J. Laskar, "Novel DC-offset cancellation techniques for even-harmonic direct conversion receivers," *IEEE transaction Microwave Theory and Techniques*, Vol. 48, No. 12, Dec 2000, pp. 2554–2559.

[30] H. Ikematsu, K. Tajima, K. Kawakami, K. Itoh, Y. Isota, and O. Ishida, "Distortion characteristics of an even harmonic type direct conversion receiver for CDMA satellite communications," *IEICE Transactions Electronics*, Vol. 82-C, No. 5, May 1999, pp. 699–707.

[31] J.J. Kim, and B. Kim, "A low phase noise CMOS LC oscillator with a ring structure," *IEEE International Solid State Circuits Symposium*, 2000, pp. 430–431.

Super-Regenerative Architectures

[32] E.H. Armstrong, "Some recent developments of regenerative circuits," *Proceedings of the IRE*, Vol. 10, Aug 1922, pp. 244–260.

[33] H.A. GLucksman, "Superregeneration-An analysis of the linear mode," *Proceedings of the IRE*, Vol. 37, May 1949, pp. 500–504.

[34] J.R. Whitehead, *Super-Regenerative Receivers*, Cambridge University Press, 1950.

[35] A. Vouilloz, M. Declercq, and C. Dehollain, "A low-power CMOS super-regenerative receiver at 1 GHz," *IEEE Journal of Solid-State Circuits*, Vol. 36, No. 3, Mar 2001, pp. 440–451.

[36] P. Favre, N. Joehl, A. Vouilloz, P. Deval, C. Dehollain, and M. Declercq, "A 2-V 600-μA 1-GHz BiCMOS super-regenerative receiver for ISM applications," *IEEE Journal of Solid-State Circuits*, Vol. 33, No. 12, Dec 1998, pp. 2186–2196.

[37] B. Otis, Y.H. Chee, and J. Rabaey, "A 400μW-RX, 1.6mW-TX Superregenerative transceiver for wireless sensor networks," *IEEE International Solid State Circuits Conference*, 2005, pp. 396–398.

Analog Trends

[38] K. Bult, "Analog design in deep sub-micron CMOS," *European Solid State Circuits Conference*, September 2000, pp. 126–132.

Hearing Aid Applications

[39] D.G. Gata, et al., "A 1.1-V 270-μA mixer signal hearing aid chip," *IEEE Journal of Solid-State Circuits*, Vol. 37, No. 12, Dec 2002, pp. 1670–1678.

[40] S. Kim, et al., "An energy-efficient analog front-end circuit for a sub-1-V digital hearing aid chip," *IEEE Journal of Solid State Circuits*, Vol. 41, No. 4, April 2006, pp. 876–882.

[41] F.S.-Graells, et al., "A true 1-V 300-μW CMOS-subthreshold log-domain hearing-aid-on-chip," *IEEE Journal of Solid State Circuits*, Vol. 39, No. 8, Aug 2004, pp. 1271–1281.

RFID Applications

[42] G.K. Balachandran and R.E. Barnett, "A 110nA voltage regulator system with dynamic bandwidth boosting for RFID systems," *IEEE Journal of Solid-State Circuits*, Vol. 41, No. 9, Sep 2006, pp. 2019–2028.

[43] C.T. Rodenbeck, "Planar miniature RFID antennas suitable for integration with batteries," *IEEE Transactions on Antennas and Propagation*, Vol. 54, No. 12, Dec 2006.

[44] Y. Ji, et al., "Analysis and design strategy of UHF micro-power CMOS rectifiers for micro-sensor and RFID applications," *IEEE Transactions on Circuits and Systems – I*, Vol. 54, No. 1, Jan 2007.

[45] B. Jiang, et al., "Energy scavenging for inductively coupled passive RFID systems," *IEEE Transactions on Instrumentation and Measurement*, Vol. 56, No. 1, Feb 2007.

[46] E. Cantatore, et al., "A 13.56-MHz RFID system based on organic transponders," *IEEE Journal of Solid State Circuits*, Vol. 41, No. 9, Sep 2006, pp. 2019–2028.

BAW Resonators

[47] M. Franosch, K.-G. Opermann, A. Meckes, W. Nessler, and R. Aigner, "Wafer-level-package for bulk acoustic wave (BAW) filters," *IEEE International Microwave Symposium*, Vol. 2, Jun 2004, pp. 493–496.

[48] R. Ruby, A. Barfknecht, C. Han, Y. Desai, F. Geefay, G. Gan, M. Gat, and T. Verhoeven, "High-Q FBAR filters in a wafer-level, chip-scale package," *IEEE International Solid State Circuits Conference*, Feb 2002.

[49] Y.H. Chee, A.M. Niknejad, and J. Rabaey, "A sub-100uW 1.9-GHz CMOS oscillator using FBAR resonator," *IEEE Radio Frequency Integrated Circuits Symposium*, Jun 2005, pp. 123–126.

[50] R. Aigner, J. Ella, H.-J. Timme, L. Elbrecht, W. Nessler, and S. Marksteiner, "Advancement of MEMS into RF-filter applications," *IEEE Electron Devices Meeting*, Dec 2002, pp. 897–900.

Ultra-Low Power Radios

[51] Low-power RF selection guide. http://www.chipcon.com.

[52] Medical Implantable RF Transceiver, ZL70101. http://www.zarlink.com.

[53] Ultra low power smart sensor interface and transceiver platform, TZ1030. http://www.toumaz.com.

[54] Amplifier sequenced hybrid radios, RF Monolithics Inc. http://www.rfm.com.

[55] J. Ryckaert, C. Desset, A. Fort, M. Badaroglu, V.D. Heyn, P. Wambacq, G.V.D. Plas, S. Donnay, B.V. Poucke, and B. Gyselinckx, "Ultra-wide-band transmitter for low power wireless body area networks: Design and evaluation," *IEEE Transactions on Circuits and Systems-I*, Vol. 52, No. 12, Dec 2005.

[56] L. Smaini, C. Tinella, D. Helal, C. Stoecklin, L. Chabert, C. Devaucelle, R. Cattenoz, N. Rinaldi, and D. Belot, "Single chip CMOS pulse generator for UWB systems," *IEEE Journal of Solid State Circuits*, Vol. 41, No. 7, Jul 2006.

Energy Scavenging/Harvesting

[57] J.A. Paradiso and T. Starner, "Energy scavenging for mobile and wireless electronics," *IEEE Pervasive Computing*, Vol. 4, No. 1, Jan 2005, pp. 18–27.

CHAPTER 7

Packaging for Integrated Communication Microsystems

INTRODUCTION

The demand for increasingly higher data rates, voice and video, and for the integration of various technologies (digital, analog, radio frequency (RF), and optical) has driven in emerging high-performance applications such as personal communication networks, wireless local area networks (WLANs), "last mile" RF-optical networks, and millimeter-wave sensors. These RF/wireless modules have defined a trend toward more flexible and reconfigurable systems, since they impose very stringent specifications never reached before in terms of low noise, high linearity, low power consumption, small size and weight, and low cost. The electronics packaging industry has proliferated to a point where its technology is at least equally, if not more, important than the semiconductor technology it is designed to serve. An increase in silicon complexity and integration has imposed demanding requirements on packaging technology for higher input–output (I/O) counts, finer lead pitch, enhanced heat dissipation, and high-speed, reliable interconnects. At the same time, the industry is also driving toward a reduced printed circuit board area and complexity. This trend puts additional strain on packaging requirements. Microprocessor, RF, and millimeter-wave packaging are undergoing major changes driven by technical, business, and economic factors. In its early evolution, the influence of the package on performance was limited; however, as the systems evolve to provide increasing performance and operation frequencies, the package must also evolve to keep up, and packaging design must ensure that it can optimally enable the systems' functionalities. From the

Advanced Integrated Communication Microsystems, edited by Joy Laskar, Sudipto Chakraborty, Manos Tentzeris, Franklin Bien, and Anh-Vu Pham

traditional role of a protective mechanical enclosure, the modern package has been transformed into a sophisticated thermal and electrical management platform. Furthermore, system architecture and design techniques have had a significant impact on the complexity and cost of packaging. The need to optimize the total solution (chip, package, board, and assembly) has never been more important to maximize performance and minimize cost.

Recent advances in high-frequency packaging indicate a migration from wirebond (where the chip or die is interconnected to the package only on the periphery of the die) to flip-chip (where the die is interconnected to the package using the entire die area); and from ceramic to organic packages, with cartridge and multichip technologies emerging as key form factors. With the "segmentation" of the high-frequency computing market (mobile, desktop, server, and associated subsegments), a significant proliferation of packaging types tailoring functionality and costs to the different applications specifications can be observed.

The RF front-end module is the core of these systems, and its integration poses a great challenge. Microelectronics technology, since the invention of the transistor, has revolutionized many aspects of electronic products. The integration and cost path has led the microelectronics industry to believe that this kind of progress can go on forever, leading to the so-called "system-on-chip" (SOC) [1] for all applications. But it is becoming clear that it is still a dream to produce a complete on-chip solution for the novel wireless communication front ends. Considering the characteristics of the RF front-end modules, such as high performance up to a 100-GHz operating frequency; large number of high-performance discrete passive components; design flexibility; reconfigurable architecture; low power consumption; compactness; customized product; short time to market; and low cost, the "system-on-package" (SOP) approach [2] has emerged as the most effective to provide a realistic integration solution. The strength of SOP is based on multilayer technology using low-cost and high-performance materials. Using this three-dimensional (3-D) topology, high-density, hybrid interconnect schemes as well as various compact passive structures, including inductors, capacitors, antennas, and filters, can be directly integrated into the substrate. Thus, high-performance modules can be implemented while achieving simultaneous cost and size reduction.

Electronic packaging has been one of the greatest challenges of the semiconductor industry. Packaging is a method for allowing electrical connection to an integrated circuit while maintaining and regulating its operating environment as well as achieving performance, reliability, and cost requirements. Because of its nature, it is a synergy of various different factors (electrical, mechanical, thermal, and material) that have to be considered during an accurate design and fabrication process, as shown in Figure 7.1.

The primary function of a package is to provide a means for electrical connectivity from the semiconductor device to a printed wiring/circuit board (PWB/PCB). It provides a path for power to be applied to the chip as well as a way for the data signals to be transmitted into and out of the chip. Its secondary function is to house and protect the fragile chip from harsh environmental conditions, such as moisture, light, and dust, that might hinder its performance. Finally, the package provides a pathway

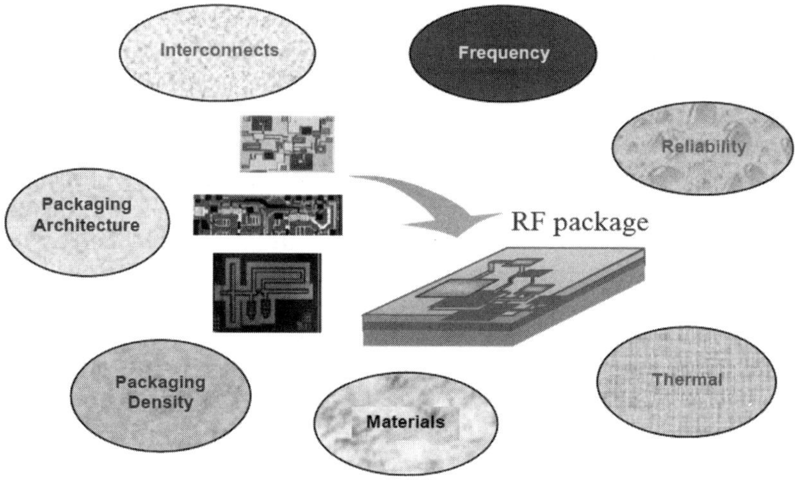

Figure 7.1. Critical factors along the design cycle of an RF package.

for dissipating the heat generated by the semiconductor device. This last function is becoming increasingly important and significantly more difficult to achieve in the beginning of the twenty-first century, as new applications require integrated circuit complexity, operating frequencies, and power consumption to reach new heights [3].

7.1 BACKGROUND

7.1.1 Trends from 1970 to 1995

The original packaging devices were typically large. They were connected to the outside world by means of long leads or pins that needed to be inserted through holes in the printed circuit boards. These devices came to be known as "through-hole" devices for that reason and were made of metal, ceramic, and plastic. The most predominant of these packages in the 1970s and 1980s was the dual-in-line package (DIP) [4], which is still in use today. The DIP's main limitation was the lead count with an upper limit of 64 pins. DIP is an early format, which used wirebond techniques while providing multiple grounds. After that point, the package simply became too large for any practical applications. A solution to this was the pin grid array (PGA) [5], which was using a two-dimensional (2-D) array of pins protruding from the bottom of the package. The PGA package could have around 200 pins on average.

The main problem with through-hole devices was their size. Furthermore, their performance at high frequencies was limited. The trend in printed circuit board manufacturing was to increase density while decreasing board area and increasing signal frequencies. Through-hole devices did not easily allow for this to happen. In the 1980s, through-hole devices began to give way to a new packaging technology called surface mount technology (SMT) [6]. The leads from these packages were mounted directly to rectangular pads on the surface of the printed circuit boards, and they did not require holes to be drilled into the PWBs. As a result, the width of the leads could be smaller and the spacing between the leads (lead pitch) could be decreased. Thus, a package with the same number of pins as a traditional through-hole device could now be significantly smaller, despite that both devices were made with the same materials. Common surface mount devices included the plastic leaded chip carrier (PLCC) and the small outline integrated circuit (SOIC). In recent years, the quad flat pack (QFP) has become a predominant fine pitch, high lead count package solution. From an RF packaging perspective, QFP and quad flat pack no lead (QFN) have become popular in recent times. QFP uses a peripheral-leaded, wire-bonded, single-layer package, and it has a more consistent lead shape in its cross section and can provide a higher pin count. QFN uses board attach technology with solder bump, and it has become popular because of its very low parasitic performance at high frequencies. Figure 7.2 illustrates most of the previous technologies, and Figure 7.3 shows some popular packaging devices used throughout the 1980s and 1990s.

Both the through-hole and surface mount devices listed above were connecting the silicon chip to the external package leads by means of a process known as wire bonding [7,8]. Very fine wires were attached directly to the silicon chip at one end and connected to the package leads at the other end. This process has been very well understood in the semiconductor industry and can be performed with very high yield for very low cost. It does, however, pose problems as I/O count, density, and operating frequencies increase and package lead pitch decreases. Another process for connecting to the outside world is tape automated bonding (TAB).

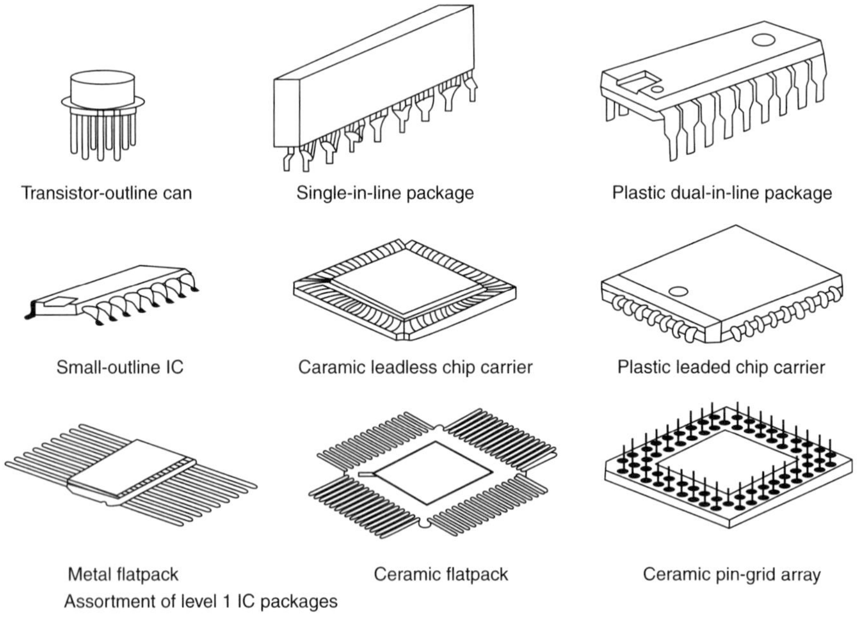

Figure 7.2. Various packaging techniques.

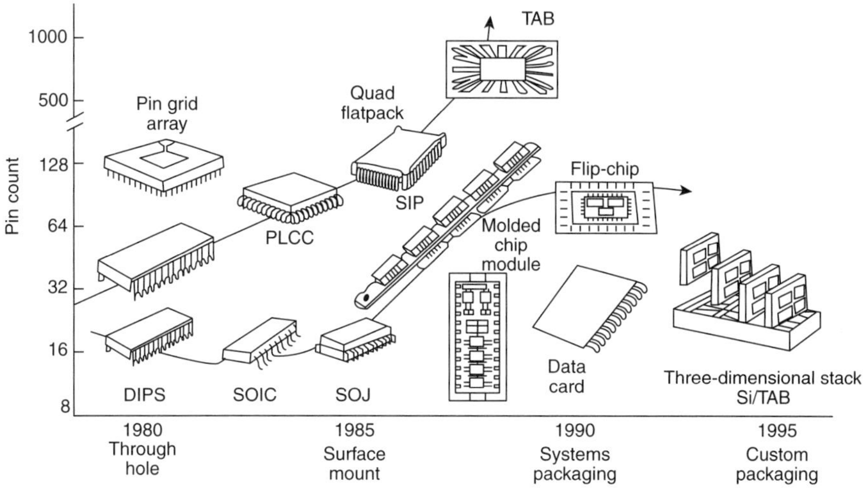

Figure 7.3. Popular packaging devices used throughout the 1980s and 1990s.

7.1.2 Trends from 1995 to Today

Today, as semiconductor technology continues the minimization trend, the level of complexity on a single silicon chip is significantly increasing. This complexity leads to more functionality in a smaller area, higher I/O counts, higher frequencies, and higher heat dissipation requirements. All of the technologies listed up to this point have been inadequate and impractical to satisfy all of these requirements. Although they still have a place in modern integrated circuit design, they will not meet the needs of leading edge technologies and, in particular, in the case of very demanding high-frequency performance, such as that of RF and millimeter-wave integrated systems.

New packaging technologies have emerged in the last ten years that are aimed at solving the I/O and thermal challenges. The first generation of these solutions continues to use the same wire bonding technology used in traditional packages, while accommodating advances in interconnect technology. Flip-chip technology (also called direct chip attach or DCA) has emerged as a possible alternative to wire bonding [9]. Regardless of the interconnect technology, the most promising packaging technology being pursued currently is the ball grid array (BGA) package, which is a descendant of the PGA package discussed earlier. Instead of through-hole pins, the BGAs have small conductive balls that are soldered directly to the surface of printed circuit boards. As the balls are located on the bottom of the package, an obvious disadvantage is the inability to inspect the connections visually between the package and the PWB. As a result, the assembly process for PWBs using BGA technology must be very precise with very low tolerance for error. Its advantages in I/O count, I/O density, and heat dissipation easily outweigh this somewhat minor disadvantage [10].

In 1997, the Semiconductor Industry Association (SIA) released "The National Technology Roadmap for Semiconductors" [11]. (All of the data in this section is taken directly from this document.) According to the SIA: "The increase in total I/O and I/O density as packaging has evolved from PGA through QFP and now to BGA has been

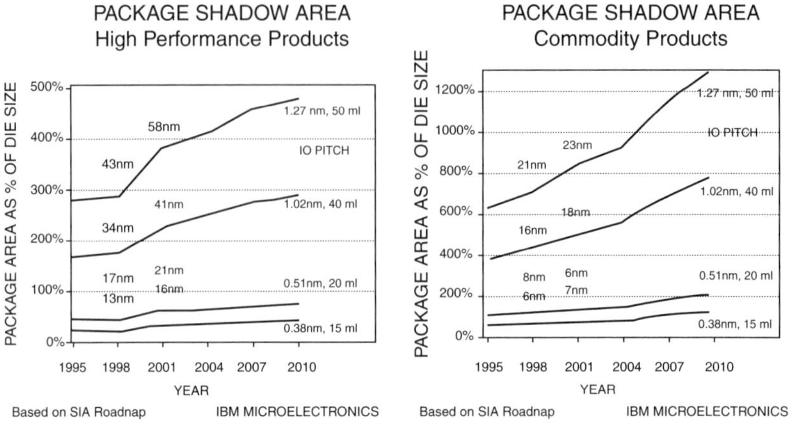

Figure 7.4. Package area evolution.

quite dramatic, but represents only a first step along the required growth path. Growing at an average compound rate of 12.6%, high performance package I/O is projected to reach 1700 by the year 2001. Commodity products, while growing at a less aggressive 8.2% rate, will require 325 package I/O in the same year" (Figure 7.4) [12].

The Assembly and Packaging section of the SIA road map claims that package design and fabrication directly affect systems applications. Device layout on the chip, chip layout on the package, and interconnects are all influenced by power and signal waveform integrity, while timing, crosstalk, and ground bounce are the factors that affect waveform integrity. Future challenges for the assembly and packaging have been sorted in two groups, as follows.

7.1.3 Before 2006

Improved organic substrates for high I/O area array flip-chip
Improved underfills for high I/O area array flip-chip
Reliability limits of flip-chip on organic substrates
Integrated design tools and simulators to address chip, package, and substrate complexity

7.1.4 After 2006

Close the gap between the substrate technology and the chip
"System-level" view and codesign of the integrated chip, package, and substrate needs

The needs of assembly and packaging are driven equally by silicon technology and marketing requirements. As always, cost will be the driving factor. It is expected that packaging costs per pin will decrease in the coming years, but the overall packaging pin count is expected to increase at a faster rate than the cost decrease. The increase in pin count is also expected to affect the substrate and the system-level costs. Also, the road map identifies thermal management as a significant challenge. As a result, handheld or portable devices that do not use forced air and rely on the operator's hand to dissipate the heat from the unit will need new heat sink technologies and materials with better thermal conductivity. On the other side, the cost-performance market (desktop processors) requires forced air-cooling. Flip-chip could be a possible enhancement to the forced air-cooling as the backside of the silicon chip provides a "direct, efficient heat path from the chip to the heat sink." Existing heat sink solutions are predicted to be ineffective above 50 W in applications where forced air is not a viable solution because of market requirements. A reduction of internal thermal resistance and better air-cooling techniques would be critical for future thermal solutions. The high-end market, with predicted power consumption between 110 and 120 W per chip, will pose an even greater challenge, necessitating a closed-loop cooling system that meets market and customer requirements. In general, flip-chip is predicted to become the predominant technology for a chip-to-next-level interconnect, since it enables the reduction of a level of interconnect. Commodity products will continue to use advanced forms of wire bonding until the cost of flip-chip becomes affordable for that market.

Another important RF packaging challenge is the matching of the coefficient of thermal expansion (CTE) between the silicon chip and the substrate. When using organic substrates, underfills will be required with high reliability, ease of manufacturability, stronger attachment at the interface, and higher resistance to moisture. Liquid-crystal polymers (LCPs) with engineered CTE could be a solution [13]. High-frequency packaging geometries are often sorted into two independent areas of focus, single-chip and multichip packages. The single-chip packages include the technologies discussed in the Background section. It is expected that devices like QFPs will reach a maximum lead count of 300 and that lead pitch will reach a minimum of 0.5 mm [7]. After this point, the package body size and the surface mount assembly complexity become cost ineffective and multichip solutions have to be investigated.

7.2 ELEMENTS OF A PACKAGE

Several elements are involved in defining a package. These elements include: (1) power/ground planes; (2) various traces, such as microstripline, buried microstripline, stripline, coplanar waveguide (CPW), and so on; (3) vias (vertical conductors) to connect signal lines between layers; (4) bondwires to connect the die pads to the package pins; (5) balls to attach the packaged die to the board; (6) package layers with various dielectric constants; and (7) heat spreaders to dissipate heat from the high-power consumption chips.

7.2.1 Power/GND Planes

Power/ground planes control the power/ground distribution, as well as the impedance levels of high-frequency performance of the conductors. Vias provide resistance and inductance, which can be modeled accurately using electromagnetic analysis, and the lumped element equivalent model can be used in circuit simulation. Bondwires are typically modeled with resistance, inductance, and capacitance parameters to be used in circuit simulations. These parameter are dependent on the cross section, and a smaller cross section leads to higher resistance and inductance values. To eliminate the effect of bondwires, they can be made short, and multiple bondwires may be used to reduce the resistance and inductance. Dielectric layers are responsible for separating conductor layers, which determines capacitive effects by determining impedance as well as delay in the signal. Heat spreaders are provided by floating metals in the package and can lead to eddy currents to the coupled signal. This phenomenon reduces the inductance of signals, but it increases the capacitance as well. At the same time, they can couple to signals to power/ground elements.

It is desired that the package should have low resistance and moderate inductance values. Fluctuation in supply voltage injects noise to the circuit and leads to functional and performance problems. High current density leads to electromigration. The voltage drop leads to the increase in switching time of the gate, and it degrades the noise margin for a digital gate. On-chip decoupling capacitors are placed on the supply routing and serve as local charge storage elements, and they reduce the peak current drawn from the supply. A typical power/ground formation is shown in Figure 7.5. The *IR* drop across the power/ground pins leads to speed degradations of the digital gates, and it may also cause a logic glitch in complicated digital logic chains. The voltage available at the circuit terminal is given by

$$V_c = V_{DD} - IR - \mathrm{L}\frac{dI}{dt} \tag{7.1}$$

Resonance in a power grid is also a major concern that needs to be addressed. In earlier package generations, the resonating frequencies in the power grid were significantly higher than the clock frequencies in the system. With the continued increase in the system clock frequencies, the resonance frequency of the power plane may become lower than the clock frequency. In this case, the system might need several clock cycles to settle down to the desired voltage levels. One may increase the resistance of the path to prevent the supply ringing, but this leads to more *IR* drop.

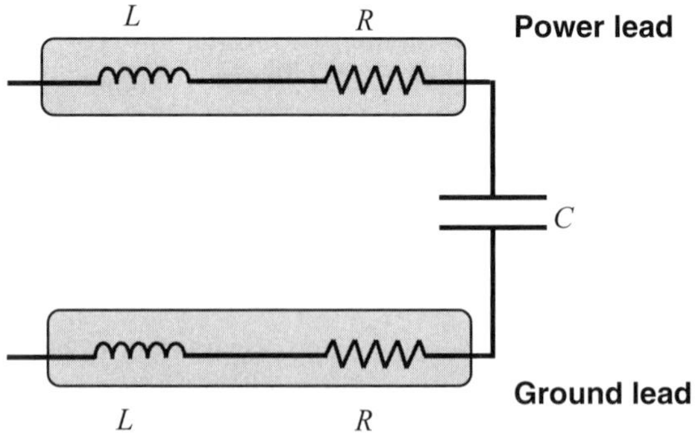

Figure 7.5. Representation of power and ground in a package.

Table 7.1. Material Conductivities

Material	Conductivity ($\Omega^{-1}m^{-1} \times 10^{-7}$)
Silver	6.3
Copper (annealed)	5.8
Gold	4.5
Aluminium	3.8
Tungsten	1.8
Solder	0.7

Table 7.2. Dielectric Constants

Material	ε
Air	1.0
Glass	2.2
Polyimide	2.5–3.5
Rogers	2.9
Goreply	3.0–3.3
GETEK	3.5
SiO_2	3.9
FR4	4.1
Epoxy glass	5.0
Alumina	9.0
Silicon	11.8

7.2.2 Package Materials

Dielectric materials and conductors determine the electrical properties of a package. Conductors are characterized by resistivity and permeability performances. Resistance impacts the current distribution because of the skin effect and dissipates heat. Table 7.1 shows a list of conductor materials and their conductivities.

Dielectric materials are characterized by dielectric constants, which determine the propagation delay in the material and are responsible for the capacitive coupling between conductors. Dielectric constants for various materials that are commonly used are shown in Table 7.2.

7.3 CURRENT CHIP PACKAGING TECHNOLOGIES

7.3.1 Ball Grid Arrays (BGAs)

BGA packages allow for PWB space savings since an array of solder bumps (or balls) is used in place of traditional package pins. The bumps have the advantage of a shorter electrical patch to the motherboard that improves the electrical performance at high frequencies. Also, they occupy a smaller area on the board for the same I/O count,

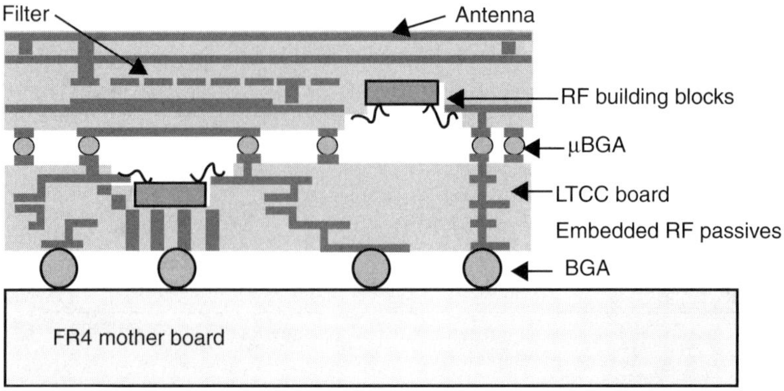

Figure 7.6. Three-dimensional Integrated RF front-end module concept view using BGA and uBGA.

while providing finer pitches and lower cost. An example of a BGA applied to a 3D integrated module concept is presented in Figure 7.6 [14]. In the proposed module concept, two stacked LTCC substrates are used and board-to-board vertical transition is ensured by μBGA balls [15]. The μBGA is a chip scale package that uses much smaller bumps than the traditional BGA. A typical BGA solder ball is 500 to 750μm in diameter and has a 40- to 50-mil pitch, whereas a typical μBGA Ni-Au ball is 50 to 100 μm in diameter and has a 20- to 30-mil pitch. It is very similar to flip-chip; the only difference is an elastomeric layer between the silicon die and the bumps for dissipating the stresses induced by the thermal expansion mismatches. Standard BGA balls ensure the interconnection of this high-density module with a motherboard such as an FR4 board. The top and the bottom substrates are dedicated, respectively, to the receiver and the transmitter building blocks of the RF front-end module. The two parts are separated by the μBGA "layer" to improve the crosstalk performance of the module.

BGAs are expected to be the solutions for packages requiring over 200 pins and will be implemented using wire bonding in the lower I/O density parts and flip-chip in the higher I/O count and higher power devices. Ultra fine pitch (UFP) wire bond technology, will allow for a larger die to be placed on smaller substrates, and an effective pad pitches below 60 microns. It is expected that increasing circuit density will enable the accommodation of more than 1000 I/Os and the die shrinkage (size reduction) by 20–50%. Placing a larger die with more functionality on a smaller substrate results in significant cost savings, as the substrate is a driving factor in the cost of the device. Recent trends indicate that I/O counts for leading-edge devices grow at a rate of approximately 10 times every 14 years. Three-dimensional or stacked packages allow a higher density of integration with thinned dies. Such configurations may be used in conjunction with wire bond or flip-chip.

Liquid encapsulant underfills are expected to be needed to relieve the stress caused by CTE differences between the chip and the package substrate in the flip-chip devices. Chip scale packages using fine pitch BGAs with a size in the order of the chip size will be the next level of advancement for applications where low weight and small package size are required. As the technology matures and processes become cheaper,

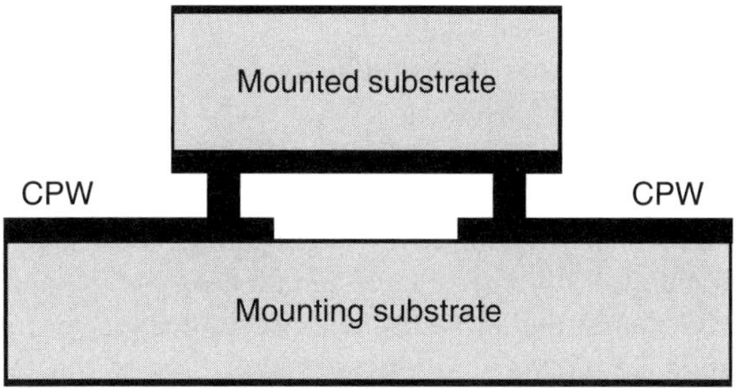

Figure 7.7. Schematic of a CPW-to-CPW flip-chip interconnection.

BGAs have the potential to become the workhorse of packaging technologies, like the DIP was for the 1970s and 1980s. BGA usually works in an array-type geometry, with a wire bond/flip chip, and it may use multiple layers and planes. The dielectric used in these packages can be ceramic or plastic. Tape BGAs use a simple construction with a possibility of one or two layers and use wire bond. Microstar BGAs are used in commercial products.

7.3.2 Flip-Chip Technology (FCT)

Flip-chip is a 30-year-old technology that has only recently matured to be widely accepted and cost effective for the semiconductor industry especially for high-frequency applications [16]. It refers to flipping a silicon die or chip and mounting it face down on a substrate. A schematic of a flip-chip transition from CPW to CPW is presented in Figure 7.7.

Of the 60 billion integrated circuits produced in 1998, approximately 1.5% of them were manufactured using flip-chip technology. From 1997 to 1998, the number of flip-chip dies grew by 40%. An average annual growth rate of 48% for flip-chip is expected over the next several years. Recently the capabilities of flip-chip were extended up to 75 GHz [17] and to optical applications using glass bumps that allow for waveguiding of optical waves [18].

7.3.3 Flip-Chip vs. Wire Bond

Although new packaging and interconnection technologies have been introduced, wire bonding is still dominant in RF/wireless products, especially in applications up to 10 GHz, since it has strong benefits in cost and reliability. However, the stringent specifications of emerging communication systems and the use of higher frequency bands accentuate the drawbacks of the wire bonding—i.e. parasitic effects and losses. When wire bonds are used as interconnects in microwave- and millimeter-wave modules, they exhibit high characteristic impedance from the high inductance of the thin wire and a small capacitance from the small dielectric constant of the air gap between the wire and the ground plane. In addition, radiation loss resulting from wire discontinuity becomes significant, particularly in the millimeter-wave frequency range.

Vertical interconnect solutions have gained a significant interest because they provide features that can eliminate the problems associated with wire bonding: the poor repeatability of the manufacturing process and a drastic increase of the losses associated with increased frequency. Along with showing better electrical performance, flip-chip technology allows several chips to be mounted together on the motherboard to increase density, improve system performance, and reduce cost. This packaging technique also allows for a combination of passive and active devices, Si and GaAs, and for analog, digital, and optical circuits on the same motherboard [19]. Furthermore, the compatibility with automatic manufacturing improves the reliability and reduces the assembly cost. However, with a judicious technical choice, wire-bonded parts can have superior performance.

In addition to these benefits, since the flip-chip die is flipped upside down, all the chip area is available for interconnect, eliminating the wire-bond restriction of having all the I/Os along the chip perimeter. This, along with the short electrical path that eliminates coupling issues, allows true chip scale packaging and, therefore, increased integration levels. Flip-chip also reduces power/ground resistance and provides better power/ground supply planes, higher I/O density, and better signal isolation. Power supply and ground inductances are reduced in wire-bond packages by connecting multiple wire bonds in parallel, whereas in flip-chip, this is realized by planes separated by a few microns that would lead to a significant reduction in inductance. In other words, to meet the similar inductance performance, bond-wired structures would require a large number of wire bonds. A flip-chip package uses the impedance of substrate, whereas a wire-bond impedance may be difficult to match to $50\,\Omega$, as the reflection from the wire bond can be significant if it is long. Two wire bonds can be placed in a ground and return path configuration in order to lead to a closer match to $50\,\Omega$. At the same time, a differential configuration can significantly reduce the electromagnetic cross-talk performance.

7.3.4 Choice of Transmission Line

Another important issue in RF packaging structures is the transmission line choice. The two most common choices for the transmission lines to be used in a flipped, monolithic microwave integrated circuit (MMIC) are microstrip and CPWs. Coplanar MMICs are more suited to flip-chip technology because of the immediate availability of all the grounds on one surface. In addition, coplanar circuitry requires no backside processing, eliminates the need for ground vias, and allows the use of a thicker, more physically robust chip. Easier matching can be achieved because of the ground–signal–ground configuration. Recently developed finite-ground coplanar waveguides (FG-CPWs) allow for a miniaturized implementation of these planar lines [20]. On the other hand, microstrip design tools are more mature and MMIC manufacturers prefer to make full use of their capabilities.

7.3.5 Thermal Issues

However, thermal performance of flip-chip packages is poor compared with wire bond. Heatsinking is more efficient when the chip is fully sitting on the vertical

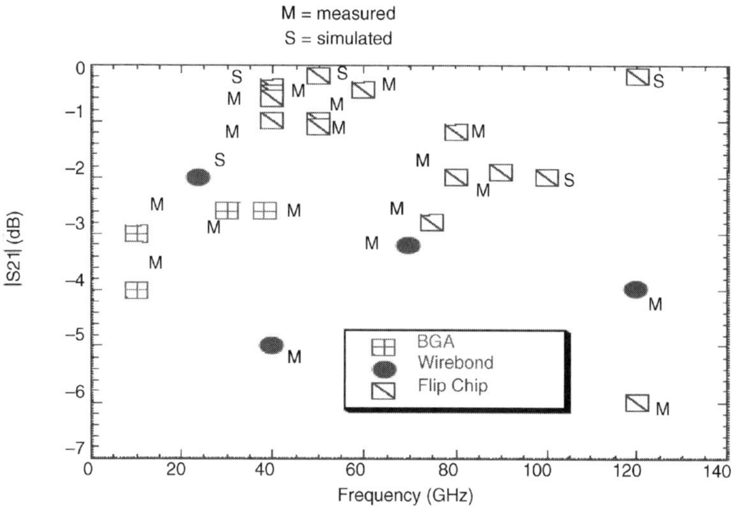

Figure 7.8. Electrical performance comparison amang flip-chip, wire bond, and BGA.

stackup rather than having no other contact with it than the interconnecting bumps. The coplanar design of the PCB eliminates the need for the vias, and shunt elements can be easily removed or added if tuning is necessary. From a mechanical reliability standpoint, flip-chip technology still needs to be improved. An underfill technology—to fill the gap between the chip die and the board with a dielectric material—has been introduced to improve the heat dissipation and the mechanical stability and to compensate the coefficient of thermal expansion (CTE) mismatch. However, it is difficult to apply underfill technology to RF modules, since the additional dielectric loss generated from the underfill material reduces the system efficiency. Moreover, the characteristic impedance of the transmission lines on an MMIC chip is significantly modified because of the higher dielectric constant of the underfill material. To avoid this problem, all MMIC circuits should be designed to compensate for the expected impedance change.

Figure 7.8 presents a review of the published electrical performance for wire bond, flip-chip, and BGA [21]. The test structures used in these analyses are not identical, and the individual characteristics are reflected in the S-parameters. However, it can be observed that flip-chip shows evidence of lower insertion loss over the entire frequency range than the wire bond. On the other hand, the BGA performance is poor compared with flip-chip, as a result of the larger dimensions and the use of more lossy substrates for the motherboard, especially for frequencies above 40 GHz. Similar results have been reported for return loss.

7.3.6 Chip Scale Packaging (CSP)

Chip scale packaging (CSP) is generally considered the fundamental evolution required to meet most of the electronic packaging needs of the twenty-first century [22]. As defined by its proponents, the CSP ranges in size from a die-equivalent area to, at its maximum, 120% of the die area. Several configurations now in development include, among others, perimeter pad and lead-frame packages, area array interposers, and μBGAs, with target market segments ranging from low-cost consumer through high-end data processing, and I/O ranges up to 1000. In addition to a smaller form factor, the CSP enjoys the benefit of very low parasitics and no lead frame, and it uses balls that are directly attached to the wafer.

7.4 DRIVING FORCES FOR RF PACKAGING TECHNOLOGY

The major driving forces for the RF and wireless packaging technology are manufacturing cost, size and weight, signal integrity, low high-frequency loss, heat dissipation, mechanical, stability, testability, and reliability. At the same time, the package parasitics play a significant role in determining the electrical characteristics at high frequencies. In detail:

Manufacturability and Cost. This depends on materials, fabrication steps, and IC cost. Costs incurred from testing, rework, yield loss, and manufacturability depend on process control, cycle time, reparability, equipment downtime, and design tolerances.

Electrical Design. Interconnect speed now plays a dominant role in determining performance limits as connection has parasitic capacitance, resistance, and inductance that limit speed, potentially distort signals, and add noise. These connections are also a source of reliability problems. Therefore, several factors need to be considered, including:

1. Lead length
2. Matched impedances
3. Low ground resistance
4. Simultaneous switching and power supply spiking.

Thermal Design

$$\text{objective} = \text{remove heat from the junctions of the ICs}$$

1. Keep dopants from moving
2. avoid self-heating effects techniques = forced air, liquid cooling, monophase cooling, dual-phase cooling

Considerations: How to remove heat (from the front or backside of the IC): air or liquid, mono or dual phase, thermal conductivity of the substrate, stresses induced due to CTE mismatches.

Mechanical Design. Susceptibility to thermal stresses must be considered in design. The tensile modulus ("stiffness") has to be thoroughly investigated.

Testability. Two types, two phases:
On wafer:

1. In-process testing (nondestructive)
2. Stress/reliability testing (destructive)

On MCMs:

1. Substrate
2. Assembled module

7.5 MCM DEFINITIONS AND CLASSIFICATIONS

The term "multichip module" (MCM) refers to the packaging of multiple silicon dies into one device [23,24]. MCMs offer the ability to reduce package pin count by combining two or more high-pin-count devices, which would normally connect to each other at the board level, into one package, where the interconnect is performed at the chip level. The resulting packaged system only needs a reduced set of pins to connect power and signals from the outside world. It can also serve to reduce PWB real estate (an expensive commodity) and presents the following advantages:

Improved performance: such as shorter interconnect lengths between dies (resulting in reduced delay time, lower RF parasitics and losses), lower power supply inductance, lower capacitance loading, less crosstalk, and lower off-chip driver power.

Miniaturization: since MCMs result in a smaller overall package size when compared with separate packaged components performing the same function; hence, the resulting I/O to the system board is significantly reduced.

Shorter time-to-market: making them attractive alternatives to application-specific integrated circuits (ASICs), especially for products with short life cycles.

Low-cost silicon sweep: allowing integration of mixed semiconductor technology, such as SiGe or GaAs.

Hybrid configurations: including surface mount devices in the form of chip scale or micro-ball grid array (μBGA) packages and discrete chip capacitors and resistors.

Simplification of board complexity by integrating several devices onto one package: thereby reducing total opportunities for error at the board assembly level.

Capability of accommodating a variety of second-level interconnects: Although BGAs are the most popular ones, lead-frame solutions can be employed for plugability, enabling reconfigurability and modularity for upgrades.

The widespread use of MCMs has been hurt by a few factors. It is a relatively high-cost process. Since not all of the signals are connected to the outside world, package level testing is difficult, so effort must be made to ensure that the die being put into the chip is good. Obtaining such "known good die" (KGD) is still a challenge. Also, the proliferation of flip-chip, BGA, and CSP for single-chip package solutions has reduced the need for chip-level system integration.

In the area of multichip packages, the SIA roadmap [11] identifies advancements in the following enabling technology solutions as requirements for the effective implementation of flip-chip:

Design tools/simulators (integrated design environment)
Flip-chip interconnect optimization
Underfill
High density substrates
Low-cost KGD

These enabling technologies are interdependent and must be selected to be compatible to provide an integrated multichip solution. Bare chips must provide the same reliability, quality, and performance as packaged chips to be considered "known good." Multichip packages containing two to three chips are in volume production today, and they could evolve over the next five years to include three–five chips. The evolution of these multichip packages into multichip modules with more than five chips will be driven by performance and cost requirements at the subsystem and system level, as well as improved chip reworkability and module testing techniques. The implementation and proliferation of multichip packages and modules will continue to be constrained by the availability of high-density substrates and "known good" die commensurate with the necessary cost and performance. Multichip modules will be driven by densification and cost reduction for low-end products and densification and performance for high-end products.

As mentioned, a key factor in the development of an MCM package is the availability of high-density substrates. Three commonly used substrate groups exist: (1) *laminated* (MCM-L)—constructed of plastic laminate-based dielectrics and copper conductors using PWB technology; lowest cost MCMs [7]; (2) *ceramic* (MCM-C)—constructed on cofired ceramic or glass–ceramic substrates using thick film (screen printing) technologies [7]; and (3) *deposited* (MCM-D)—formed by the deposition and patterning of thin films, the highest performance MCMs [7] (see Table 7.3).

To optimize cost and performance trade-offs, the high performance of MCM-D at the low cost of MCM-L are desirable, which leads to a hybrid technology, called

Table 7.3. Comparison of MCM Technologies

Characteristic	MCM-L	MCM-C	MCM-D
Minimum Line Width (mm)	60–100	75–100	8–25
Line Spacing (mm)	625–2250	125–450	25–75
Via Diameter (mm)	300–500	100	8–25
Cost ($/cm2)	3–30	50–1000	800–8000

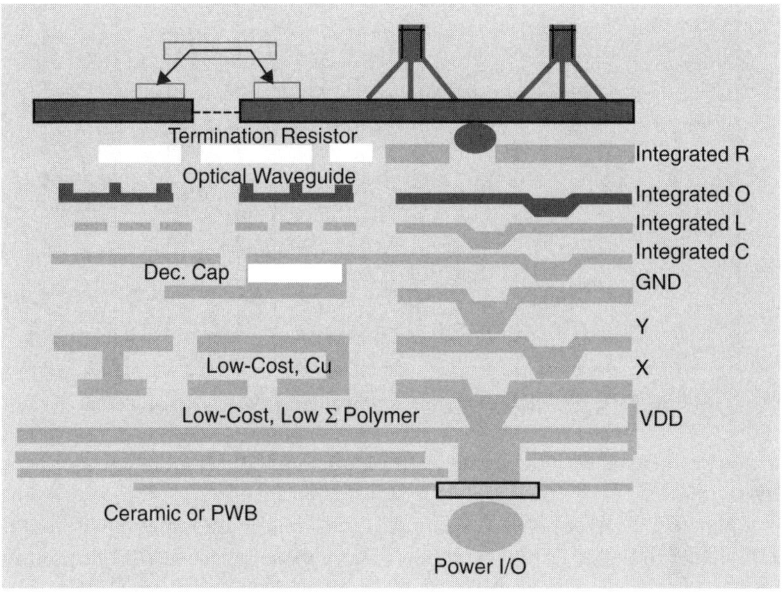

Figure 7.9. GT slim module.

MCM-L/D, where substrate structures are formed as multiple layers of metal separated by dielectric material. This is known as "single-level integrated module" (SLIM) implementation of the MCM-L/D, and it is shown in Figure 7.9 [25,26]. This technical vision integrates all the packaging levels into one, containing layers for DC power distribution; high-density, high-speed digital circuitry; embedded decoupling capacitors and termination resistors, embedded multilayer RF passives (i.e., inductors, filters, and antennas); embedded optical waveguides; built-in MEMS devices; and built-in thermal management.

7.6 RF–SOP MODULES

The future of wireless communication systems will require better performance, lower cost, and smaller size of the RF front end. To meet the critical specifications, it is common to use discrete passive components such as surface acoustic wave (SAW) filters and packaged inductors that cannot be easily integrated in a single-chip form. Although a lot of efforts have been devoted to the realization of SOC in an RF area using Si-based technology, SOC is considered as a solution for limited applications, such as Bluetooth. The strength of SOP is its multilayer technology that is using low-cost and high-performance materials. RF-SOP's mission is "to provide a complete packaging solution for RF module by integrating embedded passives components and MMIC at the package level." Using the multilayer topology, high-density hybrid interconnect schemes as well as various compact passive structures, including inductors, capacitors, filters and antennas can be directly integrated into the substrate.

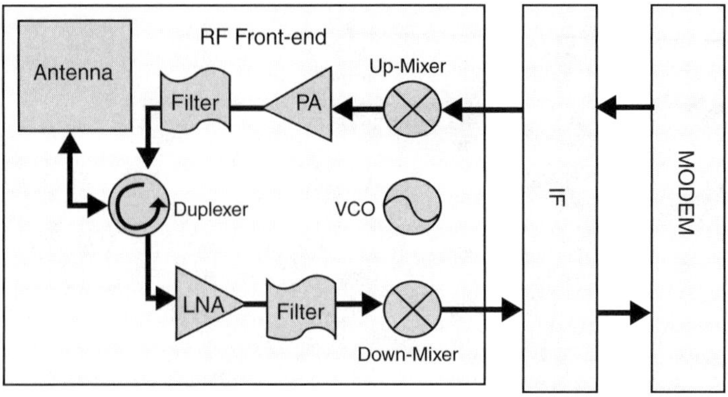

Figure 7.10. Block diagram of a typical RF front-end architecture.

Thus, high performance modules can be implemented with simultaneously lower cost and size.

To explain the SOP concept, a general system configuration of a wireless transceiver is shown in Figure 7.10. The module is composed of an MMIC chipset: power amplifier (PA), low noise amplifier (LNA), up and down mixer (MIX), voltage-controlled oscillator (VCO), and passive components: filter, antenna, and external high-Q discrete passive elements for stringent blocks such as PA and VCO. The RF-SOP approach includes replacing the discrete passives with embedded ones, adding more functional blocks, such as baluns and antennas, to the module and maximizing the performance of the MMIC chipset by replacing the on-chip passives with high-Q passives embedded in the package.

The multilayer 3D architectures have the major benefit of high integration and compactness, while adding another degree of freedom in the module design. The materials suited for this technology include low-temperature cofired ceramic (LTCC) and high-temperature cofired ceramic (HTCC) [28], multilayer organic (MLO) [27], and liquid crystal polymers (LCPs) [29]. LTCC is widely used for RF and microvawe systems, for providing a large; number of layers (up to 20–30) with high dielectric constant (5.6–20), good thermal conductivity for power amplifier applications, and low line losses at high frequencies, as well as for allowing for each layer to be manufactured and inspected before the multiple layers become laminated together. HTCC is an aluminum oxide substrate, which is also called alumina. It is cofired at a higher temperature than LTCC, using conductors with higher melting points that confer to this technology a superior stability and a better reliability in harsh environments. The disadvantages of these conductors are higher losses at high frequencies. LCP has recently received much attention as a high-frequency circuit substrate and package material. It has impressive electrical characteristics that are environmentally invariant because of extremely low water absorption [13], and it provides a nearly static dielectric constant (~3.19) across a very wide frequency range up to 110 GHz. Thermal expansion characteristics are equally desirable. For circuit applications, the controllable CTE can be engineered to match either copper, silicon, or GaAs. LCP is flexible, recyclable, impervious to most chemicals, and stable up to its high melting temperature. Most importantly, it is very cost effective in comparison with LTCC. The MLO-based process could offer another choice for next-generation SOP technology for RF-wireless, high-speed digital, and RF-optical applications. It uses a cost-effective process, while offering design flexibility and optimized integration because of its multilayer topology in which free vertical real estate is taken advantage of. And, in addition to this, it incorporates low-cost materials and processes consisting of a core substrate (FR-4, for example) laminated with two thin organic layers. The thickness of the core substrate is 40 mils, whereas the thickness of the laminate layers are 2.46 mils each. Advanced epoxy-based materials are used as laminate layers and exhibit electrical properties such as a low dielectric constant of 3.4 and a loss factor of 0.025 up to the C-band frequency range [30].

The integration of the antenna with the RF front end in compact modules is one of the major challenges for the SOP implementation. The size of the conventional patch antenna ($\lambda/2$) is often too large to be integrated with the rest of the module, because it would unnecessarily increase the size of the entire system. Various approaches have been proposed for the miniaturization, the suppresion of parasitic backside radiation (crosstalk), and the realization of multiband/multistandard antennas [31]. The schematic of a 3D integrated transceiver module with an antenna for 5.8 GHz is presented in Figure 7.11.

Three different multilayer subsystems, i.e., the transceiver, the filter, and the antenna, are vertically stacked and connected through vertical vias. The antenna and filter can be directly fabricated on the module using LTCC or LCP technology in order to reduce size and interconnection losses. The presented module has been

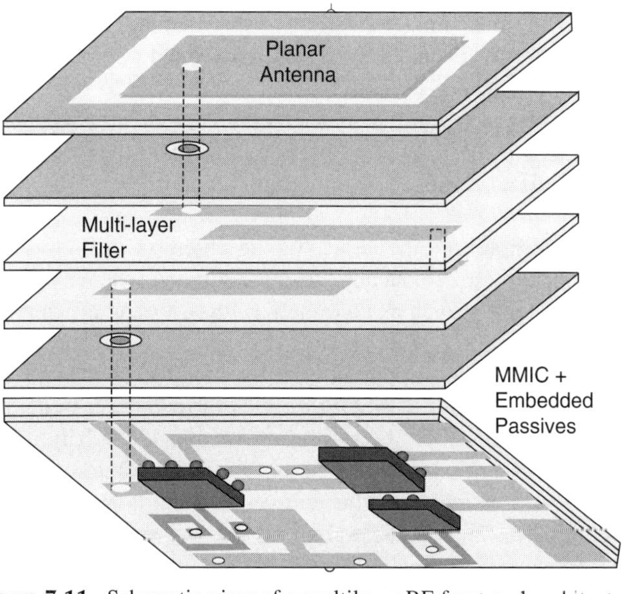

Figure 7.11. Schematic view of a multilayer RF front-end architecture.

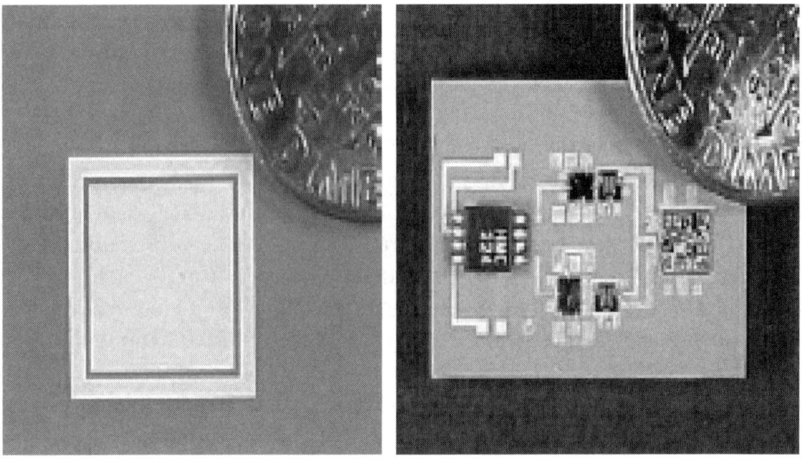

Figure 7.12. Three-dimensional integrated LTCC RF front-end module for 802.11a WLAN.

designed for 20 LTCC layers. Before designing each of the embedded passives, the layers have been properly assigned. The antenna, filter, and transceiver use 8, 10, and 2 layers, respectively. The total size of the module is $14 \times 19 \times 2$ mm, including all the RF functional blocks. The grounds are connected efficiently to suppress the unwanted parasitic modes and the photograph of the integrated module is shown in Figure 7.12.

To use the space of the module effectively, the geometries of the passive components are chosen very carefully. A cavity-backed patch antenna (CBPA) has been designed for the module. Recently, the introduction of the soft-and-hard-surfaces idea has allowed for additional miniaturization of the antennas [32]. A three-section coupled stripline filter has been designed to be embedded inside the LTCC package with its input and output ports connected to the antenna and the duplexer switch through vias. The RF functional blocks, including PA, LNA, mixers, and VCO, are attached on the top of the LTCC board. The specifications of the functional blocks are determined and verified through system simulations based on the IEEE 802.11a standard.

It is important to note that the choice of the on-chip or off-chip (on-package) passives is dependent on the frequency band, modulation scheme, available device, and packaging technology, such as the required linearity and efficiency of the PA. In VCO, the need for high-Q inductors on a package is determined by the stringent phase noise specification that is coming from the modulation scheme.

Briefly, the advantages of RF-SOP are as fallows: (1) lower cost by using embedded passives instead of discrete components, (2) flexibility for MMIC designers by embedding the high-Q passives in the package, (3) minimum loss and parasitic effects by reducing the number of interconnections, (4) size reduction by adopting multilayer packaging, (5) easy-to-realize multifunctional RF modules in a single package, and (6) better high-power handling capability than the MMIC chip.

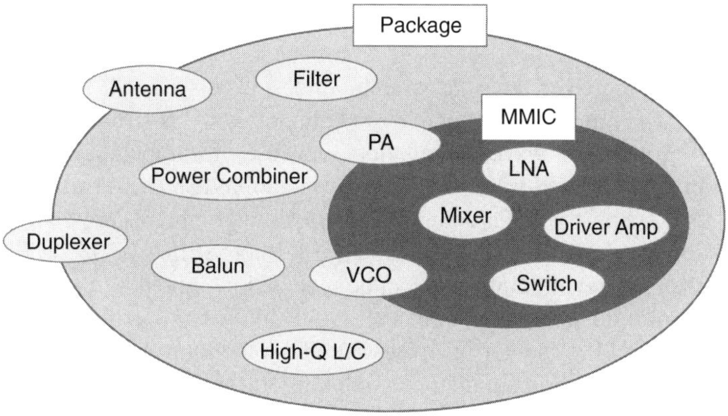

Figure 7.13. Technical mapping for RF front-end SOP implementation.

However, the SOP approach has various problems that are currently addressed by various research groups in the world, namely:

1. Interference between the blocks in the package
2. Too many degrees of freedom in multilayer packaging to build a design library
3. Size constraint if antenna is included in the package

Innovative shielding solutions are currently investigated using metalized cavities, [33] or electromagnetic band-gap (EGB) structures [34] as well as advanced simulation platforms to predict and electromagnetic interference and crosstalk within complex microsystems. In particular, EGB topologies are also used to confine the radiating field of antenna elements, which therefore leads to more efficient isolation of the rest of the module, while achieving a reduction of size of the antenna, as it has been described above. Figure 7.13 illustrates the technical mapping for RF front-end SCP implementation for various functional blocks.

Also, new design approaches based on optimization algorithms such as design of experiments (DOE) [35], feed forward neural networks [36], and genetic algorithms [37] are extensively studied and used to generate comprehensive models taking into account the fabrication and layout parameters and their impact on electrical performances in multilayer configurations.

7.7 PACKAGE MODELING AND OPTIMIZATION

The current drawbacks of most commercially available microwave- and millimeter-wave front ends are their relatively large size, heavy weight primarily caused by discrete components such as the filters, and separately located modules. Multilayer ceramic (i.e., LTCC) and organic-based SOP implementation are capable of overcoming this limitation by integrating components as part of the module package that would have otherwise been acquired in discrete form. On-package components not only miniaturize the module, but also they eliminate or minimize the need for discrete components and thereby reduce the assembly time and cost as well. The optimization of SOP structures requires the effective modeling of complex structures that involve mechanical motion and wave propagation. Because of computational constraints, many commercial simulators use various approximations to provide fast and relatively accurate results. Some popular commercial EM simulation tools are high frequency structure simulator (HFSS) [38], SONNET [39], MicroStripes [40], and IE3D [41]. Often the size or type of circuit that can be modeled is limited by these simulators. Either the approximations used limit the applicability to specific problem, or the simulation time and/or memory is excessive. To solve numerous innovative or 3D complex problems, custom simulators employing full-wave techniques can be used. Using a custom code, approximations can be made selectively and the effect on accuracy can be determined. Popular simulation techniques [42] include the method of moments (MoM), finite element method (FEM) in the frequency domain and finite-difference time domain (FDTD), transmission line matrix (TLM), and multiresolution time domain (MRTD) [43] in the time domain. Frequency domain methods are often used to simulate complex structures and can naturally handle frequency-dependent parameters such as loss. Alternatively, time domain simulation techniques allow for the use of simple grids for complex structures, parallelize well on inexpensive hardware, and through the use of a Fourier transform can give the results for a wide frequency band using a single simulation. Both types of simulators can be used on most problems, although not with the same complexity.

Modern RF 3D modules and packages demand a high level of compactness and functionality. Full-wave EM numerical tools require computational complexity that renders this kind of design approach unpractical. Also, the low-frequency RF packaging design process often requires scalable equivalent circuits for the package itself. As the problems associated with the integration involve more and more factors to be considered, the design and optimization of such systems requires more comprehensive and sophisticated tools. The current design and optimization methods, using the commercially available electromagnetic simulators, do not take into account the specific effect of each of the factors involved in the design process, the degree to which these factors interact with each other, and their ranges of values. Only this type of thorough understanding of the entire system can enable the optimization and synthesis of any module or microsystem under different given conditions. An example of successful application of a combination of DOE and response surface methods (RSM) is presented in [44]. First, the factors that affect the performance of the system and the output figures of merit have to be identified. The next step involves the design of a factorial experiment with centerpoints based on a design space for these factors to determine the effect of each of the parameters, identify their interaction, and determine which ones are significant for each of the outputs. The experiment is run using electromagnetic simulations and/or microwave measurements, and the outputs are recorded and inputed into a statistical analysis software. After the statistical analysis of the data, significant factors are identified for all figures of merit, then RSM statistical methodology is applied for optimization. The result is an explicit set of equations that show how the outputs depend on the input variables, which are used to optimize the figures of merit simultaneously. Within the design space of the experiment, the optimized figures of merit and the required design parameters are identified.

The nonlinearity of the system, combined with the lack of analytical input–output description, suggests the use of soft computing algorithms also. Genetic algorithms (GAs) can be used as an optimization method of this kind. GAs search the parameter space stochastically generating solutions that are close to the optimal. They also turn out to be very efficient for problems where small perturbations in the optimal solution lead to an abrupt increase of the error.

These techniques can be applied to any type of design, especially in complex RF microsystems and packages where the number of factors increases and it is extremely difficult to optimize using only electromagnetic simulators. It gives a thorough understanding of the system behavior and integrates geometrical, material, and functional parameters altogether. The approach is generic and independent of the choice of the electromagnetic simulator and statistical analysis software.

This approach has also been applied for RF and microwave flip-chip design rule development [45]. Previous work identified the factors to be considered in the design process and in which way they affect the assembly performance. The structure and the geometrical factors are presented in Figure 7.14. The design of experiments method gave the possibility to bring all these factors together, quantify their significance, evaluate the interaction between them, and eliminate the insignificant ones from the analysis. The significant factors in this case have been found to be the bump diameter and height, the CPW width, and the amount of transmission line overlap around the transition. Also, the transition was optimized by considering not only the individual effects of the variables, but also the interaction between them, which can be critical in determining the optimal factor combination. Finally, fully scalable lumped element models with all these factors were developed by applying regression models to the statistical data.

Package models are typically provided as lumped element components and are used in circuit simulations as such. Other approaches could include S parameter models of the package components, which can be directly included in the circuit simulator. There are several types of package models, and the commonly known are L, T, and Pi models.

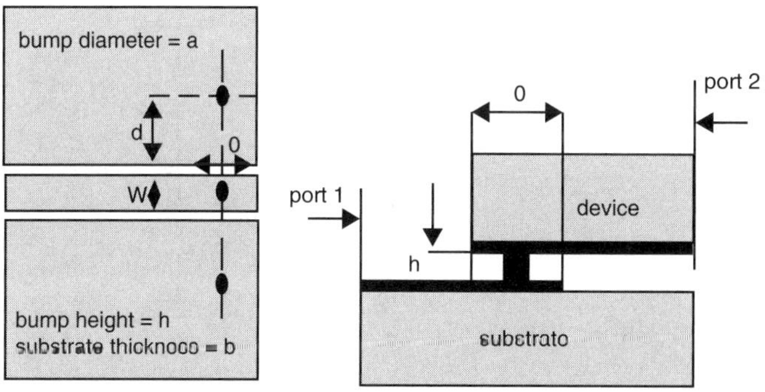

Figure 7.14. Geometrical parameters for flip-chip.

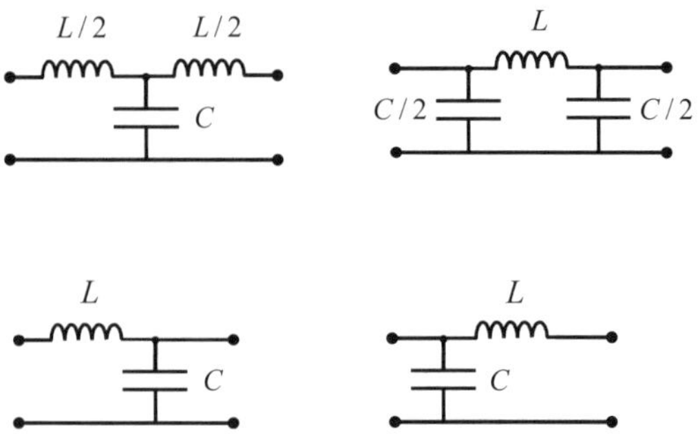

Figure 7.15. Commonly used package models: T, Pi, right L, and left L.

Figure 7.15 shows the different types of package models. It can be observed that the T and Pi models operate at higher frequencies compared with the left and right L models. Hence, while using packages for cosimulation with circuits, appropriate models with necessary bandwidth should be used. For use above such frequencies, additional effort must be spent to verify the quality of the models that are being used. Use of EM simulations can do this in an effective manner. One of the critical issues associated with packages is the pin-to-pin capacitance that can directly impact the quality of isolation between adjacent pins. Another major parasitic component is the capacitance between the pins and the ground paddle or the ground of the board. This can impact RF matching or increase loss through the pin at high frequencies.

Resistive and inductive effects are also important in modeling pins in a package. When using leadless packages, the lead inductance is typically very small, but it should still be considered during the simulations. Figure 7.16 shows an example of a package model.

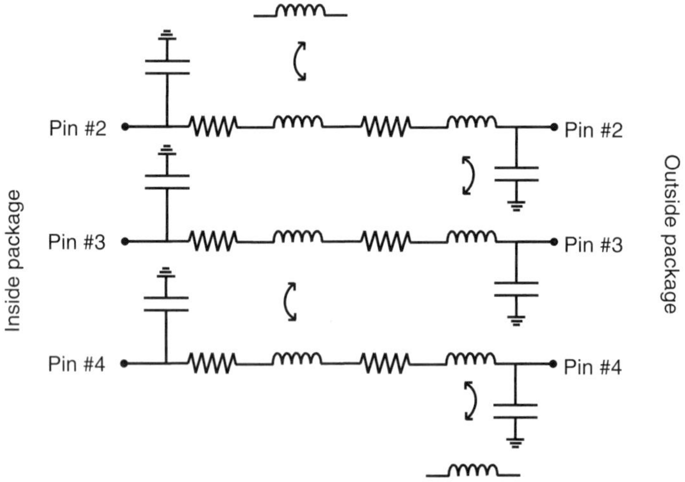

Figure 7.16. Electrical model for package pins and their pin-to-pin interaction.

Upon obtaining the package model from a vendor, it is imperative to check for its validity by exiting different ports one by one and observing the impedance and the induced voltage at all the other adjacent pins. A small signal simulation as well as a transient simulation gives more insight to this model. The components are usually of a distributed nature, but a lumped element model is usually adopted for representation in circuit simulation. The package inductance (and resistance) is determined by grounding the other end of the package line and measuring the impedance looking into it. The capacitance is measured by opening the package end and measuring the impedance looking into the package pin. By performing these two simulations, we can characterize the package in terms of its lumped element components.

7.8 FUTURE PACKAGING TRENDS

Recent developments in advanced packaging technologies such as 3D multichip modules provide an opportunity for significant reduction in mass, volume, and power consumption. Simultaneously, emerging wireless communications applications in the RF/microwave/millimeter wave regimes require miniaturization, portability, cost, and performance as key driving forces in the electronics packaging evolution. The SOP approach (versus the SOC) for module development has been proven to be the best approach for systems integration because of the real estate efficiency, cost savings, size reduction, and performance improvement potentially involved in this integral functionality, while simultaneously satisfying the specifications of the next-generation wireless communication systems. However, current RF module integration is still based on low-density hybrid assembly technologies. Embedded ICs [46] and bumpless build-up layer (BBUL) [47] packaging technology are targeted for low-cost applications and provide great opportunities for a multi-gigahertz processor and ultra-compact RF integrated system.

An example of an integration concept is illustrated in Figure 7.17. A multilayer interconnect structure is built using modified MCM-D technology and advanced photosensitive epoxy. Low-loss interconnect is fabricated using build-up technology.

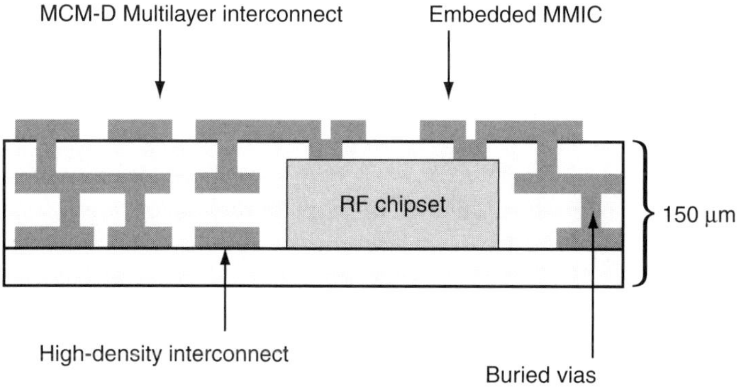

Figure 7.17. Embedded IC and BBUL packaging technology concept view.

Micro-vias with a minimum diameter of 40 μm are used to connect the different metal layers. Thus, a high-density interconnect network and integral passive components such as high-performances embedded inductors, filters, and antennas can be implemented within the multilayer wiring structure. A multi-gigahertz processor or RF commercial chipset is placed into cavities created along the MCM-D process and covered by a last dielectric layer. Cavities provide self-alignment for the dice with the interconnection structure. This approach avoids parasitics because of the wire bonding, flip-chip or BGA type of interconnection, and the parasitic interconnection length between the active circuitry and the passives components are greatly reduced.

New challenges such as RF-MEMS integration and packaging are nowadays at the leading edge of the packaging research. Various approaches have been developed by universities or worldwide packaging companies. Current techniques under investigations for the RF-MEMS packaging are as follows:

- Wafer-to-wafer anodic bonding or polymer bonding/sealing at wafer scale [48].
- Thin-film encapsulation at wafer scale [49].
- Cavity in MCM-C, L, or D substrate combined with the use of a sealing cap [50].

Nevertheless, there is still a lack of standardization that is leading to excessive cost for the development of final products. Eventually, the ideal package technology will merge the standard functions (i.e., environmental protection, hermetical sealing, accelerated testing and vacuum encapsulation skills, signal isolation, and mechanical stability) with design-specific functions, such as the RF requirement for minimized package resonance, reduced parasitic effects, and so on.

High-speed wired communication systems demand interconnect density on both the IC and the package. These ultra-high-density packages have to support digital and RF speed in the 20–50-GHz range simultaneously, posing challenges in electrical and mechanical design, fabrication of 5–10-micron structures with materials of unprecedented electrical and mechanical properties far beyond today's FR-4, BT, and other boards. All of these requirements have to be satisfied by an integrated solution for high density, high speed, low cost, and high reliability, necessitating a revolutionary interconnect methodology combining novel designs, materials, processes, structures, and test methods.

7.9 CHIP-PACKAGE CODESIGN

We would now consider a few cases of chip-package codesign examples and detailed illustration. In the beginning, we provide the readers with the important parameters about packages. Most of the commercial products in the wireless as well as wired communication domains today operate between 1 and 10 GHz, and our discussions will be very specific to these frequency bands. Chip-package codesign has become a routine practice nowadays, and while doing such, one should be careful not to generate additional crosstalk between various circuit blocks. First we will introduce various package elements, and then we will focus on modeling them.

In a package, there can be one or more power/ground planes, vias, electrical traces, and so on. In many cases, it is imperative to understand the inductance, capacitance, and so on originating from the various components on package, such as inductance and capacitance.

7.10 PACKAGE MODELS AND TRANSMISSION LINES

Package models originate in their fundamental form in understanding transmission line theory, which quantifies signal propagation on a system of two parallel conductors with cross-sectional dimensions much smaller than their length. Such structures can be classified as coaxial, stripline, microstrip, and others. Transmission lines are characterized as per-unit length capacitance C, per-unit length conductance G, per-unit length inductance L (frequency-dependent loop inductance from the skin effect), and per-unit length resistance R. Formulating the uncoupled equations in terms of voltage and current waveforms as uncoupled equations, telegrapher's equations are obtained as follows:

$$\frac{d^2V}{dZ^2} = (j\omega L + R)(j\omega C + G)V \tag{7.2}$$

$$\frac{d^2I}{dZ^2} = (j\omega L + R)(j\omega C + G)I \tag{7.3}$$

We define the characteristics impedance as $Z_0 = \sqrt{\frac{L}{C}}$, and the voltage and current waveforms are given as follows:

$$v(z,t) = f^+(z - v_p t) + f^-(z + v_p t) \tag{7.4}$$

$$i(z,t) = \frac{1}{Z_0} f^+(z - v_p t) - \frac{1}{Z_0} f^-(z + v_p t) \tag{7.5}$$

the terms represented as forward and backward waves, respectively, under the assumption that the line is lossless. The characteristic impedance determines the voltage amplitude, and any discontinuities in the characteristic impedances would lead to reflections.

7.10.1 Frequency of Operations

The type of models to be used depends on the wavelength (/frequency) under consideration. If the segment dimension d is much smaller than the wavelength (λ), then the lumped element component can be used for simplicity ($d << \lambda$). In this case, the inductance and capacitance values are obtained by simply multiplying the unit inductance and capacitance by the segment length. At larger dimensions, where $d > \lambda/10$ or $d \approx \lambda$, then a distributed transmission line model would be the most appropriate to use. In a digital system (microwaves), the frequency is interpreted as $f = \frac{0.35}{t_r}$.

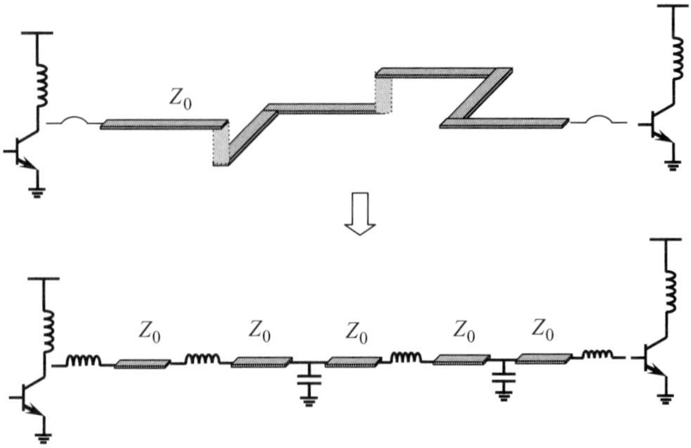

Figure 7.18. Illustrating physical bends in the transmission line.

7.10.2 Bends and Discontinuities

In a package, it is not always possible to route straight wire. Often, some amount of bent is provided in the trace. Such bends and discontinuities lead to either inductive or capacitive components along with the standard transmission line components. Wire bonds are primarily inductive in nature, whereas interconnect bends are capacitive in nature. Vias are primarily inductive. This situation is illustrated in Figure 7.18.

These unwanted additional components only add to the delay of the path, and they are dependent on the values of C, L, and Z_0. The voltages before and after the discontinuities are related by

$$V_{2i} = V_{1i}(1-e^{-t/\tau})$$

where τ is the time constant associated with C, L, and the characteristic impedance Z_0, and is given by $\tau = 0.5CZ_0$ or $\tau = 0.5(L/Z_0)$. Bends in differential traces introduce signal skew and differential reflections. Sometimes two bends can be added to compensate signal skew. Figure 7.19 illustrates the effects of inductive and capacitive discontinuities in the transmission line system.

In the presence of a capacitor, the delay is given by $V_{2i} = V_{1i}(1-e^{-\frac{\tau}{\tau_C}})$, whereas an inductive discontinuity leads to the formulation of $V_{2i} = V_{1i}(1-e^{-\frac{\tau}{\tau_L}})$. The time

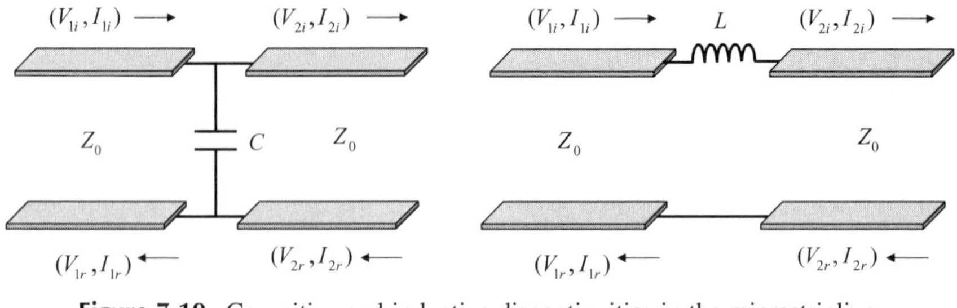

Figure 7.19. Capacitive and inductive discontinuities in the microstripline.

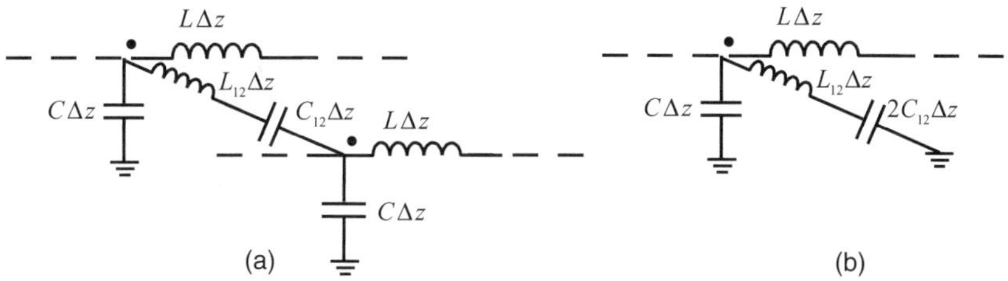

Figure 7.20. A section of common-mode and differential-mode impedance.

constants are given by $\tau_C = \frac{CZ_0}{2}$ and $\tau_L = \frac{L}{2Z_0}$. Hence, to achieve a 90% signal transmission $V_{2i} = 0.9V_{1i}$, it takes 2.3 time constants in both cases, which leads to a delay of $D = 1.15CZ_0$, and $D = \frac{1.15L}{Z_0}$. Figure 7.20 illustrates the differential mode and common mode impedance scenarios with regards to a transmission line.

7.10.3 Differential Signaling

In modern microsystems, both the circuits as well as the package structures are differential in nature. Hence, the transmission lines are also differential in nature, and the impedances are defined as even and odd modes. In differential mode, $Z_{diff} = 2Z_{odd}$. Single-ended structures are susceptible to common-mode coupling, and the potential difference at the receiver fluctuates because of the voltage drop by the other gates at the package pin, as well as because of any variation w.r.t. power supply voltages. This consideration is eliminated in the case of differential signalling, which is illustrated in Figure 7.21.

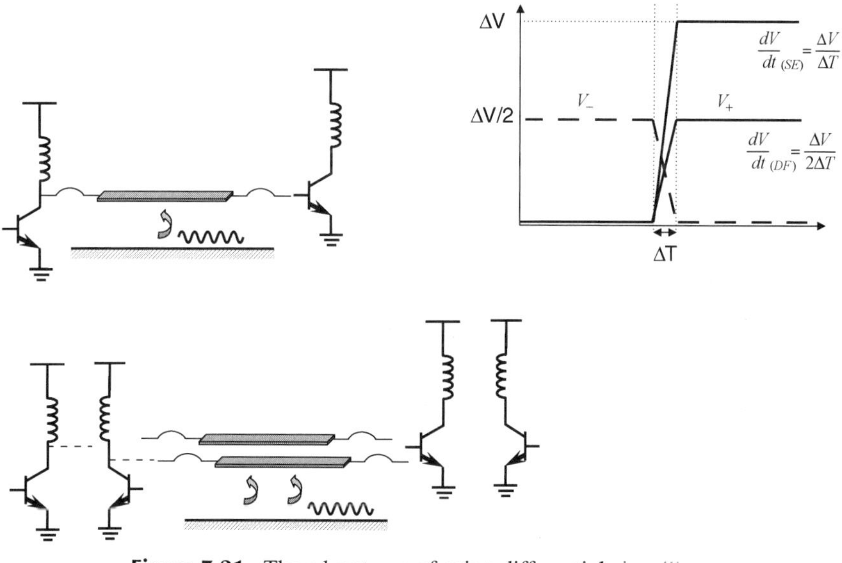

Figure 7.21. The advantages of using differential signalling.

The differential signal has another advantage in terms of delay. A differential signaling system requires half the voltage swing per output pin for the same signal swing, which leads to lower switching current to charge and discharge the capacitors at the logic gate input and to a factor of 2 improvement in speed compared with single-ended signaling. This advantage holds good for large swing digital circuits, as well as for high-speed analog circuits. Lower supply voltage also reduces the power dissipation. By reducing the susceptibility to the common-mode disturbances, differential signaling allows for a longer propagation distance as well. At the same time, differential lines radiate less as the incident and the return currents are closer to one another, and the net EM fields cancel each other. Differential structures provide better EMI performance, noise margin, and low radiation, and they are less susceptible to ground bounce.

Since every signal is a combination of differential- and common-mode signals, they need to be terminated individually (both common-mode and differential-mode termination should be used). The reflection coefficients should also be defined for a differential signal. Signal bends and so on introduce additional reflections, such as clock skews, which are similar to their single-ended counterpart. Two coupled transmission lines can be analyzed w.r.t. their even- and odd-mode impedances as shown in Figure 7.22.

It can be noted that in the even mode, no current flows through the coupling capacitor between two transmission lines, as the drive phases are the same. Hence, in odd mode, $I_1 = -I_2$. Combining voltage and current equations, the odd mode impedance for single transmission line can be given as follows:

$$Z_{odd} = \sqrt{\frac{L - L_{12}}{C + 2C_{12}}} \tag{7.6}$$

Similarly, the even-mode impedance can be given as follows:

$$Z_{even} = \sqrt{\frac{L + L_{12}}{C}} \tag{7.7}$$

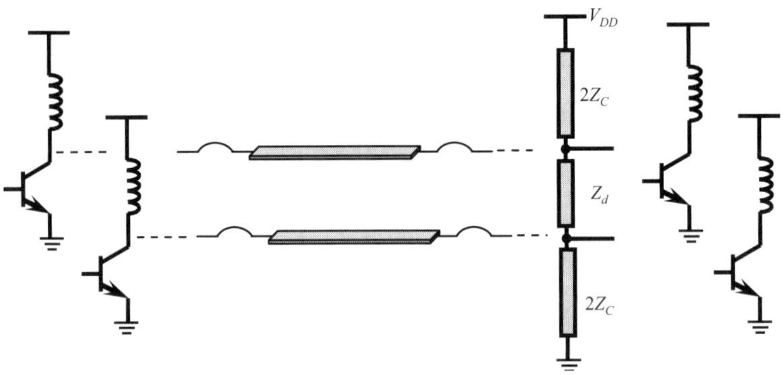

Figure 7.22. Common-mode and differential-mode termination.

The differential- and common-mode impedances are governed by

$$Z_{cm} = \frac{V_E}{2I_E} = \frac{Z_E}{2} \tag{7.8}$$

$$Z_{dm} = \frac{2V_o}{I_o} = 2Z_o \tag{7.9}$$

To ensure differential mode operation, $Z_o < Z_e \Rightarrow Z_{dm} \leq 4Z_{cm}$, equality holds when both lines are balanced and uncoupled.

7.11 CALCULATIONS FOR PACKAGE ELEMENTS

7.11.1 Inductance

Two types of inductance exist in the context of a current-carrying wire. Internal inductance of a wire is determined by the current that flows inside the conductor, which is maximum at DC (the whole current flows through an entire internal cross section), and it drops at higher frequencies. At DC, current is spread uniformly in the conductor, whereas at AC, it is confined in an outer ring as shown in Figure 7.23. However, at high frequencies, the self-inductance and resistance becomes equal $R_{AC} = \omega L_{AC}$. The transition point between the DC internal inductance and that of the AC happens when both AC resistance and DC resistance become equal, $\delta_{skin} = \sqrt{\frac{1}{\pi f \mu \sigma}} = r/2$, and the total self-inductance being given by $\frac{1}{L_{\text{int}}} = \left[\frac{1}{L_{DC}{}^2} + \frac{1}{L_{AC}{}^2}\right]^{1/2}$.

Thus, the self-inductance of a current-carrying wire/conductor is maximum at DC and drops at higher frequencies, and the rate of drop is determined by $\sqrt{f}$, where f is the operating frequency.

In many cases, the self-inductance is not of much importance, except that we are interested in the inductive impedance of the wire. Often, the Q factor becomes important. However, mutual inductance of two current-carrying lines is one of the most important parameters in integrated systems, where two wires can be placed parallel to one another and interfere significantly. Mutual inductance is determined by

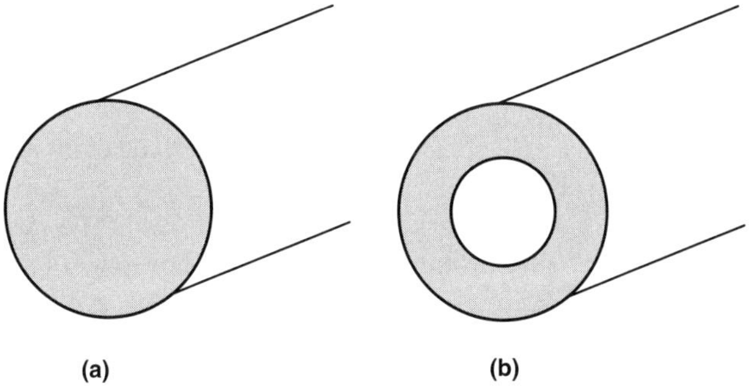

Figure 7.23. DC and AC inductance of a wire in 3D.

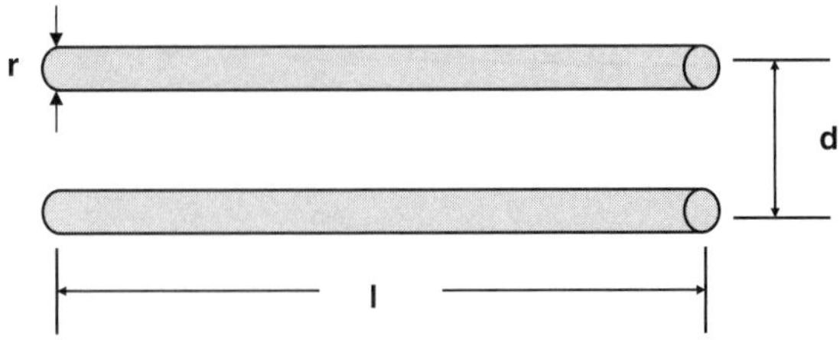

Figure 7.24. Inductance formulation of two wires.

Neumann's formula of closed form integration, which assumes the current to be flowing in the surface of the conductor.

$$M_{12} = \mu_0 \int_{l_m} \int_{l_n} \frac{1}{4\pi R_{mn}} dl_n dl_m \tag{7.10}$$

For the common case of two parallel lines with radii r, and separated by distance d, Neumann's formula in closed form leads to

$$L(l,r) = \frac{\mu_0}{2\pi} l \left(\log\left(\left(\frac{l}{r}\right) + \sqrt{1 + \left(\frac{l}{r}\right)^2} \right) - \sqrt{1 + \left(\frac{r}{l}\right)^2} - \left(\frac{r}{l}\right) \right) \tag{7.11}$$

This formula can be used to solve other complex structures as well. The actual geometry is illustrated in Figure 7.24.

7.11.2 Capacitance

Although understanding various types of coupling in package components, capacitance becomes important to consider. When two conductors carry equal and opposite charges δQ, with a potential difference of δV, the capacitance is defined as $C = \frac{\delta Q}{\delta V}$.

We can illustrate this for several cases, and geometries; however, we cover three common structures of interest: (1) Capacitance of a line to ground, (2) ball to ground, and (3) ball to infinity. These are pertinent to the package structure, and they are used commonly at the time of chip-package codesign to compute capacitance. For example, while designing an LNA, it is important to know how much the input capacitance is, while optimizing the transistors for input matching and noise performance. Fundamentally, these are computed either using method of images or isolated cases.

We would now focus on several geometries that are well known for package structures. For a parallel plate transmission line as shown in Figure 7.25, the inductance and capacitance per unit length are given by

$$L = \frac{\mu a}{w}, \ C = \frac{\varepsilon w}{a} \tag{7.12}$$

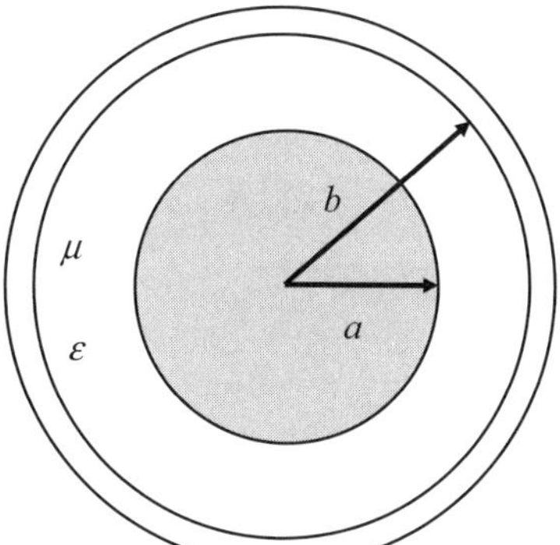

Figure 7.25. Capacitance of coaxial cable.

For a coaxial transmission line, they are given by

$$L = \mu \ln \frac{b}{a}, \; C = \frac{2\pi\varepsilon}{\ln(b/a)} \tag{7.13}$$

While designing with various types of circuits, capacitance formulation of bondpads is also very important consideration. Figure 7.26 provides an illustration of a bondpad with the key geometry parameters. The shape of the bondpad determines the fringing capacitance. In the case of wafer scale packages, the solder balls are directly attached to these bondpads, and the overall structure needs to be considered for stability. Usually bondpads require multiple metal layers stacked together by vias.

7.11.3 Image Theory

To compute inductance and capacitances of various packaging structures, image theory can be used. This is simply obtained by producing an image of the main

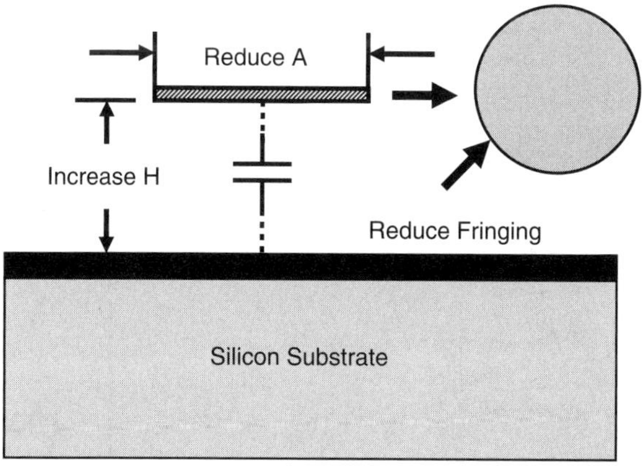

Figure 7.26. Bondpad model and the capacitive components.

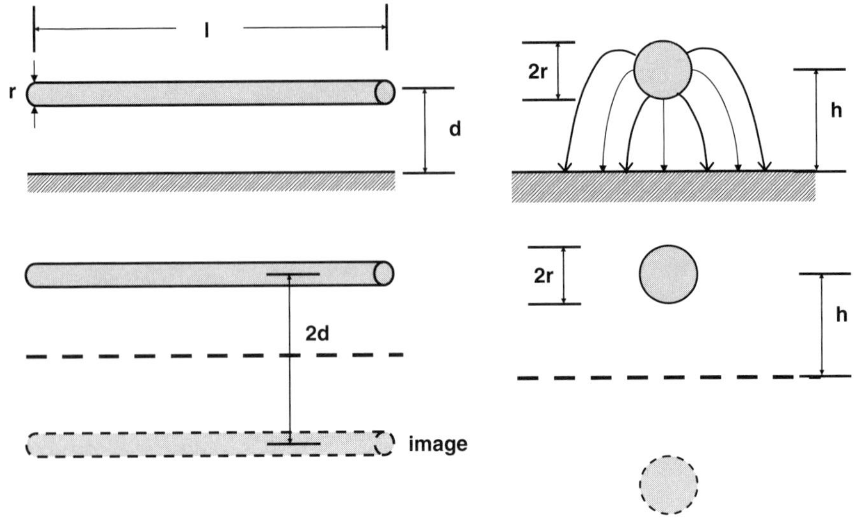

Figure 7.27. Computation of capacitance and inductance by image theory.

current/charge carrying element at a distance "d" under the ground plane. This provides computation efficiency in the case of complex geometries. Figure 7.27 illustrates the formulation of current and charge images for inductance and capacitance, calculations, respectively.

7.12 CROSSTALK

In addition to the considerations and formulations of electrical parameters such as inductance, capacitance, and so on, crosstalk plays a significant role in determining signal integrity. Crosstalk is strongly dependent on the termination of coupled lines, as shown in Figure 7.28. Unterminated lines usually exhibit worse crosstalk performance. At the same time, signal rise time impacts the crosstalk phenomenon. A faster rise time leads to more "ringing" in the signal characteristics if the line is not terminated properly.

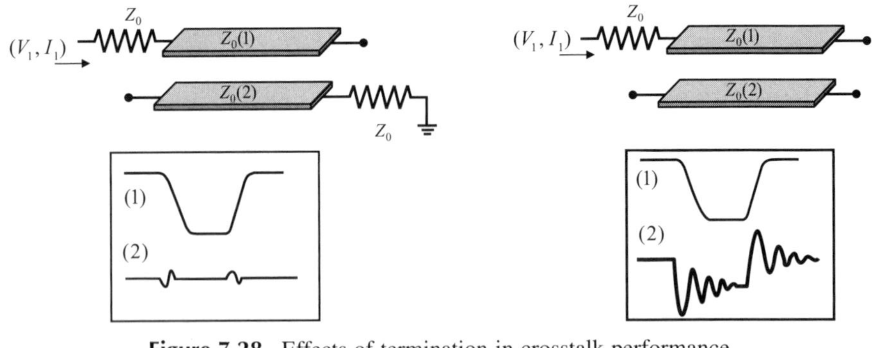

Figure 7.28. Effects of termination in crosstalk performance.

Crosstalk in coupled lines also leads to reflection, noise margin degradation, signal distortion and dispersion, attenuation (/loss), and radiation. Such issues are especially important while considering many coupled lines in a high-speed system.

7.13 GROUNDING

Grounding is critical in the successful design of any RF and microwave components. The ground reference is an inherent requirement for performing simulations, and the proper allocation and modeling of a ground path can greatly impact the accuracy of the design. Poor on-chip grounding can have deleterious effects on the fabricated IC performance, which leads to deviations in design to hardware correlation. Both types of grounds (on chip and on package) are essential to categorize the effects of ground wiring. The ground pin could provide a resistance, as well as inductance, that may result in emitter degeneration, as well as unwanted resonances at undesired frequencies. The designer must identify the location of the RF ground and carefully model any parasitic inductances that may exit in the ground path. This location is typically the ground plane of the reference board. It will translate to a nonideal ground on the IC because the path from the board ground to the ground on the chip includes the inductance and resistance of the board, package paddle or pin, and the downbond connecting the chip ground to the paddle or pin. Issues regarding the package and its type will be further discussed in the next section.

The AC ground becomes less significant when the design migrates to a differential topology. Typically, use of multiple downbonds placed in parallel can ensure availability of adequate on-chip ground for the RF components of the receiver. However, when using multiple wire bonds, additional care must be taken in modeling the mutual coupling between adjacent bonds. Although it is correct to assume that the use of multiple downbonds reduces the overall ground inductance, the mutual inductance between the bonds that are placed very close to one another can reduce the impact of this parallel combination. Figure 7.29 shows a simple model of a bondwire and the outcome of mutual coupling between the two.

Low inductance grounding is essential in reducing crosstalk. The isolation between various blocks of the receiver can be significantly improved by use of low inductance ground paths. Figure 7.30 shows a few examples for use and modeling of on-chip RF ground.

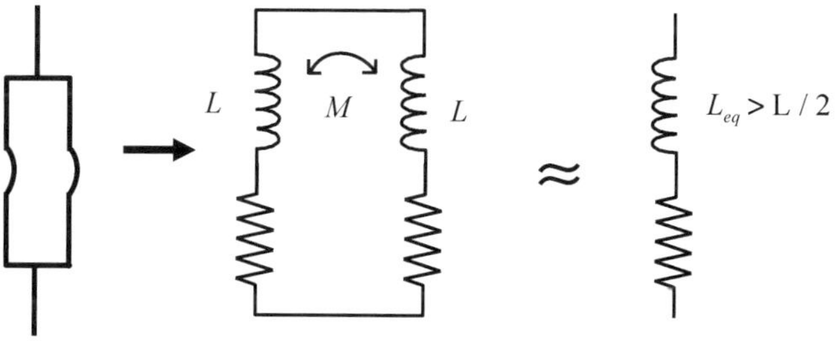

Figure 7.29. Impact of mutual coupling on adjacent wirebonds.

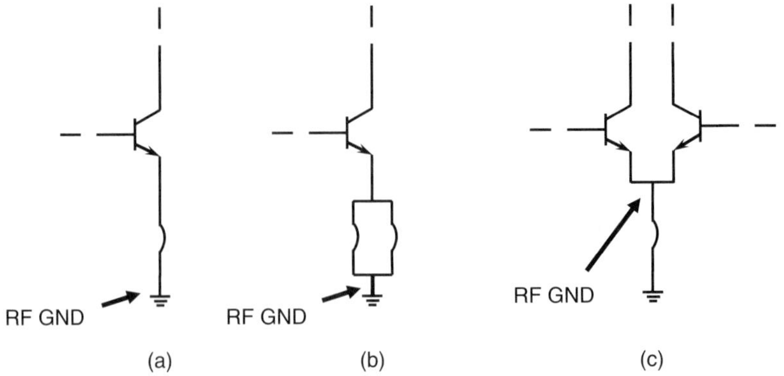

Figure 7.30. Various RF grounds: (a) high-inductance RF ground for a single-end amplifier located at the far side of the wirebond, (b) low-inductance RF ground using two wirebonds, and (c) RF ground of a differential amplifier unaffected by the wirebond inductance.

7.14 PRACTICAL ISSUES IN PACKAGING

Package modeling is very important in realizing an end-product that is feasible and can easily be integrated into a system. With migration of application into higher frequencies, the RF and microwave performance of packages are becoming more and more important and essential to success of the design process. A major problem with leaded packages is the inductance associated with the lead. This is eliminated by the use of leadless packages that reduce this inductance significantly. Another characteristic of a good package for RF applications is the paddle. To accommodate a good on-chip RF ground, the ground must be connected through downbonds to the package paddle, which is soldered on to the board assembly ground. Without a paddle, the ground connection must be routed through longer bondwires and package pins increasing the ground inductance.

7.14.1 Ground Modeling

Figure 7.31 illustrates an example schematic used to model the RF ground path from the board to circuit implemented on the chip. It is a very simplistic model that can be used with relative accuracy up to a few gigahertz. Electromagnetic simulations are necessary for modeling such a transition at higher frequencies.

The actual amount of inductance or resistances that are used is dependent on the actual dimensions of the various components. For example, a 1-mil wire bond has a typical inductance of 0.9 nH/mm, but this value changes with the diameter of the wire bond. Some minor capacitances to the paddle are also associated with the wire bond or downbond that we have ignored here in our estimates. The package pin parasitics are very much dependent on the shape and length of the pin. If a leaded package is used, the inductance and resistance values used for modeling the package pin would be significantly increased. For the case of a package with a paddle,

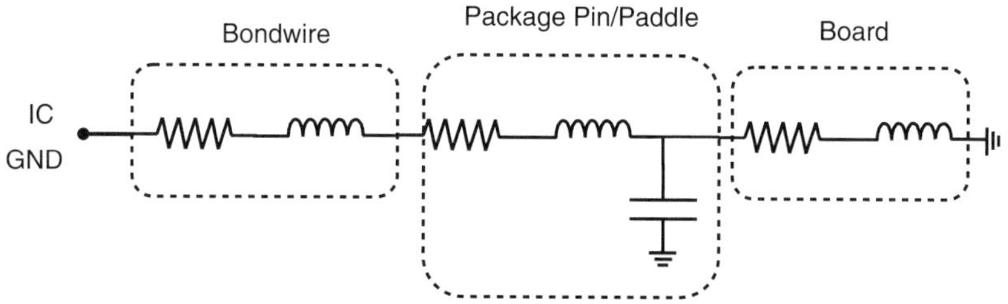

Figure 7.31. A lumped element model of the RF ground path from the circuitry on an IC to the board ground.

the parasitics are much less and depend on the dimensions of the paddle itself. Typically a few tens of pH is a good estimate for a package paddle parasitics. Modeling of the board parasitics is very much dependent on the board layout. It is desired that the board ground plane cover a large area and is well connected throughout different layers of the board with numerous ground through vias. This would ensure a relatively constant ground potential on the board and reduce any parasitic effects.

7.14.2 Isolation

As mentioned, the physical distance between various receiver components decreases by increasing the level of integration in an IC. Increasing the density of interconnection, active and passive components inside an IC, reduces the isolation between critical signals. Figure 7.32 illustrates the workings of deep trenches and isolation rings on improving isolation between two transistors.

Use of deep trench isolation and isolation rings can help to improve poor isolation, but this may not be enough. Some designers rely on careful frequency planning and unique architectures to reduce the impact of poor isolation on overall receiver performance.

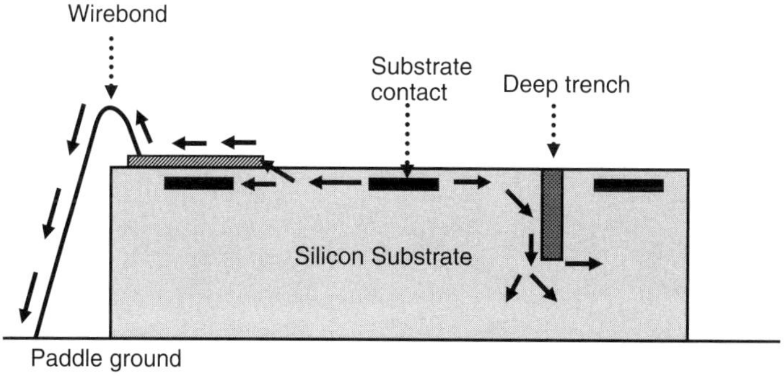

Figure 7.32. RF surface currents and the impact of deep trenches and grounded substrate contacts on the current flow and direction.

7.15 CHIP-PACKAGE CODESIGN EXAMPLES

7.15.1 Tuned Amplifier with Off-Chip Inductor

Let us now consider the issues associated with chip-package codesign. In any RF architecture, there is need for a tuned amplifier, which may use off-chip inductance. Such inductance can be very easily used from the bondwires of the package. In the case of a wafer scale package, these inductors can be patterned on the package layer itself. However, a major consideration comes as a result of the crosstalk effect. Since these bondwires radiate electromagnetic fields, the other integrated structures are susceptible to the electromagnetic interference, which may lead to unwanted spectral components and feedback loops in a receiver. It must be noted that depending on the input signal strength, the output of the LNA using bondwire inductance as loads may sometimes be significantly higher, and the field may cause significant radiation issues. As shown in Figure 7.33, the differential LNA may use two bondwires, which need to be spatially symmetrical to one another. As the currents through them flow in the opposite direction at any instant of time, the mutual inductance tends to reduce the individual inductances. This is given by the equation

$$L_{diff} = L_1 + L_2 - 2K\sqrt{L_1 L_2} \tag{7.14}$$

where L_1, and L_2 are the individual inductances and K is the coupling coefficient.

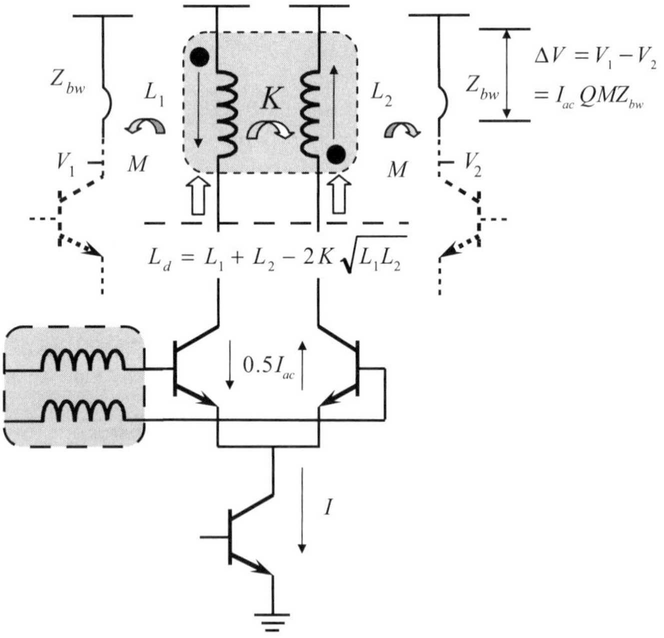

Figure 7.33. A differential LNA with bondwire inductors as load.

Hence, the individual wires need to be larger to compensate for the amount of inductance "lost" in the mutual coupling phenomenon. As the length increases, it tends to form an antenna on the package, and it picks up additional parasitic components. Such loops may also have the possibility of eddy currents on sheets of metals used for fill patterns or MIM capacitors. However, the inductance is determined by the impedance to be seen in the resonating tank, as well as in the frequency of operation.

7.15.2 LNA and Oscillator

Another common example could be taken as a resonator tank in a voltage-controlled oscillator. VCOs require a high Q inductor as part of the tank resonator for the phase noise specifications in the wireless/cellular handset solutions. Most of the wireless standards are of narrowband type, and a higher Q tank implies a lower current consumption for a given phase noise requirement. Hence, many times, high-Q off-chip components are used for VCO design. As VCOs provide significantly higher peak-to-peak swing, these inductors can significantly induce electromagnetic fields. Many of these can be handled using a differential configuration, where dipoles are formed to reduce far-field induction. Figure 7.34 illustrates these two cases of codesign. Off-chip inductors are essentially a trade-off among the Q factor, susceptibility to noise, antenna pickup, and the inductance value. These factors decide significantly on the success of integrated systems, especially in the vicinity of 1 GHz, where the on-chip Q is much lower, compared with its package counterpart.

High Q filters are also desirable for a chip-package codesign example. Any radio architecture uses high Q filters as part of the front end to provide high band selectivity.

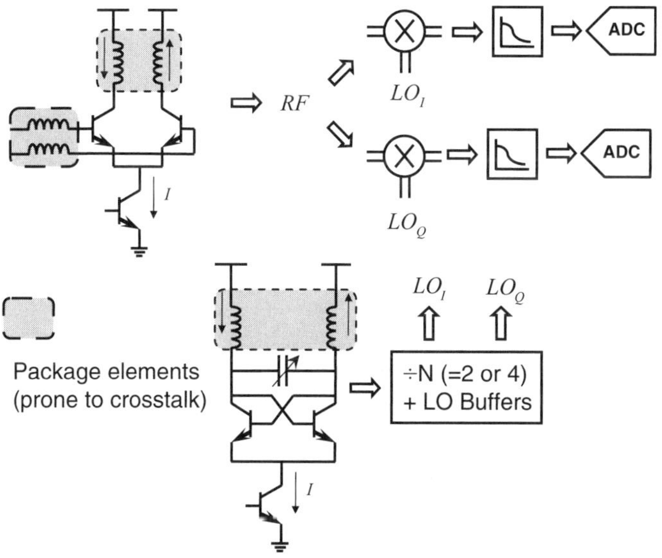

Figure 7.34. LNA and VCO circuits using codesign with bondwire.

For superheterodyne radios, SAW filters are traditionally used to provide channel selectivity. However, standard filters tend to adhere to 50 Ω impedance standards, and deviation from this impedance would lead to group delay variation in the filter. High selectivity filters can be very easily realized using high-Q off-chip inductors and on-chip capacitors. Duplexers can also be realized using such a codesign mechanism. Impedance matching is also commonly used in the case of chip-package codesign.

In integrated microsystems, where chip-package codesign is frequently used, the resonator cores should be isolated, as well as spatial quadrature should be practiced to achieve better isolation. In a bondwire-based package (e.g., BGA) pads are always placed at the periphery of the chip, and the coupling between the bondwires should be considered for final placement.

7.15.3 Magnetic Crosstalk

For an RF application, package performance is very essential, and common performance parameters are related to (1) electrical coupling and (2) magnetic coupling. Magnetic coupling is often underestimated. Let us assume that there are two bondwires leading to the differential amplifier load inductance. The inductances may have a Q of 10, and the circuit may be consuming an AC current of a 5-mA peak. This implies a 50 mA in the inductor, and with a coupling coefficient of 0.2 to the adjacent bondwire, it

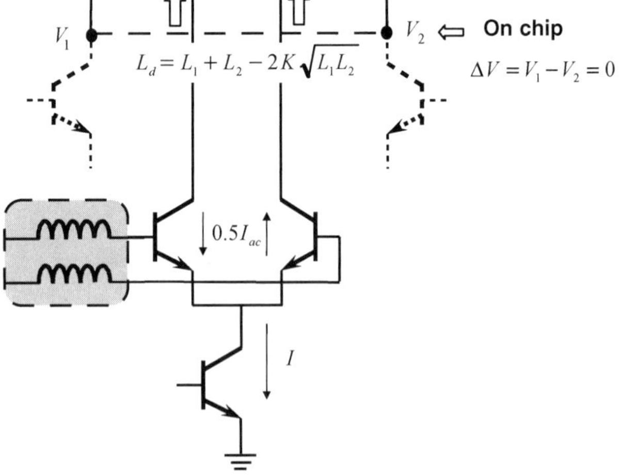

Figure 7.35. Magnetic crosstalk illustration using bondwires.

induces a current of 10 mA, which leads to a voltage difference of 50 mV across a resistor of 5 Ω, which can result from the trace resistance. As this leads to voltage imbalance, performance degradations as the circuit now is not symmeytric anymore. At the same time, it reduces headroom. To eliminate this effect, one can short the two ends of the supply wires, and there would be no voltage difference across the loop. However, in such cases, the inductor Q would be degraded by the presence of a loop around it. Connecting capacitance to alleviate the issue in fact worsens it, as the capacitance would lead to lowering the impedance of the loop. Figure 7.35 illustrates this situation.

7.16 WAFER SCALE PACKAGE

In the wafer scale package, the off-chip inductor would be a spiral-type inductor. The chip is usually placed in a package in a flip-chip fashion. However, in a wafer scale package, the pad dimensions are usually large, which leads to increased static capacitance at the circuit nodes. The wafer is covered with plastic, and then a copper layer is placed, which is used for design inductors and so on. The Q is usually lower as the thickness of the copper layer is small. A plastic is placed on top of the copper layer, and the wafers are cut into pieces to obtain packaged chips. The major problem of the wafer scale package is the lower Q and the inflexibility in placing components. There is minimal routing flexibility compared with the to ball grid array package.

Concerns of mutual inductance and magnetic crosstalk are eliminated in the case of the wafer scale package. However, the complication in wafer scale package happens because the packaged inductors are very close to the chip, and the interaction between the chip and the package would be much more compared with the BGA package. The reduced thickness of the polyimide layer increases the input capacitance at the LNA input significantly. To minimize this capacitance, the layer immediately below the ball may be removed to reduce the capacitance. The inductor may consume a large area and may lead to eddy current losses caused by interaction with the chip; it is difficult to obtain a high Q using the wafer scale package.

7.17 FILTERS USING BONDWIRE

Bondwires can be used to develop on-chip filters (see Figure 7.36) [51]. The inductance of the bondwire can be used as a filter component. However, the bondwire structure could also include the inductance of a trace, which could be either embedded into the IC carrier or a trace on the IC. Another configuration includes a bondwire to a pad on the IC or a substrate connected to another pad in the IC, or substrate connected back to another bondwire. Figure 7.37 illustrates the various types of bondwire configurations in a chip-package codesign of RF front-end filters. The capacitors could be used from standard semiconductor technologies.

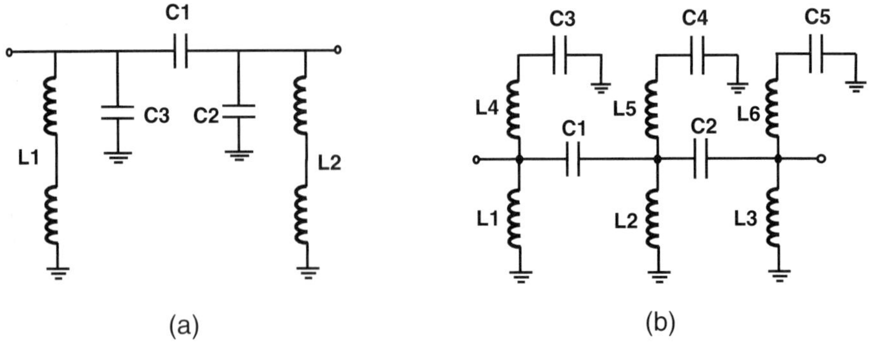

Figure 7.36. Two filters using bondwire inductors.

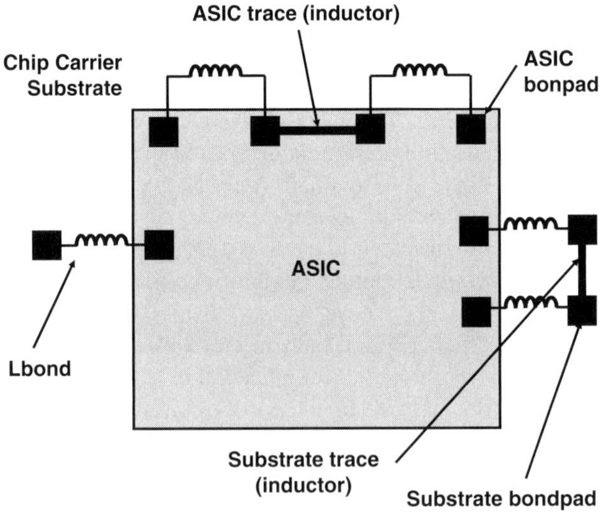

Figure 7.37. Front-end RF filters using a combination of bondwire and off-chip capacitors.

Most of the commercial applications today range between 1 and 10 GHz. Lumped element models are usually sufficient for such applications.

7.18 PACKAGING LIMITATION

Many design rules are supplied by the packaging houses that must be considered when designing integrated receivers. As mentioned, integration will typically translate into higher interconnection density in an IC. This can manifest itself into the use of larger packages that can accommodate more pins, leaving a relatively large distance between the pins and the die sitting inside the package. This can result in stretching the length of wirebonds beyond the limits defined by the packaging facilities and increasing the inductive parasitics of the wirebonds. Use of custom lead frames with multiple rows of pins is one of the ways in increasing the pin count without increasing the size of the package and the length of the wirebonds.

It must be also noted that to ensure the success of integrated microsystems, shrinkage of the semiconductor technology node is not the only aspect to be considered. Packaging form factors must shrink to accentuate the rapid shrinkage of semiconductor minimum feature sizes. It may so happen that because of the aggressive scaling of the semiconductor technology node, the entire solution may be bondpad limited. Hence, it may sometimes prove judicious to remain at a submicron node, achieving the desired performance and the form factor necessary for the solution.

CONCLUSION

Future RF and wireless communication and sensor systems will be increasingly complex, miniaturized, and reconfigurable, enhancing the importance of packaging structures. High-integration technology is expected to reduce the chip size, number of components, and total system cost. Along with the migration of ICs to integrated SOC, IC packaging is moving toward wafer-level packaging, and toward wafer-level test and burn-in. The rapid progress of scaled CMOS technology and the introduction of SiGe technology during the past few years have improved the performance of silicon RF devices and circuits to meet the requirements. However, the difficulty in lowering the power consumption of RF and analog circuits and the size of passive components and analog transistors has diminished the appeal of the SOC approach. The SOP will be the way to address these issues and lead to the successful transition from bulky board packaging to the multilayer micro-, nano-, and multifunctional boards in the future.

Useful links

http://www.chips.ibm.com/
http://www.chips.ibm.com/micronews
http://www.chips.ibm.com/products/interconnect/swg/interconnect.html
http://www.tessera.com/
http://www.flipchip.com/flipsec4/tec.htmhttp://www.semichips.org/
http://sprocket.colorado.edu/centers/yclee-group/links/
http://www.prc.gatech.edu/
http://www.smta.org/

REFERENCES

[1] A. Matsuzawa, "RF-SoC - expectations and required conditions," Invited Paper, *IEEE Transactions on Microwave Theeory and Techniques*, Vol. 50, No. 1, Jan 2002, pp. 245–253.

[2] K. Lim, S. Pinel, M.F. Davis, A. Sutono, C.-H. Lee, D. Heo, A. Obatoynbo, J. Laskar, M. Tentzeris, and R. Tummala, "RF-SOP for wireless communications," *IEEE-Microwave Magazine*, Vol. 3, No. 1, Mar 2002, pp. 88–99.

[3] C.-K. Liu and Y.-H. Cheng, "Thermal analysis of power amplifier package in cellular phones," *Proceedings of the 4th International Symposium on Electronic Materials and Packaging*, Dec 2002, pp. 415–421.

[4] M. DiOrio and S. Pinamaneni, "Material effects on the performance and reliability of high-power molded dual-in-line packages," *Proceedings of the 38th Electronics Components Conference*, May 1988, pp. 406–410.

[5] G.C. Phillips, Jr., "Planar pin grid array (PGA) ceramic packaging," *Proceedings of the 38th Electronics Components Conference*, May 1988, pp. 350–354.

[6] R. Chroneos, D. Mallik, and S. Prough, "Packaging alternatives for high lead count, fine pitch, surface mount technology," *Proceedings of Electronics Manufacturing Technology Symposium*, Sep 1991, pp. 181–186.

[7] R. Tummala, E. Rymaszewski, and A. Klopfenstein, Microelectronics packaging handbook – Part II, 2nd Edition, Kluwer Academic Publishers, 1997.

[8] M.J. Kuzawinski,"IBM' s ultra fine pitch wire bond PBGA modules," IBM Microelectronics Web Site http://www.chips.ibm.com/micronews/vol4_no4/pbga.html, 1998.

[9] M. Christensen, "Flip-chip: A Technology Reborn," Solid State Technology, Pennwell Publishing Co. , 1999.

[10] D. Dunn, "BGAs given a couple of boosts." Electronic Buyer' s News, CMP Media, Inc., Nov 29, 1999.

[11] *National Technology Roadmap for Semiconductors* 1997 edition, Semiconductor Industry Association, 1997.

[12] http://www.chips.ibm.com/products/interconnect/technology/scale.html, 1998.

[13] K. Brownlee, S. Bhattacharya, K.-I. Shinotani, C.P. Wong, and R. Tummala, "Liquid crystal polymer for high performance SOP applications," *8th International Symposium on Advanced Packaging Materials*, 2002, pp. 249–253.

[14] S. Pinel, S. Chakraborty, S. Mandal, H. Liang, K. Lim, M. Roellig, R. Kunze, R. Tummala, M. Tentzeris, and J. Laskar, "3D integrated LTCC module using μBGA technology for compact C-band RF front-end module," *Proceedings of IEEE International Microwave Symposium*, Jun 2002, pp. 1553–1556.

[15] C. Mitchell, "Assembly and reliability study for the micro-ball grid array," *Proceedings of the 16th IEEE/CPMT IEMT Symposium*, Sep 1994, pp. 344–346.

[16] M. Christensen, Flip-chip: A Technology Reborn, Solid State Technology; Pennwell Publishing Co., 1999.

[17] D. Staiculescu, J. Laskar, and M. Tentzeris, "Design of experiments (DOE) technique for microwave/millimeter wave flip-chip optimization," *International Journal of Numerical Modeling*, Vol. 16, 2003, pp. 97–103.

[18] J. Jahns, R.A. Morgan, H.N. Nguyen, J.A. Walker, S.J. Walker, and Y.M. Wong, "Hybrid integration of surface-emitting microlaser chip and planar optics substrate for interconnection applications," *Photonics Technology Letters*, Vol. 4, No. 12, Dec 1992, pp. 1369–1372.

[19] K.P. Jackson, "High-density, array, optical interconnects for multi-chip modules," *Proceedings of Multi-Chip Module Conference*, Mar 1992, pp. 142–145.

[20] G.E. Ponchak, E. Dalton, E.M. Tentzeris, and J. Papapolymerou, "Coupling between microstrip lines with finite width ground plane embedded in polyimide layers for3-DMMICs on Si," *Proceedings of IEEE International Microwave Symposium*, Jun 2002, pp. 2221–2224.

[21] D. Staiculescu, K. Lim, A. Sutono, H. Liang, M. Tentzeris, and J. Laskar, "Flip-chip vs. wirebond," *Printed Circuit Design Magazine*, June 2002, pp. 12–16.

[22] B.C. Johnson, Overview of chip-level packaging, in: *Electronic Materials Handbook Volume 1 Packaging*, ASM International, 1989.

[23] G. Vendramin and M. Weller,"MCM-L: Rethinking electronic packaging." IBM Microelectronics web site. http://www.chips.ibm.com/micronews/vol4_no4/mcm-l.html, 1998.

[24] B. Levine, "Are MCMs finished?" *Electronic News*, Cahners Publishing Co, Feb 22, 1999.

[25] R.R. Tummala, "SLIM: Third generation of packaging beyond MCM, CSP, flipchip and micro-via board technologies," *Proceedings of 2nd Electronics Packaging Technology Conference*, Dec 1998, pp. 1–8.

[26] R.R. Tummala, V. Sundaram, Fuhan Liu G. White, S. Hattacharya, R.M. Pulugurtha, M. Swaminathan, S. Dalmia, J. Laskar, N.M. Jokerst, and Y.-C. Sang, "High density packaging in 2010 and beyond," *Proceedings of the 4th International Symposium on Electronic Materials and Packaging*, Dec 2002, pp. 30–36.

[27] M.F. Davis, A. Sutono, S.-W. Yoon, S. Mandal, N. Bushyager, C.-H. Lee, K. Lim, S. Pinel, M. Maeng, A. Obatoyinbo, S. Chakraborty, J. Laskar, E.M. Tentzeris, T. Nonaka, and R.R. Tummala, "Integrated RF architectures in fully-organic SOP technology," *IEEE Transactions on Advanced Packaging*, Vol. 25, No. 2, May 2002, pp. 136–142.

[28] O. Salmela and P. Ikalainen, "Ceramic packaging technologies for microwave applications," *Proceedings of Wireless Communications Conference*, Aug 1997, pp. 162–164.

[29] S. Pinel, M. Davis, V. Sundaram, K. Lim, J. Laskar, G. White, and R. Tummala, "High Q passives on liquid crystal polymer substrates and μBGA technology for 3D integrated RF front-end module," *IEICE Transactions on Electronics*, Vol. E586, No. 8, Aug 2003, pp. 1584–1592.

[30] S. Pinel, K. Lim, M. Maeng, M.F. Davis, R. Li, M. Tentzeris, and J. Laskar, "RF system-on-package (SOP) development for compact low cost wireless front-end systems," *European Microwave Conference*, Sep 2002.

[31] M. Tentzeris, R. Li, K. Lim, M. Maeng, E. Tsai, G. DeJean, and J. Laskar, "Design of compact stacked-patch antennas on LTCC technology for wireless communication applications," *Proceedings of the 2002 IEEE AP-S Symposium*, Vol. 2, 2002, pp. 500–503.

[32] R. Li, G. DeJean, M. Tentzeris, and J. Laskar, "Integration of miniaturized patch antennas with high dielectric constant multilayer packages and soft-and-hard surfaces (SHS)," *Proceedings of the 2003 IEEE-ECTC Symposium*, May 2003, pp. 474–477.

[33] H. Jiang, Y. Wang J.-L.A. Yeh, and N.C. Tien, "On-chip spiral inductors suspended over deep copper-lined cavities," *IEEE Transactions on Microwave Theory and Techniques*, Vol. 48, Dec 2000, pp. 2415–2423.

[34] Y. Rahmat-Samii and H. Mosallaei, "Electromagnetic band-gap structures: classification, characterization, and applications," *11th International Conference on Antennas and Propagation*, Vol. 2, Apr 2001, pp. 560–564.

[35] D.C. Montgomery, *Design and Analysis of Experiments*, John Wiley & Sons, 1997.

[36] R. Hecht-Nielsen, Neurocomputing, Addison-Wesley, 1990.

[37] N. Chaiyaratana and A.M.S. Zalzala, "Recent developments in evolutionary and genetic algorithms: theory and applications," *Second International Conference on Genetic Algorithms in Engineering Systems: Innovations and Applications*, Sep 1997, pp. 270–277.

[38] http://www.ansoft.com/products/hf/hfss/.

[39] http://www.sonnetusa.com/.

[40] http://www.microstripes.com/.

[41] http://www.bay-technology.com/ie3d.htm.

[42] T. Itoh, Numerical Techniques for Microwave and Millimeter-Wave Passive Structures, John Wiley & Sons, 1989.

[43] M. Krumpholz and L.P.B. Katehi, "New time domain schemes based on multiresolution analysis," *IEEE Transactions on Microwave Theory and Techniques*, Vol. 44, Apr 1996, pp. 555–561.

[44] N. Bushyager, L. Martin, S. Khushrushahi, S. Basat, and M.M. Tentzeris, "Design of RF and wireless packages using fast hybrid electromagnetic/statistical methods," *Proceedings of the IEEE-ECTC Symposium*, May 2003, pp. 1546–1549.

[45] D. Staiculescu, J. Laskar, and M.M. Tentzeris, "Design rule development for microwave flip-chip applications," *IEEE Transactions on Microwave Theory and Techniques*, Vol. 48, No. 9, Sep 2000, pp. 1476–1481.

[46] S. Pinel, C.-H. Lee, S.-W. Yoon, K. Lim, and J. Laskar, "Embedded IC packaging technology for ultra-thin and highly compact RF modules," *Proceedings of the IEEEECTC Symposium*, May 2003, pp. 24–29.

[47] S. Towle, H. Braunisch, C. Hu, R.D. Emery, and G.J. Vandentop, "Bumpless build-up layer packaging" *Intel Corporation, Components Research Intel Corp.*

[48] A. Margomenos, D. Peroulis, J.P. Becker, and L.P.B. Katehi, "Silicon micromachined interconnects for on-wafer packaging of MEMS Devices," *Topical Meeting on Silicon Monolithic Integrated Circuits in RF Systems*, Sep 2001, pp. 33–36.

[49] Y.K. Park, Y.K. Kim, C.J. Kim, B.K. Ju, and J.O. Park, "Innovative ultra thin packaging for RF-MEMS devices," *12th International Conference on Solid-State Sensors, Actuators and Microsystems*, 2003, pp. 903–906.

[50] http://global.kyocera.com/prdct/semicon/ic_pkg/mems.html.

[51] M.A.I. Mostafa, J. Schlang, and S. Lazar, "On-chip RF filter using bond wire inductors," *Proceedings of IEEE International ASIC/SOC Conference*, 2001, pp. 98–102.

[52] J. Laskar, B. Matinpour, S. Chakraborty, "Modern Receiver front-ends, systems, circuits and integration", chapter 8, John Wiley, 2004. US Patent 6815961, issued on Nov 9, 2004.

CHAPTER 8

Advanced SOP Components and Signal Processing*

INTRODUCTION

Since the early 1970s, microelectronics has been a driving force behind the production of electronic devices in various industries. Some of these industries include the consumer markets, communications, military applications, and automotives [1]. The integration of large numbers of transistors as well as of passive components and mixed-signal active devices into a working system started with the design approach of system-on-chip (SOC) [2]. This method allowed the development of high-performance integrated circuits (ICs) to be produced with a high volume. Some drawbacks from this approach were the high production costs from the use of expensive silicon-based discrete chips, low functional reliability from manufacturing, thermal and mechanical problems, and the limited capabilities of integrating transistors and active (and passive) components on the same chip [3]. In the early 1980s, the advent of the multichip module (MCM) offered an alternative to the restrictions of integrating components on the same chip by integrating multiple ICs (optical, radio frequency (RF), and digital) on a substrate using ball grid arrays (BGAs). This system approach alleviates issues of crosstalk between ICs by allowing the chips to perform their functions independent of each other. The packaging of ICs in MCMs is typically designed in a one- and two-dimensional (1-D and 2-D) platform [4]. The system-in-a-package (SIP) approach introduced approximately a decade ago allowed some similar

*This chapter has been contributed by Dr. Rong Lin Li of Georgia Institute of Technology

Advanced Integrated Communication Microsystems, edited by Joy Laskar, Sudipto Chakraborty, Manos Tentzeris, Franklin Bien, and Anh-Vu Pham

features as MCMs at a lower cost. The big advantage that SIP enjoys over MCMs is the ability to package chips vertically on different layers, thus allowing a three-dimensional (3-D) module. Some have called this technique of system integration, "SIP-based MCMs" based on the addition of the third dimension in the module. This vertical placement of the ICs allows for more compact modules to be produced. In today's consumer market and for years to come, the integration approach that has the most promise is system-on-package (SOP) technology [5]. The SIP approach uses discrete, bulky passive and active components in its integration [4,5]. The advantage of SOP over SIP modules is the use of thin-film-embedded components. This usage allows for the maximal compactness of full-scale integrated modules that can be placed in third-generation (3G) and fourth-generation (4G) cellular phones and other wireless modules. SOP technology consists of a multifunction, multichip package that enables the integration of many system-level functions, such as digital, optical, analog, and micro-electro-mechanical systems (MEMS) [6,7]. The integration of the antenna into a compact SOP module continues to be a limiting factor in the development of a complete RF SOP device. Much research have been done in the field of designing antennas that have a large array of functions such as broad impedance bandwidths, multifrequency operations, and beam steering capabilities. The challenge is to maximize the performance characteristics of antennas while maintaining a compact size.

Microstrip antennas have received much consideration for implementation in SOP technology [8]. An illustration of a microstrip patch antenna is shown in Figure 8.1. Additionally, SOP enables 3D compact architectures to be realized. Planar antennas can be integrated into 3-D modules for millimeter-wave short-range broadband communications and reconfigurable sensor networks.

These structures enjoy many advantages over their counterparts, such as low manufacturing cost via modern printed circuit technology, low profile, ease of integration with monolithic microwave integrated circuits (MMICs) and integrated passives, and the ability to be mounted on planar, nonplanar, and rigid exteriors [9].

Low-temperature cofired ceramics (LTCC) multilayer technology is becoming more and more popular for producing complex multilayer modules and antennas

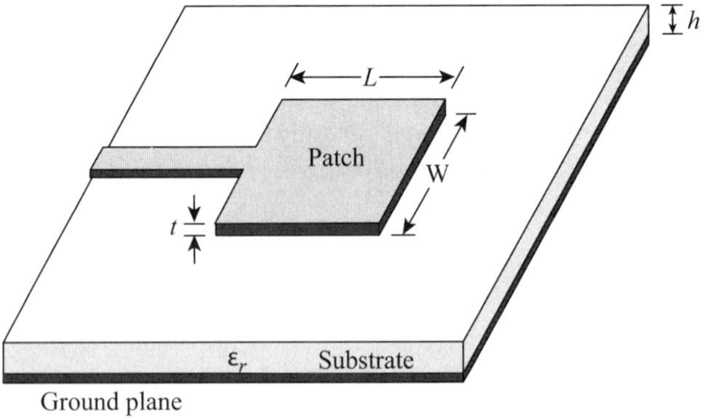

Figure 8.1. Microstrip antenna [8].

because of its flexibility in realizing a variable number of laminated layers [10,11]. The major advantages of using LTCC are lower dielectric loss, size reduction from a high dielectric constant, hermeticity, ability to integrate surface mount devices (SMDs) and ICs, and ability to incorporate embedded passives and interconnect circuitry to be sandwiched between the substrate layers. The high dielectric constant of LTCC is significant in realizing more compact 3-D architectures because of the inversely proportional relationship between the dimensions and the dielectric constant. Designing microstrip antennas on LTCC layers offers a desirable approach for integration with RF devices. To increase the physical area of the antenna, one can place air vias in the structure that will, in turn, lower the effective dielectric constant and produce a larger physical area. Some disadvantages associated with LTCC are the shrinkage of the material and the surface wave excitation caused by the dielectric constant of the substrate. Alternative organic materials, such as the liquid crystal polymer (LCP), offer certain advantages over LTCC. These advantages include a lower cost, engineered transverse coefficient of thermal expansion (CTE), and flexibility, although this is a less mature fabrication technology [12].

When designing planar antennas for wireless communications, it becomes necessary to have a microstrip antenna configuration that is compact in size and able to be integrated with other devices. With the physical area of the antenna being inversely proportional to the frequency, it is sometimes difficult to achieve a compact size for WLAN applications for acceptable efficiency and isolation values. Maintaining a compact antenna that can be optimized for impedance bandwidth, radiation pattern characteristics, high gain, and low front-to-back (F/B) ratios can be used for SOP applications as well as for RF certificates of authenticity (COAs) for consumer applications. There is often a trade-off in realizing compact antennas while maintaining performance characteristics.

The topic of this chapter is the full-scale design of compact antenna structures using SOP techniques for modern, commercial, mixed-signal RF systems such as cellular phones for PCS applications, Bluetooth and 2.4-GHz ISM applications, RFIDs, WLAN (802.11a,b,g,n), LMDS, remote sensing applications, and millimeter-wave applications at 60 GHz. These compact designs will have many performance characteristics such as a broad bandwidth, multifrequency operation, surface-wave suppressing features, and high gain "quasi"-endfire radiation properties. The performance characteristics of these antennas will vary based on the design application. This research will also consist of theoretical investigations of the electromagnetic phenomena, equivalent circuit modeling of the scattering parameters, and measurement of the fabricated prototypes. Most of the compact antenna configurations will consist of microstrip patch designs, and compact optimization will be applied to the structures when necessary.

Compact microstrip antenna design has been a pivotal topic of discussion in order to meet the miniaturization requirements of portable communications equipment [13]. Many attempts have been made to decrease the size of antennas from reducing the substrate thickness to using a substrate with a high permittivity. The first part will focus on a review of compact microstrip antennas with an illustration of how compact antennas can be applied to achieve specific performance characteristics, such as enhanced bandwidth and gain, dual-frequency operation, and circularly polarized radiation. We also provide the readers with some complexities and modeling issues associated with designing compact antennas for integrated transceivers and wireless applications.

8.1 HISTORY OF COMPACT DESIGN

The simplest method of antenna size reduction is to use a high dielectric constant substrate [8,13,14]. Microstrip antennas are approximately half-wavelength structures, as the resonant length is half of a guided wavelength ($\lambda_g/2$). One guided wavelength can be expressed as the ratio of the phase velocity (v_p) to the frequency (f). Additionally, the phase velocity can be expressed as the ratio of the speed of light ($c \approx 3 \times 10^8$) to the square root of the effective dielectric constant. The length of a microstrip antenna can be approximately given by

$$L = \frac{c}{2f\sqrt{\varepsilon_{eff}}} \tag{8.1}$$

The effective dielectric constant increases with increasing dielectric constant [9]. Depending on the process and application, the dielectric constant may remain a fixed parameter; a high dielectric constant will maintain the compactness of the structure, whereas a low dielectric constant will result in a more efficient radiator. If this dielectric constant has a high value, surface modes may be launched at the interface of the air and dielectric material. Surface waves are transverse magnetic (TM) and transverse electric (TE) modes that propagate into the substrate outside the microstrip patch [15]. These modes have a cutoff frequency that is different than the resonant frequency for the dominant mode of the antenna. Surface waves become a problem when their cutoff frequency is lower than the resonant frequency of the antenna causing overmoding (more than one propagating mode at a given frequency). The cutoff frequency of a surface wave is inversely proportional to the dielectric constant of the substrate as follows:

$$fc = \frac{nc}{4h\sqrt{\varepsilon_r - 1}} \tag{8.2}$$

where c is the speed of light ($c \approx 3 \times 10^8$), h is the substrate thickness, ε_r is the dielectric constant, and $n = 1,3,5\ldots$for TE_n modes and $n = 2,4,6\ldots$for TM_n modes. If surface waves are present, the total efficiency of the antenna will be reduced. There is often a trade-off between compact size and efficiency. Therefore, other methods have been proposed to reduce antenna dimensions with fixed substrate properties. One method is the use of a meandered patch Figure 8.2. The meandering is done by cutting slots in the nonradiating edges of the patch [16,17]. This effectively elongates the surface current path on the patch and increases the loading, which results in a decrease in the resonant frequency.

The trade-off is a decrease in impedance bandwidth and antenna gain, which causes a severe limitation in practical applications [18]. Additionally, high levels of cross-polarization may originate from sections of the meandered patch [19]. Another

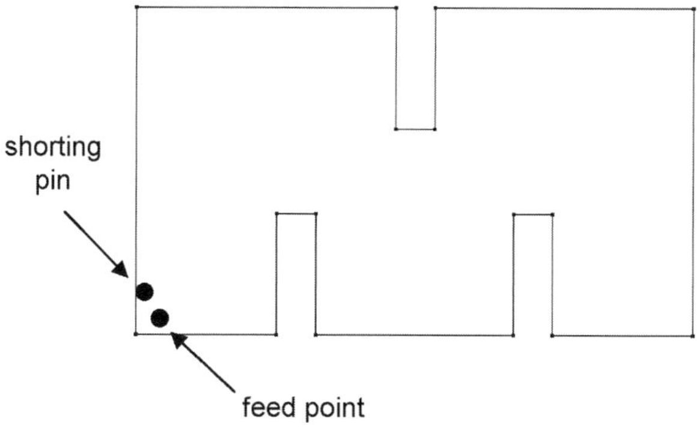

Figure 8.2. Meandered patch by inserting slots in nonradiating edges [17].

method includes the meandering of the ground plane [18]. In a similar approach, the insertion of slots in the ground plane can reduce the resonant frequency for a given length (Figure 8.3).

The slots in the ground plane may cause unwanted levels of backside radiation [20], which potentially leads to high absorption of energy from the human head when the antenna is used in PCS applications, specifically when the antenna is placed in cellular phones [21]. This absorption of radiation to the head is characterized by the specific absorption ratio (SAR). An acceptable SAR level required by the Federal Communications Commission (FCC) for public exposure is 1.6 W/kg [22]. Another popular technique involves a shorted plane that is placed along the middle of the patch parallel to the radiating edge between the patch and the ground plane. With the presence of the shorted plane, half of the patch can be omitted. The patch now has a resonant length of a quarter-wavelength ($\lambda/4$). Theoretically, the position of the shorted plane is selected where the electric field normal to the patch is nonexistent. Therefore, the fields parallel

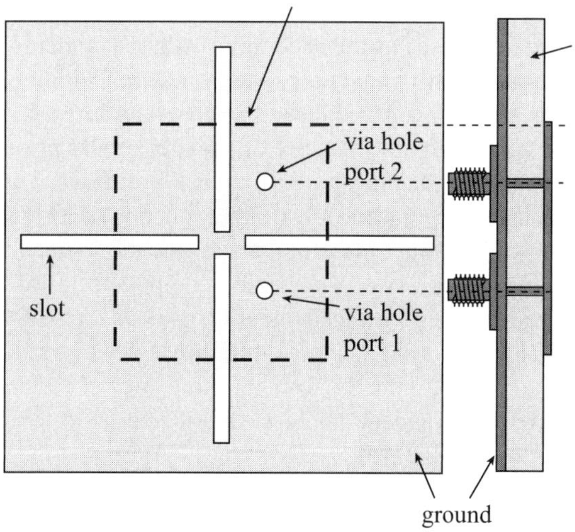

Figure 8.3. Meandered patch by inserting slots in the ground plane [18].

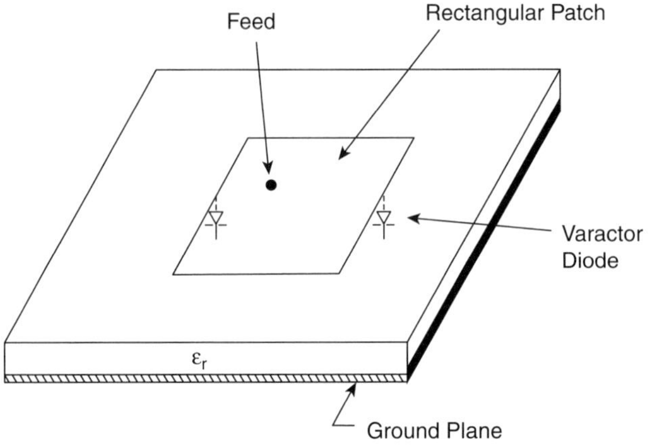

Figure 8.4. Patch antenna loaded with varactor diodes [25].

to the shorted plane are undisturbed. The major disadvantage of this method is a narrower impedance bandwidth for some applications, such as digital European cordless telephones (DECTs) [23]. Also, punching vias through the substrate to create the shorted patch may not be suitable for some materials, such as LCP, because of the alignment inaccuracies of the vias. Targonski and Waterhouse attempted to alleviate this problem by using a thick foam substrate with a low dielectric constant [24], but this affected the compactness of the antenna. The low-cost use of foam may not be suitable for RF packaging applications that involve a high-temperature environment. Finally, the use of varactor diodes (Figure 8.4) has been shown to contribute to the compact operation by means of tuning the resonant frequency [25]. Varactor diodes contain a capacitance that can be adjusted by changing their voltage. This additional capacitance helps to decrease the resonant frequency making for a more compact geometry. This structure could also support circularly polarized radiation.

One concern from this approach is that the close distance between a varactor diode and the coaxial probe can cause unwanted coupling, whereas widening this distance by moving the probe may destroy the impedance match. Additionally, although a large bandwidth can be achieved, reduced efficiency and increased levels of cross-polarization are present. Ultimately, the use of varactor diodes presents a problem in terms of integrating the antenna into an RF module. Many times, devices operating in the very high frequency (VHF) and the ultra high frequency (UHF) require antennas to be completely passive elements. Therefore, a varactor diode (an active component) would have to be realized in terms of a printed component instead. du Plessis and Cloete proposed a solution of using a metallic pad at the radiating edges of a rectangular patch [26]. This design, shown in Figure 8.5, is completely passive and has the similar feature of changing the resonant frequency of the antenna. When the size of the pads is determined and the antenna is fabricated, a trimming device could be used to trim off metal from the pads.

Using this technique, the antenna's resonant length is shorter because of the shorter surface current path; hence, the resonant frequency can be increased. With this

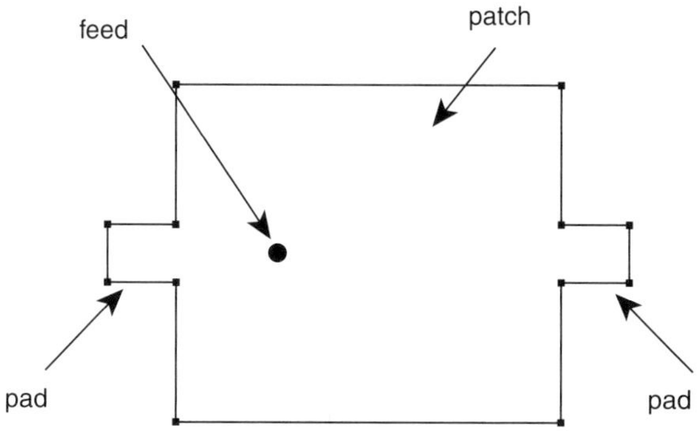

Figure 8.5. Patch antenna loaded with metallic pads.

method, the frequency of operation cannot be decreased. Also, the trimming of the antenna may affect the performance of the design. The idea of modifying a structure once it has been fabricated is often not practiced in RF packaging and antenna design.

8.2 PREVIOUS TECHNIQUES IN PERFORMANCE ENHANCEMENT

Compact antenna design can be used to achieve circular polarization (CP), enhanced gain, and wideband operation for many applications, such as GPS, Bluetooth, and WLAN applications. Recent attempts have been made to realize these performance characteristics. One of the simplest ways to attain circular polarization is to insert a cross-shaped slot in the patch [27,28]. This tends to excite two orthogonal modes with a 90° phase difference between them, a necessary condition for CP. This method is useful because it only requires a single feed point. Trimming off the corners of the patch along the same diagonal direction (Figure 8.6) is another means of achieving CP while maintaining a compact design [29].

An additional simple approach can be taken where one can slightly increase the length (or width but not both) of a square patch. This makes the patch "nearly square."

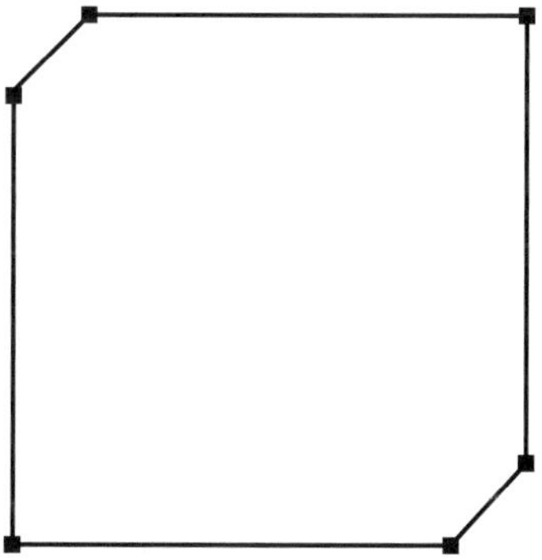

Figure 8.6. Patch antenna with trimmed corners.

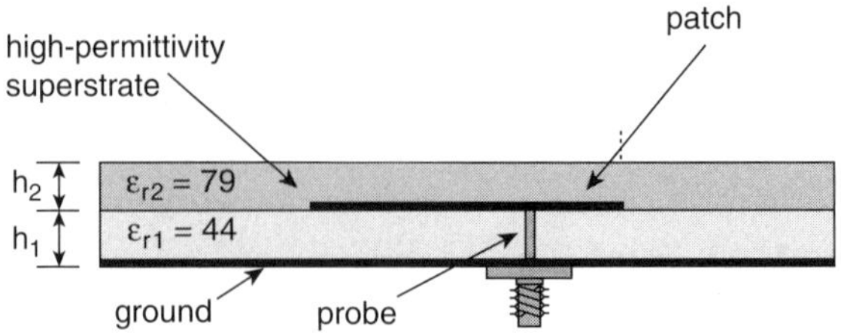

Figure 8.7. Patch antenna with low and high permittivity substrate [33].

Using this technique as well as exciting the patch along the diagonal, achieves CP by obtaining two modes with slightly different resonant frequencies. One mode can "lead" by 45°, whereas the other mode can "lag" by 45°; hence, a 90° phase difference is produced while maintaining electric field amplitudes that are equal [30,31]. An advantage to this design is the ease in which CP can be achieved. The circuit modeling and radiation characteristics for this approach remain unchanged. Despite the advantages, there has been no formulated approach of choosing the correct length perturbation to achieve CP. There are few designs that have been reported for producing enhanced gain while maintaining compact operation. One of these designs incorporates two substrate layers with the patch antenna embedded between them Figure (8.7). The lower substrate (between the patch and the ground plane) has a low dielectric constant ($\varepsilon_r < 5$), whereas the other substrate layer (above the patch) has a high dielectric constant ($\varepsilon_r > 15$) [32,33]. This high dielectric constant will excite substrate modes, thus lowering the efficiency and bandwidth, which may not be suitable for a desired application [20].

Another technique that can be used for enhancing the gain of microstrip antennas involves placing parasitic elements next to the radiating patch [34]. The radiating patch will capacitively couple energy to the nearby parasitic elements creating a wider aperture. Although this wider aperture will increase the gain of the structure, the effect may not be significant. A possible drawback using this approach is the increased lateral area of the design, which may prohibit the compactness of the structure. Careful placement of the parasitic elements must also be taken into account. Placing these elements too close to the radiating patch can greatly decrease the resonant frequency of the antenna, whereas positioning the elements too far from the radiating patch will exhibit no effects at all. Finally, the achievement of the wideband frequency operation has been reported in [13]. In addition to stacking, one design incorporates an aperture-coupled shorted patch with a slot in the ground plane. The uniqueness of this design is the thick air substrate employed under the patch [35]. With the length and width of the patch chosen to resonate at two frequencies that are close to each other, the use of the air substrate helps to widen the bandwidth to a point where it combines to cover the bandwidth of both resonant frequencies. With this design, a total impedance bandwidth of 26% can be achieved. A second compact design with wideband operation uses a chip resistor that is placed between the patch and the ground plane at one radiating edge of the structure [36]. The wideband effect can be observed by considering the decrease in the quality factor, or Q factor, when additional resistance is introduced into the circuit. This decreased Q factor greatly increases the bandwidth of the antenna as observed in the equation below:

$$BW = \frac{S-1}{Q\sqrt{S}} \tag{8.3}$$

where BW is the bandwidth and S represents the maximum voltage standing wave ratio (VSWR) value that is desired for an acceptable impedance match. For antennas, this value is usually equal to 2. The major disadvantage of this design is the reduced efficiency since a large portion of the input power is dissipated in the resistor, which takes away available power that can be radiated by the antenna.

8.3 DESIGN COMPLEXITIES

Despite the fact that much work has been done in the area of compact antenna design, some design complexities exist that must be taken into account. One complexity originates from the use of a high dielectric constant substrate. High dielectric constant substrates are favored for miniaturization of structures, but surface modes are launched into the substrate, which reduces the radiated power, thus significantly reducing the efficiency of the antenna [20]. Another complexity stems from the feeding structure. Some designers prefer to use microstrip lines printed on the same layer as the antenna for excitation. Ease of fabrication and simplicity in circuit modeling are two advantages of using these lines. Unfortunately, radiation loss of microstrip lines increases as the ratio of the square of the length to the square of the free space wavelength (L^2/λ^2) increases [37]. The radiation from the microstrip line may also tilt the main beam a few degrees in the direction of the feed line [38]. A coaxial cable may be suitable for excitation, but depending on the substrate, this may not be possible to manufacture. Additional feeding methods have been used when a planar design is necessary. In particular, proximity coupling and aperture coupling are two of the more popular feeding methods caused by the decreased levels of cross-polarization and the shielding of the feedline radiation by the ground plane (applicable only in aperture coupling) [39]. The proximity-coupled feeding is capacitive in nature, whereas the aperture-coupled feeding is inductive. The lack of design rules can cause the analysis of both feeding techniques to be complex. Moreover, these feeding methods can only be used in a multilayer environment, but they can take full advantage of the 3-D integration. This places an additional restriction on the design, which may not be suitable for the desired application. It is important to maintain the effectiveness of the antenna by taking into account all complexities that are associated with a design.

8.4 MODELING COMPLEXITIES

Many electromagnetic modeling tools (simulators) have been introduced in the last ten years for modeling complex structures and high-frequency effects of vias and

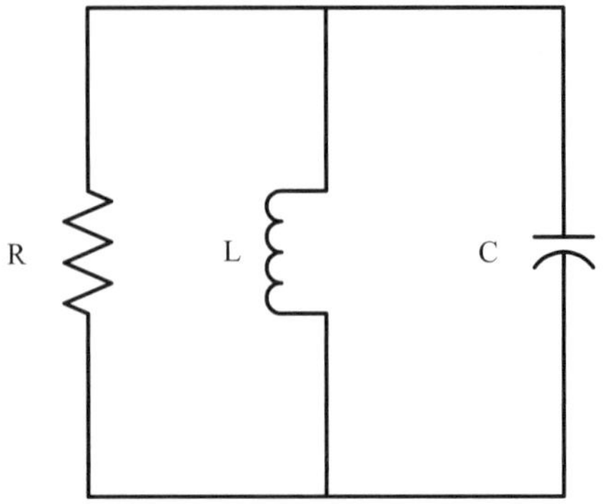

Figure 8.8. Parallel RLC circuit to simulate a $\lambda/2$ microstrip patch antenna.

high-impedance feedlines. These simulators have been based on many computational algorithms such as method of moments (MoM), finite element method (FEM), and finite difference time domain (FDTD) schemes. They have been useful in the analysis of the electromagnetic phenomena of passive and active devices. When integrating components together on the SOP module, it is necessary to simulate the electromagnetic characteristics of the entire IC. Because of the physical size of the components as well as the multifrequency operation of the modules, simulating these structures simultaneously may not be feasible on computing systems that have insufficient memory. To obtain a macroscopic insight on the resonance of microstrip patch antennas, an equivalent circuit of a parallel resistor–inductor–capacitor (RLC) network has been proposed in many publications (Figure 8.8). This circuit representation gives a designer an idea of the resonant frequency and Q factor of the antenna.

Design equations for determining the R, L, and C values are presented in [9] based on the physical dimensions of the antenna (length and width of the patch and the height of the substrate) and the dielectric constant of the substrate. This circuit is only capable of analyzing single-layer microstrip patch designs. For more complex, multilayer antenna structures, there is a need for additional passive components that can represent the higher order effects of the antenna such as parasitic resonances in the scattering parameter data. Additionally, a more complex passive circuit could be used in conjunction with the electromagnetic analysis to form a complete representation of the design.

The stacked patch antenna is a common approach for achieving a wider impedance bandwidth. One problem with using a stacked structure is that the distance between the two antennas may possibly shift the design frequency. In addition, many parameters need to be adjusted for an optimal bandwidth performance, such as the length and width of each patch, the thickness of the substrate, as well as the position of the feed point. With so many parameters that need to be accounted for, to date, there has been no control over adjusting all variables, simultaneously, to achieve optimal bandwidth performance. Therefore, there is a need for a set of design rules to guide antenna engineers in designing wideband antennas.

8.5 COMPACT STACKED PATCH ANTENNAS USING LTCC MULTILAYER TECHNOLOGY

The integration concept of 3D modules that is considered in this chapter is illustrated in Figure 8.9. A stacked-patch antenna is embedded on the top of an RF frond-end module in an LTCC multilayer package. The input of the antenna comes from the output of an embedded band-pass filter that is connected with a block of RF-active devices by processes called "flip-chipping" and "wire bonding." The vertical integration capabilities in the LTCC technology provide the space for the embedded RF block. The LTCC cavity process also provides integration opportunities for RF-passive components such as switches and/or off-chip matching networks. The vertical board-to-board transition of two LTCC substrates is implemented using a micro ball grid array (μBGA) ball process.

The standard BGA balls ensure the interconnection of the highly integrated LTCC module with a mother board such as an FR-4 substrate.

The antenna structure for this design is shown in Figure 8.10. It consists of two square patches (lower and upper) of length L that are stacked on a grounded LTCC substrate. Square patches were used in this design for the purpose of two orthogonal modes, TM_{10} and TM_{01}, with resonant frequencies that are in close proximity to each other, therefore, obtaining a wider bandwidth. The total thickness of the substrate is denoted h. This thickness can be divided into two smaller thicknesses, h_1, the distance between the lower patch and the ground plane, and h_2, the distance between the lower and upper patch where $h = h_1 + h_2$. The lower patch of the antenna structure is excited through a via that is connected to the output port of a filter. A via is a slender piece of metal that vertically connects components on different layers. Then, electromagnetic coupling of energy is transferred from the lower patch to the upper patch. The position of the (via) is placed at the center of the radiating edge in order to match a 50-Ω coaxial line.

In the 3D RF front-end module, the antenna is integrated with other RF circuits. It is essential to prevent any unwanted radiation from other RF components in the integrated module. Therefore, a metal-backed cavity is introduced in order to shield the RF signals of components surrounding the antenna from the separate antenna signals to preserve functionality. In LTCC packaging technology, a continuous metal

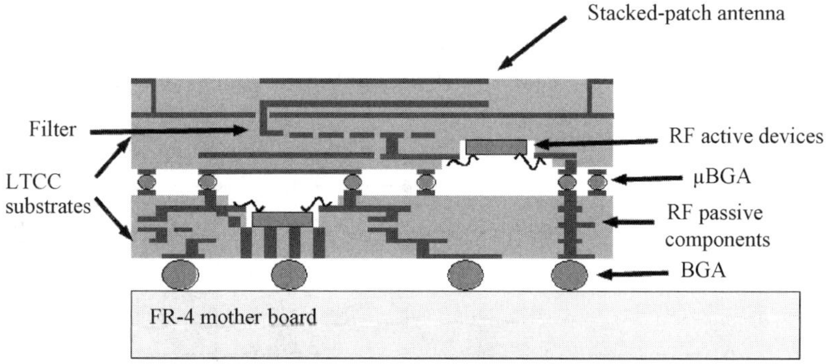

Figure 8.9. 3D intcgrated module.

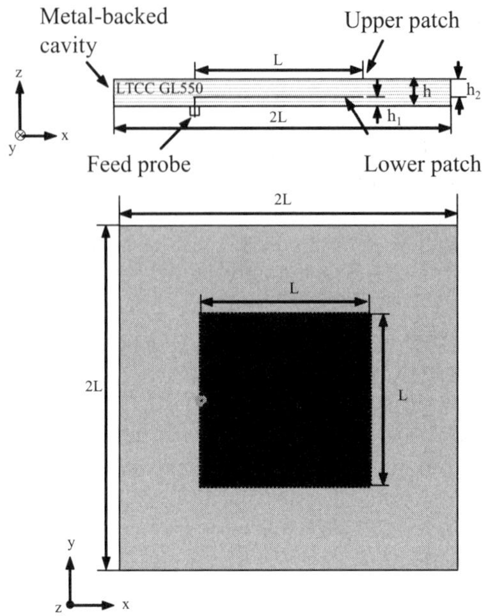

Figure 8.10. Stacked-patch antenna architecture on LTCC multilayer substrate.

wall cannot be realized in fabrication. This obstacle is overcome through the use of an array of vertical vias. It is recommended to choose the lateral dimension of the cavity to be twice that of the stacked patch. A smaller dimension may result in reflections from the walls, which may affect the impedance characteristics of the antenna and, hence, the bandwidth of the structure. A larger dimension may hinder the compactness of the structure.

Besides the use of square patches in increasing the bandwidth, the major contribution in the wide-bandwidth performance of the stacked-patch antenna is achieved through the combination of two close resonant frequencies that respond, respectively, to the lower patch and the upper patch. The combination is made by an electromagnetic coupling between the two patch resonators, which can be modeled by the equivalent circuit shown in Figure 8.11.

This circuit consists of two electromagnetically coupled parallel resonant circuits where L_1 and L_2 are equivalent inductances, C_1 and C_2 are equivalent capacitances, and

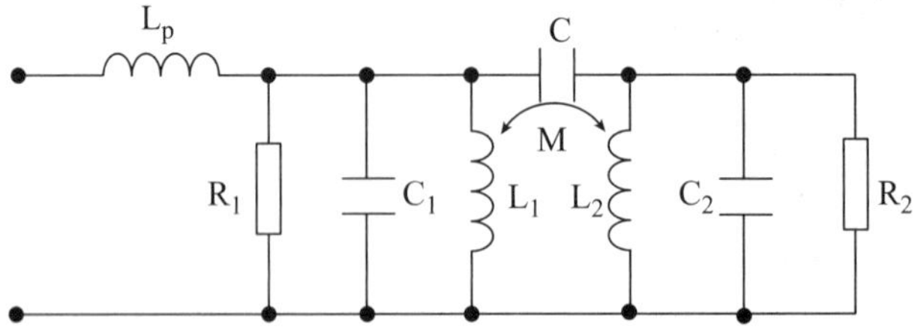

Figure 8.11. Equivalent circuit of the probe-fed stacked-patch antenna.

R_1 and R_2 are radiation resistances. (Subscript 1 refers to the lower patch, and subscript 2 refers to the upper patch.) A series inductance, L_p, is included to model the inductance of the feed probe [4]. Two resonant frequencies depend on L_1C_1 and L_2C_2. This is shown in the formula below:

$$f_n = \frac{1}{2\pi\sqrt{(L_n C_n)}} \tag{8.4}$$

where $n = 1, 2$. Furthermore, the tightness of the electromagnetic coupling is decided by the coupling capacitance C and mutual inductance M. By adjusting the heights of the lower and upper patches, the corresponding resonant frequencies and the coupling tightness can be varied, thus resulting in an optimal impedance performance.

Simulations of this structure using Microstripes 5.5 were performed, and initial results were taken and analyzed. Microstripes 5.5 is a 3-D fullwave simulator by Flomerics Ltd. that uses transmission line matrix (TLM) modeling for analysis. First, a comparison was done to show the variation of input impedance as a function of frequency for five values of relative dielectric constant: $\varepsilon_r = 2, 4, 6, 8$, and 10. This is shown in the form of a Smith chart Figure 8.12. A circle labeled "vswr 2:1" represents a reflection coefficient of one third and a −10-dB return loss. The plot of the input impedance inside this circle shows the −10-dB return loss bandwidth that can be achieved for the particular frequency band. The more input impedance points that lie in the "vswr 2:1" circle, the wider the bandwidth of operation. The horizontal center line of the Smith chart represents a purely resistive impedance. An impedance above

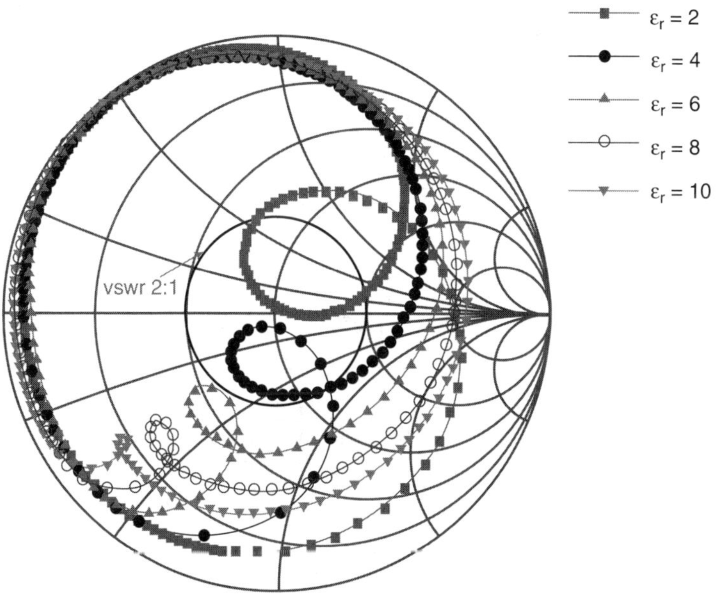

Figure 8.12. Smith chart of input impedance for variable values of ε_r.

this line is resistive and inductive, whereas an impedance below this line is resistive and capacitive. The impedance loop tends to move downward as the dielectric constant increases. This is a result of the increased parallel plate capacitance from the proportional relationship between the capacitance and the dielectric constant. Additionally, the parallel plate capacitance is inversely proportional to the plate separation.

The formula for the capacitance C of two finite size plates with a finite separation distance from each other is shown below:

$$C = \frac{\varepsilon_r \varepsilon_o A}{d} \tag{8.5}$$

where ε_r is the dielectric constant, ε_o is the permittivity of free space, A is the lateral area of the plate, and d is the separation distance of the plates. For this simulation, the length and width of both patches is each 10 mm. The distance from the parasitic patch to the excited patch is 1 mm. The distance from the excited patch to the ground is also 1 mm.

A second simulation was done to analyze the effect of changing the position of the excited patch. By doing this simulation, the input impedance bandwidth can be optimized. Figure 8.13 shows the input impedance as a function of frequency for a fixed dielectric constant ($\varepsilon_r = 7$) in the form of a Smith chart. It can be observed that a position of 0.5 mm above the ground plane gives an impedance loop that is totally inside the "vswr 2:1" circle, hence, an optimal bandwidth. It is also clearly shown that

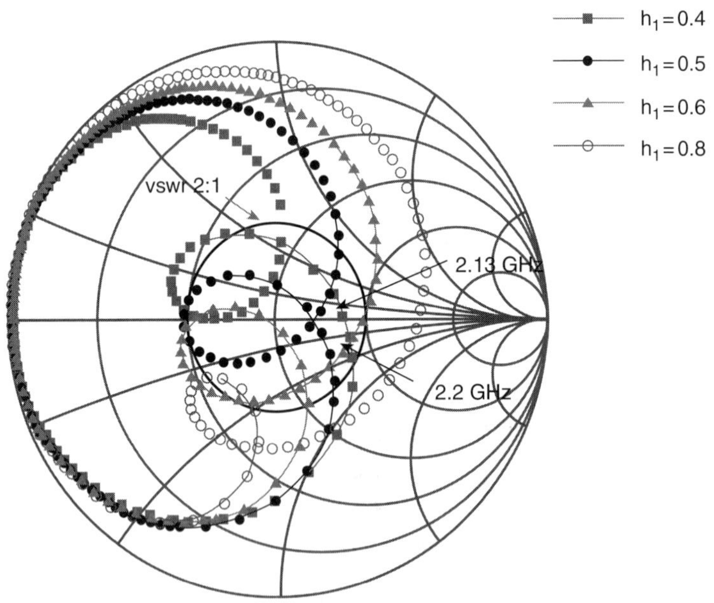

Figure 8.13. Smith chart of input impedance versus frequency for variable values of lower patch height at $\varepsilon_r = 7$.

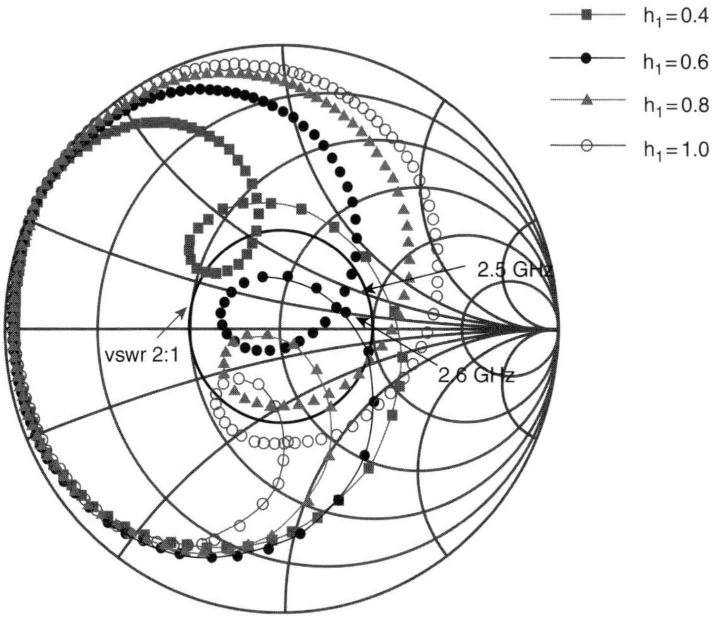

Figure 8.14. Smith chart of input impedance versus frequency for variable values of lower patch height at $\varepsilon_r = 5$.

the impedance loop moves downward as the height of the excited patch (distance from the ground plane) increases. This simulation was done for a stacked patch structure where each patch has a length and width of 10 mm. The total substrate thickness is 2 mm.

Another simulation was done to also analyze the effect of changing the positioning of the excited patch. Figure 8.14 shows the input impedance as a function of frequency for a fixed dielectric constant ($\varepsilon_r = 5$) in the form of a Smith chart. It can be observed that a position of 0.6 mm above the ground plane gives an impedance loop that is totally inside the "vswr 2:1" circle and, therefore, a bandwidth that is optimal. It is again clearly shown that the impedance loop moves downward as the height of the excited patch above the ground plane increases. This simulation was done for a stacked patch structure where each patch has a length and width of 10 mm. The total substrate thickness is 2 mm.

A final simulation was done to examine the effect of changing the positioning of the excited patch. Figure 8.15 shows the input impedance as a function of frequency for a fixed dielectric constant ($\varepsilon_r = 3$) in the form of a Smith chart. The Smith chart shows that a position of 1 mm above the ground plane will give an impedance loop that is totally inside the "vswr 2:1" circle and, therefore, an optimal bandwidth. It is again clearly shown that the impedance loop moves downward as the height of the excited patch increases. This simulation was done for a stacked patch structure where each patch has a length and width of 10 mm, and the total substrate thickness is 2 mm.

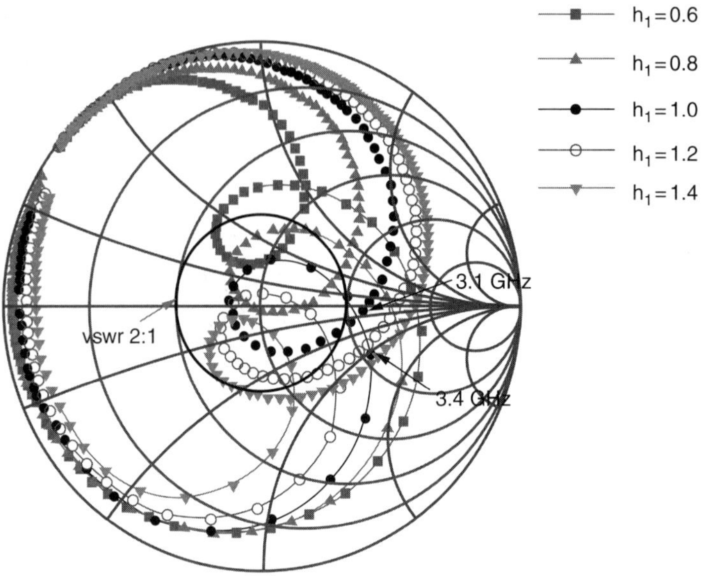

Figure 8.15. Smith chart of input impedance versus frequency for variable values of lower patch height at $\varepsilon_r = 3$.

Upon examination of the simulations and results presented in the last section, a major point of interest can be deduced. When stacked patch antennas were designed on an LTCC Kyocera-GL550 multilayer substrate with a layer thickness of 4 mils per layer, dielectric constant (ε_r) = 5.6, and loss tangent (tan δ) = 0.0012, a relationship between the bandwidth and the substrate thickness for a vertically compact structure (substrate thickness = 0.01–0.03 λ_0) is obtained. This is shown in Figure 8.16, where

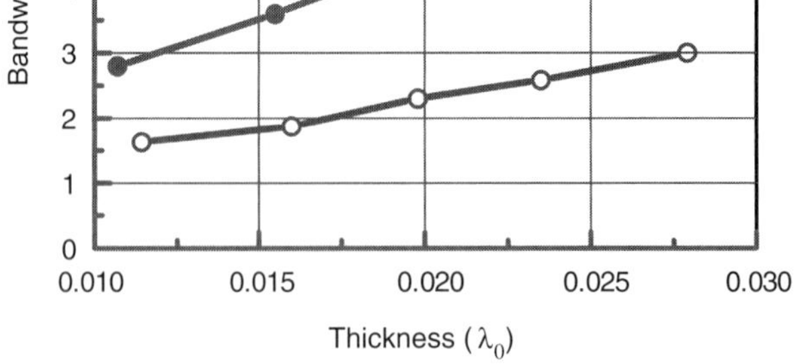

Figure 8.16. Impedance bandwidth versus total thickness of patch antennas on LTCC multilayer substrate.

the relative 10-dB return-loss bandwidth (normalized to the resonant frequency f_r) is plotted as a function of the thickness of the antenna (normalized to the free-space wavelength λ_0 at f_r).

The bandwidth of a single-patch antenna using the same type of substrate and thickness is presented in this figure as well to show the improvement in the bandwidth when using a stacked configuration. From the plot, it is observed that the compact stacked-patch antenna can achieve a bandwidth of up to 7%. This is 60–70% wider than that obtained from a single patch antenna. It is worth noting that the bandwidth of a patch antenna is mainly limited by the dielectric constant and total thickness of the substrate (i.e., the total volume occupied by the antenna). This is understandable by making note of the inversely proportional relationship between the bandwidth and the Q factor. Moreover, a higher volumetric structure will have a lower Q factor and, therefore, a greater bandwidth.

In lieu of this point and the results obtained in the last section, a set of design rules can be established for the design of stacked-patch antennas on LTCC multilayer substrates with ε_r and tan δ close to 5.6 and 0.0012, respectively. The steps are as follows:

Step 1: Choose an initial value for the total design thickness of the substrate h ($h = h_1 + h_2$). This thickness is usually less 0.05 λ_0 for a compact design.

Step 2: Select the lower substrate thickness h_1. Through analyzing the results of many simulations using the LTCC Kyocera-GL550 multilayer, it is observed that the impedance loop will be totally inside the 2:1 vswr circle when plotted on a Smith chart, if $h_1 \approx h/4$, which is optimal for an enhanced bandwidth.

Step 3: Design the length L (which is also the width) according to the appropriate resonant frequency f_r required for the application. The equation below is suggested for designing the length L:

$$L = \frac{c}{2f_r\sqrt{\varepsilon_r}} \tag{8.6}$$

where all variables have been previously defined.

Step 4: Determine the upper-substrate thickness h_2 for an optimal return loss. The initial value of h_2 can be chosen as $3h_1$ according to Step 2. The final value of h_2 may be obtained by simulation. Upon our observation, it is found that the impedance loop in the Smith chart will move from the upper (inductive) portion of the Smith chart to the lower (capacitive) portion as the distance between the upper patch and the lower patch is shortened. The upper substrate thickness h_2 is determined when the center of the impedance loop moves closest to the center of the Smith chart, which corresponds to a minimum return loss and a better matched circuit.

Step 5: Lastly, adjust the length L slightly to cover the desired frequency band. The simulations will assist in determining the optimal length for the design.

It is possible that the optimized bandwidth of the structure is unsatisfactory for the desired application. To overcome this dissatisfaction, simply increase the lower substrate thickness h_1 and repeat steps 4 and 5 until the required design specifications are met.

The information obtained from initial simulations and analysis as well as from the design rules that have been postulated will be applied to three emerging wireless communication bands. These are as follows: the 2.4-GHz ISM band, the IEEE 802.11a 5.8-GHz band, and the 28-GHz LMDS band. The substrate used in these applications is an LTCC Kyocera-GL550 multilayer laminate.

The 2.4-GHz ISM band has a 3.4% bandwidth with the center frequency f_c at 2.4415 GHz. Referring back to Figure 8.16, a stacked-patch antenna with a bandwidth of 3.4% should have an electrical substrate thickness of about 0.015 λ_0. The physical substrate thickness at the specified center frequency is 72 mils. At 4 mils per layer, the total requirement is 18 LTCC layers. By selecting the lower substrate thickness h_1 to be one quarter the total thickness, four or five layers should be a suitable selection. By using Eq. (8.6), the length L is 1022 mils. The upper substrate thickness h_2 can be selected through simulation. The simulated input impedance is plotted for different values of h_2 (in layers) on the Smith chart shown in Figure 8.17.

The Smith chart shows that the impedance loop tends to move downward in the capacitive region as h_2 decreases. As explained earlier, a higher capacitance is exhibited as the lower and upper patches are closer to each other. When the impedance loop moves closest to the center of the Smith chart and the loop is totally inside the 2:1 vswr circle, a minimum return loss and an optimized bandwidth can be achieved. This is the case for $h_2 = 14$ layers. Therefore, a total substrate thickness h of 19 layers (not 18 layers) is necessary for an optimized bandwidth

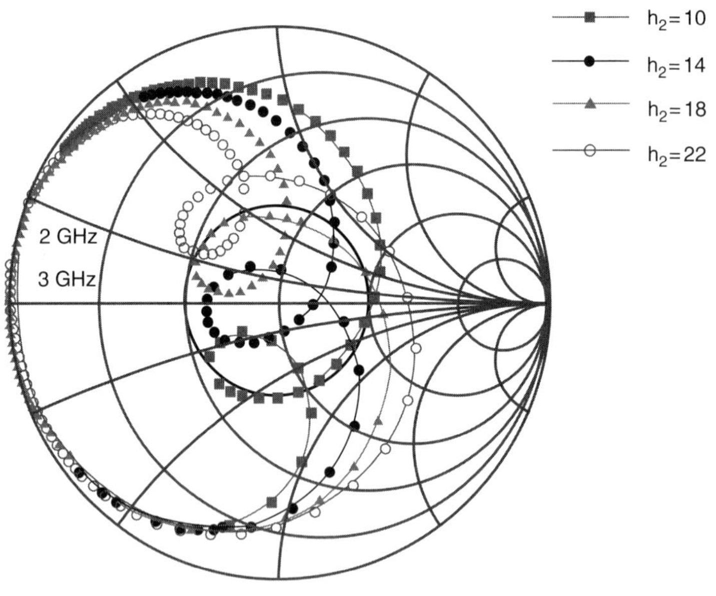

Figure 8.17. Smith chart of input impedance versus frequency for variable values of upper patch height at 2.4415 GHz.

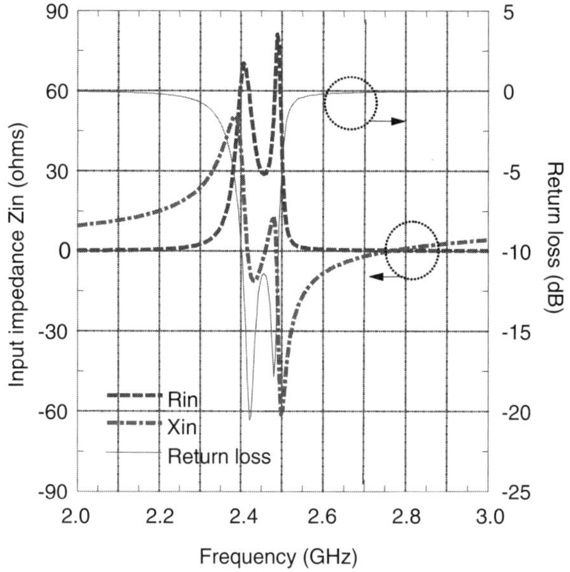

Figure 8.18. Input impedance and return loss versus frequency of a stacked-patch antenna at 2.4415 GHz.

design with h_1 and h_2 equal to 5 and 14 layers, respectively. This coincides closely to the one-quarter ratio of h_1 to h. Lastly, the antenna length L has to be modified to meet the band specification. Upon simulation, this value is reduced to 966 mils. The input impedance and the return loss versus frequency are plotted in Figure 8.18. This graph shows the two resonances that are close to each other in the return loss, which contributes to a wider bandwidth.

A similar approach was taken for this application. The selected center frequency is around 5.8 GHz. Refering to Figure 8.8, the electrical substrate thickness for this band should be approximately 0.015 λ_0, which corresponds to a physical substrate thickness of about eight layers. Initially h_1 and h_2 are chosen to be 2 and 6 layers, respectively. Additionally, the length L is set at 400 mils using the formula given in step 3 of the design rules. Once again, the simulated input impedance is plotted for different values of h_2 (in layers) on the Smith chart that is shown in Figure 8.19. Coincidentally, the optimized value of h_2 from the Smith chart is six layers. This value agrees perfectly with the one-quarter ratio of h_1 to h proposed in step 2. It is worth noting that the one-quarter condition can only be met for total layer thicknesses that are multiples of 4.

Figure 8.20 shows the simulated and measured return loss versus frequency of this structure (when $h_2 = 6$ layers) that was fabricated at Kyocera Industrial Ceramics, Corp. The measured return loss is in good agreement with the simulated results, and a bandwidth of 3.5% is observed. To measure the radiation pattern of the antenna, a modification had to be made in the feeding structure that consisted of a microstrip line connected to the lower patch with a via that passed through the substrate to the top layer and terminated to the surface of a metallic pad. The signal line of an SMA connector

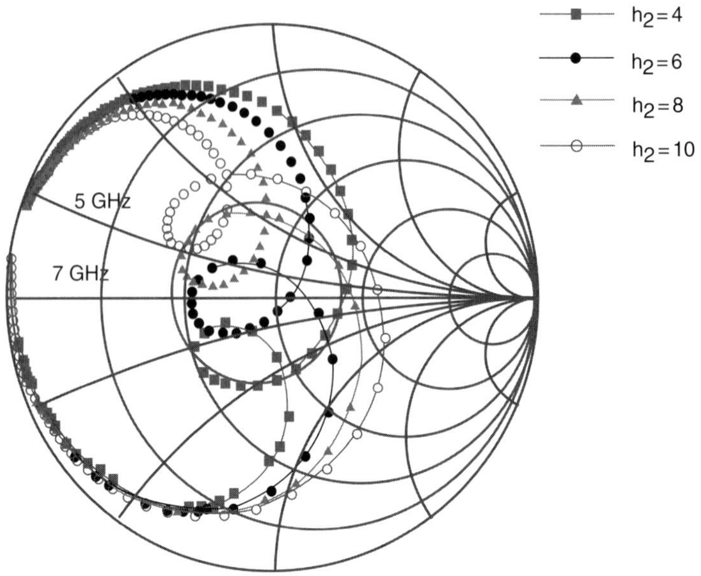

Figure 8.19. Smith chart of input impedance versus frequency for variable values of upper patch height at 5.8 GHz.

was then connected to the pad. The simulated and measured radiation patterns taken at 5.8 GHz are illustrated in Figure 8.21. The backside radiation level is about 10 dB lower than the maximum gain, which is about 4.5 dBi. The low gain of this antenna resulted from the high dielectric constant and the thin substrate thickness (0.015 λ_0).

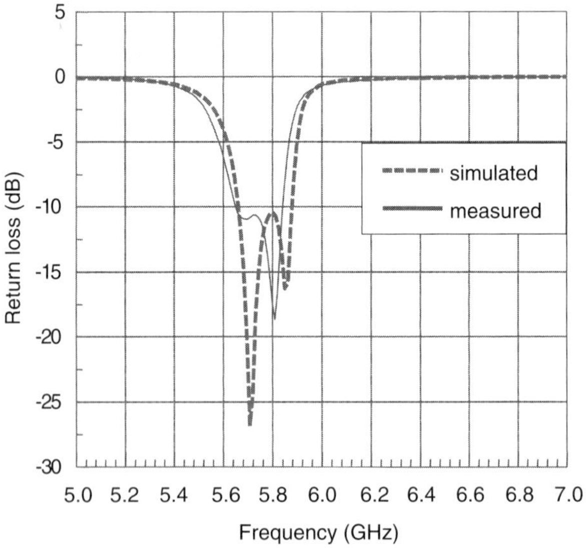

Figure 8.20. Return loss versus frequency of a stacked-patch antenna at 5.8 GHz.

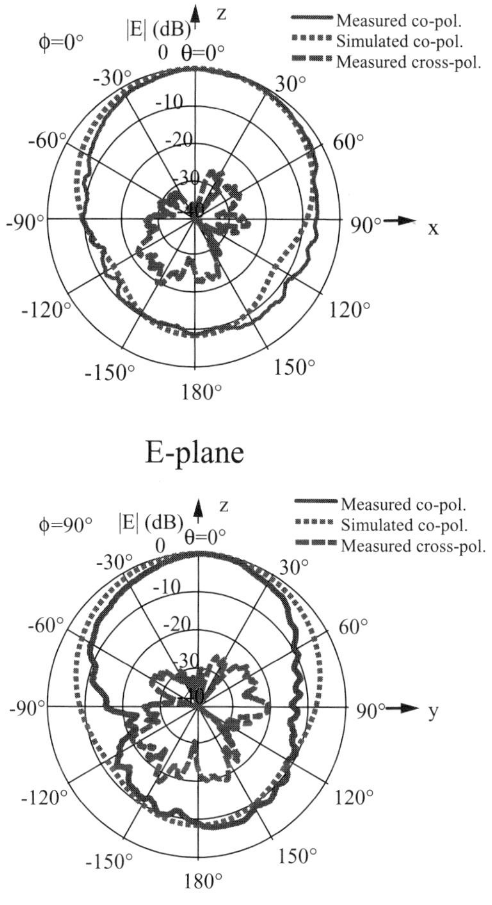

Figure 8.21. Simulated and measured radiation patterns of a stacked-patch antenna at 5.8 GHz.

The simulated cross-polarization was less than −40 dB in the E- and H-planes, whereas the measured cross-polarization was less than −20 dB, which is acceptable for this design. The modified feeding structure caused degradation in the cross-polarization performance of the antenna.

Once again, a similar approach was applied to the LMDS band. The required bandwidth is 7%, and the center frequency is 28 GHz. Based on Figure 8.16, the electrical thickness is about 0.03 λ_0, which corresponds to a physical thickness of only three LTCC layers because of the high operational frequency. The height h_1 is chosen to be one layer. Because of the low number of layers that is necessary to achieve an optimal bandwidth at this frequency, the one-quarter ratio rule of h_1 to h starts to breakdown. The simulated input impedance is plotted on the Smith chart, shown in

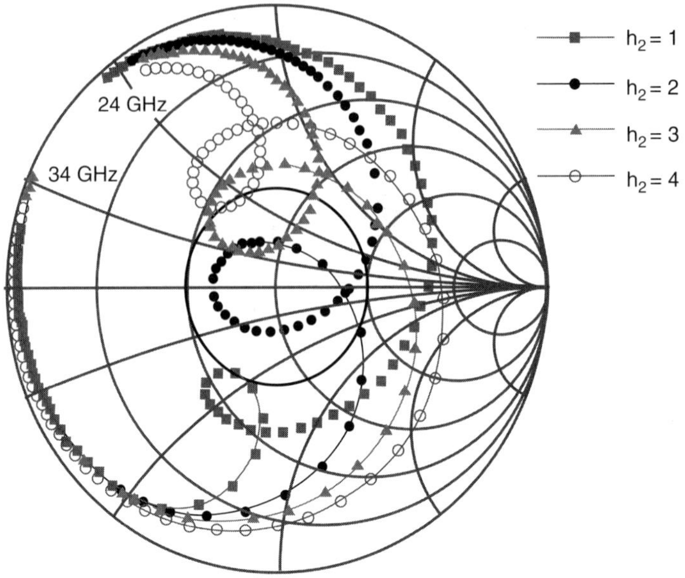

Figure 8.22. Smith chart of input impedance versus frequency for variable values of upper patch height at 28 GHz.

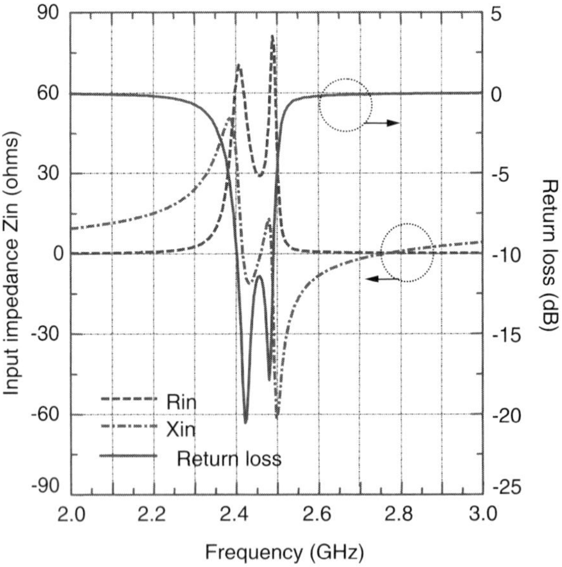

Figure 8.23. Input impedance and return loss versus frequency of a stacked-patch antenna at 28 GHz.

Figure 8.22, for various values of h_2. Since this structure has a thin substrate and a fixed layer thickness, a more extreme variation will exist in the movement of the impedance loops as the value of h_2 changes. As is predicted by Figure 8.16, this value should be set at two layers to obtain a minimum return loss and an optimized bandwidth. The length L of the patch is tuned to 80 mils to cover the required band fully. The input impedance and return loss versus frequency, simulated in MicroStripes 5.5, is plotted in Figure 8.23 when $h_2 = 2$ layers. For comparison, the return loss versus frequency, simulated in an "in house" FDTD code, is also shown in Figure 8.23. From the plots, the two simulators are in good agreement with each other. The return loss is below -15 dB for both plots. The bandwidth for both simulations is close to 7%.

The radiation patterns for the stacked-patch structure as well as a single patch with the same total thickness h is illustrated in Figure 8.24. The copolarized (co-pol.)

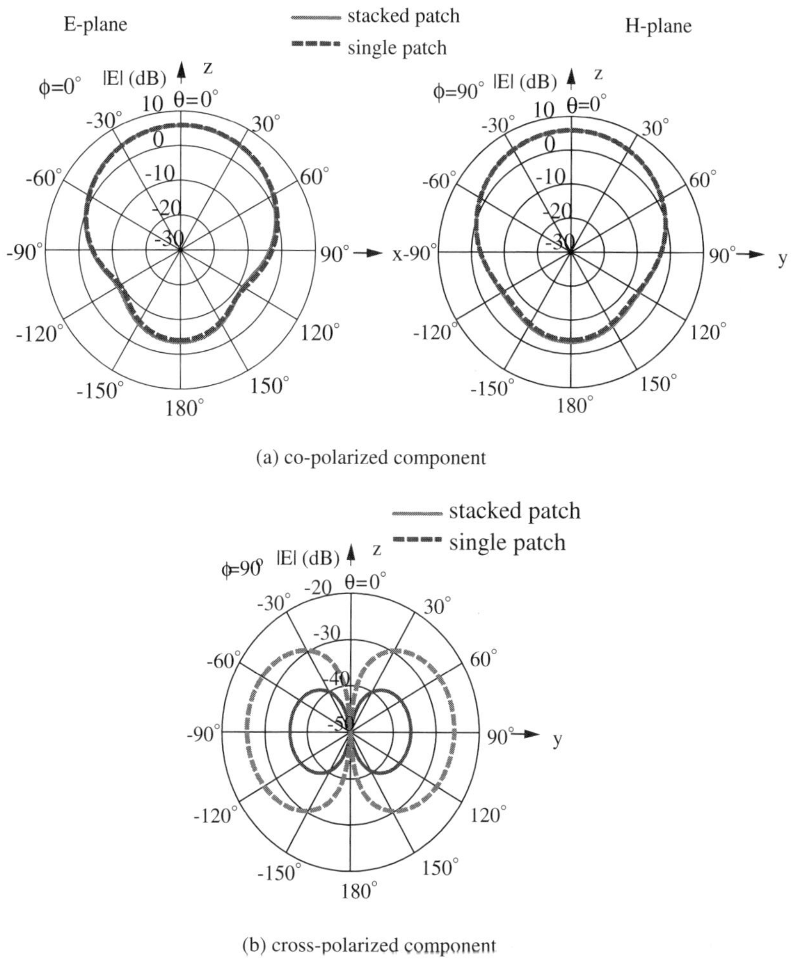

Figure 8.24. Radiation pattern performance comparison of stacked-patch versus single-patch antenna at 28 GHz.

components of the two designs show similar performance for the E- and H-planes. The stacked-patch design has a much lower cross-polarization than the single-patch design. This is mainly because the single patch has a feed probe that is four times longer than the stacked-patch structure, therefore, contributing to higher cross-polarization. All of the designs presented have a radiation efficiency greater than 85%. The power lost in these structures is from conductor loss (conductivity σ equal to 5.8×10^7 siemens per meter, S/m), dielectric loss ($\tan\delta = 0.0012$), surface wave loss ($\varepsilon_r = 5.6$), and feedline radiation. The first three types of loss are properties of the metal (copper, Cu) and the substrate.

Designing thin feeding structures and transitions can circumvent the problem of feedline radiation. It is worth noting that the size of a single-patch design would have to be doubled to achieve the same bandwidth as the stacked-patch antenna. Therefore, the stacked-patch antenna is a great solution for vertical integration of wireless transceivers using multilayer substrates such as LTCC, LCP, and multilayer organic (MLO).

8.6 SUPPRESSION OF SURFACE WAVES AND RADIATION PATTERN IMPROVEMENT USING SHS TECHNOLOGY

As multilayer materials such as LTCC, LCP, and MLO are being used for the integration of antennas into SOP modules, some concerns need to be addressed to maintain the performance of the antenna. One of the drawbacks of designing antennas on LTCC is the high dielectric constant of the substrate ($\varepsilon_r > 5$), which facilitates the propagation of surface waves that may be a larger problem at higher frequencies (millimeter-wave range). Designing antennas on high dielectric constant substrates can severely degrade the performance of the antenna's radiation characteristics as well as reduce the efficiency of the radiator. Another disadvantage of the surface-wave propagation from antennas on high dielectric constant substrates is the unwanted coupling of energy between the antenna and other active devices on the module. Although placing the antenna at a distance of about three wavelengths (3λ) away from the active devices can reduce the crosstalk between devices, this approach is not feasible for maintaining a compact module. Additionally, the use of vertical integration capabilities associated with SOP technology has allowed the lateral size (length and width) of the substrate to be decreased considerably. Despite this innovation, planar antennas that are designed on small-size substrates contribute to backside radiation below the ground of the structure. There is a need for a surface-wave suppressing mechanism that can be integrated to the patch antenna in order to maximize the performance of the antenna and to minimize the degradation of other devices on the module. The most common method of suppressing surface waves is the use of a periodic band-gap structure (PBG). Sometimes integrating PBG structures into SOP-based devices may not be suitable for maintaining the compactness of the design because PBG structures can be large because of the rows of via holes needed to realize the band gaps. A new implementation using the soft-surface properties of a soft-and-hard surface (SHS) structure is applied to a patch antenna on LTCC multilayer substrates.

The ideal SHS conditions can be characterized by the following symmetric boundary conditions for the electric and magnetic fields [10]:

$$\hat{h}\cdot\vec{H} = 0,\ \hat{h}\cdot\vec{E} = 0 \tag{8.7}$$

where $\hat{h}$ is a unit vector tangential to the surface. This boundary is called "ideal" since the complex Poynting vector $\vec{S} = \frac{1}{2}\vec{E} \times \vec{H}^*$ has no component normal to the boundary on the surface. This can be seen through the expansion

$$\begin{aligned} \hat{n}\cdot(\vec{E} \times \vec{H}^*) &= [\hat{h} \times (\hat{n} \times \hat{h})]\cdot(\vec{E} \times \vec{H}^*) \\ &= (\hat{h}\cdot\vec{E})[(\hat{n} \times \hat{h})\cdot\vec{H}^*] - [(\hat{n} \times \hat{h})\cdot\vec{E}](\hat{h}\cdot H)^* \\ &= 0 \end{aligned} \tag{8.8}$$

where $\hat{n}$ is the unit vector normal to the SHS. In a similar way, it can be proved that the complex Poynting vector has a zero component in the direction (say $\hat{s}$) transverse to $\hat{v}$ on the SHS, which means that the SHS in direction $\hat{s}$ can be considered a soft surface, a concept originated from acoustics. We will take advantage of the characteristics of the soft surface to block the surface waves propagating outward along the high dielectric-constant substrate, thus alleviating the diffraction at the edge of the substrate. Shown in Figure 8.1, a patch antenna with arbitrary configuration lies on a finite substrate with a dielectric constant of ε_r and is surrounded by an SHS that is formed as a soft surface in the outward direction. As a result, it would be difficult for surface waves to propagate from the microstrip patch to the substrate edge. Such an SHS can be realized using several via rings whose height h must be equal to $\lambda_0/(4\sqrt{\varepsilon_r})$, where λ_0 is the free-space wavelength. Based on this basic configuration, we will in the next two sections investigate two antenna structures. One is a patch antenna on a large-size substrate, and the other is a CP antenna.

The antenna structure for this implementation is shown in Figure 8.25. It consists of two square patches that are stacked for bandwidth enhancement. Each side of the patches is 0.75 mm. The LTCC substrate has a dielectric constant ε_r of 5.4 and a loss tangent of 0.0015. The substrate thickness is 500 μm, which comprises five total layers that are each 100 μm thick. The upper patch is placed on the top layer of the substrate, whereas the lower patch is positioned three layers below the upper patch. One layer separates the intermediate ground plane and the lower patch. A 100-μm cavity has been etched out between the intermediate ground plane and the total ground plane of the structure for the purpose of burying the MMICs. Surrounding the patch antennas are square metallic rings that mimic the perfect electric conducting (PEC) regions of the SHS. These rings are separated by a distance of 0.5 mm. This distance is chosen arbitrarily. Although a smaller distance does not have a significant effect on the suppressing ability of the SHS, a larger distance allows the fields to continue to propagate to the edges of the substrate. The substrate that is between the rings represents the perfect magnetic conducting (PMC) regions of the SHS. In simulation, continuous metal walls can be used for the PEC regions, whereas in fabrication, it is necessary to use vias that are separated by 0.5 mm. The diameter of the vias is 0.13 mm. A requirement for an effective surface wave suppressing SHS is that the substrate thickness maintains a

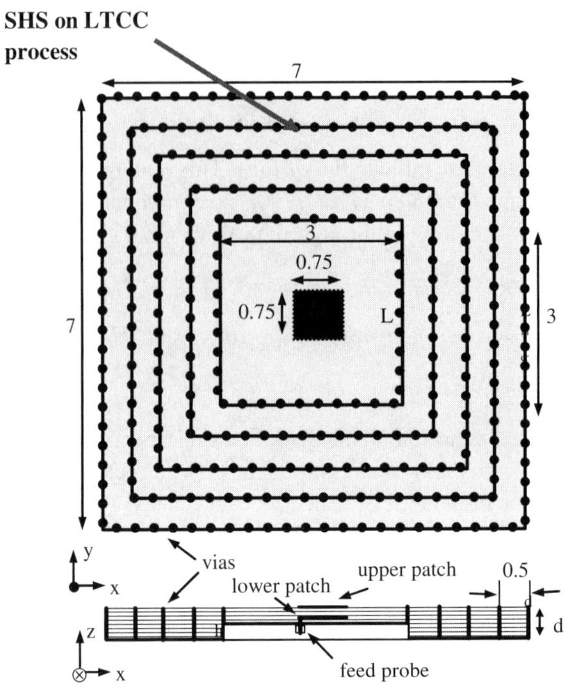

Figure 8.25. Stacked-patch antenna surrounded by the SHS structure.

height of $\lambda_o/(4 \times \varepsilon_r^{1/2})$, where λ_o is the free space wavelength. Becasue of this limitation, the resonant frequency for this design is chosen to be 64.55 GHz. The metal for the conductors is silver (Ag). The total area for the antenna design is 7 mm $\times$ 7 mm.

The antenna was simulated using MicroStripes 5.6. A similar design without the SHS structure was also simulated in the software to compare the performance of the two structures. The total size of both designs is 7 mm $\times$ 7mm $\times$ 0.5 mm. The radiation patterns at the resonance are shown in Figure 8.26. Two major effects can be observed from the E-plane and H-plane radiation patterns. The first effect is the reduction in the backside radiation (lower hemisphere of the plots), which is less than 5 dB below the main beam for both patterns in the design without the SHS. In contrast, the design with the SHS exhibits F/B ratios around 15 dB in the E- and H-planes. The cross-polarization level, which can be seen in the H-plane, is slightly higher in the SHS implemented structure in comparison with the structure without the SHS, but its value (below -18 dB) is suitable for this design. The second effect that can be seen in the two plots is the increased gain in the design with SHS. Both the E-plane and the H-plane radiation patterns exhibit on-axis ($\theta = 0°$) gain enhancements of 10 dB. The major reason for this enhancement is the edge diffraction of the finite-size substrate. As the finite-size substrate increases, the edge diffraction effects contribute increased radiation at angle between 30° and 45° away from the broadside direction (normal to the surface of the patch). This is only observed in the E-plane plot in which the resonant length is along the horizontal axis.

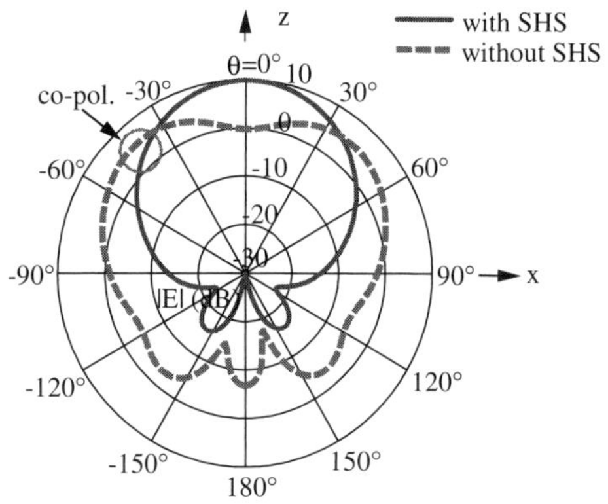

(a)

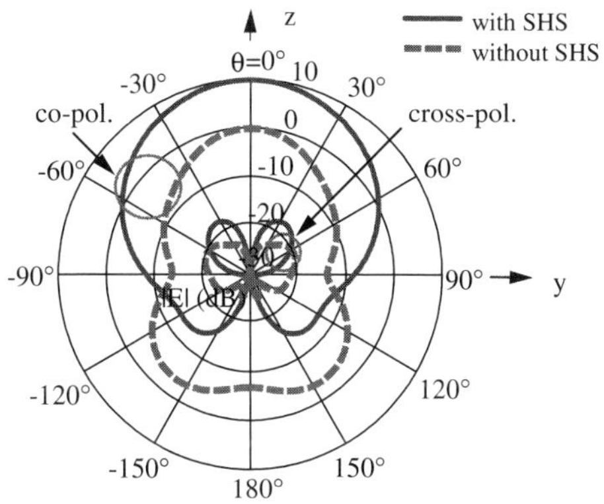

(b)

Figure 8.26. Two-dimensional radiation patterns on (a) E-plane and (b) H-plane for designs with and without SHS.

The return loss versus frequency plots for the two designs are shown in Figure 8.27. It is observed that one TM_{10} resonance is present in the structure at the resonant frequency of 64.55 GHz. The impedance bandwidth of the SHS, implemented structure is smaller that that of the design without the SHS, which results, from the contribution of the currents on the metals that adds inductance to the antenna. The on-axis gain versus frequency is also displayed in Figure 8.27. It can be observed that the gain of the antenna with the SHS is around 10 dB, and this value is constant over the entire bandwidth from approximately 61.5 to 66 GHz. The on-axis gain for the antenna structure without the SHS is 0 dB.

Plots of the near-field distributions for both designs are presented in Figure 8.28. As the fields propagate away from the radiating edges of the patch, significant field intensities are present at the edges of the substrate in the design without the SHS. On the other hand, when the SHS is implemented to the patch antenna design, the electromagnetic fields become more and more weak as they pass the SHS. At the edges of the substrate, the fields are significantly suppressed. Because of the results

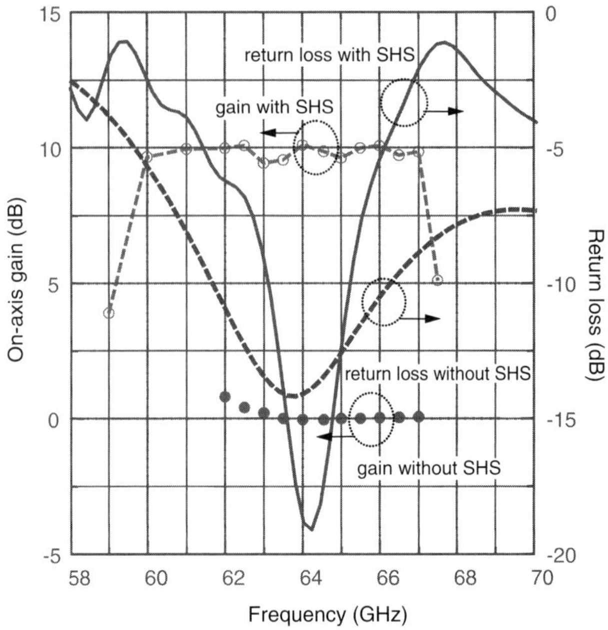

Figure 8.27. Return loss and on-axis gain versus frequency for antenna designs with and without SHS.

of the near-field distributions, it is believed that an SHS structure using a smaller number of rings could be used to obtain the same effect of suppressing substrate modes as they propagate away from the radiating edges of the patch. This will also reduce the overall size of the structure, thus, allowing for a more compact device for integration to RF SOP modules.

8.7 RADIATION-PATTERN IMPROVEMENT USING A COMPACT SOFT-SURFACE STRUCTURE [40]

In the previous section, it has been demonstrated that the concept of artificial SHS can be introduced to improve the radiation patterns for different types of antennas. This technique can be implemented on standard board materials or using emerging multilayer technologies, such as LCP, MLO, and LTCC processes. However, the conventional SHS concept based on corrugated structures requires a substrate thickness of one-quarter guided wavelength as well as a large area. Recently, a modern realization of artificial soft-surface structures has been proposed in [41]. The new proposed soft surface consists of several quarter-wavelength metal strips that are short-circuited to the ground plane. The operating frequency for this soft surface is determined by the strip width, not by the thickness of the substrate, thus allowing for its implementation in a substrate with arbitrary thickness.

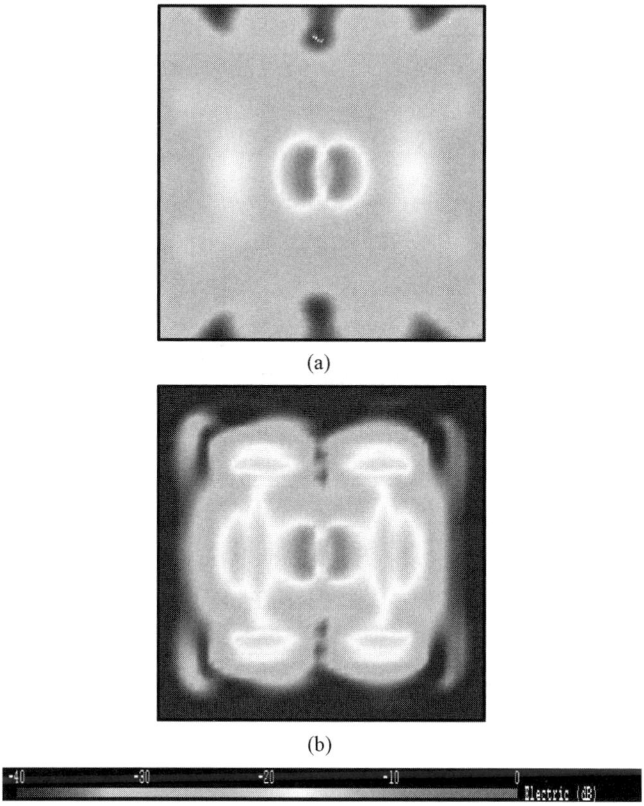

Figure 8.28. Near-field distributions on electric field for antenna design (a) without SHS and (b) with SHS.

In this section, we apply the concept of the modern soft surface to improve the radiation pattern of patch antennas. A single square ring of the shorted quarter-wavelength metal strips is employed to form a soft surface and to surround the patch antenna for the suppression of outward propagating surface waves, thus alleviating the diffraction at the edge of the substrate. Since only a single ring of metal strips is involved, the formed soft-surface structure is compact and easily integrable with 3-D modules. A numerical investigation is presented for a patch antenna surrounded by an ideal compact soft-surface structure for three different dielectric substrates. The effectiveness of the compact soft surface in terms of radiation pattern improvement is verified by its implementation on LTCC multilayer technology.

For the sake of simplicity, we consider a probe-fed square-patch antenna on a square grounded substrate with thickness H and a dielectric constant ε_r. The patch antenna is surrounded by a square ring of metal strip that is short-circuited to the ground plane by a metal wall along the outer edge of the ring, as shown in Figure 8.29. The shorted square ring represents an ideal soft-surface structure when the width of the metal strip

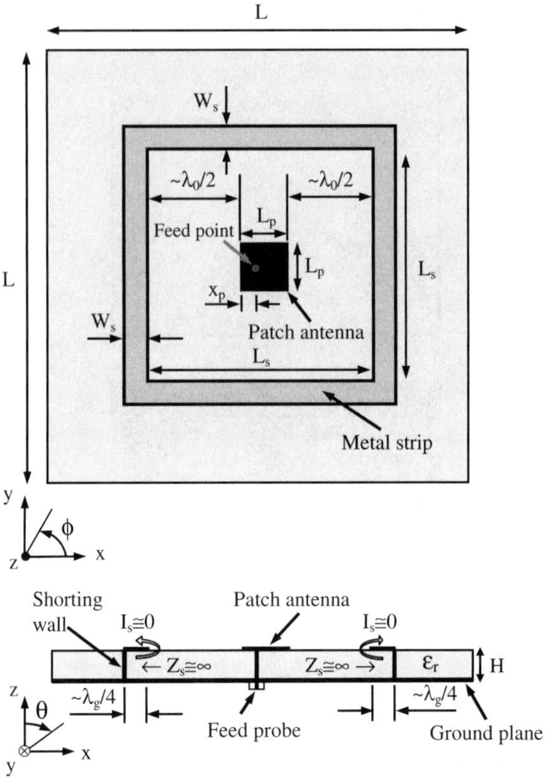

Figure 8.29. Patch antenna surrounded by an ideal compact soft-surface structure that consists of a ring of metal strip and a ring of shorting wall (I_s = the surface current on the top surface of the soft surface ring, Z_s = the impedance looking into the shorted metal strip).

(W_s) is approximately equal to a quarter of the guided wavelength:

$$W_s \cong \frac{\lambda_g}{4} = \frac{c}{4f_0\sqrt{\varepsilon_r}} \tag{8.9}$$

where f_0 is the operating frequency, λ_g is the guided wavelength, and $c = 3 \times 10^8$ m/s. Taking into account the effect of fringing field at the inner edge of the shorted ring, we obtain the following design value for W_s:

$$W_s = \frac{c}{4f_0\sqrt{\varepsilon_{re}}} - \Delta W_s \tag{8.10}$$

where ε_{re} is the effective dielectric constant given by

$$\varepsilon_{re} = \left[\frac{\varepsilon_r + 1}{2} + \frac{\varepsilon_r - 1}{2} F(W_s/H)\right] \tag{8.11}$$

with

$$F(W_s/h) = (1 + 6H/W_s)^{-1/2} \tag{8.12}$$

ΔW_s is an empirical correction factor obtained as

$$\Delta W_s = H \cdot \left[0.882 + \frac{0.164(\varepsilon_r - 1)}{\varepsilon_r^2} + \frac{\varepsilon_r + 1}{\pi \varepsilon_r} \left\{ 0.758 + \ln\left(\frac{2W_s}{H} + 1.88 \right) \right\} \right] \tag{8.13}$$

The design value for W_s is optimized by simulation for a maximum directivity at broadside (the z direction).

To demonstrate the effectiveness of the compact soft-surface structure on the radiation pattern improvement, we simulated the soft-surface-surrounded patch antenna for three different dielectric constants: $\varepsilon_r = 2.9$, 5.4, and 9.6, which correspond to the typical values of LCP and to two types of LTCC materials, respectively. The operating frequency is assumed to be $f_0 = 15$ GHz. The substrate thickness H is assumed to be 0.5 mm ($0.025\lambda_0$, λ_0 = the free-space wavelength), and the lateral size ($L \times L$) of the substrate is 40 mm × 40 mm ($2\lambda_0 \times 2\lambda_0$), which is much larger than the size ($< 0.5\lambda_g \times 0.5\lambda_g$) of the square patch. The numerical simulation tool used is the TLM-based software: *Microstripes 6.0*. For comparison, we also simulated the patch antenna without the soft surface. The patch lengths were found to be $L_p = 5.4, 3.94$, and 2.9 mm, respectively, for $\varepsilon_r = 2.9$, 5.4, and 9.6 at $f_0 = 15$ GHz. The optimized (for a maximum directivity) values of W_s are 2.65, 1.88, and 1.36 mm, respectively, for $\varepsilon_r = 2.9, 5.4$, and 9.6, close to the analytically calculated design values: 2.75, 1.93, and 1.35 mm. An important observation after the simulation is that the maximum directivity is dependent on the inner length L_s of the soft-surface ring. Figure 8.30 shows the directivity as a function of L_s. We can see the maximum directivity values of 9.6, 9.2, and 9.1 dBi obtained at $L_s = 24$, 22, and 20 mm for $\varepsilon_r = 2.9$, 5.4, and 9.6, respectively. The reason for the dependence will be explained later in this section. Note that the maximum directivity was found to be 8.8 dBi for a patch antenna (on a substrate with $\varepsilon_r = 10$) surrounded by a PBG ring with a ring width of $\sim 2.5\lambda_g$, ten times of the width of the soft-surface ring ($\sim 0.25\lambda_g$).

Since both the patch antenna and the soft-surface ring are resonant structures, the directivity is definitely related to the frequency. The relationship is plotted in Figure 8.31 for an optimized soft-surface structure. The variation of directivity with frequency is slightly smaller for $\varepsilon_r = 2.9$ than those for $\varepsilon_r = 5.4$ and 9.6. A comparison of return loss between the patch antennas with and without the soft surface is shown in Figure 8.32 for $\varepsilon_r = 2.9$. It is observed that the soft-surface structure has no significant effect on the impedance performance of the patch antenna since the soft-surface ring is almost half a wavelength away from the edge of the patch. This observation also applies to the cases for $\varepsilon_r = 5.4$ and 9.6.

Next, we proceed to the evaluation of the radiation-pattern improvement achieved by the soft-surface structure. The radiation patterns for $\varepsilon_r = 9.6$ plotted in two principal planes (E-plane and H-plane) are compared in Figure 8.33 between the patch

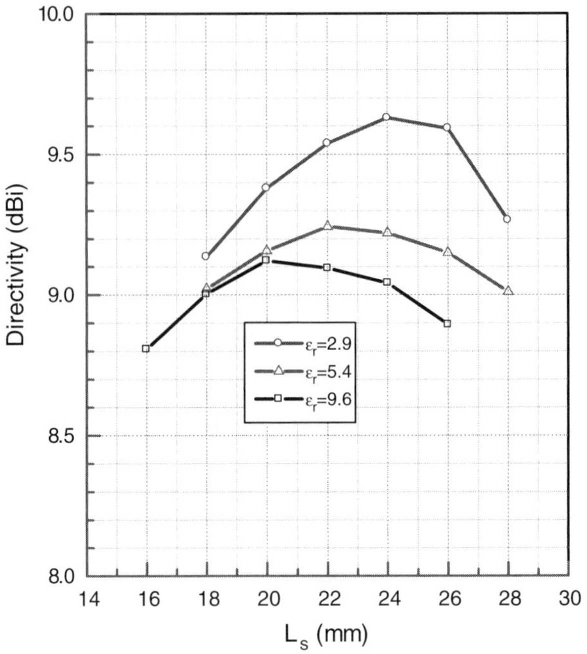

Figure 8.30. Directivity of the soft-surface-surrounded patch antenna as a function of the inner length (L_s) of the soft-surface ring with $W_s = 2.65$, 1.88, and 1.36 mm for $\varepsilon_r = 2.9$, 5.4, and 9.6, respectively.

antennas with and without the soft surface. It is observed that without the soft surface the direction for the maximum radiation is no longer in the z-direction because of the contribution from the surface-wave diffraction. By introducing the soft-surface structure, the radiation pattern is significantly improved, including an enhanced radiation field in the z-direction, narrowed beamwidth in the E plane, and reduced backside radiation. The directivity at broadside is increased by more than 5 dB (from 4.0 dBi to 9.1 dBi), whereas the backside radiation level gets reduced by about 8 dB. A similar radiation pattern improvement by the soft surface is observed for $\varepsilon_r = 2.9$ and 5.4, where the directivity at broadside increases from 5.7 dBi and 4.5 dBi to 9.6 dBi and 9.2 dBi, respectively.

The mechanism for the radiation pattern improvement achieved by the introduction of a compact soft-surface structure can be understood by considering two factors. First the quarter-wave shorted metal strip serves as an open circuit for the TM_{10} mode (the fundamental operating mode for a patch antenna). Therefore, it is difficult for the surface current on the inner edge of the soft-surface ring to flow outward (also see Figure 8.1). As a result, the surface waves can be considerably suppressed outside the soft-surface ring, hence reducing the undesirable diffraction at the edge of the grounded substrate. This explanation can be confirmed by checking the field distribution in the substrate. Figure 8.34 shows the electric field distributions on the top surface of the substrate for the patch antennas with and without the soft-surface. We

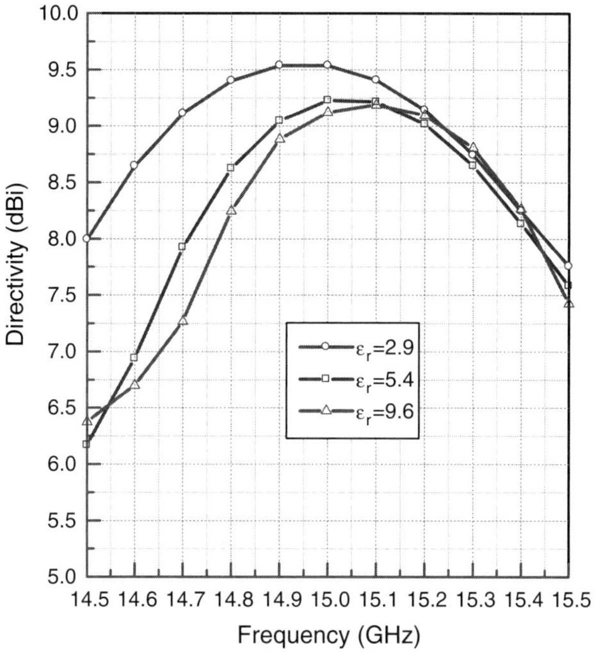

Figure 8.31. Directivity of the soft-surface-surrounded patch antenna as a function of frequency with optimized $W_s = 2.65$, 1.88, and 1.36 mm and optimized $L_s = 24$, 22, and 20 mm for $\varepsilon_r = 2.9$, 5.4, and 9.6, respectively.

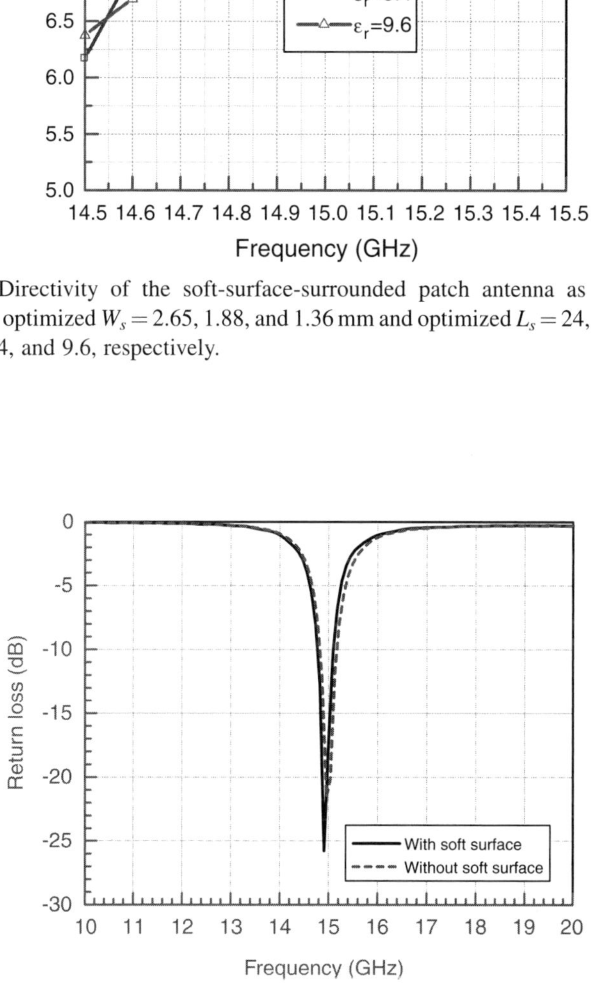

Figure 8.32. Comparison of return loss between the patch antennas with and without the soft surface ($\varepsilon_r = 2.9$, $x_p = 1.9$ mm).

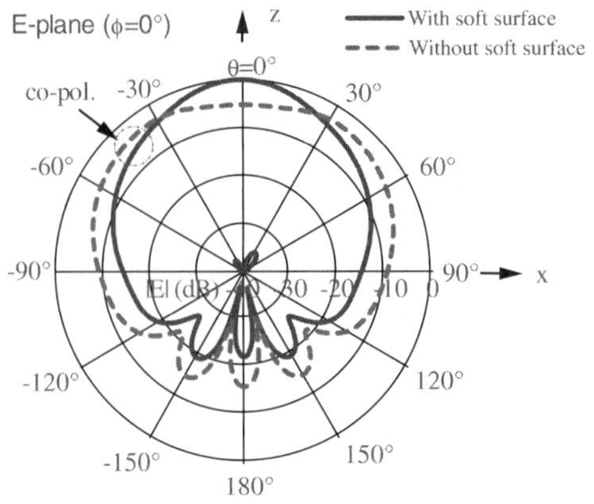

(a) E-plane (ϕ=0°)

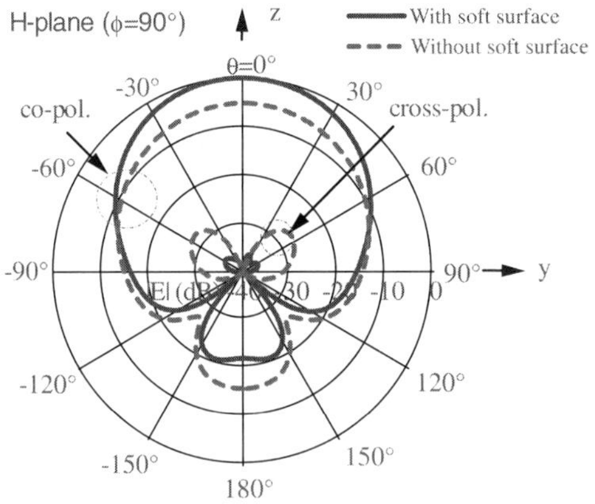

(b) H-plane (ϕ=90°)

Figure 8.33. Comparison of radiation-patterns in principal planes between the patch antennas with and without the soft surface ($\varepsilon_r = 9.6$, $f_O = 15$ GHz).

can see that the electric field is indeed contained inside the soft-surface ring. It is estimated that the field magnitude outside the ring is approximately 5 dB lower than that without the soft surface.

The second factor contributing to the radiation pattern improvement is the fringing field along the inner edge of the soft-surface ring. This fringing field along with the fringing field at the radiating edges of the patch antenna forms an antenna array in the E-plane. The formed array acts as a broadside array with minimum radiation in the x–y plane when the distance between the inner edge of the soft-surface ring and its nearby radiating edge of the patch is roughly half a wavelength in free space. Even though the magnitude of the fringing field along the soft surface may be much lower compared with that at the radiating edges, the size of the soft-surface ring is much larger than the patch. As a result, the contribution from the soft-surface ring to the radiated field at broadside may be significant. Therefore, the radiation pattern can be considerably

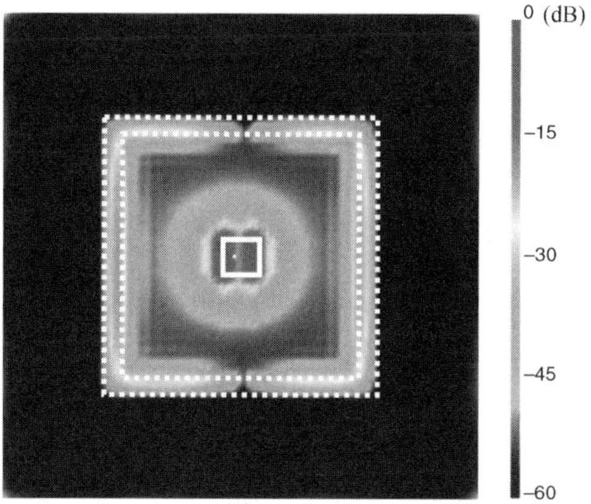

(a) With soft surface

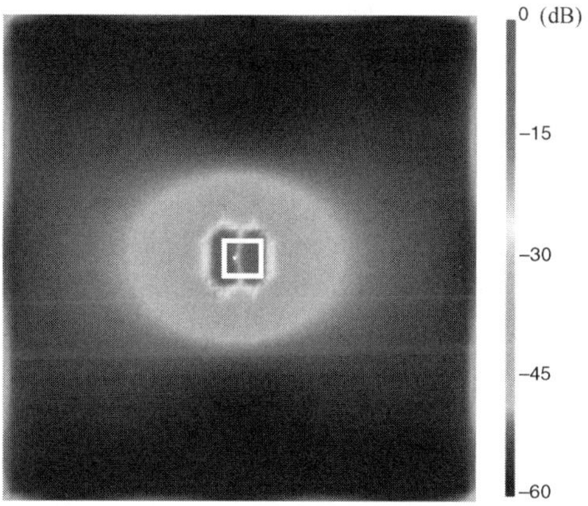

(b) Without soft surface

Figure 8.34. Simulated electric field distributions on the top surface of the substrate for the patch antennas with and without the soft-surface ($\varepsilon_r = 9.6$).

improved by the soft-surface ring. Note that if the inner length of the soft-surface ring (L_s) is further increased, the fringing field will become weaker. Therefore, there is an optimized value for L_s. By simulation, it is found that the optimized L_s is approximately equal to one wavelength in free space plus the length of the patch antenna (L_p). In this sense, the soft-surface ring is also designed as a radiating element that launches the impinging surface wave into free space and in doing so effectively increases the size of the radiation aperture. By simulation, it is observed that the radiation pattern cannot be further improved by adding more soft-surface rings, which supports the claim that the soft-surface ring indeed acts as a radiating element.

The LTCC laminated technology is getting more and more popular in wireless and millimeter-wave applications because of its easy realization of highly integrated, complex multilayer 3-D modules and circuits. This technology is appreciated for its flexibility in connecting an arbitrary number of layers with via holes (or via metallization). By using the LTCC technology, it is easy to realize the shorting wall required by the compact soft-surface structure, minimizing the parasitic surface currents. To demonstrate the feasibility of this technology on the implementation of the soft surface, we first simulated a benchmarking prototype that was constructed to replace the shorting wall with a ring of vias. Note that the LTCC substrate is hidden in order to see the via-related structures clearly. The used LTCC material had a dielectric constant of 5.4. The whole module consists of a total of 11 LTCC layers (layer thickness = 100 μm) and 12 metal layers (layer thickness 10 μm). The diameter of each via was specified by the fabrication process to be 100 μm, and the distance between the centers of two adjacent vias was 500 μm. To support the vias, a metal pad is required on each metal layer. To simplify the simulation, all pads on each metal layer are connected by a metal strip with a width of 600 μm. Simulation showed that the width pad metal strips has little effect on the performance of the soft-surface structure as long as it is less than the width of metal strips for the soft-surface ring (W_s).

The size of the LTCC board was 30 mm × 30 mm. The operating frequency was set within the K_u band (the design frequency $f_0 = 16.5$ GHz). The design value of the distance (i.e., W_s) between the center of the vias and the inner edge of the soft-surface ring was estimated by Eq. (8.10). The optimized values for L_s and W_s were, respectively, 22.2 mm and 1.4 mm, which led to a total via number of 200 (51 vias on each side of the square ring). Including the width (300 μm) of the pad metal strip, the total metal strip width for the soft-surface ring was found to be 1.7 mm.

Since the substrate was electrically thick at $f_0 = 16.5$ GHz ($> 0.1\lambda_g$), a stacked configuration was adopted for the patch antenna to improve its input impedance performance. By adjusting the distance between the stacked square patches, a broadband characteristic for the return loss can be achieved. For this case, the upper and lower patches (with the same size 3.4 mm × 3.4 mm) were, respectively, printed on the first LTCC layer and the seventh layer from the top, leaving a distance between the two patches of Six LTCC layers. The lower patch was connected by a via hole to a 50-Ω microstrip feed line that is on the bottom surface of the LTCC substrate. The ground plane was embedded between the second and third LTCC layers from the bottom, which means that among the total 11 LTCC layers, two were used for the microstrip feed line, whereas nine layers were used for the patch antenna and the soft-surface structure. Figure 8.35 shows the layout and a prototype of the stacked-patch antenna surrounded by a compact soft-surface structure implemented on the LTCC technology. The inner conductor of an SMA (semi-miniaturized type-A) connector was connected to the microstrip feed line while its outer conductor was soldered on the bottom of the LTCC board to a pair of pads that were shorted to the ground through via metallization. Note that the microstrip feed line was printed on the bottom of the LTCC substrate to avoid its interference with the soft-surface ring and to alleviate the contribution of its spurious radiation to the radiation pattern at broadside. For comparison the same stacked-patch antenna on the LTCC substrate but without the soft-surface ring was also built.

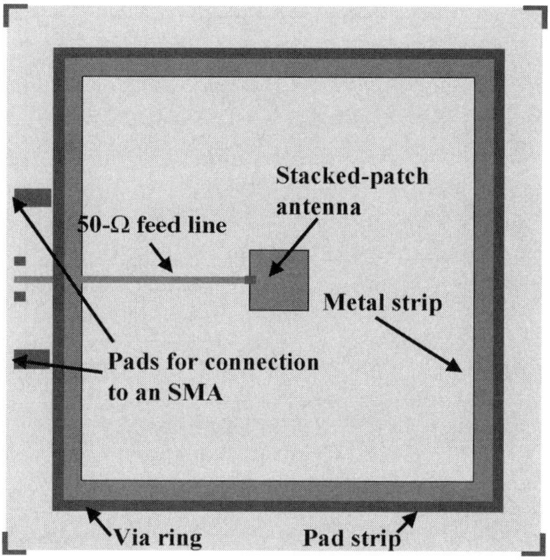

(a) layout

(b) Prototype

Figure 8.35. Layout and prototype of a stacked-patch antenna surrounded by a compact soft-surface structure implemented on LTCC technology.

The simulated and measured results for the return loss are shown in Figure 8.36, and good agreement is observed. Note that because the impedance performance of the stacked-patch antenna is dominated by the coupling between the lower and upper patches, the return loss for the stacked-patch antenna seems more sensitive to the soft-surface structure than that for the previous thinner single-patch antenna. Fortunately the bandwidth is indeed improved both for the stacked patches with and without the soft surface. The measured return loss is close to −10 dB over the frequency range 15.8–17.4 GHz (about 9% in bandwidth). The slight discrepancy between the

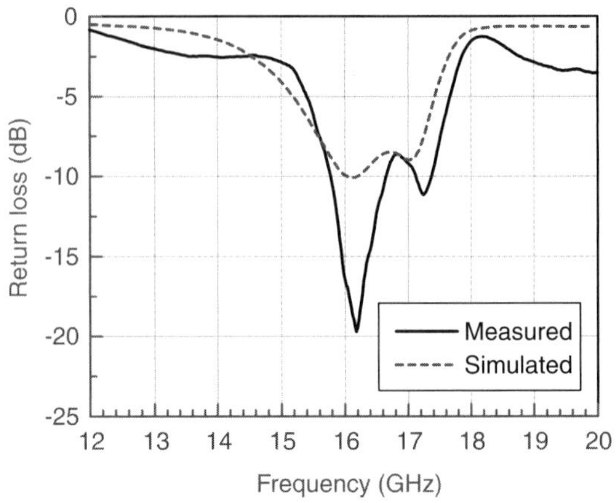

(a) With soft surface

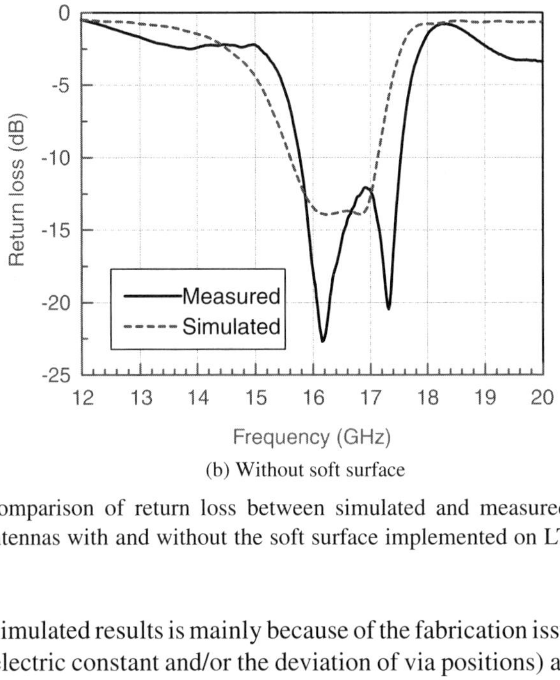

(b) Without soft surface

Figure 8.36. Comparison of return loss between simulated and measured results for the stacked-patch antennas with and without the soft surface implemented on LTCC technology.

measured and simulated results is mainly because of the fabrication issues (such as the variation of dielectric constant and/or the deviation of via positions) and the effect of the transition between the microstrip line and the SMA connector. It is also noted that there is a frequency shift of about 0.3 GHz (about 1.5% up). This may be caused by the LTCC material, which may have a real dielectric constant a little bit lower than the overestimated design value. Note that it is normal for practical dielectric substrates to have a dielectric constant with $\pm 2\%$ deviation.

The radiation patterns measured in the E- and H-planes are compared with simulated results in Figure 8.37. The radiation patterns compared here are for a frequency of 17 GHz where the maximum gain of the patch antenna with the soft surface was observed. From Figure 8.37, we can see good agreement for the

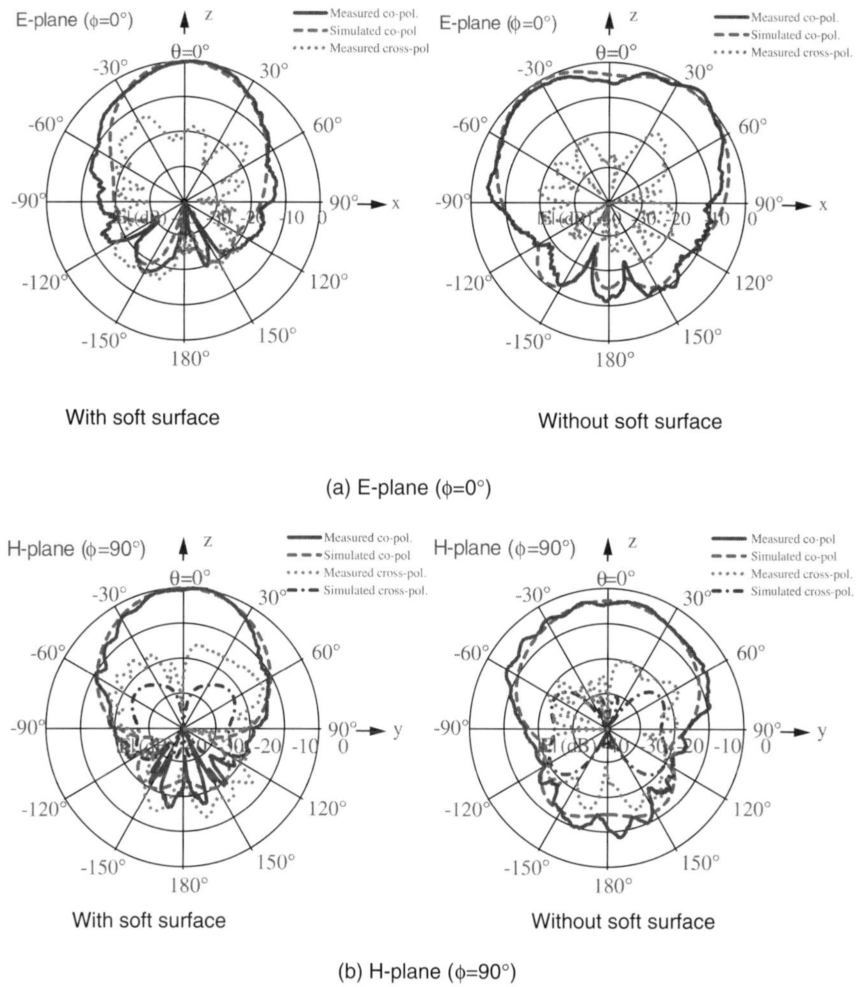

Figure 8.37. Comparison between simulated and measured radiation patterns for the stacked-patch antennas with and without the soft surface implemented on LTCC technology ($f_0 = 17$ GHz).

copolarized components. It is confirmed that the radiation at broadside is enhanced and that the backside level is reduced. Also the beamwidth in the E-plane is significantly reduced by the soft surface. It is noted that the measured cross-polarized component has a higher level and more ripples than the simulation result. This is because the simulated radiation patterns were plotted in two ideal principal planes, i.e., $\phi = 0°$ and $\phi = 90°$ planes. From simulation, it was found that the maximum cross-polarization may happen in the plane $\phi = 45°$ or $\phi = 135°$. During measurement, a slight deviation from the ideal planes can cause a considerable variation for the cross-polarized component since the spatial variation of the cross-polarization is quick and irregular. Also, a slight polarization mismatch and/or some objects nearby the antenna (such as the connector and/or the connection cable) may considerably contribute to the high cross-polarization.

Finally let us compare the gains between the stacked-patch antennas with and without the soft surface. The simulated and measured gains at broadside (i.e., the z-direction) are shown in Figure 8.38. The simulated gain was obtained from the numerically calculated directivity in the z-direction and the simulated radiation efficiency, which is defined as the radiated power divided by the radiation power plus the ohmic loss from the substrate and metal structures ($\tan\delta = 0.002$ and $\sigma = 5.8 \times 10^7$ S/m were assumed). It is found that the simulated radiation efficiency at the design frequency of the stacked-patch antenna with the soft surface is about 88%, only 2% lower than that of the same stack-patch antenna without the soft surface. The measured gain was obtained by calibrating the stacked-patch antennas with a standard gain horn antenna. From Figure 8.38, we can see that the simulated broadside gain around the design frequency of the stacked-patch antenna with the soft surface is more than 9.0 dBi, about a 6.5-dB improvement as compared with the stacked patch without the soft surface. This gain improvement is also nearly 2 dB higher than that obtained for the single patch on a thinner substrate as described in the previous section. This is due to the contribution from the thicker substrate with the help of a soft surface. The thicker substrate excites stronger surface waves, whereas the soft surface blocks and transforms the excited surface waves into space waves. Because of the frequency shift,

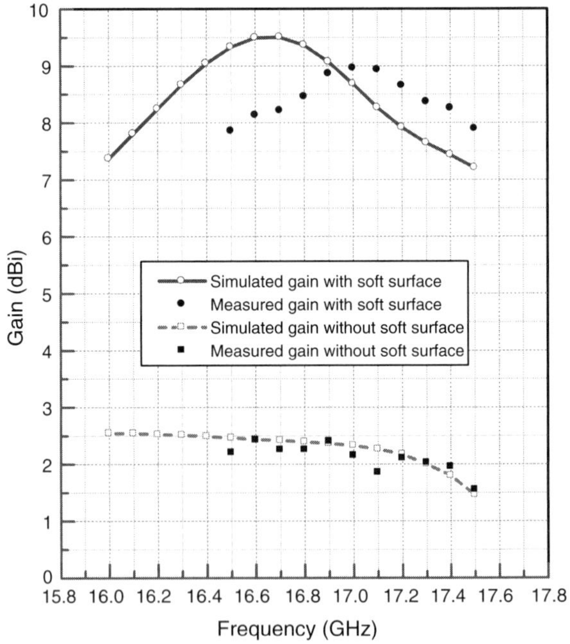

Figure 8.38. Comparison of simulated and measured gains at broadside between the stacked-patch antennas with and without the soft surface implemented on LTCC technology.

the maximum gain measured for the stacked patch with the soft surface is observed around 17 GHz and its value is near 9 dBi, almost 7 dB higher than the gain at broadside for the stacked-patch antenna without the soft surface.

8.8 A PACKAGE-LEVEL-INTEGRATED ANTENNA BASED ON LTCC TECHNOLOGY [42]

In recent years, numerous studies have focused on the development of highly integrated transceiver systems, including the SOP approach. The RF SOP technique can provide a single package solution by integrating embedded passives, MMICs, and CMOS bare chips using advanced packaging technology. As an example, a package-level integration of an LTCC antenna is investigated in this section for a highly integrated 5-GHz wireless transceiver module. The LTCC process is an attractive solution for RF applications, as low-loss transmission lines, high Q passives, and three dimensional stack-up geometries can be easily realized. Furthermore, regarding the overall circuit size, the high relative permittivity of LTCC material is advantageous. Recent studies have focused on the suitability of LTCC technology for the development of antennas, filters, baluns, and modules.

In general, a model number and a manufacturer name are printed on the top surface of IC packages. Thus, antenna integration on the top surface of a package would be very space efficient. The proposed LTCC package has a stacked-patch antenna integrated on top of a package together with an inner packaging space in the middle layer for the integration of other passive components. This antenna works as a package as well as simultaneously works as a radiator. Moreover, to reduce the undesirable coupling effect, one metallic layer is inserted between the antenna and the inner passives, and this internal ground plane is connected directly with the system ground plane through multiple via holes. Figure 8.39 shows the design concept of the package-integrated antenna. The printed radiating element can be situated on the top package layer, whereas the internal space under the antenna can be used to host other chip-scale packaged components, including a film bulk acoustic resonator (FBAR) filter, a diplexer, and lumped elements. Moreover, transmission lines and metallic pads for the interconnection are printed on the bottom layer. Those chip-scale packaged components can be mounted on the bottom layer using Au/Ni solders.

Most 5-GHz commercial wireless communication systems currently deployed require more than 100 MHz of bandwidth. The stacked-patch antenna topology is employed to achieve the required specifications of a broad bandwidth and a nearly omnidirectional radiation pattern. A schematic diagram of the proposed package with a stacked-patch antenna fed by a strip line is given in Figure 8.40. As shown in the figure, it comprises of the LTCC stacked-patch antenna, a feeding network, an antenna ground plane, a package side wall with a cavity, and a system ground plane. The LTCC multilayer structure used in the current research consists of DuPont 951 sheets with a thickness of 0.05 mm after sintering and with a dielectric constant of 7.8.

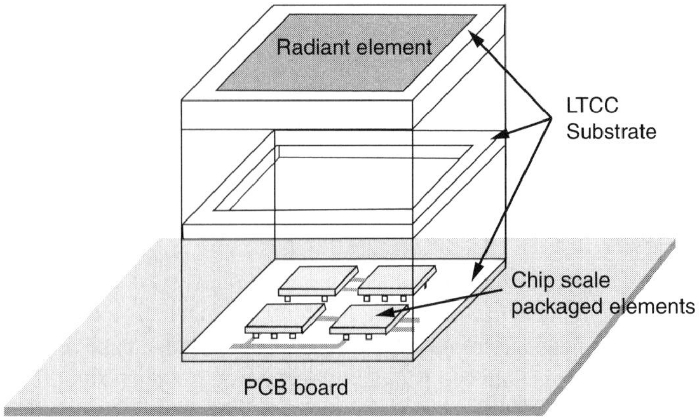

Figure 8.39. Design architecture of the package-level integration of an antenna.

To fit the needs of next-generation wireless communication systems with compact transceiver modules, the package height (with antenna) should be less than 1.5 mm, whereas the system ground size is assumed to be $20 \times 20\,\text{mm}^2$. The thicknesses of each dielectric layer used in this work are 0.1, 0.2, 0.3, and 0.7 mm. There are four Ag screen-printed metallic layers on the cavity structure. The upper two metallic layers are designed to host two radiating elements of the stacked patch antenna, whereas the lowest metallic layer works as an antenna ground. When Ag is screen-printed onto the sheets, the dielectric layers require a guiding area (0.2 mm for each side). These

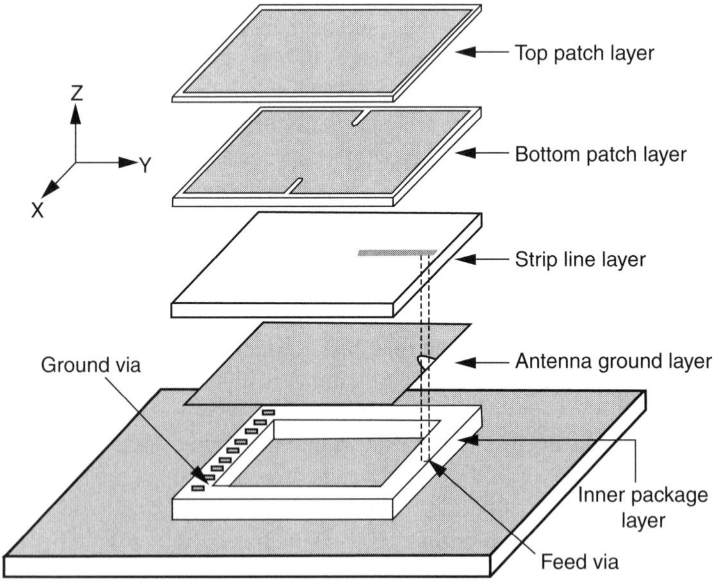

Figure 8.40. Expanded view of the proposed package.

metallic layers are then designed slightly smaller than the substrate layer. The above-mentioned three layers each have dimensions of $9.9 \times 9.9 \times 0.01\ mm^3$. For the first resonant frequency, the bottom patch employs two short slits (width = 0.2 mm, length = 3.9 mm) in the middle of nonradiating edges. The third metallic layer contains a strip line providing proximity feeding to the bottom patch. The length and width of the feeding strip line play an important role in the input impedance performance. This line is vertically connected through a via structure that is embedded in the inner package layer. The lowest metallic layer is designated as an antenna ground and is interconnected with the system ground plane using nine via holes. This is a part of the system ground, and it prevents any unwanted radiation from the antenna into the innerintegrated elements. The inner package structure has a dimension of 10.3×10.3 $0.7\ mm^3$, and it works as a package side wall and provides $8.3 \times 8.3 \times 0.7\ mm^3$ of internal space. The width of the wall is 1 mm, and the height of the tallest of the bare components is assumed to be less than 0.7 mm.

Figure 8.41 shows the detailed configuration of the proposed antenna-integrated package topology as well as the design parameters of each layer. The back side of the system ground plane has a microstrip line connected directly with the feed via. Its width is chosen to be 0.61 mm so that the line to has the characteristic impedance of nearly 50. The length is defined as 6 mm from the edge of the system ground plane.

Based on the calculated results, the proposed package integrated antenna is fabricated using LTCC technology. Ag screen printing on the dielectric sheet as well as the low-loss via process and high conductivity metallization are useful for high-frequency applications. Because of the size difference between the LTCC package and the system ground layer, they are fabricated separately. After fabricating each element, the package is mounted on the system ground plane. Figure 8.42 shows a photograph of the implemented LTCC package on the system ground plane. In the figure, the metal layer has a golden color because of the Au/Ni plating after sintering.

Figure 8.43 shows scanning electron microscope (SEM) images of the fabricated antenna. It is interesting to note that Figure 8.5(a) shows a cross-sectional view in the middle of the YZ plane, where the via on the right side is connected with a feeding strip line. It is also interesting to note that there is a slight via misalignment caused by a shrinkage difference in the firing process between the inner package and the strip line layers. The cross-sectional view of the ground vias in the XZ plane is shown in Figure 8.5(b). Most via holes are extremely well fabricated and are connected with a square pad placed in the bottom. These pads have a slightly larger dimension than the diameter of the via holes; this is in order to provide a simple connection between the package and the system ground plane. As shown, all dielectric layers were fabricated as expected and the metallic layers were successfully screen printed onto the sheets.

Figure 8.44 compares the calculated and measured return loss characteristics of the antenna. The antenna has two resonant frequencies, coming from the bottom and top patches. The calculated resonant frequencies of the designed geometry are 5.264 and 5.355 GHz for the top and bottom patches, respectively, and the 10-dB impedance bandwidth is 140 MHz. Despite the two neighboring resonant frequencies, the bandwidth is relatively narrow because of the relatively small ground plane as well as the finite substrate size. The measured impedance bandwidths are 40 and 100 MHz

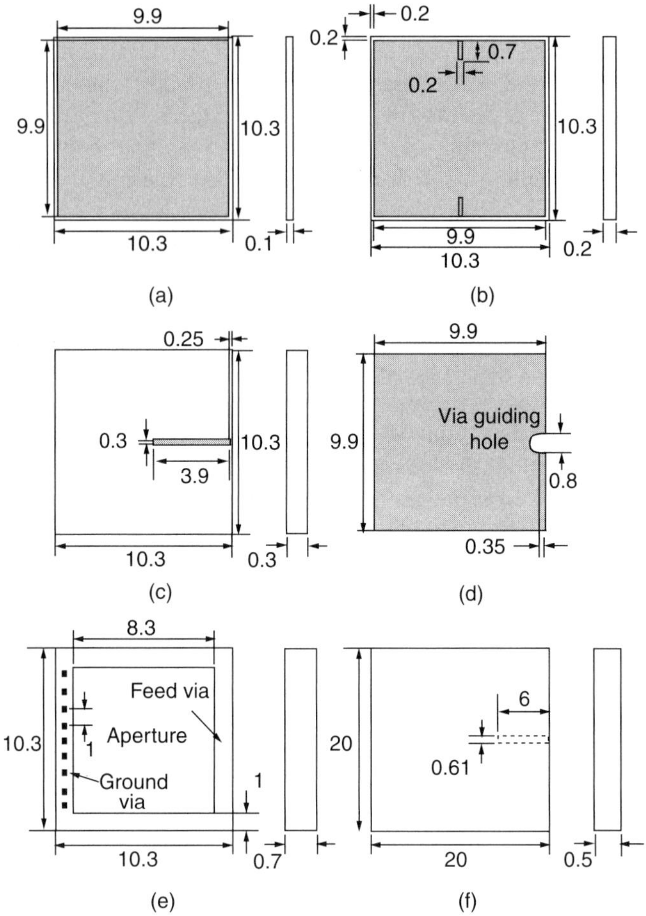

Figure 8.41. Design parameters of each layer: (a) top patch layer, (b) bottom patch layer, (c) strip line layer, (d) antenna ground layer, (e) inner package layer, and (f) system ground layer.

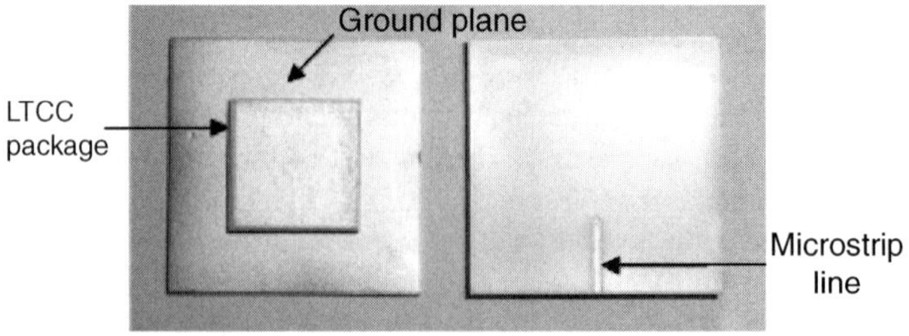

Figure 8.42. The top and bottom view of the implemented LTCC package.

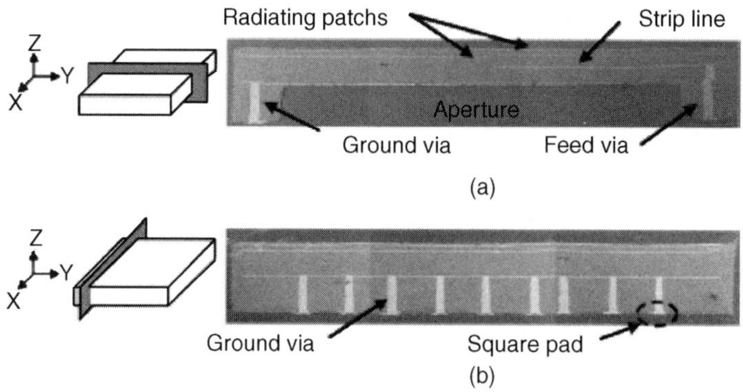

Figure 8.43. SEM image of the fabricated antenna: (a) cross-sectional view in the middle of the YZ plane, and (b) cross-sectional view of the ground vias in the XZ plane.

at the resonant frequencies of 5.280 and 5.450 GHz, respectively. The discrepancy in the resonant frequencies may be caused by fabrication tolerances (shrinkage differences), the finite metal thickness, or other loss mechanisms. The dimensions of the fabricated antenna differ slightly from the designed dimensions because of fabrication tolerances and the dicing process. The package size (length and width) is 0.05 mm, which is 0.1 mm smaller than that of the designed value. The height of the package is approximately 0.04 mm, which is 0.05 mm thicker than the designed height. In addition, the distance between the package and the system ground plane is approximately 0.03 mm because of the mounting process and because the SMA connector affects the antenna return loss characteristics. For these reasons, the simulation is

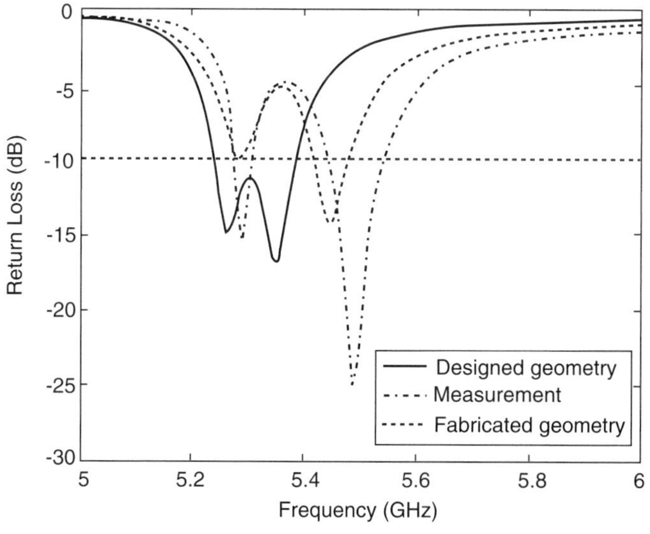

Figure 8.44. Comparison of the calculated and measured return loss characteristics.

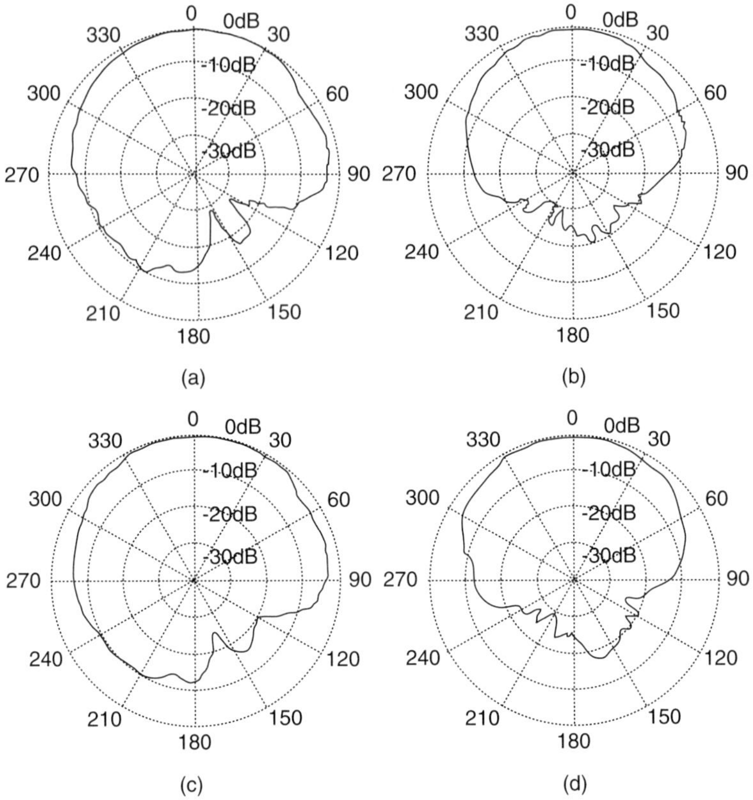

Figure 8.45. Measured radiation patterns: (a) YZ plane at 5.280 GHz, (b) XZ plane at 5.280 GHz, (c) YZ plane at 5.480 GHz, and (d) XZ plane at 5.480 GHz.

conducted with fabricated antenna dimensions attached to the SMA connector model, and the result is inserted into Figure 8.44. The simulated result shows that the two resonant frequencies are similar to the measurement results.

The measured radiation patterns at 5.280 GHz and 5.480 GHz are given in Figure 8.45. The E- and H-planes of the radiation patterns were measured in an anechoic chamber. It can be observed that the radiation patterns of the two resonant frequencies are nearly identical. Moreover, the proposed antenna reveals radiation characteristics nearly identical to the conventional patch antenna at each resonant frequency. The nonsymmetrical YZ patterns are caused by the SMA connector. The gain is measured by comparing the receiving power of the antenna under test (AUT) with a standard gain horn antenna. The measured gains of the proposed antennas are 0.08 dBi and 0.17 dBi at the first and second resonant frequencies, respectively. It is important to note that the small system ground plane, compact antenna dimension, and high dielectric constant material influence the reduced antenna gain.

CONCLUSION

The integration of passive components into a working system-on-package (SOP) allows the development of high performance integrated circuits (ICs) and signal processing. Antenna is one of the most important passive components in SOP technology, which places tremendous challenges on their performance characteristics while maintaining a compact size, such as broad impedance bandwidths, multi-frequency operations, and beam steering capabilities. This chapter provides the readers with some complexities and modeling issues associated with designing compact antennas for integrated transceivers and wireless applications. A number of approaches to the compact design and performance enhancement have been reviewed. Different design and modeling complexities are discussed. The design procedure for compact stacked patch antennas based on LTCC multilayer technology is described. The soft-and-hard surface (SHS) technology is proposed for surface-wave suppression and radiation-pattern improvement. A package-level integration of antenna is implemented on LTCC Technology. The advanced SOP technologies presented in this chapter may find a variety of applications in consumer markets, communications, military, and automotives.

REFERENCES

[1] R.R. Tummala and V.K. Madisetti, "System on chip or system on package," *IEEE Design & Test of Computers*, Vol. 16, No. 2, Apr-Jun 1999, pp. 48–56.

[2] S.K. Lim, "Physical design for 3D system on package," *IEEE Design & Test of Computers*, Vol. 22, No. 6, Nov-Dec 2005, pp. 532–539.

[3] R.R. Tummala, M. Swaminathan, M.M. Tentzeris, J. Laskar, G.-K. Chang S. Sitaraman, D. Keezer, D. Guidotti, Z. Huang, K. Lim, L. Wan, S.K. Bhattacharya, V. Sundaram, F. Liu, and P.M. Raj, "The SOP for miniaturized, mixed-signal computing, communication, and consumer systems of the next decade," *IEEE Transactions on Advanced Packaging*, Vol. 27, No. 2, May 2004, pp. 250–267.

[4] R.R. Tummala, "SOP: what is it and why? A new microsystem-integration technology paradigm-Moore's law for system integration of miniaturized convergent systems of the next decade," *IEEE Transactions on Advanced Packaging*, Vol. 27, No. 2, May 2004, pp. 241–249.

[5] R.R. Tummala and J. Laskar, "Gigabit wireless: System on package technology," *Proceedings of the IEEE*, Vol. 92, No. 2, Feb 2004, pp. 376–387.

[6] K. Lim, S. Pinel, M. Davis, A. Sutono, C. Lee, D. Heo, A. Obatoynbo, J. Laskar, M. Tentzeris, and R. Tummala, "RF-system-on-package (SOP) for wireless communications," *IEEE Microwave Magazine*, Vol. 3, No. 1, Mar 2002, pp. 88–99.

[7] R. Tummala, *Fundamentals of Microsystems Packaging*, The McGraw-Hill Companies, Inc., 2001.

[8] C. Balanis, *Antenna Theory: Analysis and Design*, John Wiley & Sons, Inc., 1997.

[9] R. Garg, P. Bhartia, I. Bahl, and A. Ittipiboon, *Microstrip Antenna Design Handbook* Artech House, Inc., 2002.

[10] S. Consolazio, K. Nguyen, D. Biscan, K. Vu, A. Ferek, and A. Ramos, "Low temperature cofired ceramic (LTCC) for wireless applications," *IEEE MTT-S Symposium on Technologies for Wireless Applications*, Feb 1999, pp. 201–205.

[11] L. Devlin, G. Pearson, and B. Hunt, "Low-cost RF and microwave components in LTCC," *Proceedings of MicroTech* Jan 2001, pp. 59–64.

[12] D.C. Thompson, O. Tantot, H. Jallageas, G.E. Ponchak, M.M. Tentzeris, and J. Papapolymerou, "Characterization of liquid crystal polymer (LCP) material and transmission lines on LCP substrates from 30 to 110 GHz," *IEEE Transactions on Microwave Theory and Techniques*, Vol. 52, No. 4, Apr 2004, pp. 1343–1352.

[13] K. Wong, *Compact and Broadband Microstrip Antennas*, John Wiley & Sons, Inc., 2002.

[14] W. Stutzman and G. Thiele, *Antenna Theory and Design*, John Wiley & Sons, Inc., 1998.

[15] I. Bahl and P. Bhartia, *Microstrip Antennas*, Artech House, Inc., 1980.

[16] S. Bokhari, J. Zurcher, J. Mosig, and F. Gardiol, "A small microstrip patch antenna with a convenient tuning option," *IEEE Transactions on Antennas and Propagation*, Vol. 44, No. 11, Nov 1996, pp. 1521–1528.

[17] S. Dey, R. Mittra, T. Kobayashi, M. Itoh, and S. Maeda, "Circular polarized meander patch antenna array," *IEEE Antennas and Propagation Society International Symposium*, Vol. 2, Jul 1996, pp. 1100–1103·

[18] K. Wong, J. Kuo, and T. Chiou, "Compact microstrip antennas with slots loaded in the ground plane," *IEEE Antennas and Propagation Society International Symposium*, Vol. 2, Jul 2001, pp. 732–735·

[19] H. Wang and M. Lancaster, "Aperture-coupled thin-film superconducting meander antennas," *IEEE Transactions on Antennas and Propagation*, Vol. 47, No. 5, May 1999, pp. 829–836·

[20] M. Vaughan, K. Hur, and R. Compton, "Improvement of microstrip patch antenna radiation patterns," *IEEE Transactions on Antennas and Propagation*, Vol. 42, No. 6, Jun 1994, pp. 882–885.

[21] R. Mittra and S. Dey, "Challenges in PCS design," *IEEE International Symposium, Antennas and Propagation Society*, Vol. 1, Jul 1999, pp. 544–547.

[22] http://www.fcc.gov/cgb/sar/.

[23] R. Waterhouse, S. Targonski, and D. Kokotoff, "Design and performance of small printed antennas," *IEEE Transactions on Antennas and Propagation*, Vol. 46, No. 11, Nov 1998, pp. 1629–1633.

[24] S. Targonski and R.B. Waterhouse, "An aperture coupled stacked patch antenna with 50% bandwidth," *IEEE Antennas and Propagation Society International Symposium*, Vol. 1, Jul 1996, pp. 18–21·

[25] P. Bhartia and I. Bahl, "A frequency agile microstrip antenna," *IEEE Antennas and Propagation Society International Symposium*, Vol. 20, May 1982, pp. 304–307·

[26] M. du Plessis and J. Cloete, "Tuning stubs for microstrip-patch antennas," *IEEE Antennas and Propagation Magazine*, Vol. 36, No. 6, Dec 1994.

[27] C. Huang, J. Wu, and K. Wong, "Cross-slot-coupled microstrip antenna and dielectric resonator antenna for circular polarization," *IEEE Transactions on Antennas and Propagation*, Vol. 47, No. 4, April 1999, pp. 605–609.

[28] H. Iwasaki, "A circularly polarized small-size microstrip antenna with a cross slot," *IEEE Transactions on Antennas and Propagation*, Vol. 44, No. 10, Oct 1996, pp. 1399–1401.

[29] P. Sharma and K. Gupta, "Optimized design of single feed circular polarized microstrip patch antennas," *Antennas and Propagation Society International Symposium*, Vol. 19, Jun 1981, pp. 19–22.

[30] K. Carver and J. Mink "Microstrip antenna technology," *IEEE Transactions on Antennas and Propagation*, Vol. 29, No. 1, Jan 1981, pp. 2–24.

[31] W. Richards, "Microstrip antennas," Chapter 10 in *Antenna Handbook: Theory, Applications and Design*, Van Nostrand Reinhold Co., 1988.

[32] Y. Hwang, Y. Zhang, G. Zheng, and T. Lo, "Planar inverted F antenna loaded with high permittivity material," *Electronic Letters*, Vol. 31, No. 20, Sep 1995, pp. 1710–1712·

[33] C. Huang, J. Wu, and K. Wong, "High-gain compact circularly polarised microstrip antenna," *Electronic Letters*, Vol. 34, No. 8, Apr 1998, pp. 712–713·

[34] L. Zaid, G. Kossiavas, J. Dauvignac, J.-Y. Cazajous, and A. Papiernik, "Dual- Frequency and broad-band antennas with stacked quarter wavelength elements," *IEEE Transactions on Antennas and Propagation*, Vol. 47, No. 4, Apr 1999, pp. 654–660·

[35] M. El Yazidi, M. Himdi, and J. Daniel, "Aperture coupled microstrip antenna for dual frequency operation," *Electronic Letters*, Vol. 29, No. 17, Aug 1993, pp. 1506–1508·

[36] K. Wong, and Y. Lin, "Small broadband rectangular microstrip antenna with chip-resistor loading," *Electronic Letters*, Vol. 33, No. 19, Sept 1997, pp. 1593–1594·

[37] E. Levine, G. Malamud, S. Shtrikman, and D. Treves, "A study of microstrip array antennas with the feed network," *IEEE Antennas and Propagation*, Vol. 37, No. 4, April 1989, pp. 426–434·

[38] J. James and P. Hall, *Handbook of Microstrip Antennas*, Volumes 1 and 2, Peter Peregrinus, 1989.

[39] R.L. Li, G. DeJean, M.M. Tentzeris, and J. Laskar, "Development and analysis of a folded shorted-patch antenna with reduced size," *IEEE Transactions on Antennas and Propagation*, Vol. 52, No. 4, Feb 2004, pp. 555–562·

[40] R. Li, G. DeJean, M.M. Tentzeris, J. Papapolymerou, and J. Laskar, "Radiation-pattern improvement of patch antennas on a large-size substrate using a compact soft-surface structure and its realization on LTCC multilayer technology," *IEEE Transactions on Antennas and Propagation*, Vol. 53, No. 1, Jan 2005, pp. 200–208.

[41] G. Ruvio, P.-S. Kildal, and S. Maci, "Modal propagation in ideal soft and hard waveguides," *Proceedings of IEEE Antenna and Propagation Society International Symposium*, Vol. 4, June 2003, pp. 438–441.

[42] S.-H. Wi, Y.-B. Sun, I.-S. Song, S.-H. Choa, Il-S. Koh, Y.-S. Lee, and J.-G. Yook, "Package-level integrated antennas based on LTCC technology," *IEEE Transactions of Antennas and Propagation*, Vol. 54, No. 8, Aug 2006, pp. 2190–2197.

CHAPTER 9

Simulation and Characterization of Integrated Microsystems

INTRODUCTION

Computer-aided analysis is powerful and necessary while dealing with complex circuit blocks and integrated systems. It is important to interpret the results of computer simulations correctly in order to achieve the design target in fewer steps. At the same time, characterization of mixed-signal communication systems provides meaningful insights into the design and performance validation. In this chapter, we will illustrate the various methods of computer-aided analysis and their interpretations. We will also provide fundamental considerations toward various characterization methods and standards. These considerations are applicable to active and passive devices as well as to integrated circuits and communication systems. The general mindset of silicon-based designs can be set aside in this chapter, and the building blocks, however complicated, may as well be treated as a "black box." Although from a functional and specification perspectives accurate simulators provide design confidence and design robustness, laboratory measurements provide meaningful insights (and often more learning compared with the design phase). In the design phase itself, the circuit needs to be analyzed in the same way it is to be measured in the laboratory.

9.1 COMPUTER-AIDED ANALYSIS OF WIRELESS SYSTEMS

In recent years, semiconductor technology platforms have advanced aggressively to reduce the form factor and power dissipation of communication system solutions.

Advanced Integrated Communication Microsystems, edited by Joy Laskar, Sudipto Chakraborty, Manos Tentzeris, Franklin Bien and Anh-Vu Pham

To represent these transistor devices accurately, multiple parameters are required in the model, and it becomes extremely difficult to perform a "hand calculation" of integrated circuits, however simple their topology may be. At the same time, one cannot simply apply heuristics and obtain a performance optimization of multiple devices in a circuit without an intuitive design approach. In addition, we must be able to understand the interpretation of these analyses quickly and to obtain the right design direction. This is performed iteratively, and in this section, we illustrate a few details of such considerations.

9.1.1 Operating Point Analysis

The DC operating point is the first step to observe in order to design/analyze any integrated circuit. Performances of circuits operating under small-signal approximation (such as low noise amplifier (LNA), radio frequency (RF) input port of the mixer, low-frequency amplifier, etc.) depend heavily on the DC operating point, as the input signals "ride" on top of the DC operating point. This is typical for "small signal" circuits where the AC or transient signal excursion is 10% or less than the DC operating point. However, in the case of many large signal circuits, the DC operating point merely sets the operating conditions, whereas the large signal excursions determine the circuit performance. This is typical of voltage-controlled oscillators (VCOs), antiparallel diode pair-based passive mixers, field-effect transistor (FET) ring mixers, digital circuits (with the exception that they consume current only at the instant of switching), and so on. Many circuits, including power amplifiers, are operated in switching modes to achieve high efficiency. While analyzing for DC, the following parameters should be observed:

1. Transconductance g_m determines the voltage-to-current transfer ratio of amplifiers. This is dependent on the bias current, voltages, and the device geometries.
2. Output conductance g_{ds} determines the output impedance of an active element. Current reference circuits and amplifiers usually require a low value of g_{ds} (high impedance) compared with the load conductance $1/Z_L$.
3. Bias current I_{DC} determines the operating point of transistors. In the case of hard-switched transistors, such as current-mode logic circuits, the voltage swing is given by the product of I_{DC} and by the load impedance Z_L. This is usually the case for large signal circuits, such as VCO, switched-mode power amplifiers (PAs), frequency dividers, and so on.
4. Voltage headroom: In terms of voltage headroom, both gate-to-source and drain-to-source voltages are important design parameters. The overdrive voltage $V_{OV} = V_{GS} - V_t$ is responsible for noise, linearity, and matching. A low value of V_{OV} indicates a low input-referred noise, whereas a high V_{OV} is preferred for device matching and to improve linearity. Hence, in an LNA design, noise and linearity trade-off at a given bias current provides a design challenge. Usually, the degeneration impedance helps to obtain

simultaneous noise and input match, while improving the linearity performance.

In the case of switching transistors, such as passive FET ring mixers, or any hard-switched state, the overdrive voltage is set to zero or slightly negative to provide maximum utilization of a driving signal.

The drain-to-source headroom is obtained by the difference between the drain and source voltages ($V_{DS} = V_D - V_S$). To ensure that a transistor operates in the saturation region, we require $V_{DS} > V_{D,SAT}$. In the case of large signal circuits, both drain and source would swing large, and this condition must be ensured in the steady-state time domain operation. If this is not maintained, then the transistor would come out of saturation and provide nonlinearity, as well as reduce the speed of the circuit. In the case of switching transistors, $V_{DS} \sim 2V_{D,SAT}$ is a good starting point. In voltage/current reference circuits, high $V_{D,SAT}$ is desired for better device matching, and a larger length can be used to obtain lower drain-referred flicker noise as well.

5. Current headroom: Similar to the above consideration, when the current excursion is large compared with the DC bias current, the individual branch may "run out of current" and cause nonlinearity. Hence, while biasing the circuits, it must be ensured that the branch DC currents are higher than the transient peak value of the current waveform. However, the bias current should be optimized to provide the desired linearity.
6. ON resistance and OFF capacitance: In the case of a transistor being used as a switch, the ON resistance and OFF capacitance metrics are important. In reality, these switches contribute to the nonlinrearity of the system, which is governed by the voltage variation of the OFF capacitance and the current variation of the ON resistance. These are obtained by DC operating points for small-signal circuits.
7. Common-mode voltages: At fairly low frequencies (below 100 MHz), on-chip capacitors to AC couple functional blocks prove to be area inefficient and impractical. Hence, these stages are directly coupled to one another. Hence, the common-mode input and output ranges play a critical role in determining the proper compliance in their DC input–output voltages. This is another reason for using complementary stages (NMOS input stage followed by PMOS input stage, etc.) for a cascade of baseband amplifiers and so on.

   In reality, the boundary of "small" signal versus "large signal" is just a conceptual one. An LNA much farther from the base station would be operating under the limits of a small signal operation, and it should provide maximum gain in order to provide the highest possible sensitivity. However, when it operates near the base station, the signal excursions are significantly higher, and the circuit operates under the limits of a large signal. In this case, reliability of the devices (LNA and ESD, etc.), as well as nonlinear terms resulting from the large signal, prove to be more critical. Large signal excursions can be conceptually thought of as traversing through multiple DC operating points within the limits of device time constants.

9.1.2 Impedance Matching

To evaluate the maximum power transfer in the receiver/transmitter front end, scattering parameters (S-parameter) simulation w.r.t. reference source impedance is performed. Inside the chip, impedance matching is often not used, whereas at the interfacing points to off-chip components such as filters and duplexers, we use an impedance match. Performances of the off-chip components can vary significantly depending on the matching impedance. In a "mostly monolithic" integrated system, matching performances are critical at the LNA input and PA driver output. If an on-chip switch is used (in TDD systems), the matching should be obtained at the switch output. The matching network provides voltage gain, which leads to a lower noise figure of the LNA (or a lower current consumption in the LNA as well).

S-parameter analysis is essentially a small signal analysis, and it must not be employed in circuits where frequency conversion with respect to a large signal is performed (i.e., it is meaningless to calculate conversion gain of mixers by simply performing S-parameter analysis form the RF to IF port). In many large signal circuits (e.g., APDP mixers and FET ring mixers), the input impedance is a strong function of the drive signal (obtained from another source). At the same time, these impedances may vary significantly between their ON and OFF states. Hence, the DC biasing and any large signal stimuli should be present while characterizing the input matching of the circuit.

S-parameter simulation indicates how closely the circuit's input or output impedance matches with a standard impedance (50 or 75 Ω). Not only do we observe the S-parameter in decibel scale, but also we note the actual output impedance of the circuit, which provides better physical understanding of the circuit. A 10-dB match indicates a 90% power transfer from the test equipment to the circuit, and a 20-dB match indicates a 99% power transfer from the test equipment to the circuit.

S-parameter performance trades off with the noise figure in the case of a receiver, and with output power in the case of a transmitter. Also, while obtaining a desired matching performance, care should be taken such that any unwanted stability degradations of the circuit are avoided. Q factors of matching network components indicate the sensitivity of input matching to component mismatches. In wideband systems, the design considerations may be complicated.

9.1.3 Tuning at Resonance

In their basic nature, RF architectures employ tuning and frequency selectivity. This is obtained by $L-C$ resonance circuits. The Q factor of the $L-C$ circuit along with the inductance and bias current determines the gain and center frequencies, whereas the loaded Q determines the frequency selectivity. It must be ensured that the tank is tuned to the desired frequency. This is usually obtained by small signal simulation of the $L-C$ resonance circuits. This is illustrated in Figure 9.1.

The solid curve at the LNA/mixer interface is the desired frequency response. The dashed curve is the result of the AC simulation when the LNA and mixer interface is not

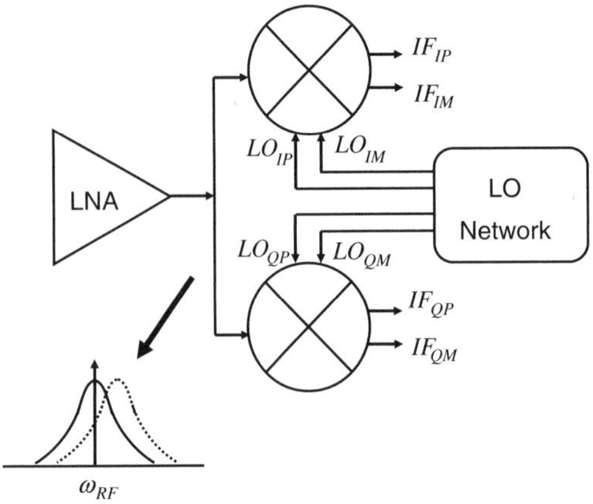

Figure 9.1. Resonance and peak considerations.

tuned at the desired RF frequency ω_{RF}. This situation becomes worse for a high Q interface, as the frequency response exhibits a sharp peak at a different frequency, and falls rapidly in both sides. Because of the process variation of components in an integrated circuit (IC), very high Q circuits are often avoided. As is evident from the figure, if one performs a transient simulation prior to doing an AC simulation, the voltage gain might appear to be small and this might give rise to a false indication that the current consumption of the circuit needs to be increased for higher voltage gain, while it is indeed the tuning at the LNA/mixer interface, which needs to be optimized to obtain the maximum available gain for a given amount of power consumption. This is the reason why a small signal AC simulation must be performed before doing a transient simulation.

9.1.4 Transient Analysis

Transient (or time domain) simulation provides detailed insight into the actual large signal behavior of any circuit and system. This is used for the following purposes:

1. Signal excursions in circuit nodes should be observed to ensure enough voltage and current headroom.
2. This is the most suitable analysis technique for circuits and systems with multiple, widely varied time constants. This is typical for receivers, transmitters, and the signal generation path (PLL). One cycle of RF frequency is of the order of nS, whereas the same for baseband frequency is of the order of mS.
3. Any concerns of reliability in the circuit operation under worst-case circumstances. Large voltage swings in deep submicron transistors may pose significant hot carrier degradations, and so on. This is prevalent in the case of PA drivers, and so on.

4. Any stability issues, common/differential-mode oscillations, VCO startup, and so on. These are well captured using transient simulations. Small signal stability analyses using a phase margin or Middle-Brooke test may not lead to accurate results in the case of a large signal circuit operation.
5. A large signal 1-dB compression performance of circuits. Transient simulation can analyze the circuit under the real compression region. Transient simulation can accurately capture the effects of envelope variation in the case of modulated signals and systems.
6. Large signal interaction between common-mode and differential-mode signals. This is extremely critical in modern RF systems. Often a conventional current source may provide headroom and signal swing issues in large signal circuits (in the case of PA drivers, VCOs, etc.). The complicated interplay of common-mode to differential-mode conversion may lead to unwanted oscillations. Although in our previous discussions we have been paying close attention to all the differential-mode operations, common-mode performance is quite critical.

Transient simulation should be carried out to find out the voltage gain of the integrated receiver front end. This is especially useful to observe the effects of any harmonics resulting from the front end, which are not captured by the small signal evaluation, as described before. As the receiver front ends provide differential outputs, the voltage gain should also be observed by taking the difference between the output signals, in both in-phase and quadrature. An I/Q imbalance can also be observed by transient simulation. However, in the simulation, we do not usually observe I/Q imbalance, unless we deliberately use some nonidealities. The source of I/Q imbalance includes layout asymmetry and any mismatch in amplitude and phase from the circuits at very high frequency.

Important insight can be obtained by performing a discrete Fourier transform (DFT) of various time domain signals (branch currents and node voltages) and can be interpreted in terms of intermodulations, and so on. However, setting convergence tolerance for transient simulation is important, especially in cases of high-dynamic-range circuits. A time step used in transient simulation indicates the simulation noise floor due to truncation, and should be reduced by performing the transient analysis at a fixed, predefined timestep. The simulation may be much slower, and memory intensive, but it will lead high accuracy result.

9.1.5 Noise Analysis

In integrated transceivers, mostly two types of noise generation mechanisms are prevalent: (1) linear noise, and (2) nonlinear noise. In the case of linear noise, no frequency translation occurs, and this is the situation with amplifiers and other single-frequency circuits. Nonlinear noise is important whenever frequency translation occurs, which is inherent to a up/downconversion process in mixers and VCOs. In the frequency translation systems, noise is down/upconverted from various sidebands.

Device noise is fundamentally of broadband or flicker in nature. The spectral profile of output noise is dependent on the specific circuit functionality and transfer functions from various noise sources to the output.

Linear noise is usually simpler to analyze. Noise is analyzed similar to AC analysis in the most part. Various uncorrelated noise sources resulting at the circuit output by means of circuit transfer function are added in terms of noise power, and the noise power is divided by the gain of the circuit to obtain input-referred noise.

Nonlinear noise is always analyzed after the circuit stabilizes, and the periodicity is obtained at the desired node of analysis. It is analyzed in different ways in the transmitter and the receiver. The analysis mechanism in the transmitter occurs under the assumption of large signal operating conditions, and in the receiver, it is obtained under a small signal operating condition. In both cases, the frequency of interest is chosen (sideband), and at a frequency offset from that sideband, all the noise contributions are obtained. In any transceiver system, we obtain degradations caused by broadband noise and phase noise. Broadband or "thermal" noise is associated with the white noise generated by the circuit and is dependent on the g_m of the devices. However, phase noise is associated with the switching of the transistors and is associated with the slope of the switching waveform at zero crossings, as well as with the bias condition of the transistors (bias current and overdrive). In the case of the receivers, the broadband noise figure implies signal-to-noise (SNR) degradation, whereas the phase noise in the case of the transmitters determines the spectral leakage to the other adjacent frequency bands.

For a direct conversion receiver, the output noise spectral density should be integrated over the entire bandwidth of interest, and then averaged out. For a low intermediate frequency (IF) receiver front end, the spot noise (output spectral density at a particular frequency offset) should be observed. Noise can be hand-calculated using basic circuit configurations. Assuming a reasonable input match level, when the circuit is fully noiseless, the input source resistor (of 50 Ω) is the only contributor of noise and provides a voltage spectral density of 0.9 $\text{nV}/\sqrt{Hz}$. This is divided equally, and at the input of the circuit, we would have 0.45 $\text{nV}/\sqrt{Hz}$. After the noise simulation, the input-referred noise can be calculated by referring the output-referred noise to the input by dividing it by the voltage gain. The input-referred noise can be formulated as $V_{n,in} = \frac{V_{n,out}}{A_v}$. The noise figure can be formulated as $NF = 20\log(\frac{V_{n,in}}{0.45})$. The noise figure obtained by such an equation gives a double sideband (DSB) noise figure estimate for the receiver front end. Also, while integrating the output noise spectral density, one must integrate it over the required message bandwidth of a channel under consideration.

9.1.6 Linearity Analysis

As illustrated before, 1-dB compression and third-order intercept points are used to obtain linearity performance. The third-order intercept point provides a conceptual picture of linearity, but the 1-dB compression point provides the actual power level where the compression occurs. At the compression point, signal energy from the fundamental component gets distributed to various harmonics.

Linearity simulations must be performed using input power sources in the case of matched terminals (input of receiver or output of transmitter). The need for input matching requirements can be emphasized once more, as the interpretation on the power levels is dependent on the matching levels. Once again, we analyze the circuit until circuit periodicity is reached, and then we obtain the relative magnitudes of the fundamental and the intermodulation tones.

Let us consider a simple architecture using an LNA and a mixer, where both generate intermodulation terms, as illustrated in Figure 9.2. These intermodulation terms add/subtract depending on the relative phase relationship. Usually intermodulations are obtained in transconductance stages (V–I converter), assuming the current-mode switching to be linear in nature. All the simulations on linearity must be performed at the linear region of the receiver. A typical power level is −50 dBm for these simulations, as it falls well in the dynamic range of the receivers (usually starts from ∼−100 dBm and goes up to −20 dBm). Three different linearity simulations are important: *IIP*3 (input third-order intercept point), input *P*1*dB* (1-dB compression point), and *IIP*2 (input second-order intercept point). All of these parameters should be obtained separately in the simulation. Because of the differential nature of the circuits used in modern front ends, the even-order nonlinearity terms (*IIP*2, *IIP*4, etc.) are expected to be much higher, and in a simulation, they may not be observed without introducing some deliberate mismatches (3–5% mismatch and mismatch applied to one component only). In the specific cases, the even-order intermodulation terms generated from the LNA are not important for our consideration, as they are shifted up in frequency by the mixer, and subsequently filtered out by the frequency response of the baseband filter. In the linear region of the circuit, the fundamental and the *n*th-order intermodulation (*IIPn*) tones at the output usually follow a slope of 1 and *n*,

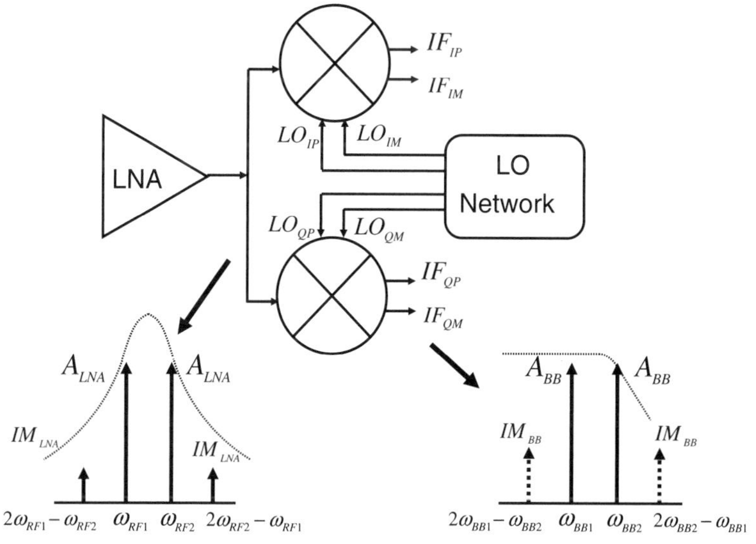

Figure 9.2. Interpreting intermodulation terms.

respectively, when plotted in logarithmic scale w.r.t. the input power. However, the 3× slope of the IM3 products w.r.t. fundamental is not true for all circuits, and one must be very careful while interpreting the results.

In the case of high dynamic range blocks, a fixed timestep can be used to ensure accurate simulation results. The choice of individual frequencies is important in determining the simulation speed. The entire system is simulated w.r.t. one cycle of the beat frequency, and the smaller the beat frequency, the time duration of the simulation increases. A higher beat frequency leads to faster simulation and analysis time, but one needs to be careful in the following considerations:

1. Components of odd-order intermodulation should not overlap the even-order intermodulation terms. Considering a choice of $\omega_{RF} + 30M$ and $\omega_{RF} + 40M$ as fundamental tones would lead to third-order intermodulation terms at (20M, 30M, 40M, 50M), and the fifth-order intermodulation term would be located at 10M, which is also the second-order intermodulation term. Hence, our interpretation of the *IIP*2 number would be incorrect. Even in a perfectly matched circuit, *IIP*2 may appear much worse because of this superimposition effect.
2. The baseband bandwidth and resonant tank circuit would provide finite rejection to the intermodulation terms because of their frequency-dependent nature of impedance. (Baseband stages are represented by $R - C$ and tank circuits are represented by $L - C$). To obtain an accurate estimate of the intermodulation tones, the differential output current may be observed. A broadband load circuit can be used (with the same gain as the real circuit) to accurately estimate intermodulation caused by the active circuitry. Figure 9.3 illustrates an illustration of a two-tone frequency separation and the impact on intermodulation performance.

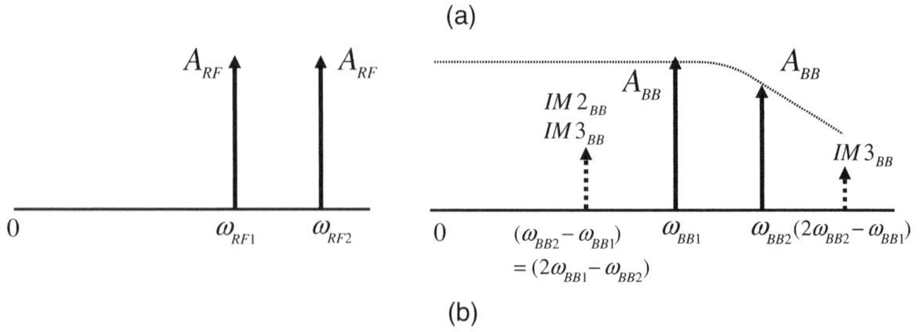

Figure 9.3. Frequency separation and intermodulation.

9.1.7 Parasitic Elements

Parasitc elements (capacitor, inductor, resistor) are very important to RF performance. After performing the layout of the entire receiver front end, it is necessary to find out the parasitic capacitances and resistances at various circuit node points and interconnect traces. Because of their reactive nature, the parasitic inductor and capacitor can be tuned out but not the parasitic resistors. Parasitic resistors affect the circuit performance by introducing noise. At extremely high frequencies (of the order of tens of gigahertz), these parasitics can be used for matching purposes as well. Parasitic elements depend on geometry and can be well calculated depending on the 3D nature of actual geometries. However, most extraction tools extract these capacitors as simple capacitors or as "capacitors with infinite Q factor". This leads to gross simulation inaccuracy. Substrate resistors should always be associated with parasitic capacitor networks. Simulations performed with extracted parasitics deviate from the circuit performance w.r.t. matching, noise figure, and gain (to a very little extent). Parasitic inductors may also lead to unwanted oscillations at high frequencies.

9.1.8 Process Variation

At highly scaled geometries (180 nm and beyond), because of variations in lithography, the circuit geometries become highly prone to fabrication inaccuracies, which leads to process variation. Hence, any design at these nodes must consider a process spread over which the parameters would vary, in order to maintain design robustness. Typical parameters include the oxide thickness, t_{OX}, V_t, and so on. In the case of a digital switching gate, these parameters lead to the variation in the $R-C$ time constant, which results in "slow" and "fast" corners. In practice, the process corners are statistically distributed within these limits.

Process variation is also modeled to ensure the design robustness of the circuit at various foundries. In a specific foundry, the essential quantities such as doping profile can be well controlled. However, these quantities will vary from one foundry to the other, and when such process robustness is ensured, it saves significant time to market for commercial products.

9.2 MEASUREMENT EQUIPMENTS AND THEIR OPERATION

In this subsection, we illustrate the operating principles of test equipments used in the characterization phase. The device under test may be a single transistor or a complicated integrated transceiver. Many measurements are performed to obtain an accurate device model.

9.2.1 DC/Operating Point

Device modeling starts with DC characterization. DC characterization is performed using a source-monitor unit (SMU), which allows the characterization to be performed

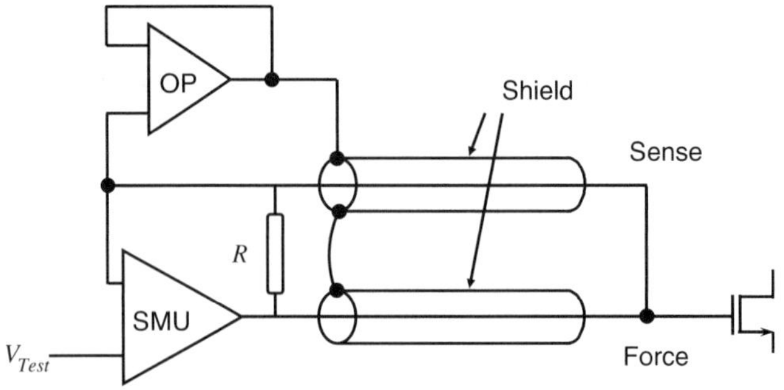

Figure 9.4. A DC measurement technique.

over a wide current and voltage range over all quadrants. Figure 9.4 illustrates the DC characterization method using a four-wire configuration using "force" and "sense" lines. The OP-Amp used in the voltage follower mode eliminates ohmic losses in the "force" lines. However, it should be noted that the DC signal lines experience capacitance to the "shield," which would limit the speed of measurement. This problem is solved using a double-shielded method as shown in Figure 9.4. It is desired that the force and sense connections are made close to the device under test.

9.2.2 C–V Measurement

Various capacitors associated with a transistor must be obtained to model its behavior accurately. This is obtained using a capacitance meter, which can use one of the two methods to provide the capacitance value. These methods are illustrated in Figure 9.5.

The first method, known as the "$I - V$" method, is a reference current injected into a network consisting of a reference impedance and the device under test in

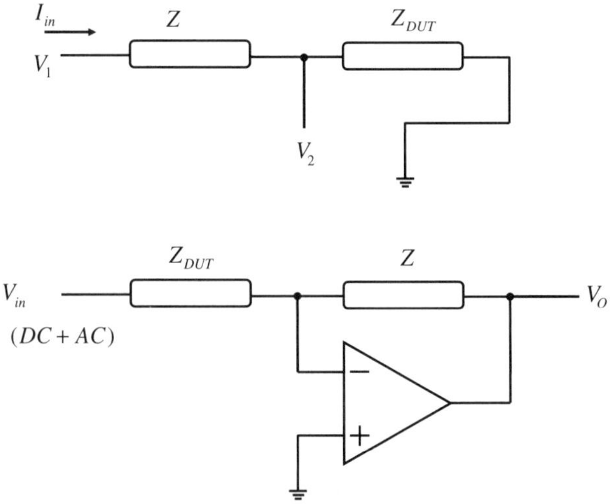

Figure 9.5. An impedance measurement using the autobalancing method.

series. Z_{DUT} is calculated w.r.t. the ratios of voltage drop across the reference impedance and the impedance under test, and it is given by

$$Z_{DUT} = (\frac{V_1}{V_2} - 1) \times Z \tag{9.1}$$

This method can easily measure structures connected to the ground, and in practice, a transformer is used in place of a low-valued resistor. This limits the usable frequency ranges over which the measurement can be performed.

In the second method, called the "autobalancing" method, input current flows through the reference resistance and device, connected at the virtual ground of the OP-Amp. The measurement range can be easily changed by adjusting the value of reference impedance Z. A practical implementation includes a zero detector and a modulator, and care must be taken to ensure that the OP-Amp is not saturated by a low-impedance device (diodes and so on under a forward bias condition).

$$Z_{DUT} = -\frac{V_{in}}{V_o} \times Z \tag{9.2}$$

Where the voltages in these equations represent a "complex" voltage, it is determined by magnitude and phase.

9.2.3 Vector Network Analyzer and S-Parameter Measurements

Figure 9.6 illustrates the block diagram of a test system to measure the S-parameter. A vector network analyzer measures the complex power transmission at the two ports. The device under test (DUT) is connected between the two ports.

The RF synthesizer provides a signal to the S-parameter test set and is applied first to port 1 and then to port 2, and finally the complex power signal is measured. The block diagram of a simple four-sampler, two-port vector network analyzer is shown in Figure 9.7, which is used for S-parameter measurements.

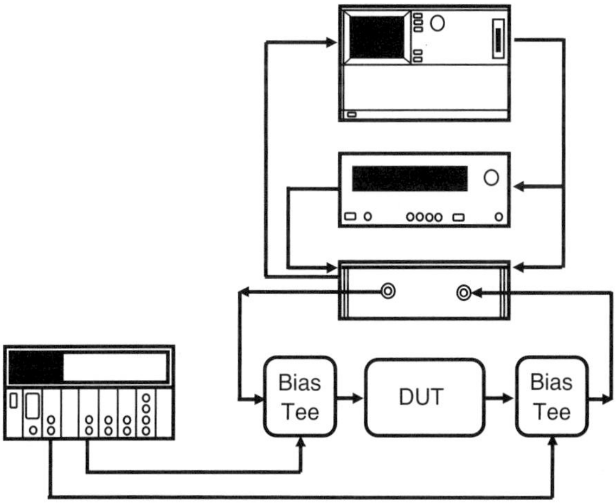

Figure 9.6. A measurement using VNA.

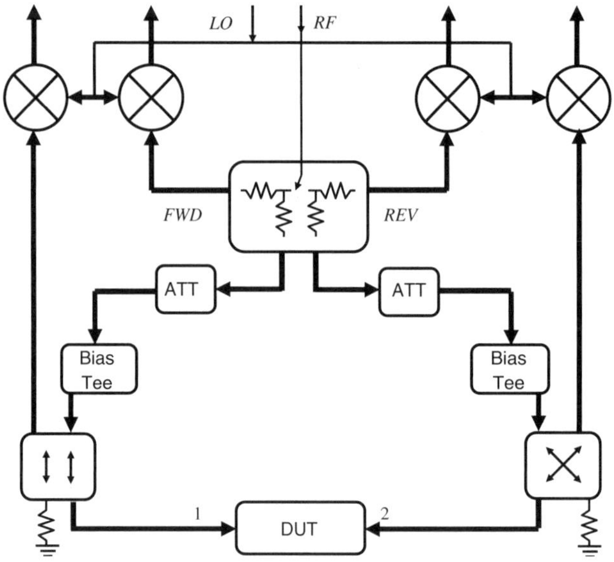

Figure 9.7. A VNA block diagram.

The reflected and transmitted complex power (magnitude and phase) is measured by changing the position of the switch. The four downconverted IF signals are used by the back-end signal processor to provide S-parameters. This simple technique is widely used for frequencies $\sim$26 GHz. Figure 9.7 illustrates an integrated measurement setup to characterize the S-parameter for a device under bias. The bias is applied through a bias Tee, which provides the sum of the DC bias and the AC signal from the S-parameter test set.

9.2.4 Spectrum Analyzer (SA)

A spectrum analyzer is used to analyze the frequency content of an RF signal under consideration. It represents the frequency contents of a time-varying signal. As the time and frequency domain representations are the dual of one another, a simple sinusoidal waveform provides a delta function in the frequency domain and rotates as the time domain signal changes amplitude and phase. Because of various modulations on RF signals, spectrum analysis provides an easy and convenient representation method.

Figure 9.8 illustrates a simple functional block diagram of a spectrum analyzer. The spectrum analyzer system is a superheterodyne AM receiver. It can be used to measure intermodulation distortion, noise, and to spectrum analysis. Fundamentally it is of two types: (1) swept tuned and (2) Fourier analysis. In a Fourier analysis, the spectrum analyzer samples the time domain signal and displays its fast fourier transform (FFT). Thus, they can be used to display frequency and phase information. A swept tune analyzer is commonly used, and it sweeps across a frequency range and displays the amplitude of the frequency components. It cannot display the phase information.

In a swept tune analyzer, as illustrated in Figure 9.8, the input frequency is swept using an upconversion mixer, and the filter selects the amplitude of the

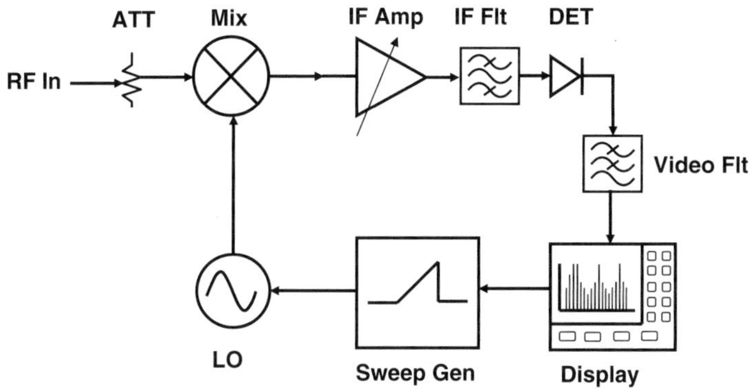

Figure 9.8. A spectrum analyzer block diagram.

upconverted frequency. An input attenuator is used to attenuate the input signal to a desired range to set its operation at optimum dynamic range. Local oscillator (LO) frequency is selected using a sweep generator. It controls the horizontal display of the spectrum analyzer (frequency axis). The LO operates under the control signal from the sweep generator. The nonlinearity of the mixer creates various harmonics at its output. The IF filter (bandpass) is centered at a fixed frequency, and it provides finite rejection to the out-of-band signals. Bandwidth of the IF filter is set by the "resolution bandwidth" parameter. After the IF filter, the diode detector is used to detect the downconverted signal amplitude. Resolution bandwidth denotes the 3-dB bandwidth of the IF filter. The IF gain adjusts the input level of the input signal, and it controls the level of the displayed signal. The video filter averages the output signals, and smoothens out the display, and it helps observe the low-level signals, which are located close to the noise floor. The bandwidth is adjusted by the video bandwidth (VBW).

LO feedthrough and phase noise provide unwanted components around DC. Instead of a sharp response at the output, they exhibit the frequency-dependent shape of the IF filter.

The performance of a spectrum analyzer is dependent on (1) the frequency range of the sweep, (2) the resolution of two closely spaced frequencies, (3) the detection of in-band distortion terms, and (4) the ability to detect very low-level signals. Often the harmonics of fundamental tone needs to be measured, in addition to the low frequency tones. To resolve closely spaced signals, filters with a smaller bandwidth should be preferred. The resolution bandwidth quantifies the amount of frequency by which two equal amplitude signals can be displayed on the screen with a 3-dB dip in between them. Video bandwidth is usually kept 10 times smaller compared with the resolution bandwidth. Noise sideband and residual FM (short-time LO instability) effects impact the resolution of the analyzer. Video bandwidth determines the display of a low-level signal and separates it from the noise floor. It does not impact the noise floor of the spectrum analyzer.

A narrow resolution bandwidth leads to an increased response time from the IF filter, and the sweep time of the filter increases. The sweep time is given as $T_{sw} = \frac{SP}{RBW^2}$.

This leads to a quadratic reduction of the sweep time with an increase in the resolution bandwidth.

Like any communication system, spectrum analyzers provide internal distortions, and the mixer is the prime component of harmonic and intermodulation distortions. The level of input signal is limited by these considerations. Typically the desired levels are 50 dBc for the third harmonic and the third-order intermodulations. The second harmonic distortion can be 40 dB below the fundamental tone. The internal noise of the spectrum analyzer is characterized by DANL (displayed average noise level), and it determines the lowest possible signal level that can be detected. Typically, the spectrum analyzer dynamic range varies between 90 and 145 dB. The noise level is dependent on the resolution bandwidth, and if RBW increases by factor of 10, the noise floor increases by 10 dB. The input attenuator causes SNR degradations, and SNR degrades with increased attenuation. The analyzer's dynamic range is also limited by the phase noise of the LO generator.

9.3 NETWORK ANALYZER CALIBRATION

9.3.1 Overview of Network Analyzer Calibration

When obtaining measurements, calibration must be performed to remove systematic errors, or to shift the measurement reference planes to a DUT. In other words, calibration is the process of characterizing systematic errors or determining error terms of a linear model that mathematically represents the physical connections (cables, connectors, etc.) between a network analyzer and a DUT. The errors are then used in calculation to remove the effects of physical connections in order to shift measurement reference planes. The error terms are determined by a set of measurements of standards or known devices (short, open, load, and thru). From these sets of measurements, a network analyzer will then determine the error terms from a set of linear equations. Let's take a look at an example of calibrating a one-port network analyzer to understand the concept of calibration.

Example 9–1

Develop equations to extract error terms for a one-port calibration model. The linear model is predetermined as shown in Figure 9.9.

Solution

Figure 9.9 illustrates the error model for a one-port calibration. The error terms e_{00}, e_{10}, e_{01}, and e_{11} will be determined during calibration. These error terms represent the responses of connectors, adapters, and cables connecting between a DUT and a network analyzer. If these error terms are known, then the one-port S-parameters of a DUT is given as follows:

$$S_{11} = \Gamma_D = \frac{\Gamma_M - e_{00}}{e_{11}(\Gamma_M - e_{00}) + (e_{10}e_{01})} \tag{9.3}$$

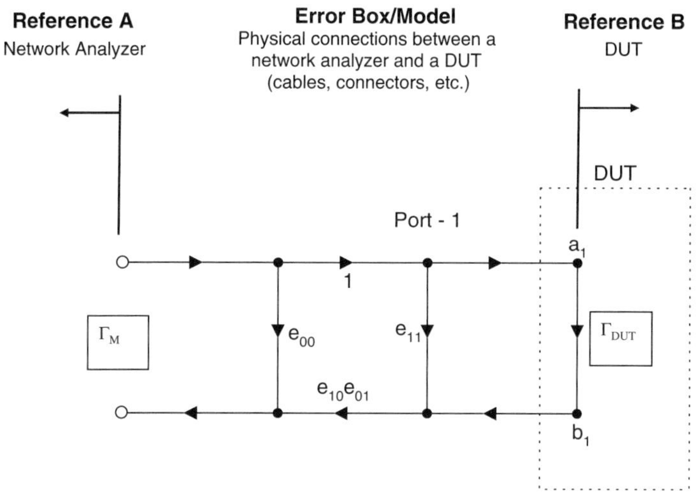

Figure 9.9. An error box for one-port calibration.

where Γ_M is the measured reflection of coefficient at the reference plane A. Γ_M includes the responses of the physical connection (cables, adapters, and connectors) and the DUT. Γ_D is the desired reflection coefficient of the DUT that users would like to obtain. The terms e_{00} (directivity), $e_{10}e_{01}$ (frequency response), and e_{11} (port match) are the error terms of the linear model that represent the physcial connection. Calibration is the process of determining these error terms so that the responses of the DUT can be obtained using Eq. (9.3). Equation (9.3) is transparent in network analyzer measurements. After calibraiton, users will get the measured responses of a DUT at a calibrated reference plane.

To dertermine the error terms e_{00}, e_{10}, e_{01}, and $e_{11,}$ a set of linear equations and known conditions are required. As shown in Eq. (9.3), the reflection coefficient of a one-port measurement only requires known error terms e_{00}, e_{11}, and $e_{10}e_{01}$. The error terms e_{10} and e_{01} are not required to be known separately, but their product is required. Therefore, the number of unknowns is reduced to three and requires three linear equations for a full solution. From a mathematical point of view, one can understand why one-port calibration only needs measurements of a short, an open, and a load. Let's see how these measurements can assist in determining the error terms.

Let's set

$$\begin{aligned} a &= e_{10}e_{01} - e_{00}e_{11} \\ b &= e_{00} \\ c &= -e_{11} \end{aligned} \tag{9.4}$$

Substituting a, b, and c into Eq. (9.3) and rearranging Eq. (9.3) yields

$$\Gamma_D a + b - \Gamma_D \Gamma_M c = \Gamma_M \tag{9.5}$$

There are three unknowns in Eq. (9.5), which require three measurements to solve for a, b, and c. Each measurement will establish one condition. For three measurements (short, open, and load), we will obtain three equations, as follows:

$$\begin{aligned}\Gamma_{D1}a+b-\Gamma_{D1}\Gamma_{M1}c &= \Gamma_{M1}\\ \Gamma_{D2}a+b-\Gamma_{D2}\Gamma_{M2}c &= \Gamma_{M2}\\ \Gamma_{D3}a+b-\Gamma_{D3}\Gamma_{M3}c &= \Gamma_{M3}\end{aligned} \tag{9.6}$$

The terms a, b, and c are unknowns. Γ_{M1}, Γ_{M2}, and Γ_{M3} are the measured quantities each time the network analyzer is connected to a short, open, and load. Again, Γ_{M1}, Γ_{M2}, and Γ_{M3} will include the responses of each standards (short, open, and load) and the physical connection. Γ_{A1}, Γ_{A2}, and Γ_{A3} are the mathematically defined reflection coefficients of the short, open, and load. Bascially, Γ_{A1}, Γ_{A2}, and Γ_{A3} are 1, $1\angle 108°$, and 0 for a short, open, and load, respectively. Hence, Eq. (9.6) has three unknowns to solve.

The network analyzer will use the information preentered to define the standards in Eq. (9.6). The information entered into the network analyzer to define these standards is called a calibration kit. As one can see, it is crtical that the standards are accurately defined and measured. The network analyzer will heavily rely on the measurements of these standards and their definitions to solve for Eq. (9.6) to obtain the error terms. Any errors in standards will result in uncertainties in the measurements. After calibration, e_{00}, $e_{10}e_{01}$, and e_{11} are determined and stored in the network analyzer for each setup with its own cables and connectors. For each measurement after calibration, the network analyzer will rely on Eq. (9.6) to provide users with the S-parameters defined at the measurement reference planes.

9.3.2 Types of Calibration

As shown in the example above, a variety of models can be developed to calibrate the network analyzer. Depending on the model, different standards will be required to determine error terms sufficiently. Before we discuss the accuracy of calibration with respect to standards, we will review existing conventional calibration techniques and their evolution. Basically, there are two widely used types of calibration, namely short-open-thru-load (SOLT) and thru-reflect-line (TRL). The TRL calibration has a couple of its derivatives including thru-reflect-match (LRM) and thru-reflect-reflect-match (LRRM).

9.3.3 SOLT Calibration

The SOLT technique was one of the initial calibration techniques developed for network analyzers in the 1980s. SOLT calibration uses a 12-term error model as shown in Figure 9.10, and requires measurements of an open circuit, a short circuit, a 50-Ω circuit, and a thru connection (back-to-back cable connection). As mentioned, the error terms will be determined based on the measurements of the SOLT standards and on the mathematical definitions or circuit models of the standards.

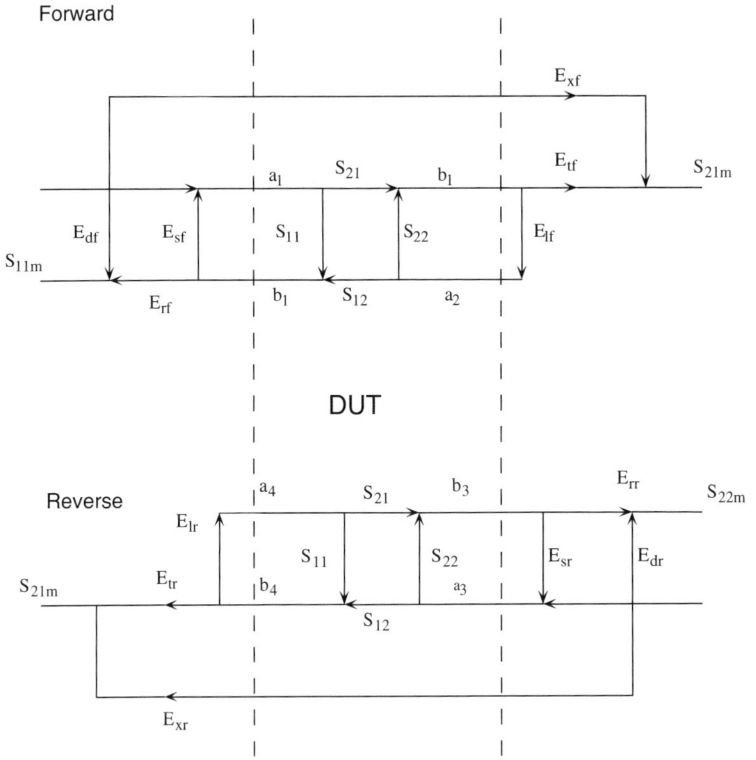

Figure 9.10. A complete 12-term error model for a two-port signal flow graph used in SOLT. The forward and reverse error terms consist of the directivity (E_{df}, E_{dr}), the source match (E_{sf}, E_{sr}), the isolation (E_{xf}, E_{xr}), the transmission tracking (E_{tf}, E_{tr}), the reflection tracking (E_{rf}, E_{rr}), and the load match (E_{lf}, E_{lr}).

Table 9.1 provides an example of an on-wafer SOLT calibration kit. Each standard circuit is defined to represent its behavior mathematically for the calibration algorithm to calculate error terms. When the definition of Table 9.1 is entered into the network analyzer, the network analyzer will use the information in calculating the error terms from a set of linear equations. This is why each set of calibration standards must have a calibration kit (cal kit) that includes the information represented in Table 9.1. The accuracy of the calibration and measurements greatly depend on the accuracy of the standard definitions. For example, the load is mathematically defined as 50 Ω, which means the value of exact 50 Ω will be used in the calibration algorithm. However, the measured value of the load may not be exactly 50 Ω over a range of frequency. Parasitic elements would be associated with the resistor load, which brings challenges to SOLT calibration since the standards must be accurately defined and modeled and the accuracy of SOLT calibration at high frequency is questionable.

The error terms of a two-port calibration are determined similarly to the one-port technique. In the two-port calibration, the thru' and the load are used to determine the transmission and isolation characteristic, respectively. Initially, one-port measurements

Table 9.1. SOLT Calibration Standards for On-wafer Measurements

STANDARD		C0 pF	C1 pF	C2 pF	C3 pF			OFFSET			FREQUENCY (GHz)[4]		
NO.	TYPE	L0 pH	L1 pH	L2 pH	L3 pH	FIXED OR[2] SLIDING	TERMINAL IMPE Ω[3]	DELAY ps	Z_0 Ω	LOSS GΩ/s	MIN	MAX	STND LABEL
1	OPEN	−9.3	0	0	0			0					
2	SHORT	2.4	−3.5	0	0			0					
3	LOAD					FIXED		0.015	50	—	0	999	
4	THRU							1	50	0	0	999	
5	OPEN	−9.3	0	0	0			0	50	0	0	999	
6	SHORT	2.4	−3.5	0	0			0	50	0	0	999	
7	LOAD					FIXED		0.015	500	0	0	999	

1. Open, short, load, delay/thru, or arbitrary impedance.
2. Load or arbitrary impedance only.
3. Arbitrary impedance only. Device terminating impedance.
   (Defaults:
   Short = 0 Ω
   Open = ∞ Ω
   Load = 50 Ω
4. For waveguide, the minimum frequency is the same as F_{co}.

(at each port) on short, open, and load standards are conducted to determine source match (E_{sf}, E_{sr}), directivity (E_{df} and E_{dr}), and reflection tracking (E_{rf}, E_{rr}). Connecting each port to a load (Z_o), the isolation terms (E_{xf} and E_{xr}) are determined. When a thru connection, which is typically a back-to-back cable connection, is measured, the load match and transmission tracking are determined

$$E_{lf} = \frac{S_{11M} - E_{df}}{S_{11M}E_{sf} - E_{df}E_{sf} + E_{rf}}$$

and

$$E_{lf} = (S_{21M} - E_{xf})(1 - E_{sf}E_{lf})$$

Once the error terms are determined, the measured S-parameters of a DUT can be calculated as follows:

$$S_{11} = \frac{\left\{\left(\frac{S_{11m}-E_{df}}{E_{rf}}\right)\left[1+\left(\frac{S_{22m}-E_{dr}}{E_{rr}}\right)E_{sr}\right]\right\} - \left[\left(\frac{S_{21m}-E_{xf}}{E_{tf}}\right)\left(\frac{S_{12m}-E_{xr}}{E_{tr}}\right)E_{lf}\right]}{\left[1+\left(\frac{S_{11m}-E_{df}}{E_{rf}}\right)E_{sf}\right]\left[1+\left(\frac{S_{22m}-E_{dr}}{E_{rr}}\right)E_{sr}\right] - \left[\left(\frac{S_{21m}-E_{xf}}{E_{tf}}\right)\left(\frac{S_{12m}-E_{xr}}{E_{tr}}\right)E_{lf}E_{lr}\right]}$$

$$S_{21} = \frac{\left[1+\left(\frac{S_{22m}-E_{dr}}{E_{r}}\right)(E_{sr}\text{-}E_{lf})\right]\left(\frac{S_{21m}-E_{xf}}{E_{tf}}\right)}{\left[1+\left(\frac{S_{11m}-E_{df}}{E_{rf}}\right)E_{sf}\right]\left[1+\left(\frac{S_{22m}-E_{dr}}{E_{rr}}\right)E_{sr}\right] - \left[\left(\frac{S_{21m}-E_{xf}}{E_{tf}}\right)\left(\frac{S_{12m}-E_{xr}}{E_{tr}}\right)E_{lf}E_{lr}\right]}$$

$$S_{12} = \frac{\left[1+\left(\frac{S_{11m}-E_{df}}{E_{rf}}\right)(E_{sf}\text{-}E_{lf})\right]\left(\frac{S_{12m}-E_{xr}}{E_{tr}}\right)}{\left[1+\left(\frac{S_{11m}-E_{df}}{E_{rf}}\right)E_{sf}\right]\left[1+\left(\frac{S_{22m}-E_{dr}}{E_{rr}}\right)E_{sr}\right] - \left[\left(\frac{S_{21m}-E_{xf}}{E_{tf}}\right)\left(\frac{S_{12m}-E_{xr}}{E_{tr}}\right)E_{lf}E_{lr}\right]}$$

Ideal open circuits do not exist at microwave frequencies. An open circuit, such as an open-ended coax and a probe lifted in air, has both radio loss and capacitive "fringing" fields. In coaxial transmission media, shields techniques are used to reduce radiation loss. In addition, the fringing capacitance that causes resultant phase shift is modeled as follows:

$$C = C_0 + C_1 \times F + C_2 \times F^2 + C_3 \times F^3 \tag{9.7}$$

where C_0 (F), C_1 (F/Hz), C_2 (F/Hz^2), and C_3 (F/Hz^3) are the coefficients for a cubic polynomial that best fits the actual capacitance of the open. Similarly, the short circuit will have residual inductance that will cause phase shift. The inductance as a function of frequency can be modeled as

$$L = L_0 + L_1 \times F + L_2 \times F^2 + L_3 \times F^3 \tag{9.8}$$

where L_0 (H), L_1 (H/Hz), L_2 (H/Hz^2), and L_3 (H/Hz^3) are the coefficients of a third-order polynomial that best fits the actual inductance of the short.

Obviously, there are assumptions made in defining SOLT standards and the models are simple to characterize the behavior of a wide-band open or a short. As frequency increases, the fringing fields and residual inductance become complex. As frequency increases, shielding becomes ineffective, which makes the assumption of the open magnitude $(\rho) = 1$ invalid for all frequencies. The model does not account for the frequency-dependent magnitude of the open and short circuits. Furthermore, the models of the open and short circuits as a polynomial function are not adequate to characterize the complex fringing capacitance and residual inductance. Since SOLT calibration heavily relies on the definition and accuracy of the standards, these nonideal open and short circuits impact the accuracy of SOLT at high-frequency and high-reflection coefficient measurements. If one would like to measure devices that have a high reflection coefficient ($\Gamma \sim 1$), the measurement results become highly uncertain. One can extend the models of open and short circuits to improve the accuracy of the calibration. However, this will require a new algorithm to interface with a network analyzer and complex modeling of these circuits. Since SOLT has major drawbacks in measurement accuracy, this book will not discuss the detail of SOLT calibration but will provide the derivation and analysis of TRL calibration.

9.3.4 TRL Calibration

Ideally, a desired calibration technique is one that relies the least on the models or definitions of standards. If even calibration relies on the definitions of standards, these circuit parameters can accurately be defined and remain constant with respect to frequency. The thru-reflect-line (TRL) calibration is developed based on these basic concepts. Even better, the TRL can characterize standards during calibration. These measured frequency-dependent parameters of standards are used in the TRL algorithm to calculate the error terms. To date, TRL is the most accurate calibration technique used in modern network analyzers.

The standards for TRL calibration include a thru, reflects (short or open circuits), and lines (transmission lines of different lengths). The goal of the TRL calibration algorithm is to eliminate mathematical definitions of open and short standards. These two standards are complex to model at a high frequency (up to 110 GHz). In addition, raw S-parameters will be used to extract the circuit parameters of transmission line standards during calibration. Therefore, this type of calibration is also referred to as self-calibration, for which the standards are being characterized empirically and accurately. Table 9.2 illustrates the definition of standards for a TRL calibration kit. Figure 9.11 illustrates the eight-term error model for the TRL calibration. In this model, the cables and connectors are represented by Error Boxes 1 and 2. Therefore, the raw S-parameters will include

$$[T_{\mathrm{raw}}] = [E_1][\mathrm{DUT}][E_2] \tag{9.9}$$

Table 9.2. TRL Calibration Standards

STANDARD		C0 pF	C1 pF	C2 pF	C3 pF			OFFSET			FREQUENCY (GHz)[4]		
NO.	TYPE	L0 nH	L1 nH	L2 nH	L3 nH	FIXED OR[2] SLIDING	TERMINAL IMPE Ω[3]	DELAY ps	Z_0 Ω	LOSS $G\Omega/s$	MIN	MAX	STND LABEL
1													
2													
3													
4													
5													
6													
7													
8	THRU							1	50	0	0	999	
9	REFLECT							0	50	0	0	999	
10	REFLECT							0	50	0	0	999	
11	LINE							10	50	0	5.64	45.1	
12	LINE							25	50	0	2.25	18	
13	LINE							40	50	0	0	11.3	

1. Open, short, load, delay/thru, or arbitrary impedance.
2. Load or arbitrary impedance only.
3. Arbitrary impedance only. Device terminating impedance. (Defaults:
   Short = 0 Ω
   Open = ∞ Ω
   Load = 50 Ω
4. For waveguide, the minimum frequency is the same as F_{co}.
5. TRL lines are 1375, 3325, and 5275 μm.

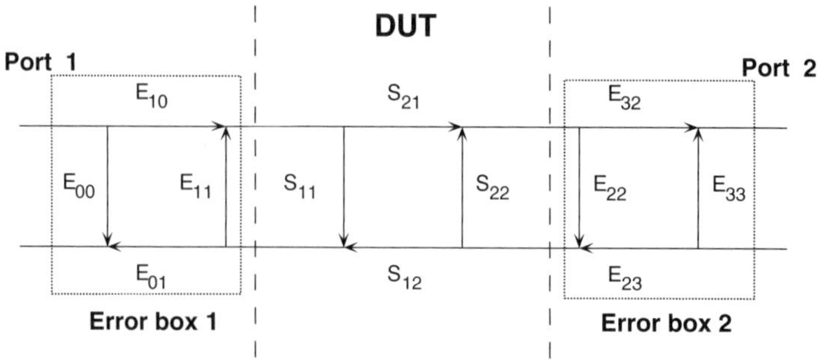

Figure 9.11. A complete eight-term error model for a two-port signal flow graph used in TRL. The cascading error terms consist of the directivity (E_{00}, E_{33}), the frequency responses [(E_{10}, E_{01}) and (E_{23}, E_{32})], and the port match (E_{11}, E_{22}).

where E_1, DUT, E_2, and T_{raw}, are the ABCD matrices of error box 1, device under test, error box 2, and the total responses. Once error terms or error boxes are determined after calibration, the ABCD matrix of the DUT is given as follows:

$$[\text{DUT}] = [E_1]^{-1}[T_{\text{raw}}][E_2]^{-1} \tag{9.10}$$

As shown in Eqs. (9.9) and (9.10), TRL calibration will provide error terms E_{00} to E_{33} or the cumulative S-parameters of physical connections at ports 1 and 2. In other words, the S-parameters of physical connections at ports 1 and 2 can be determined during calibration. This feature is very useful in obtaining the S-parameters of fixtures in load-pull measurements.

Although short and open circuits are used in TRL, models of open and short circuits are not required to solve for error terms. As shown in the calibration kit, the reflects (short or open) are intentionally left blank and are not defined. This is why they are called "reflects" rather than short and open. However, TRL calibration requires that these reflects at two ports must be identical in TRL calibration. This equality is used as one condition to solve a set of equations that lead to error terms. As shown in Eqs. (A.17)–(A.19) in the Appendix, the unknown reflection coefficient of the reflects is eliminated by equating the measured reflection coefficients at ports 1 and 2. For two-port calibration, this condition can be easily achieved by using the same reflect standard at two ports. Since the reflection coefficient of the reflect is theoretically 1, which is the largest value, any uncertainty or changes in the magnitude will be negligible. Identical reflects are adequate to determine the error terms and to eliminate the definition of the short or open circuits that rely on the lumped circuit models. By not defining short or open circuits, the calibration will be very accurate at high-frequency and high-reflection coefficient measurements.

For transmission line standards, the two parameters to describe transmission lines are delay and characteristic impedance. As shown in the TRL calibration kit (Table 9.2), the required definitions of the transmission line standards are the delay and characteristic impedance. Basically, the delay is to provide the physical lengths of transmission line standards to a network analyzer. Assumptions made in TRL calibration are that all transmission lines have transmission electron microsope (TEM) wave propagation characteristics and are dielectric lossless. When designing standards for TRL calibration, these two conditions must be met to achieve accuracy. For coaxial measurements, air-line coaxes are used as transmission line standards. Obviously, the TEM propagation mode of the standards ensures that the transmission lines maintain 50-Ω characteristic impedance over the range of calibration frequency as users input exactly 50 Ω into the network analyzer. The characteristic impedance is the only the model on which TRL calibration relies. Hence, TRL calibration relies very little on the models of the standards. Furthermore, the characteristic impedance can even be determined over the range of frequency using the measured propagation constant.

By measuring two transmission lines with unknown reference planes, the propagation constant can be empirically determined as shown in the Appendix. The measured frequency-dependent propagation constant can be used to determine the frequency-dependent characteristic impedance of the transmission line standards. This feature is critical in using transmission line standards that have a quasi-TEM propagation mode and slight variations in characteristic impedance with respect to frequency. Furthermore, as frequency rises, the variation of the impedance will increase. For a TEM transmission line, the characteristic impedance can be calculated from the propagation constant [4]:

$$Z_o = \frac{\gamma}{j\omega C + G} \tag{9.11}$$

where γ is the measured propagation constant, C is the capacitance per unit length, and G is the shunt conductance per unit length. The transmission line standards used in TRL calibration are assumed to be dielectric lossless. Therefore, the transconductance G is negligible. For a low-loss dielectric medium, G is negligible since G is much less than ωC ($G \ll \omega C$). Equation (9.11) becomes

$$Z_o = \frac{\gamma}{j\omega C} \quad \text{or} \quad C = \frac{\gamma}{Z_o j\omega} \tag{9.12}$$

The capacitance per unit length of the TEM transmission lines is very well behaved and is constant with respect to frequency [4]. Since C depends weakly on frequency and conductivity, C can be well approximated by the DC-capacitance C_{dc} [5]. C_{dc}, the DC-capacitance assuming perfect conductors, can be calculated with known geometry of the transmission line. Figure 9.12 demonstrates the measured capacitance of a TEM transmission line over a frequency range. The capacitance value is fairly constant with respect to frequency. Therefore, by providing the constant capacitance C_{dc}, the frequency-dependent characteristic impedance can be empirically calculated using the measured propagation constant obtained during calibration. As mentioned, TRL only relies on the model of the characteristic impedance. However, the characteristic impedance is indirectly determined from actual measurements, which makes TRL a very accurate technique. Typically, the network analyzer first uses the input characteristic impedance and measured propagation constant at low frequency to determine

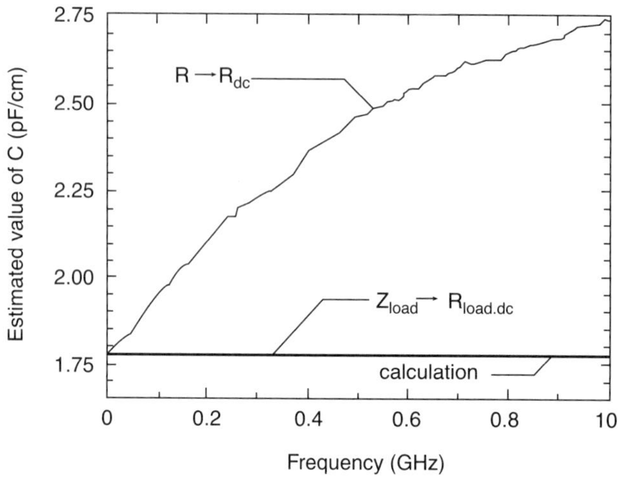

Figure 9.12. The measured frequency-dependent capacitance of a lossless transmission line [5].

the capacitance per unit length since the only calibration kit information is the impedance. Since the capacitance is constant with respect to frequency, the calculated capacitance can be used to calculate the frequency-dependent Z_o Figure 9.13 using the measured propagation constant. In the NIST TRL calibration, the software requires the value of the capacitance up front along with the characteristic impedance.

In summary, TRL standards include reflects and transmission lines of different lengths. The reflects need not to be defined but must be identical. This condition is easily achieved by using the same standard for both ports. The transmission line

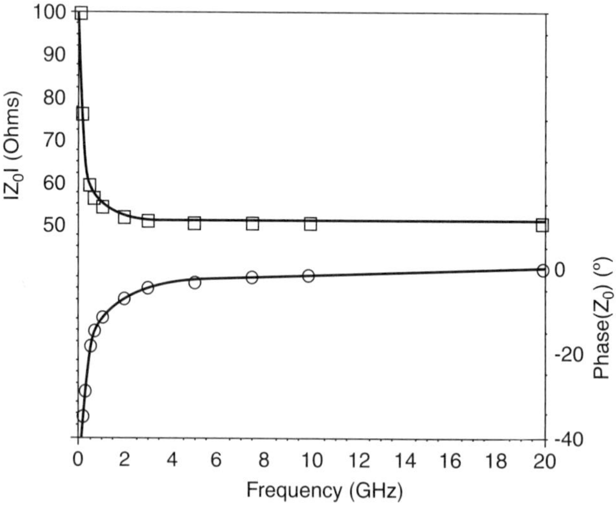

Figure 9.13. The measured frequency-dependent characteristic impedance.

standards are defined by characteristic impedance and delay. The characteristic impedance of the transmission line standards are indirectly measured during calibration. As a result, TRL calibration is the most accurate provided that the transmission lines have a TEM propagation mode.

9.4 WAFER PROBING MEASUREMENT

Modeling of active and passive devices is performed using on-wafer measurement techniques. During the 1990s, on-wafer measurement techniques were developed to measure microwave monolithic integrated circuits directly using coplanar waveguide probes. Progress in this development has resulted in mature on-wafer measurement techniques that are widely used today. On-wafer measurements employ coplanar waveguide micro-probes that provide two grounds and a signal contact on the same contact plane. Figure 9.14 demonstrates coplanar waveguide probes that are used to make direct contact to integrated circuits. The concepts of calibration for on-wafer measurements are similar to those for coaxial transmission lines. For on-wafer measurements, the calibration standards are fabricated on a substrate that includes short, open, load, and transmission lines in coplanar waveguide configurations.

9.4.1 Calibration Quantification of Random Errors

Most coaxial and waveguide calibration techniques have NIST accuracy tracibility. However, on-wafer calibration does not have any tracibility for accuracy. One technique that has been developed at NIST is to check the repeatability of calibration

Figure 9.14. A zoomed view of coplanar waveguide probes.

[10]. For on-wafer calibration, random errors are contributed by probe tip contact on a DUT, tolerances of on-wafer standards, and repeatability of probe contact. Random errors cannot be corrected. A calibration verification technique developed at NIST can be used to determine the largest possible vector error in measured S-parameters resulting from drift and random errors [10]. This quantification is determined using the concept of two-tier calibrations. Initially, an automatic network analyzer (ANA) is calibrated using a set of TRL standards to set measurement reference planes at probe tips. This is the first-tier calibration. If [T] is the transmission matrix of a transmission line, the measured results can be expressed as follows:

$$[\mathrm{T}] = [\mathrm{Y}]^{-1}[\mathrm{DUT}]_{\mathrm{Meas.}}[\mathrm{X}]^{-1} \tag{9.13}$$

where [X], [Y], and $[\mathrm{DUT}]_{\mathrm{Meas.}}$ are the error boxes at ports 1 and 2 and the overall measured responses, respectively. With the ANA error correction on, the same TRL standards are measured and these S-parameters are used as raw data to perform the second correction. It is the second tier calibration. It is an analogy to the first tier calibration. However, the internal reference planes of the second tier calibration are at the probe tips instead of inside the ANA as for the first tier. The measured transmission matrix of the transmission line is

$$[\mathrm{T}] = [\mathrm{Y}]^{-1}[\mathrm{DUT}]_{\mathrm{Meas.}}[\mathrm{X}]^{-1} = [\mathrm{Y}_2]^{-1}[\mathrm{T}]_{\mathrm{Meas.}}[\mathrm{X}_2]^{-1} \tag{9.14}$$

where $[\mathrm{X}_2]$ and $[\mathrm{Y}_2]$ are the error boxes of the second tier calibration at ports 1 and 2. $[\mathrm{T}]_{\mathrm{Meas}}$ is the measured transmission matrix of the transmission line using the second calibration. Ideally, the measurement reference planes of the second calibration are the same as that of the first. Therefore, $[\mathrm{Y}_2]^{-1}$ and $[\mathrm{X}_2]^{-1}$ are supposed to be identical matrices and $[\mathrm{T}] = [\mathrm{T}]_{\mathrm{Meas.}}$. Practically, the two calibrations differ because of instrument drift, random errors, and probe contact. Therefore, the error boxes $[\mathrm{Y}_2]$ and $[\mathrm{X}_2]$ determined by the second tier calibration are used to quantify the differences between the scattering parameters measured by the two calibrations. This quantification determines the worst-case error differences in measured S-parameters.

9.4.2 On-Wafer Measurement at the W-Band (75–110 GHz)

Although on-wafer measurements have become popular for the microwave community, phenomena such as surface wave phenomena, cross-talk, and repeatability of measurements at very high frequencies are still new to several users. One question is what the behavior of these coplanar waveguide transmission lines at W-band (75–110 GHz) is. These questions raise the suspicion of the validity of on-wafer techniques at high frequencies. In this section, we present experimental studies of on-wafer measurements and calibration standards at the W-band [11]. The W-band is currently the highest frequency band that commercial network analyzers can support.

9.4.2.1 Measurement Setup. The measurement system is a vector network analyzer HP85109C whose W-band test set has a WR-10 waveguide output.

Figure 9.15. A W-band on-wafer measurement setup.

The measurement system is shown in Figure 9.15. The two multiplier modules are used as W-band test sets. The WR-10 waveguide to 1-mm coaxial connector adapters are used to establish connections from the test set to 1-mm coax-fed microprobes. The WR-10 waveguide input microprobes are connected directly to the W-band test set. The 1-mm connector input probe consists of air coplanar tips attached to a 1-mm cable that has a 1-mm female coaxial connector. The WR-10 waveguide input probe has an absorber between the waveguide input and the 1-mm cable where the probe tips are attached. Today, both WR-10 waveguide and 1-mm coaxial connection network analyzers are commercially available. Figures 9.15 and 9.16 illustrate the measurement setup for both types of CPW microprobes.

Figure 9.16. A measurement setup for the 1-mm coaxial input CPW microprobes.

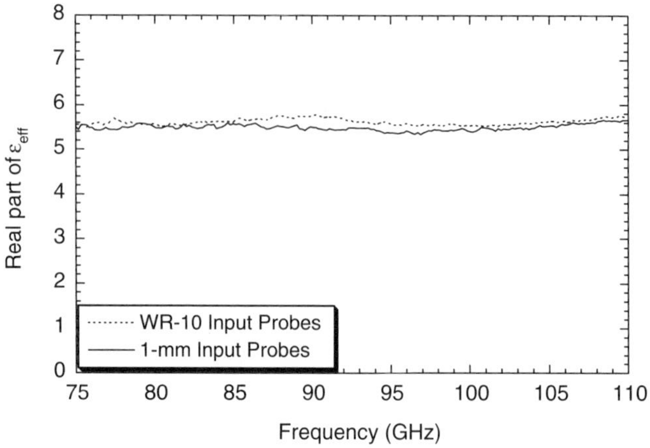

Figure 9.17. The real part of ε_{eff} measured by the TRL calibration on a W-band ISS substrate from Cascade Microtech, Inc.

9.4.2.2 On-Wafer Calibration at the W-Band (75–110 GHz). TRL calibration is performed on a W-band ISS (Figure 9.17) using two types of probes, the 1-mm coaxial connector and WR-10 input microprobes, respectively. The two on-wafer coplanar waveguide (CPW) transmission lines used in the calibration are a 51-μm-wide center conductor separated from 270-μm-wide ground planes by 22-μm gaps. The lengths of the two CPW lines are 127 μm and 448 μm. The ISS substrate is 635 μm thick and has a relative dielectric constant of 9.9. The quasi-static effective dielectric value for the CPW mode can be approximated as follows [12]:

$$\varepsilon_{eff} = \frac{\varepsilon_r + 1}{2} = 5.45 \tag{9.15}$$

Equation (9.15) calculates the effective dielectric constant for a CPW line with an infinitely thick substrate. The formula is derived using the conformal mapping technique. This technique employs the Schwarz–Christoffel transformation to map the CPW geometry into a parallel plate capacitor one. The capacitance per unit length of the CPW line is easily calculated from the new topology so that the quasi-static approximation can be used to determine the effective dielectric constant. Equation (9.15) is also a good approximation for a 635-μm-thick CPW line. The TRL algorithm determines the propagation constant of the transmission line standards during calibration. The effective dielectric constant can be derived from measurements as follows:

$$\varepsilon_{eff} = -\left(\frac{c\gamma}{\omega}\right)^2 \tag{9.16}$$

where c is the speed of light in vacuum, $\gamma = \alpha + j\beta$ is the measured propagation constant, and ω is the angular frequency.

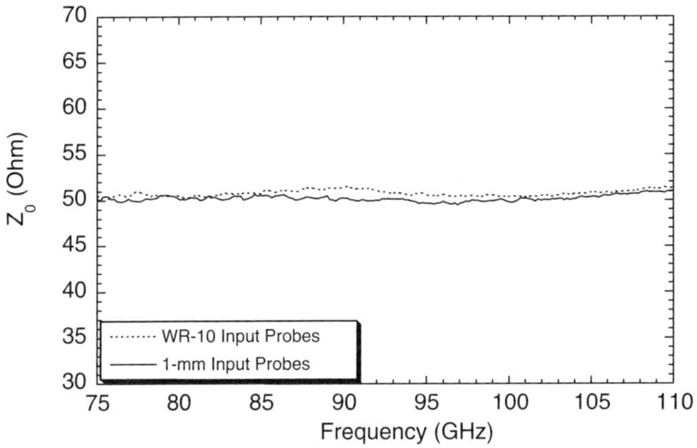

Figure 9.18. The real part of the characteristic impedance measured by the TRL calibration using both types of probes.

Figure 9.17 illustrates the measured effective dielectric constant determined during TRL calibration using both types of probes [13–15]. The measured effective dielectric constants correspond closely to the quasi-static value 5.45 of the CPW mode. This correlation demonstrates that the TRL calibration measures the quasi-TEM CPW mode for both probes. As shown in the Appendix, the surface-wave mode is dispersive and causes loss. If the surface-wave mode were dominant, the propagation constant would exhibit a nonlinear behavior with respect to frequency. As a consequence, the linear relation (eq. 9.13) would calculate the effective dielectric constant to be different from 5.45. However, the experimental results demonstrate that the measured effective dielectric constant determined by the linear Eq. (9.14) is approximately 5.45. This result demonstrates that the calibration has not been affected significantly by the surface-wave mode. In addition, the calibration also determines the characteristic impedance of the CPW lines. Figure 9.18 illustrates that there is a good correlation between the measured and the designed 50-Ω characteristic impedance of the CPW lines.

9.4.2.3 Repeatability Study. The repeatability tests are performed on two CPW transmission lines at two different days using both types of probes in TRL calibrations. The lengths of the two CPW lines are 900 μm and 1800 μm, respectively. Using the NIST software [13], two-tier calibration based on concepts described previously is performed to study the repeatability of on-wafer W-band measurements. Figures 9.19 and 9.20 illustrate the differences in measured S-parameters of the two CPW lines using the 1-mm connector input probes. In comparison, Figures 9.21 and 9.22 illustrate the repeatability of the WR-10 input probes. The bounds $|S_{ij}-S'_{ij}|$ for $ij \in \{11, 12, 21, 22\}$ are the largest differences or worst-case errors in the measured S-parameters. The maximum bounds in S-parameters for the WR-10 probes are superior than that of 1-mm connector input probes. However, the actual measured differences by both types of probes are similar. The actual differences are determined by subtracting two

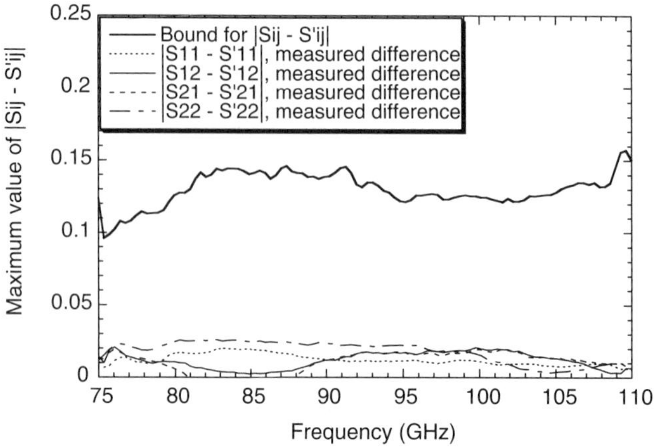

Figure 9.19. The worst-case measurement differences of a 900-μm CPW line at two different times using TRL calibrations with 1-mm connector input probes.

sets of measured S-parameters at the same reference planes. In all cases, the actual measured differences do not exceed the maximum bounds. Both types of probes are highly repeatable for measurements at the W-band.

9.4.2.4 Cross-Talk between Two CPW Microprobes [11]. The coupling energy from probe-tip-to-probe-tip can change the characteristics of a DUT. An experiment is conducted to examine the coupling energy between two probes separated by various distances. Two on-wafer 50-Ω resistors separated by 100 μm and 6 mm are measured using both types of probes. These resistors are printed on a 635-μm-thick alumina

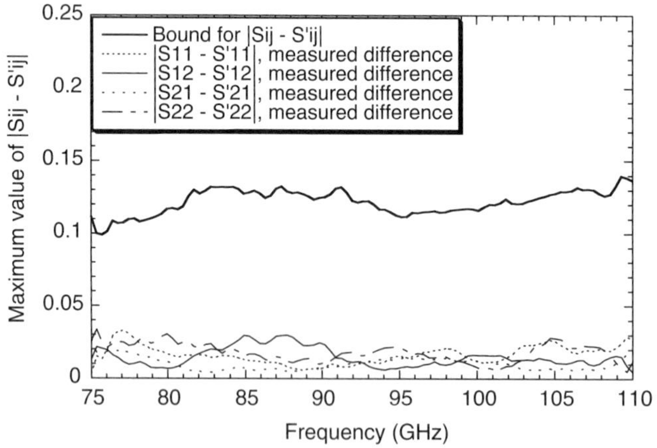

Figure 9.20. The worst-case measurement differences of a 1800-μm CPW line at two different times using TRL calibrations with 1-mm connector input probes.

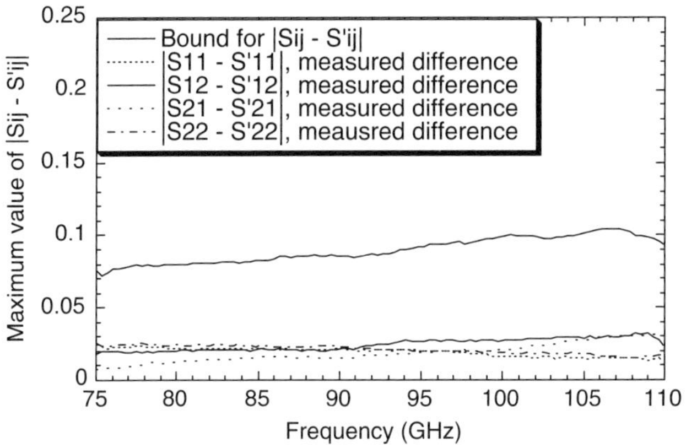

Figure 9.21. The worst-case measurement differences of a 900-μm CPW line at two different times using TRL calibrations with WR-10 waveguide input probes.

substrate. TRL calibration is performed for the measurements. Ideally, there is no energy transmitting from port 1 to port 2. For separations of 100 μm and 6 mm, the transmission coefficients obtained from measurements are −35 dB and −50 dB, respectively, as shown in Figure 9.23. The coupling energy of the coaxial input probe is 2 dB less than that of the WR-input probe for a separation of 100 μm. With 100-μm separation, the isolation is around 40 dB, which is adequate for most measurements.

9.4.3 On-Wafer Microstrip Characterization Techniques

Probing at microwave/millimeter-wave frequency offers an accurate and convenient way to characterize devices. The challenges for on-wafer as well as for coaxial

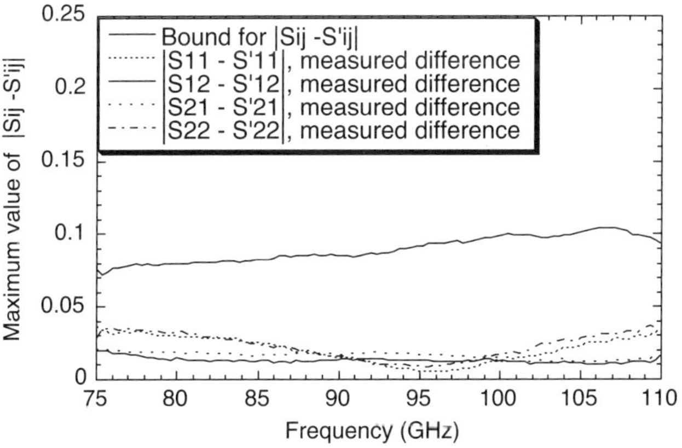

Figure 9.22. The worst-case measurement differences of a 1800-μm CPW line at two different times using TRL calibrations with WR-10 input probes.

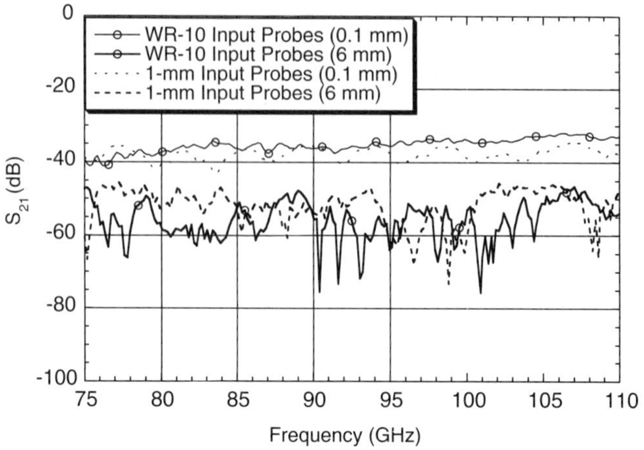

Figure 9.23. The coupling energy between two probes as a function of separation distance.

measurements originate when the signal input and output of a DUT are not compatible with network analyzer connectors. For on-wafer measurements, the microprobes that make contact with a DUT have coplanar waveguide configurations. The coplanar waveguide guide configurations provide both signal and ground contact for microwave measurements. Adapters are required for measurements of a DUT that has microstrip input and a ground plane on the backside of a substrate.

Coplanar waveguide-to-microstrip (CPW/MS) adapters provide a solution to measurements of microstrip launch devices [11]. Figure 9.24 illustrates the schematic representation of the on-chip CPW/MS adapter manufactured by J Microtechnology®. Most often, these adapters and launches are not compatible with the available standard calibration kit. For on-wafer measurements, the reference planes are calibrated at the

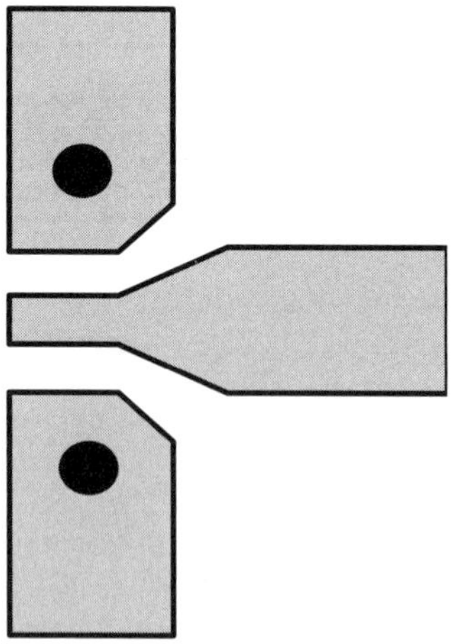

Figure 9.24. A coplanar-to-microstrip adapter.

probe tips. The CPW/MS adapters will contribute unwanted responses to the measured data of a DUT. De-embedding these transitions is tedious and requires significant effort in characterizing the launches. Time domain gating is often employed to move the reference planes. However, the accuracy of this technique is questionable and a low-loss assumption is made. The following three sections will describe techniques measuring devices with any launches that can be automatically removed through a custom design calibration kit.

9.4.3.1 CPW/MS Calibration Kit. Figure 9.24 demonstrates the coplanar-to-microstrip adapter for on-wafer probing. The input has a ground–signal–ground configuration that is compatible with standard CPW microprobes. The two ground pads are connected to the ground at the substrate backside using 7-mil diameter-plated vias. The signal and the gap at the input are designed to fit standard probe pitch (50-to 200-μm probe pitch). The signal line is tapered to a larger width microstrip line designed to achieve 50-Ω characteristic impedance. The use of CPW/MS standards causes unwanted responses in measured S-parameters of a DUT. It is important that the adapter responses can be removed efficiently.

To remove the effects the CPW/MS adapters efficiently, a custom calibration kit is required to calibrate the measurement reference planes beyond the probe tips, into the microstrip input of a DUT. Figure 9.25 illustrates the calibration kit that can be used to remove the effects of the adapters in the S-parameter measurements. After the design, CPW/MS adapters will be added to each calibration standard. In addition, both LRM and SOLT can be implemented using these adapters. One thing to consider is the parasitic of the resistor since vias are required to connect the resistor to ground. This feature will limit the frequency range of the calibration. Table 9.3 demonstrates the calibration kit of the CPW/MS adapters. Note that the length of the transmission lines excludes the length of the adapter as shown in Figure 9.25.

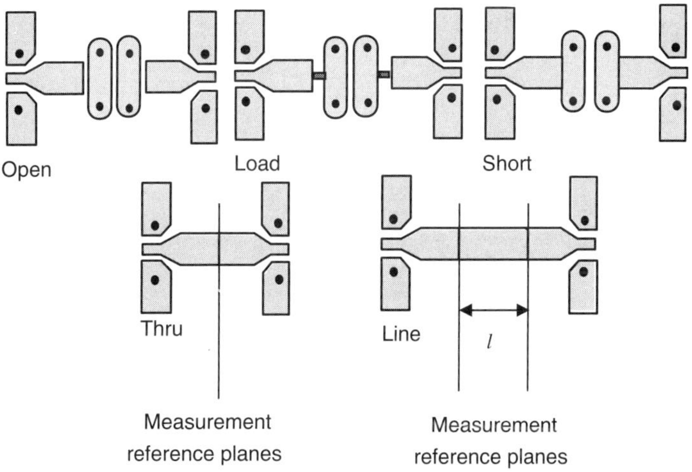

Figure 9.25. The CPW/MS calibration standards thru, line, short, open, and load.

Table 9.3. Calibration Kit for the CPW/MS Adapters

STANDARD		C0 pF	C1 pF	C2 pF	C3 pF			OFFSET			FREQUENCY (GHz)[4]		
NO.	TYPE	L0 pH	L1 pH	L2 pH	L3 pH	FIXED OR[2] SLIDING	TERMINAL IMPE Ω[3]	DELAY ps	Z_0 Ω	LOSS GΩ/s	MIN	MAX	STND LABEL
1	OPEN	25						0			0	999	OPEN25
2	SHORT	43	160					0			0	999	SHORT
3	LOAD					FIXED		0.09	500		0	5	LOAD
4	THRU							5.2	50		0	999	LINE5
5	OPEN	25						0	50		0	999	OPEN25
6	SHORT	43	160					0	50		0	999	SHORT
7	LOAD					FIXED		0.09	500		0	999	LOAD
8	THRU							0	50		0	999	LINE0
9	OPEN							0	50		0	999	OPEN
10	SHORT							0	50		0	999	SHORT
11	THRU							5.2	50		10.6	84.6	LINE5
12	THRU							15.6	50		3.5	28	LINE15
13	THRU							31.2	50		1.7	13.8	LINE37
14	THRU							10.4	50		5.2	41.5	LINE10

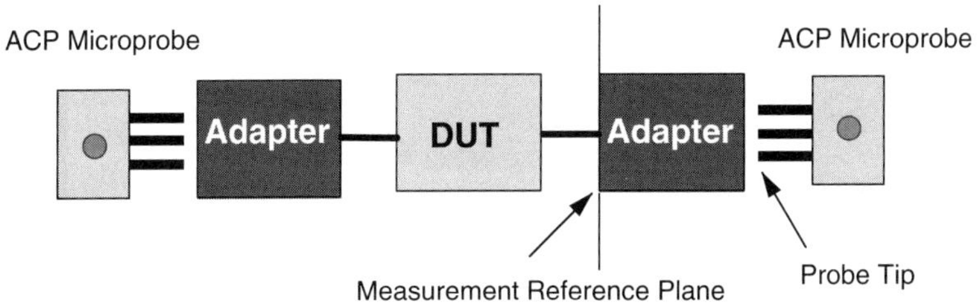

Figure 9.26. A schematic representation of a measurement setup.

The calibration kit for the CPW/MS includes LRM, SOLT, and TRL. As mentioned, the vias connecting the thin-film resistor to the ground will induce parasitics. Therefore, the frequency coverage of the load is only up to 5 GHz. Standards 5 through 14 will be used for TRL calibration. The longest line (31.2-ps delay) covers from 1.7 to 13.8 GHz. Therefore, the low-frequency option in the ANA must be used to cover the lower frequency band. In this case, TRL calibration is first performed using the chosen lines to cover the frequency of interests. Once the TRL calibration is finished, the ANA can be switched to a lower frequency calibration, in which LRM or SOLT can be executed. The combined calibration steps will cover from the lower up to 84.6 GHz, which is the maximum frequency of this calibration kit. At the lower frequency (less than 5 GHz), the resistor will accurately represent the 50-Ω characteristics with negligible parasitics.

Figure 9.26 illustrates the custom calibration technique. Normally, when a system is calibrated using a conventional impedance standard substrate (ISS) from Cascade Microtech, Inc., the measurement reference planes are at the coplanar waveguide microprobe tips. When calibration is performed on CPW/MS standards, the measurement reference planes are set and shifted to the interface of the adapter (shown in Figure 9.25). When measurements are taken, only S-parameters of a DUT are obtained. The adapters are automatically de-embedded. The assumption made for such techniques is that the adapter on each standard must be identical. Each time a measurement is made during calibration, the adapter is considered to be part of the cables and connectors that are included in the error box. Therefore, when a measurement is made on a DUT, the same adapter must be present and is automatically removed.

Calibration verification is performed using the VERIFY program from NIST to investigate the repeatability of the adapters [13]. An HP85109C with CPW microprobes is initially calibrated using on-wafer CPW/MS standards. With the error correction on, the second tier calibration is performed on the same standards. The data of error boxes determined by the second tier calibration is saved and stored in the computer. The NIST VERIFY program uses the error boxes to calculate the worst-case measurement differences in S-parameters as shown in Figure 9.27. The worst-case differences in measured S-parameters are 0.02 up to 20 GHz. The calibrated system is highly repeatable. This demonstrates that the assumption of identical adapters is valid. Otherwise, the worst-case measurement differences can be very large.

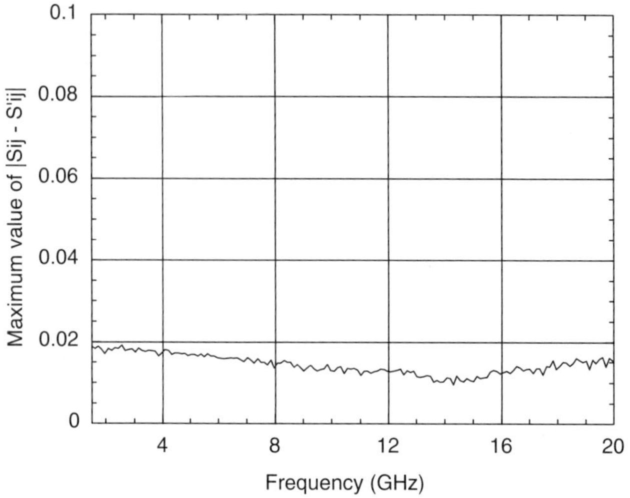

Figure 9.27. The worst-case measurement differences calibrated on CPW/MS standards.

9.4.4 On-Wafer Package Characterization Technique

9.4.4.1 On-Wafer Package Adapters. Similarly, other on-wafer adapters can be designed to characterize microwave packages that require test fixtures to make the transition to CPW microprobes. The challenges again are the design of test fixtures that provide resonant-free measurements and the need to de-embed them. The use of on-wafer adapters provides a high degree of reproducibility and repeatability. The on-wafer technique is accurate since the adapters and calibration standards are fabricated with tight tolerances. This section will describe in detail a technique that can be used to characterize surface mount packages. This technique can be applied to other types of devices that require an adapter transition to CPW microprobes.

To obtain resonant-free S-parameters of a package under test, a variety of thin-film alumina substrates as illustrated in Figure 9.28 has been designed to mount shrink small-outline packages (SSOPs). The package is mounted on the substrate using conductive epoxy. These adapters are specifically designed to make the transition from package pins to the coplanar waveguide configuration. The adapters are tapered coplanar waveguide transmission lines that allow for the package pin to be soldered at one end and the CPW microprobe contact at the other. The gaps and width of the tapered CPW are designed to achieve 50-Ω characteristic impedance. As shown from Figure 9.28, the measurement reference planes of interest are at the package pins. Therefore, on-wafer calibration using CPW microprobes must result in reference planes shifted to the package pins. Conventional standard CPW substrates (ISS substrate) are not suitable in these measurements since the reference planes are at the probe tips.

9.4.4.2 On-Wafer Package Adapter Calibration Kit. To establish the measurement reference at the package pins, custom calibration standards are designed to result

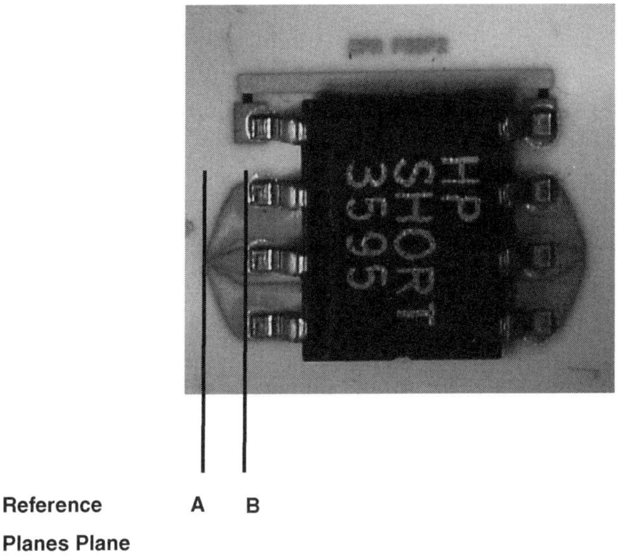

Figure 9.28. Package measurements with coplanar waveguide-to-package adapters and reference planes. Reference A is at the CPW probe tips, and reference plane B is the desired S-parameter measurements.

in shifted reference planes. These calibration standards are sometimes called off-set standards. Offset CPA calibration standards for LRM calibration are fabricated including a thru, an offset short, an offset open, a 50-Ω offset load, and a delay line Figure 9.29. In this design, the resistors are directly connected to the ground in a coplanar waveguide configuration. Therefore, minimal parasitic effects take place in this structure to allow the calibration validity to reach a high frequency. Table 9.4 illustrates the calibration kit that defines these custom standards. The use of custom calibration standards allowed the measurement reference planes to be shifted from probe tips at reference plane A to package pins at reference plane B. During the calibration and measurements, the network analyzer will see the consistent transition that is considered an integral part of the error box. Therefore, the measured S-parameters will only include the responses of the DUT at reference plane B, excluding the

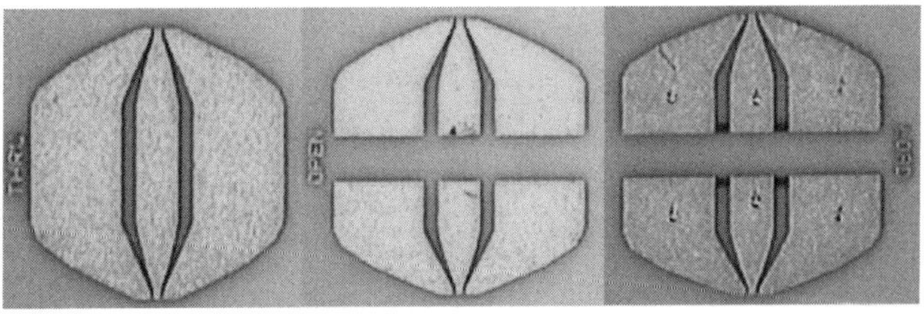

Figure 9.29. Custom coplanar-to-package calibration kit for an LRM technique that requires short, open, and match.

Table 9.4. Calibration Kit for the Custom CPA LRM Structures

STANDARD		C0 pF	C1 pF	C2 pF	C3 pF			OFFSET			FREQUENCY (GHz)[4]		
NO.	TYPE	L0 nH	L1 nH	L2 nH	L3 nH	FIXED OR[2] SLIDING	TERMINAL IMPE Ω^3	DELAY ps	Z_0 Ω	LOSS GΩ/s	MIN	MAX	STND LABEL
8	THRU							0	50	0	0	999	THRU
9	REFLECT							0	50	0	0	999	REFLECT
10	REFLECT							0	50	0	0	999	SHORT
11	LINE							0	50	0	0	999	MATCH

transitions. Practically, these sets of calibration kits must be carefully designed prototypes to achieve accuracy. By establishing the measurement reference plane at the end of the package pins in the calibration, the adapter response is removed from the measurement. By including the adapters as part of the calibration standards, the measurement reference plane is set to the end of the adapters. An LRM calibration is performed with the offset calibration structures, and the package is measured with CPW on-wafer probes. After calibration, only the S-parameters of the package are collected. When the reference plane is set to point B, the responses of the adapters are de-embedded from the measurement.

9.4.4.3 Experiment and Packaging Modeling. The previously described calibration procedure allows us to collect data that can be used to develop equivalent circuits for the package. Accurate modeling expedites the design process by accounting for parasitic elements during the design of the circuit and by predicting the packaged component's characteristics. The packages are made by encapsulating a copper lead-frame in plastic. The heat slug, which extends from the die pad inside the package to the mounting substrate, provides a ground plane as well as a thermal path to the second-level printed circuit board (PCB). To characterize the package, known circuits (50-Ω load, short, thru line, and open) fabricated on 125-μm-thick sapphire substrates are to be mounted on the copper die pad of the package using conductive epoxy. These known circuits are to assist the modeling and extraction of package pin parasitic elements. In addition, the test CPW transmission line is used to provide transmission from port 1 to 2. Wire bonds are used to attach the test circuits to the lead frame pins that provide transitions to the CPA adapters. An example of a coplanar waveguide transmission line mounted in a package is shown in Figure 9.30. These known passive structures are measured with on-wafer CPW probes using an LRM calibration on a Cascade Microtech impedance standard substrate (ISS), thereby generating accurate S-parameters representing their characteristics.

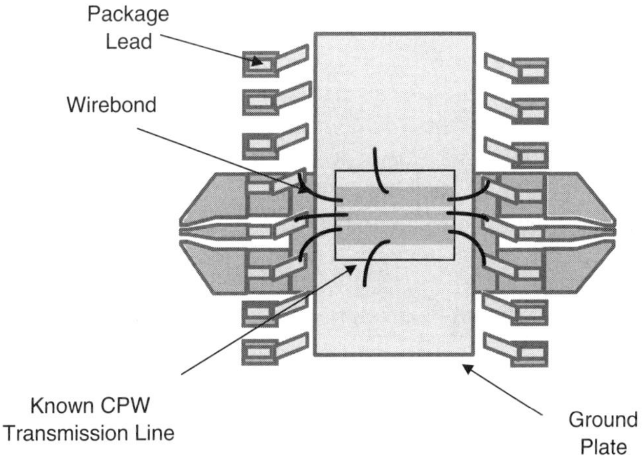

Figure 9.30. A package configuration for microwave characterization.

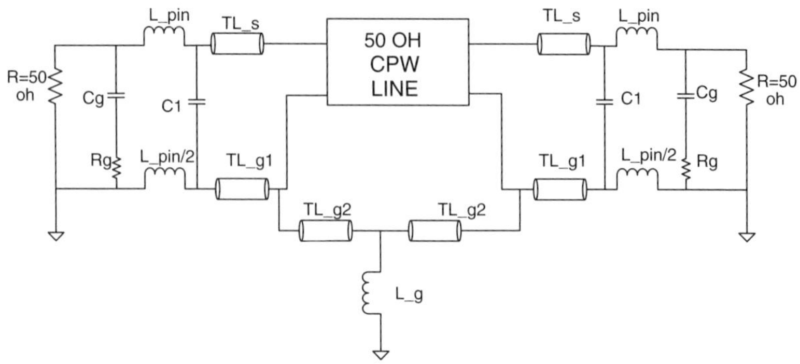

Figure 9.31. An equivalent circuit model of the package based on measured S-parameters.

After an LRM calibration using the CPA standards, the package carrying the test circuits is measured to obtain S-parameters. Based on the measured S-parameters, an equivalent circuit model for the package is constructed with a CAD tool. Figure 9.31 demonstrates the equivalent circuit model of the package. In this model, the measured S-parameters of the test structures are imported as a two-port box. The unknown elements are the parasitic of the package pins. The parasitic elements of the equivalent circuit model are optimized until the modeled and measured S-parameters are well correlated.

In the arrangement of the package model, the mutual inductance from the signal path to the two surrounding ground pins is assumed to be negligible because the ground–signal–ground (GSG) input approximates a transmission-line feed. Three basic components (the leads, the coupling of the leads, and the ground plane) are put together to form the pseudo-physical model. The parasitic element of the lead are modeled the series inductors L_pin. The capacitor C_1 connected between the pins models the coupling between adjacent pins. The inductance associated with the ground plane is represented by the inductor L_g. For an improved parasitic model, a capacitance, C_g, in series with a small resistance, R_g, is shunted across the signal pin to ground. Since the bond wires are no longer surrounded by air, the bond wires are replaced with transmission lines and are optimized with the parasitic parameters of the equivalent circuit. Because of the GSG configuration of the CPW probes, there are two ground paths, each with parasitic elements (L_pin, C_1, TL_g1). Since these two paths are identical and in parallel, they are combined together to form a single path and the corresponding model values are adjusted accordingly. The ground pads of the

Table 9.5. Equivalent Circuit Parameters

Transmission line	$Z_o(\Omega)$	Length (mil)	R (Ω/mm)
TL_s	82.0	74.5	0.001
TL_g1	41.0	74.5	0.0005
TL_g2	41.0	46.5	0.0005

Table 9.6. Equivalent Circuit Parameters

L_pin	L_g	C_g	C_1	R_g
0.85 nH	0.022 nH	0.22 pF	0.044 pF	3.7 Ω

passive standards are also wire-bonded to the slug of the package, which is represented in the equivalent circuit by the transmission lines labeled TL_g2.

Conventional models of bondwires are not used because the inherent assumption of air surrounding the wire is incorrect for the encapsulated package. The characteristic impedance of the bondwires needs to be optimized because the encapsulating epoxy affects the bondwire's characteristic response; however, the bondwire lengths are constrained to the physical dimensions of the actual bondwire to isolate the effect of our epoxy bondwire model from that of the package model. Tables 9.5 and 9.6 summarize the value of the parasitic elements in the model. In the case of the SSOP16, 50-Ω load, short, open, and thru standards are used to generate the equivalent-circuit values. Each standard is assembled in three packages to help remove any deviations caused by the reliability of package placement and connection. Figures 9.32, 9.33, and 9.34 illustrate the reflection and insertion loss of three different packaged thru lines.

9.4.4.4 Application of Package Model in Active Devices. As a demonstration of the usefulness of this electrical model, the characteristics of a packaged amplifier are predicted. A two-stage heterojunction bipolar transistor (HBT) GaAs MMIC amplifier is used to validate the developed package model. Figure 9.35 demonstrates the effects of package parasitic elements to the performance of an amplifier. Also, the need to predict the performance of a packaged MMIC is very critical. The package model is incorporated into the unpackaged HBT amplifier circuit to predict the

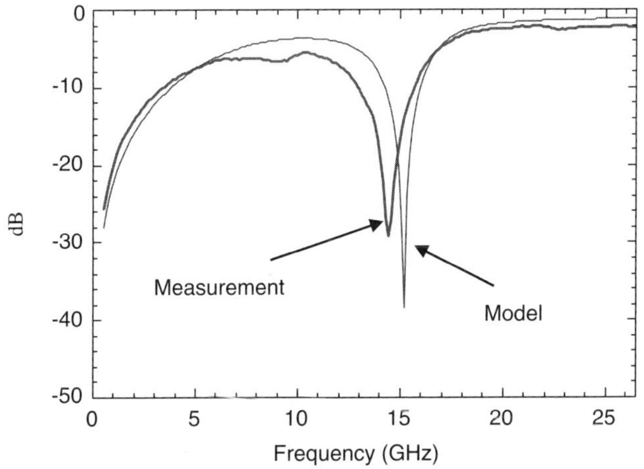

Figure 9.32. Comparison of measured magnitude of S11 of the packaged thru line.

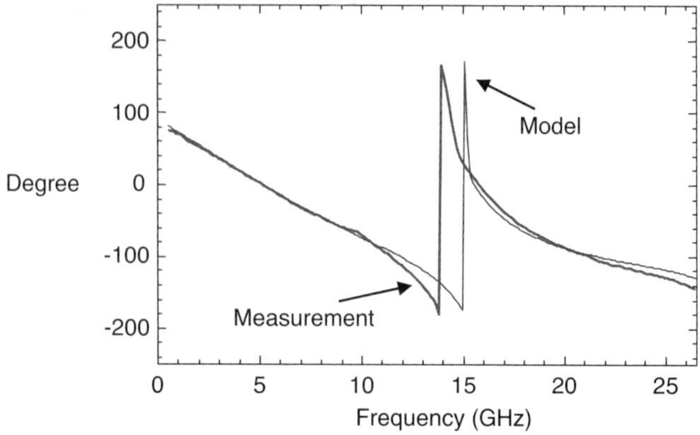

Figure 9.33. Comparison of measured phase of S11 of the packaged thru line.

cumulative performance when an amplifier is housed in a package. The unpackaged amplifier circuit shows a 3-dB cutoff frequency of greater than 15 GHz, but packaging the chip reduces the cutoff frequency to about 8 GHz. The amplifier is wire-bonded into an SSOP16 package and measured with CPAs on alumina and CPW probes to substantiate the predicted results. The gain of the simulated unpackaged, simulated packaged, and the measured packaged amplifiers are compared in Figures 9.35, 9.36, and 9.37, and it confirms the usefulness of the model. Other characteristics of a package can be studied using on-wafer probing techniques. Figures 9.38–9.40 illustrate the arrangement for isolation studies of surface mount packages.

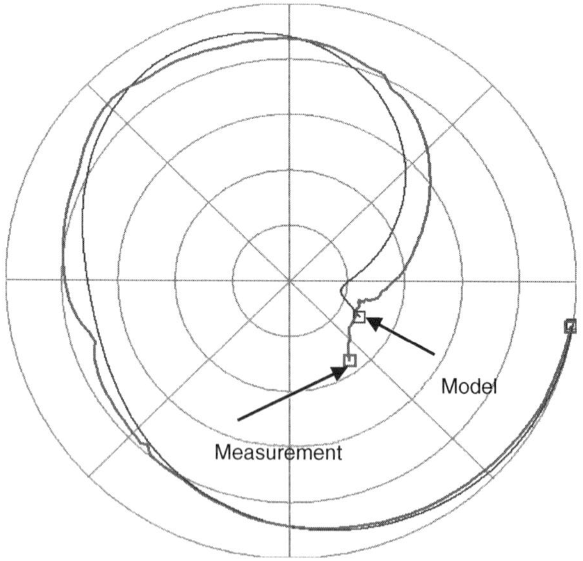

Figure 9.34. Comparison of measured S21 of the packaged thru line (0–26.5 GHz).

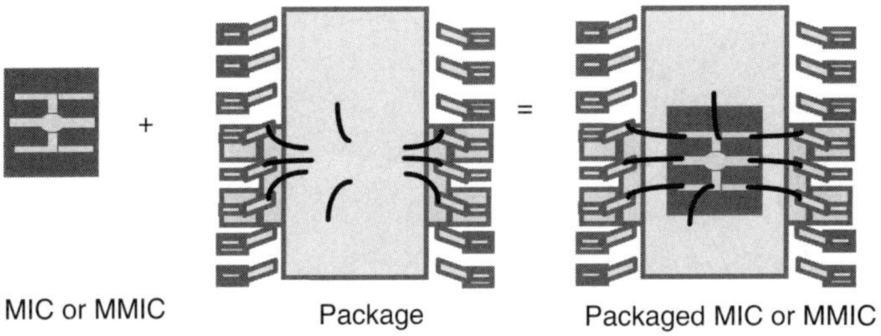

Figure 9.35. Utilization of the package model for packaged integrated circuits.

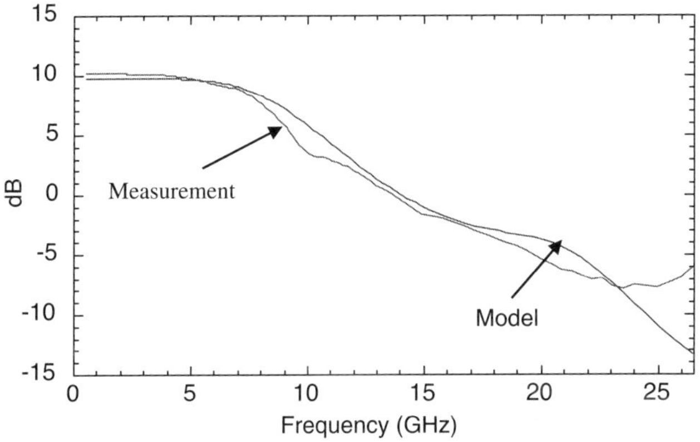

Figure 9.36. Measured and modeled magnitude of S21 of the packaged amplifier.

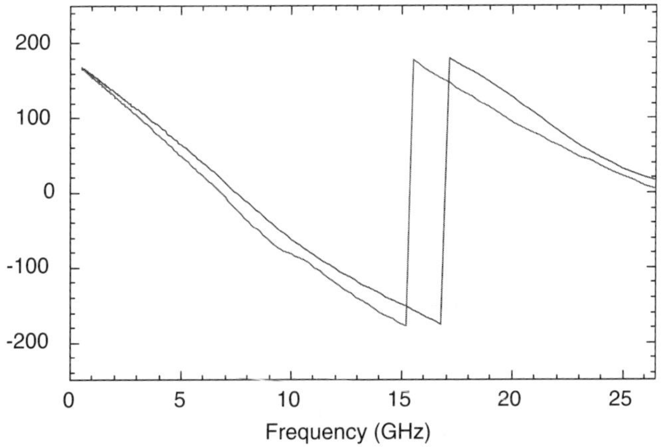

Figure 9.37. Measured and modeled phase of S21 of the packaged amplifier.

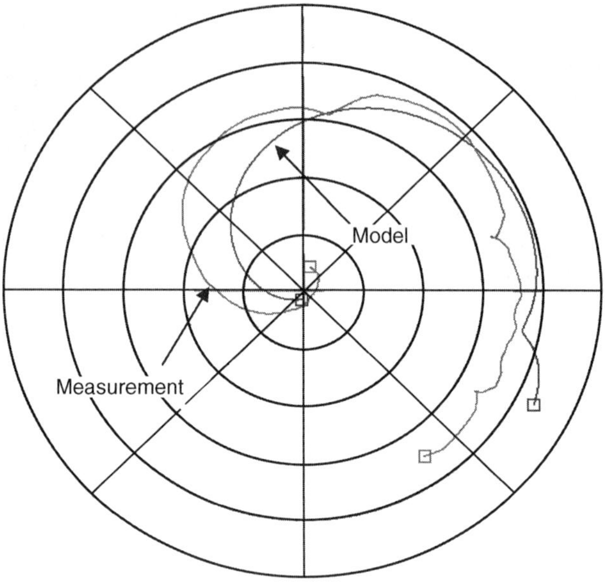

Figure 9.38. Measured and modeled phase of S11 of the packaged amplifier.

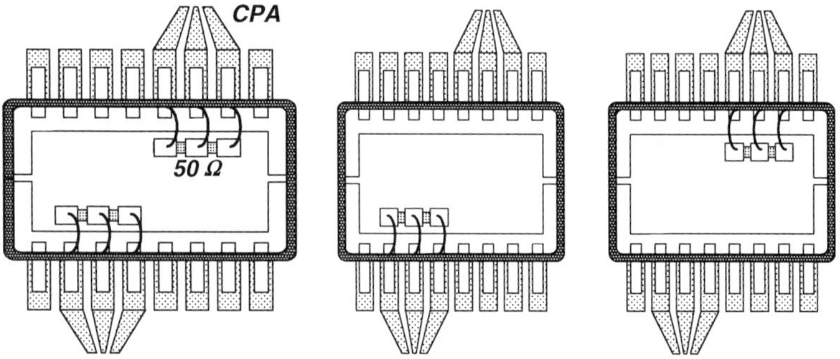

Figure 9.39. Isolation measurement configuration for packages.

9.5 CHARACTERIZATION OF INTEGRATED RADIOS

In this section, we will cover the various steps necessary for characterizing integrated radios. In most cases, the front end is characterized with single tone, as the specifications of any radio standard can be referred to a single tone. This may also lead to obtaining key performances with the minimum possible expensive hardware. Debugging is a major task of any test engineer, and the success is dependent on the complexity of the test setup. In the initial phase of the chip bring-up, many issues need careful attention up-front. In addition to that, a test or product engineer needs to

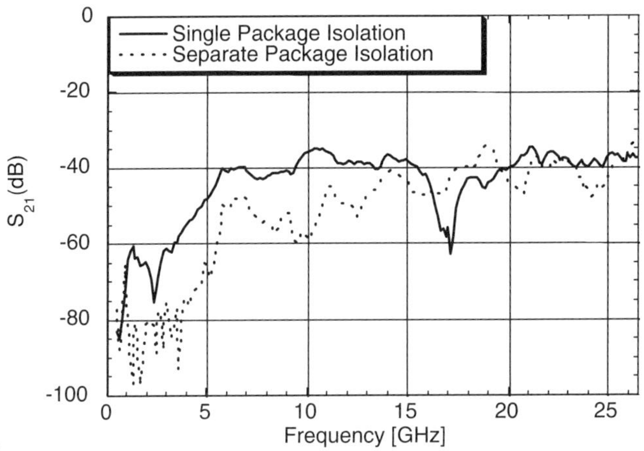

Figure 9.40. Measured isolation of packages.

understand circuit functionalities in order to obtain meaningful conclusion from the test results. Key performances of integrated receivers include:

1. Voltage gain/power gain
2. Input-referred linearity (P1dB, IIP3, IIP2)
3. Noise figure (linear and nonlinear) and noise figure slope
4. Input matching performance (S11)
5. I/Q imbalance (amplitude and phase)
6. DC offset
7. Blocker rejection and degradation on noise figure
8. Group delay
9. Stability of LO signal locking
10. Calibration modes

Key performance of transmitter include:

1. Output power and compliance with FCC spectral mask requirements
2. Linearity (output-referred IP3)
3. Output matching
4. Transmitter output power range and steps
5. LO leakage at the output
6. Out-of-band phase noise
7. Error vector magnitude (constellation degradation)
8. LO harmonics

The above performances are also measured w.r.t. temperature variations and process variations (various lots are fabricated in the foundry). Various test ports are also inserted in the IC chain, and the outputs are also monitored for key specification considerations.

There are two fundamental approaches for characterizing integrated communication systems. They can be characterized by on-wafer probing techniques by (1) probe card and (2) wedge probes. The advantages of this approach are in the possibility of fast characterization by elimination of packaging and PC board fabrication steps, which are both time and resource consuming, and the parasitics from the on-wafer probing system are minimal. In fact, device modeling at the multi-gigahertz frequency as well as characterization of high-speed communication systems are performed using on-wafer techniques. For a well-designed and correctly fabricated circuit, performance obtained from the wafer probing system should match closely with the simulated performance. Although this approach is attractive, it does not capture real effects from package components, which are extremely critical in nature, irrespective of the frequency of operation. At the same time, the end user needs to operate with a "packaged" chip. For these reasons, wafer probing is not very effective for building products.

The test equipments usually include:

1. High-resolution DC sources, power supplies, and ammeters.
2. Signal sources (single-tone output) should have a low phase noise and be capable of providing high output levels, and they should have large bandwidth in terms of output frequency.
3. Spectrum analyzers should have phase noise capability, high resolution bandwidth, and high dynamic range.
4. High-speed oscilloscope, to obtain the phasing of the output signals, as well as "eye diagrams" in case of wired communication systems. Should have low jitter and the option of input impedance selection (1 MΩ/50 Ω). Should have more input channels in order to display I/Q waveforms.
5. Network analyzers, to obtain the input matching performance of a receiver, and the output matching of a transmitter, should have a high-frequency capability.
6. Calibrated noise source, usually capable of generating a precalibrated noise figure at the output, is used for noise figure measurement.
7. Noise figure meter is used to obtain the receiver noise figure.
8. Power attenuators, used to provide sufficient attenuation for operating in linear region of amplifiers, are important for small-signal linearity tests, such as IP3.
9. Waveform generators, which usually operate at low frequencies, can generate signals of various waveforms with desired offset values that are used mostly for debugging purposes.
10. External components, such as low-loss connectors, cables, high-linearity mixers, and amplifiers provide additional flexibility in design/debug capabilities.

Both receivers and transmitters can be characterized in either time or the frequency domain. Depending on the frequency of interest, appropriate equipment would be chosen. At high frequencies, a voltage measurement using an oscilloscope would be limited by the bandwidth of the oscilloscope, whereas at low to moderately high frequencies, both the oscilloscope and the spectrum analyzer (sometimes also a dynamic signal analyzer) can be used, and the results can be verified w.r.t. one other. Spectrum analyzers provide an estimation of the noise floor, which is effective for calculating the output-referred noise voltage or the noise figure degradation with the noise injected at the input of the receiver.

9.6 IN THE LAB

Let us now consider the various types of tests for mixed-signal communication systems.

9.6.1 Operating Point

The first step in any characterization is a DC test, which indicates the power consumption of the circuit by biasing every circuit blocks one by one. In the case of dynamically biased systems (circuits operating under large signal limits), the appropriate power sources should be turned "ON" to obtain the average current flow through the circuit block. DC testing also indicates a potential "short" (large current flow) or "open" (zero current flow) in the circuit, package, or PCB. It is performed by monitoring current through an ammeter (/multimeter) connected between the test-bench supply and the supply node of the circuit. In integrated circuits, individual blocks are provided with the "enable/disable" option, and the power consumption of each block can be observed separately. During the DC test, it is essential to provide inputs to the circuit in the correct sequence to prevent any unwanted breakdown of the devices caused by current/voltage spikes. Although capacitors are provided in the circuit/PCB to eliminate this effect, care should be taken in the testing phase as well.

9.6.2 Functionality Test

In the receiver and transmitter, the functionality of the entire chain includes down/upconversion of signals, which can be observed by providing signals at various test ports, and by observing the output. In the receiver, a single tone can be provided, and the output should be observed to confirm the downconversion functionality (Figure 9.41). A low IF (1–2 MHz) can be used for this test. Similarly, a transmitter could be characterized in terms of a single-tone test. Input matching is not very important in functionality test performance.

9.6.3 Impedance Matching

Input matching indicates the amount of power transfer to the circuit under considerations, and prior to any performance test on the receiver, the input matching test needs

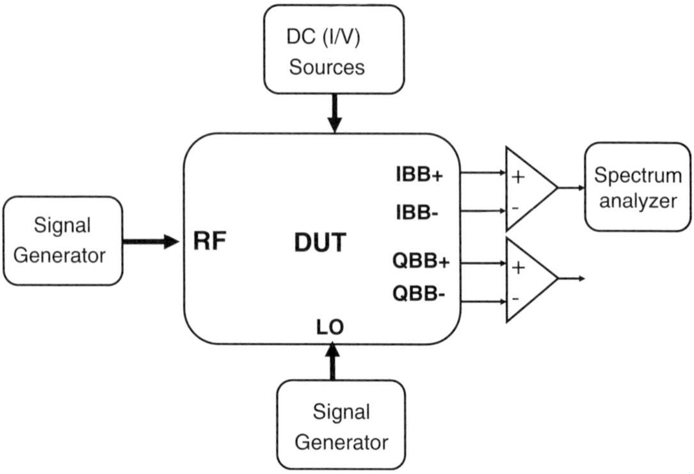

Figure 9.41. Receiver functionality test.

to be performed to ensure that the signal level from the signal generator output is close to that provided to the receiver (because of maximum power transfer). In integrated transceivers, usually this needs to be performed at the input of the receiver, output of the transmitter, and other test ports, as appropriate. However, as the signal levels in the LO chain can be increased, reflections at this port may not be critical in terms of functionality. To obtain the input/output matching level, it is ensured that the network analyzer provides a small-signal output (ensured by setting the output power level), thus performing the operation within the limits of small signal input. Figure 9.42 illustrates input matching test setup for a direct conversion receiver.

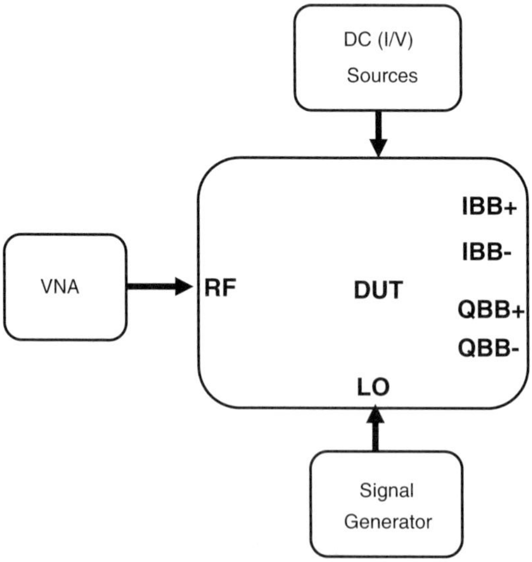

Figure 9.42. S-parameter test for input/output matching.

9.6.4 Conversion Gain

Integrated receivers employ some gain control mechanisms to handle various input signal levels. Input to the receiver can be usually an input power, and the output is usually a voltage (can be current too). Gain can be measured in terms of power or voltage, and both quantities are the same when the output impedance is also 50 Ω. However, driving a 50 Ω impedance requires high current consumption in the baseband buffers at the output. This single-tone test should be performed to evaluate the end-to-end functionality of a receiver, and the gain range is also observed (highest to lowest). Off-chip accessories, such as differential to single-ended converters, could be used to interface with the single ended test equipments. Test equipment could be a spectrum analyzer (in frequency domain) or an oscilloscope (in time domain, to obtain both I and Q channels). The voltage gain of the additional buffers should be subtracted to provide power/voltage gain values.

9.6.5 Linearity

We first consider the small-signal linearity (Figure 9.43). Two closely spaced single tones at frequencies f_1 and f_2, where $(f_2 - f_1) \sim 100$ kHz are summed using a combiner and are applied to the RF port of the receiver through a variable attenuator. The outputs corresponding to IM2 and IM3 are observed in the spectrum analyzer. In most circuit configurations, IM2 and IM3 follow a slope of 2 and 3, respectively, w.r.t. input power when plotted in a log–log scale. This test must be performed in the linear region of the receiver, and hence, a variable attenuator is used to control the level of input power. Once again, the off-chip OP-Amp buffers must have more linearity to ensure that no additional distortions crop up because of their presence. Also, the frequency source for the LO should provide a very low phase noise, such that the IM-tree is clearly observed at the output. At increased input power levels, IM5 terms will also appear at the output.

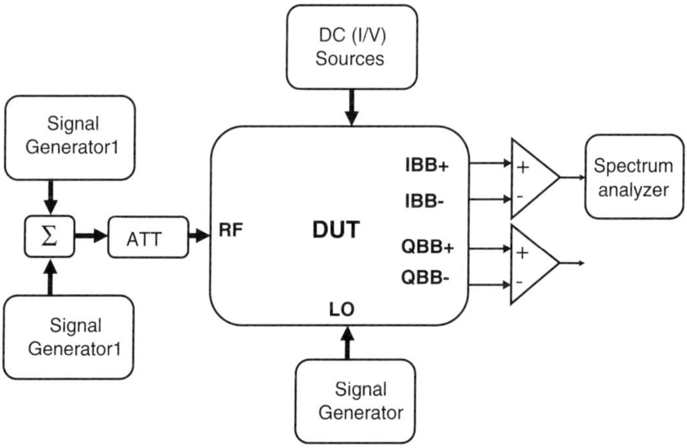

Figure 9.43. Receiver two-tone linearity test.

Since IM2 is a function of mismatch and even-order effects, it usually varies from one die to another.

9.6.6 Nonlinear Noise Figure

Three types of characterization can be applicable for noise figure measurements, such as (1) noise figure meter method, (2) gain method, and (3) Y-factor method. The first type is related to actually calibrating the noise figure meter using a predefined noise source. The noise figure of the system is computed by the ratio of total output noise power to the source noise power. The noise figure meter is precalibrated w.r.t. a known noise source, and the input noise and the SNR of the noise source are known in advance. This method is accurate for very low noise figure systems, and the frequency range is limited.

In the noise figure meter method, a calibrated noise source is provided at the input of the receiver, and the output of the differential IF terminals are interfaced with a transformer to obtain the necessary impedance conversion (as the output impedance of the IF stage is $\sim$200–300 Ω). Transformer-based circuits do not provide any power gain, but they keep the noise contribution to a minimal value. Also, the NF measured by the NF meter is a double sideband (DSB) noise figure, which is relevant for a direct conversion receiver. This noise figure can also be measured versus frequency by sweeping the RF frequency (Figure 9.44).

In the gain method, the noise figure is calculated by the ratio of total output noise power to the output noise caused by input sources. In this method, the gain of the device under test is characterized beforehand, and the device is terminated while calculating output noise spectral density. The noise figure is determined by

$$NF = P_{N,OUTD} + 174\ \text{dBm/Hz} - G$$

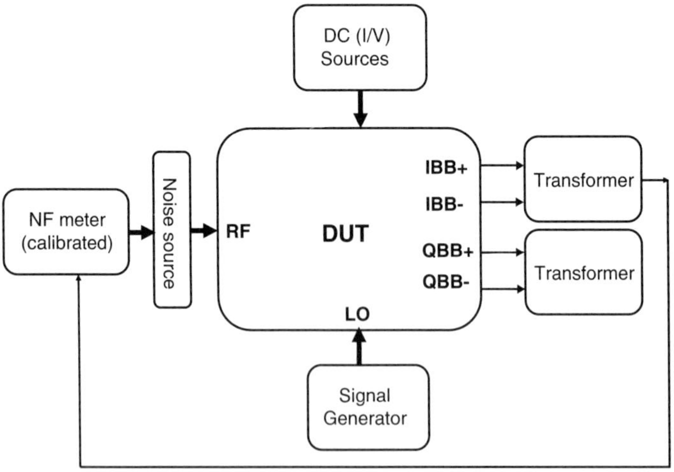

Figure 9.44. Receiver noise figure test.

The limitation of this method originates from the noise floor of the spectrum analyzer. If the system gain is high, then this method can produce accurate results, and it works until it reaches the frequency limits of the spectrum analyzer. The gain method is well suited for high gain and high NF.

The third method is known as the "excess noise ratio" and is obtained by connecting a noise source at the input of the device, turning ON and OFF the noise generator. The output noise spectral density is observed with the source turned ON versus OFF.

$$NF = 10\log_{10}\frac{10^{(ENR/10)}}{10^{(\Delta/10)}-1}$$

ENR denotes the noise figure of the noise source, and Δ denotes the difference between the output noise spectral density and the source ON versus OFF. The excess noise ratio method can be applicable for a wide range of noise figure values.

9.6.7 I/Q Imbalance

This is a time domain test, performed to observe the gain and phase imbalance of quadrature signals. Identical cables from the receiver outputs are interfaced to an oscilloscope. The test setup for the I/Q measurement is shown in Figure 9.45. The high input impedance of an oscilloscope can be used for measurement. If the oscilloscope is capable of providing the waveform statistics (amplitude and phase information, along with their mean and variance values), they would provide the imbalance information directly.

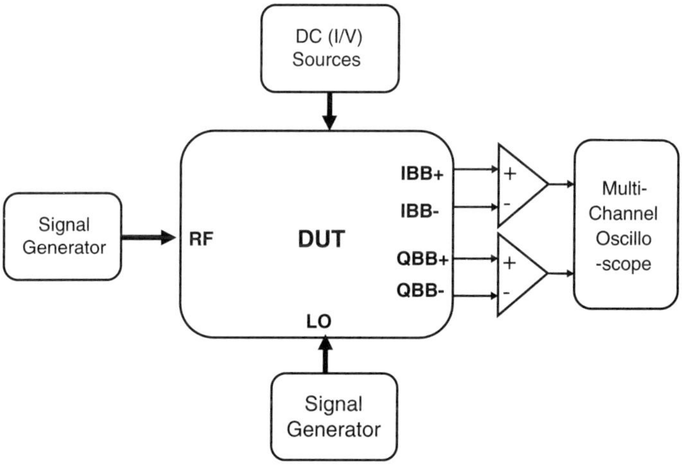

Figure 9.45. Receiver time domain test.

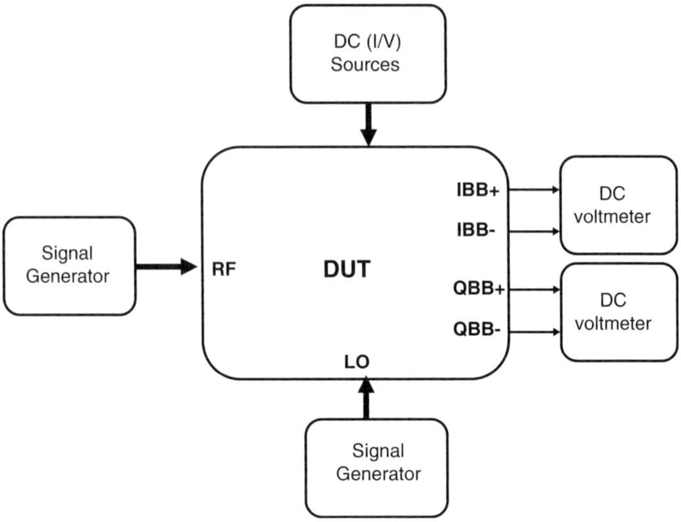

Figure 9.46. Receiver DC offset test.

9.6.8 DC Offset

DC offset can be measured by multimeter, at the output of the receiver. The test setup for DC offset measurement is shown in Figure 9.46. For the static DC offset, LO and RF sources are turned OFF and the DC offset is measured. For the dynamic offset, either RF or LO, or both should be turned ON one by one to observe the dynamic DC offset value. Static DC offset is also dependent on device mismatch, and varies from one chip to another. In any integrated radios, DC offset must be trimmed prior to obtaining any meaningful data, as a large value of DC offset may lead to the saturation of the baseband stages.

Although the basic concepts are illustrated in the above subsections, in a real-life scenario, most of these tests are automated. However, because of the diverse nature of RF testing, sometimes the right test vector generation for RF test may pose a significant challenge to the test engineer. The RFIC should be designed for maximum test coverage and for flexibility of signal injection at different points in the signal path. Typically, buffers or transmission gates as shown in Figure 9.47 can be placed at different signal nodes. The major considerations of such a circuit include:

1. Loading of the main signal path
2. Linearity, noise degradations, and frequency shift
3. Large signals in the chain may accidentally turn the test buffer ON when it is intended to be OFF

Loading of the signal path would occur by the input capacitance of the test circuit, most importantly, the Q of the OFF state impedance. They certainly lead to

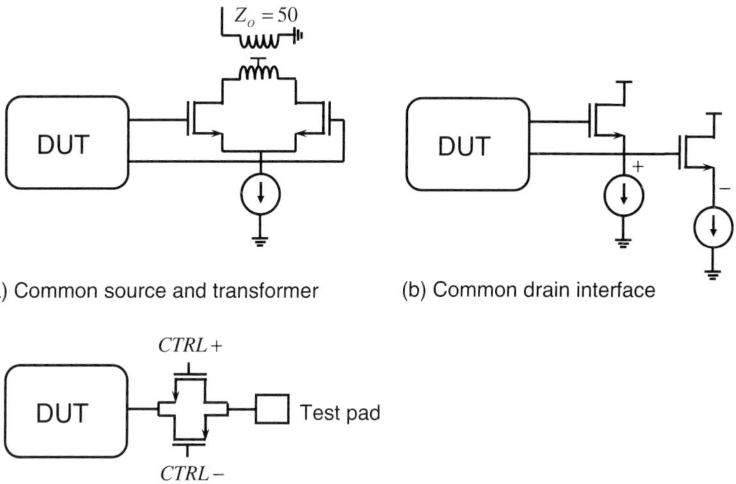

(a) Common source and transformer (b) Common drain interface

(c) T-gate interface for voltage and current

Figure 9.47. Interface circuits/test port hardware.

noise figure degradations (which must be minimized) and degrade the linearity. Thus, tapping RF signals at various points of the receiver may be limited. However, since performance of an RF circuit can be diagnosed by observing the DC voltages across critical transistors, the bias voltages and currents can be monitored to obtain the RF performance of the circuit. Even in real-life scenarios, when an interface point is not accessible (not brought out to the IC pad), injecting various signals provides much insight to the operation of the circuit. Because the pad restrictions on the IC, many signals are "MUX"ed to the same output testing port to reduce testing complexity.

CONCLUSION

In this chapter, we have illustrated the simulation and characterization techniques required for integrated communication microsystems. Accurate characterization, modeling, right interpretation of simulation results, and correlating them with measurement results include a full circle of validation. This is, of course, a cost- and resource-intensive process. Recent advances in silicon technology lead to integration of various functionalities in the same die, and the measurement techniques become complicated for the entire system. Because of the limitations in the number of bond pads many signals are usually multiplexed and tested from the same pad with digital control. Some amount of built-in self-test can be placed as a part of a mixed-signal system in order to perform automatic calibration, process compensation, and so on.

REFERENCES

[1] "Exploring the architectures of network analyzers," Application Note 1287-2. Hewlett Packard and Agilent Technologies, 1997.

[2] D.F. Williams, R.B. Marks, and A. Davidson, "Comparison of on-wafer calibrations," *38th IEEE ARFTG Symposium Digest*, Dec 1991, pp. 68–81.

[3] D. Rytting, *An Analysis of Vector Measurement Accuracy Enhancement Techniques*, Hewlett-Packard Co., 1980.

[4] R.B. Marks and D.F. Williams, "Characteristic impedance determination using propagation constant measurement," *IEEE Microwave and Guided Wave Letters*, Vol. 1, No. 6, 1991, pp. 141–144.

[5] D.F. Williams and R.B. Marks, "Transmission line capacitance measurement," *IEEE Microwave and Guided Wave Letters*, Vol. 1, No. 9, Sept 1991, pp. 243–245.

[6] G.F. Engen and C.A. Hoer, "Thru-Reflect-Line: An improved technique for calibrating the dual six-port automatic network analyzer," *IEEE Transactions of Microwave Theory and Techniques*, Vol. MTT-277, Dec 1979, pp. 987–993.

[7] D. Rytting, *An Analysis of Vector Measurement Accuracy Enhancement Techniques*, Hewlett-Packard Co., 1980.

[8] "Applying the HP 8510B TRL calibration for non-coaxial measurements," Product Note 8510-8, Hewlett-Packard Co., Oct 1987.

[9] D.F. Williams and R.B. Marks, "LRM probe-tip calibrations using nonideal standards," *IEEE Transactions of Microwave Theory and Techniques*, Feb 1995, pp. 466–496.

[10] D.F. Williams, R.B. Marks, and A. Davidson, "Comparison of on-wafer calibrations," *38th IEEE ARFTG Conference Digest*, Dec 1991, pp. 68–81.

[11] A. Pham,"On-wafer characterization," M.S Thesis, Georgia Institute of Technology, Atlanta, 1997.

[12] D. Pozar, *Microwave Engineering*, Addison-Wesley Publishing Co., 1993.

[13] NIST/Industrial MMIC Consortium Software Manuals, *The National Institute of Standards and Technology*, Feb 1992.

[14] A. Pham, J. Laskar, S. Basu, and J. Pence, "Comparison of air coplanar microprobes for on-wafer measurements at W-band," *IEEE International Microwave Symposium Digest*, Vol. 3, Jun 1997, pp. 1659–1662.

[15] D.F. Williams, J. Belquin, G. Dambrine, and R. Fenton, "On-wafer measurements at millimeter wave frequencies," *IEEE International Microwave Symposium Digest*, Jun 1996, pp. 1683–1686.

[16] A. Pham, J. Laskar, and J. Schappacher, "Development of on-wafer microstrip characterization techniques," *IEEE ARFTG Conference Digest*, June 1996, pp. 85–94.

[17] D.F. Williams and R.B. Marks, "LRM probe-tip calibrations using non-ideal standards," *IEEE Transactions on Microwave Theory and Techniques*, Vol. 43, Feb. 1995, pp. 466–469.

[18] Maxim Application notes. http://www.maxim-ic.com/an2875.

[19] F. Sischka,"Characterization Handbook. http://eesof.tm.agilent.com/docs/iccap2002/iccap_ mdl_handbook.html.

APPENDIX

A Compendium of the TRL Calibration Algorithm

$[T_x]$ and $[T_y]$ are defined as transmission matrices modeling the error boxes in TRL calibrations. The standards can be defined in terms of transmission matrix as follows:

$$[T_t] = \begin{bmatrix} 1 & 0 \\ 0 & 1 \end{bmatrix} \tag{A.1}$$

$$[T_d] = \begin{bmatrix} e^{-\gamma l} & 0 \\ 0 & e^{\gamma l} \end{bmatrix} \tag{A.2}$$

where $[T_t]$ and $[T_d]$ represent the characteristics of a thru and a delayed line with a propagation constant of γ. With a matched condition, $S_{d11} = S_{d22}$ is equal to 0. Therefore, $T_{d12} = T_{d21} = 0$.

For a thru connection, the measurement can be expressed as

$$[T_{mt}] = [T_x][T_t][T_y] \tag{A.3}$$

where $[T_{mt}]$ is the overall measurement. Since $[T_t]$ is an identity matrix, $[T_{mt}]$ becomes

$$[T_{mt}] = [T_x][T_y] \tag{A.4}$$

$$[T_y] = [T_x]^{-1}[T_{mt}] \tag{A.5}$$

Advanced Integrated Communication Microsystems, edited by Joy Laskar, Sudipto Chakraborty, Manos Tentzeris, Franklin Bien, and Anh-Vu Pham

For a delayed line connection, the measurement can be expressed as

$$[T_{md}] = [T_x][T_d][T_y] \tag{A.6}$$

Substituting Eq. (A.5) into Eq. (A.6), the following equation is obtained:

$$[T_{md}][T_{mt}]^{-1} = [T_x][T_d] \tag{A.7}$$

Defining $[T_m] = [T_{md}][T_{mt}]^{-1} = \begin{bmatrix} m_{11} & m_{12} \\ m_{21} & m_{22} \end{bmatrix}$, Eq. (A.7) becomes

$$\begin{bmatrix} m_{11} & m_{12} \\ m_{21} & m_{22} \end{bmatrix} \begin{bmatrix} x_{11} & x_{22} \\ x_{21} & x_{22} \end{bmatrix} = \begin{bmatrix} x_{11} & x_{22} \\ x_{21} & x_{22} \end{bmatrix} \tag{A.8}$$

Expanding Eq. (A.8), the four equations with five unknown are written as

$$m_{11}x_{11} + m_{12}x_{21} = x_{11}e^{-\gamma l} \tag{A.9}$$

$$m_{21}x_{11} + m_{22}x_{21} = x_{21}e^{-\gamma} \tag{A.10}$$

$$m_{11}x_{12} + m_{12}x_{22} = x_{12}e^{\gamma l} \tag{A.11}$$

$$m_{21}x_{12} + m_{22}x_{22} = x_{22}e^{\gamma l} \tag{A.12}$$

Eliminating $e^{-\gamma l}$ from Eqs. (A.9), (A.10) and (A.11), (A.12), quadratic equations with x_{11}/x_{21} and x_{12}/x_{22} as unknowns are established

$$m_{21}\left(\frac{x_{11}}{x_{21}}\right)^2 + (m_{22}-m_{11})\left(\frac{x_{11}}{x_{21}}\right) - m_{12} = 0 \tag{A.13}$$

$$m_{21}\left(\frac{x_{12}}{x_{22}}\right)^2 + (m_{22}-m_{11})\left(\frac{x_{12}}{x_{22}}\right) - m_{12} = 0 \tag{A.14}$$

The solutions can be found by solving the quadric equations with variables of $\left(\frac{x_{11}}{x_{21}}\right)$ and $\left(\frac{x_{12}}{x_{22}}\right)$. Since Eq. (A.13) and (A.14) are identical, the root choices are

$$\left(\frac{x_{11}}{x_{22}}\right) = a_1 = E_{00} - (E_{10}E_{01})/E_{11} \tag{A.13}$$

$$\left(\frac{x_{12}}{x_{22}}\right) = a_2 = E_{00} \tag{A.14}$$

Duality applies for port 2, the error terms can be expressed as

$$\left(\frac{y_{11}}{y_{21}}\right) = b_1 = (E_{23}E_{32})/E_{22} - E_{33} \tag{A.15}$$

$$\left(\frac{y_{21}}{y_{22}}\right) = b_2 = -E_{33} \tag{A.16}$$

For the unknown reflect connections, the measurement at port 1 and 2 can be expressed as

$$\Gamma_{mx} = E_{00} + \frac{(E_{10}E_{01})\Gamma_A}{1-E_{11}\Gamma_A} \tag{A.17}$$

$$\Gamma_{my} = E_{33} + \frac{(E_{23}E_{32})\Gamma_A}{1-E_{22}\Gamma_A} \tag{A.18}$$

where Γ_A is the reflection coefficient of the unknown reflect. Eliminating Γ_A from Eqs. (A.17) and (A.18), the result is

$$\frac{1}{E_{22}} = \frac{1}{E_{11}}\left(\frac{a_2-\Gamma_{mx}}{a_1-\Gamma_{mx}}\right)\left(\frac{b_1+\Gamma_{my}}{b_2+\Gamma_{my}}\right) \tag{A.19}$$

From the thru measurement, the reflection can be derived as

$$\Gamma_{m1} = E_{00} + \frac{(E_{10}E_{01})E_{22}}{1-E_{11}E_{22}} \tag{A.20}$$

Using Eqs. (B.19) and (B.20), E_{11} is found as

$$E_{11} = \left[\left(\frac{a_2-\Gamma_{mx}}{a_1-\Gamma_m}\right)\left(\frac{b_1+\Gamma_{my}}{b_2+\Gamma_{my}}\right)\left(\frac{a_2-\Gamma_{ml}}{a_1-\Gamma_{ml}}\right)\right]^{1/2} \tag{A.21}$$

Also,

$$E_{10}E_{01} = (a_2-a_1)E_{11} \tag{A.22}$$

For duality, the error terms E_{33}, E_{22}, and $(E_{32}E_{23})$ are also determined. Therefore, the transmission matrices are computed. The propagation constant can be determined as

$$e^{2\gamma l} = \frac{m_{11}+m_{22} \pm R}{m_{11}+m_{22} \mp R} \tag{A.23}$$

where

$$R = \sqrt{(m_{11}-m_{22})^2 + 4m_{12}m_{21}} \tag{A.24}$$

APPENDIX A

A(1) Single Sideband Combination

Upper sideband

$$I_{BB} \times I_{LO} - Q_{BB} \times Q_{LO}:$$
$$S_{TX}(t) = A_{BB}\cos\omega_{BB}t \times \cos\omega_{LO}t - A_{BB}\sin\omega_{BB}t \times \sin\omega_{LO}t = A\cos(\omega_{LO} + \omega_{BB})t$$

$$I_{BB} \times Q_{LO} + Q_{BB} \times I_{LO}:$$
$$S_{TX}(t) = A_{BB}\cos\omega_{BB}t \times \sin\omega_{LO}t + A_{BB}\sin\omega_{BB}t \times \cos\omega_{LO}t = A\sin(\omega_{LO} + \omega_{BB})t$$

Lower sideband

$$I_{BB} \times I_{LO} + Q_{BB} \times Q_{LO}:$$
$$S_{TX}(t) = A_{BB}\cos\omega_{BB}t \times \cos\omega_{LO}t + A_{BB}\sin\omega_{BB}t \times \sin\omega_{LO}t = A\cos(\omega_{LO} - \omega_{BB})t$$

$$I_{BB} \times Q_{LO} - Q_{BB} \times I_{LO}:$$
$$S_{TX}(t) = A_{BB}\sin\omega_{BB}t \times \cos\omega_{LO}t - A_{BB}\cos\omega_{BB}t \times \sin\omega_{LO}t = -A\sin(\omega_{LO} - \omega_{BB})t$$

While analyzing an upconversion mixer for nonlinearity, it is to be noted that the third harmonic nonlinearity of the baseband signal appears in the opposite side of the desired sideband; i.e., if we use the current combination to provide upper sideband $(\omega_{LO} + \omega_{BB})$, the third harmonic tone would be present at $(\omega_{LO} + 3\omega_{BB})$.

A(2) Fourier Transformations

Fourier transforms are used extensively in RF systems, and a few important ones can be used as follows:

(a) Rectangular pulse in time domain (often encountered in baseband)

$$A \times \mathrm{rect}\left(\frac{t}{T}\right) \leftrightarrow AT \times \mathrm{sinc}(fT)$$

Most signal energy is confined within the first null point, and the zero crossings occur at $f = \pm n/T$.

This formulation is useful to understand pulse-based and multiband OFDMs-based ultrawideband communication systems.

(b) Exponentially decaying pulse, symmetric around zero:

$$e^{-a|t|} \leftrightarrow \frac{2a}{a^2 + (2\pi f)^2}$$

(c) Time and frequency scaling

$$F[s(at)] \leftrightarrow \frac{1}{a}S\left(\frac{f}{a}\right)$$

Hence, compressing a function in the time domain leads to expansion in the frequency domain and vice versa. Thus, the time and frequency domain descriptions of a signal change in inverse relationship with one another. The product of time and bandwidth is a constant.

(d) Frequency shifting:

$$F[e^{(j2\pi f_c t)} s(t)] \leftrightarrow S(f-f_c)$$

This is an extremely important result in communication systems, and it is used frequently in formulation of mixer and modulator functionalities.

(e) Convolution:

$$\text{If } s_1(t) \leftrightarrow S_1(f), \quad \text{and} \quad s_2(t) \leftrightarrow S_2(f), \text{ then}$$

$s_1(t)s_2(t) \leftrightarrow \int_{-\infty}^{\infty} S_1(f)S_2(f-u)du$; this can be used to calculate complicated problems like second- and third-order intermodulation terms in the case of nonlinear systems. When multicarrier signals are passed through nonlinear network, second- and third-order product terms is difficult to hand-calculate. In this case, convolution terms provide a very useful tool in order to evaluate the spectrum of output signal.

(f) Sampling:

$$\sum_{k=-\infty}^{\infty} s(t-kT_0) \leftrightarrow f_0 \sum_{p=-\infty}^{\infty} S(pf_0)\delta(f-pf_0)$$

This is very useful in order to analyze a sampled data system, which is common in communication systems. Examples include switch capacitor circuits, ADCs, and so on.

(g) Hilbert transformation:

$$\tilde{s}(t) \leftrightarrow \frac{1}{\pi} \int_{-\infty}^{\infty} \frac{s(\tau)}{t-\tau} d\tau, \quad \tilde{S}(f) \leftrightarrow -j\,\text{sgn}(f)S(f)$$

This property can be associated with a network that can provide a 90° phase shift for negative frequency, and a −90° phase shift for the positive frequencies, whereas the magnitude remains the same across frequency. This provides the mathematical basis for circuits operating in single sideband fashion, such as polyphase filters.

A(3) Interconversion between Two-Port Network Parameters

1. Y parameter to Z parameter conversion and vice versa

$$Y^{-1} = Z \quad \text{and} \quad Z^{-1} = Y$$

2. S parameter to H parameter conversion

$$H_{11} = \frac{(1+S_{11}) \times (1+S_{22}) - S_{21} \times S_{12}}{(1-S_{11}) \times (1+S_{22}) + S_{21} \times S_{12}} \times Z_0$$

$$H_{21} = \frac{-2S_{21}}{(1-S_{11}) \times (1+S_{22}) + S_{21} \times S_{12}}$$

$$H_{12} = \frac{-2S_{12}}{(1-S_{11}) \times (1+S_{22}) + S_{21} \times S_{12}}$$

$$H_{22} = \left[\frac{(1-S_{11}) \times (1-S_{22}) - S_{21} \times S_{12}}{(1-S_{11}) \times (1+S_{22}) + S_{21} \times S_{12}}\right] \Big/ Z_0$$

This is useful for evaluating f_T and f_{MAX} of transistors. Usually S parameter measurements are taken and H_{21} roll-off is plotted to find f_T.

3. H parameter to Y parameter conversion

$$\begin{pmatrix} Y_{11} & Y_{12} \\ Y_{21} & Y_{22} \end{pmatrix} = \frac{1}{H_{11}} \begin{pmatrix} 1 & -H_{12} \\ H_{21} & \det(H) \end{pmatrix}$$

A(4) Three-Port *S* Parameters to Two-Port *S* Parameter Conversion

In this case, we assume that all the ports are terminated in 50-Ω load; the two-port S parameter can be obtained by terminating any of the three ports to ground (however, the network configuration does not change). Such a configuration leads to the reflection coefficient at that port to be -1 (ratio of output wave amplitude to the input amplitude)

$$[S_{3p}] = \begin{vmatrix} S_{11} & S_{12} & S_{13} \\ S_{21} & S_{22} & S_{23} \\ S_{31} & S_{32} & S_{33} \end{vmatrix}, \quad \frac{B_3}{A_3} = -1$$

$$[S_{2p}] = \begin{vmatrix} S_{11} - \dfrac{S_{13} \times S_{31}}{1+S_{33}} & S_{12} - \dfrac{S_{13} \times S_{32}}{1+S_{33}} \\ S_{21} - \dfrac{S_{31} \times S_{23}}{1+S_{33}} & S_{22} - \dfrac{S_{23} \times S_{32}}{1+S_{33}} \end{vmatrix}$$

A one-port S parameter can be obtained from a two-port S parameter by the same method as before, which leads to

$$S_{11,1p} = S_{11} - \frac{S_{12} \times S_{21}}{1+S_{22}}$$

A(5) *S*-Parameter of Commonly Used Networks

(a) Bias-Tee

$$[S] = \begin{vmatrix} 0 & 1 & 0 \\ 1 & 0 & 0 \\ 0 & 0 & 1 \end{vmatrix}$$

(b) Transformer

$$[S] = \begin{vmatrix} \frac{N^2}{N^2+1} & \frac{N}{N^2+1} & \frac{-N}{N^2+1} & \frac{1}{N^2+1} \\ \frac{N}{N^2+1} & \frac{1}{N^2+1} & \frac{N^2}{N^2+1} & \frac{-N}{N^2+1} \\ \frac{-N}{N^2+1} & \frac{N^2}{N^2+1} & \frac{1}{N^2+1} & \frac{N}{N^2+1} \\ \frac{1}{N^2+1} & \frac{-N}{N^2+1} & \frac{N}{N^2+1} & \frac{N}{N^2+1} \end{vmatrix}$$

(c) Isolator

$$[S] = \begin{vmatrix} \frac{Z_1-Z_0}{Z_1+Z_0} & 0 \\ \sqrt{\left(1-\frac{Z_1-Z_0}{Z_1+Z_0}\right)^2 \times \left(1-\frac{Z_2-Z_0}{Z_2+Z_0}\right)^2} & \frac{Z_2-Z_0}{Z_2+Z_0} \end{vmatrix}$$

(d) 90° hybrid

$$[S] = -\frac{1}{\sqrt{2}} \begin{vmatrix} 0 & j & 1 & 0 \\ j & 0 & 0 & 1 \\ 1 & 0 & 0 & j \\ 0 & 1 & j & 0 \end{vmatrix}$$

(e) 180° hybrid

$$[S] = -\frac{j}{\sqrt{2}} \begin{vmatrix} 0 & 1 & 1 & 0 \\ 1 & 0 & 0 & -1 \\ 1 & 0 & 0 & 1 \\ 0 & -1 & 1 & 0 \end{vmatrix}$$

A(6) *T* to π Conversion and Vice Versa

This is often used in RF circuits, also known as "star to Delta" conversion in classic electrical engineering in the context of three phase circuits.

$$\begin{aligned}
Z_{1T} &= \frac{Z_{1\pi} \times Z_{2\pi}}{Z_{1\pi} \times Z_{2\pi} + Z_{2\pi} \times Z_{3\pi} + Z_{3\pi} \times Z_{1\pi}} \\
Z_{2T} &= \frac{Z_{2\pi} \times Z_{3\pi}}{Z_{1\pi} \times Z_{2\pi} + Z_{2\pi} \times Z_{3\pi} + Z_{3\pi} \times Z_{1\pi}} \\
Z_{3T} &= \frac{Z_{1\pi} \times Z_{3\pi}}{Z_{1\pi} \times Z_{2\pi} + Z_{2\pi} \times Z_{3\pi} + Z_{3\pi} \times Z_{1\pi}}
\end{aligned}$$

Similarly, the reverse conversion is related as follows:

$$\begin{aligned}
Y_{1\pi} &= \frac{Y_{1T} \times Y_{2T}}{Y_{1T} \times Y_{2T} + Y_{2T} \times Y_{3T} + Y_{3T} \times Y_{1T}} \\
Y_{2\pi} &= \frac{Y_{2T} \times Y_{3T}}{Y_{1T} \times Y_{2T} + Y_{2T} \times Y_{3T} + Y_{3T} \times Y_{1T}} \\
Y_{3\pi} &= \frac{Y_{1T} \times Y_{3T}}{Y_{1T} \times Y_{2T} + Y_{2T} \times Y_{3T} + Y_{3T} \times Y_{1T}}
\end{aligned}$$

A(7) Derivation of Miller's Theorem

Using nodal equations,

$$-V_{in} \times \left(\frac{1}{Z_s} + \frac{1}{Z_f} + \frac{1}{Z_{in}}\right) + V_s \times \left(\frac{1}{Z_s}\right) + V_o \times \left(\frac{1}{Z_f}\right) = 0 \qquad \text{(I)}$$

$$V_o \times \left(\frac{1}{Z_f} + \frac{1}{Z_o}\right) - A_v V_{in} \times \left(\frac{1}{Z_o}\right) + V_{in} \times \left(\frac{1}{Z_f}\right) = 0 \qquad \text{(II)}$$

From (II), $V_o \times \left(\frac{1}{Z_f} + \frac{1}{Z_o}\right) = -A_v V_{in} \times \left(\frac{1}{Z_o} - \frac{1}{A_v Z_f}\right) \Rightarrow V_o = \frac{-A_v V_{in}(Z_f - Z_o/A_v)}{Z_f + Z_o}$

Assuming that this effect causes an impedance Z_L at the output, which is the equivalent output impedance, it is given as follows:

$$Vo = \frac{-A_v V_{in} Z_L}{Z_L + Z_o}, \text{ equating the two, we obtain}$$

$$Z_L = \frac{Z_f - Z_o/A_v}{1 + 1/A_v} \approx \frac{Z_f}{1 + 1/A_v}$$

For an equivalent input-referred impedance Z_G, we obtain

$$-V_{in} \times \left(\frac{1}{Z_s} + \frac{1}{Z_G} + \frac{1}{Z_{in}}\right) + V_s \times \left(\frac{1}{Z_s}\right) = 0 \qquad \text{(III)}$$

Equating (I) and (III), we obtain

$$\frac{1}{Z_f} = \frac{1}{Z_G} - \frac{A_v V_{in}(Z_f - Z_o/A_v)}{Z_f + Z_o} \Rightarrow Z_G = \frac{Z_f(1 + Z_o/Z_f)}{(1 + A_v) + (Z_o/Z_f)(1 + A_v)} \approx \frac{Z_f}{(1 + A_v)}$$

A(8) Composite Beat Formation in Wideband Systems

In modern communication systems, multicarrier-based wideband systems have become commonplace. This can lead to a constant power level across the spectrum, and it can be represented by

$$g(t) = A_1\cos(\omega_1 t) + A_2\cos(\omega_2 t) + A_3\cos(\omega_3 t) + \ldots]$$

RF blocks are often represented as cubic nonlinearity, such as

$$y(t) = \alpha_1 V_{in}(t) + \alpha_2 V_{in}{}^2(t) + \alpha_3 V_{in}{}^3(t) + \ldots$$

$$\alpha_3 V_{in}{}^3(t) \rightarrow \alpha_3 \left[\frac{1}{4}A_1{}^3\cos(3\omega_1 t) + \frac{1}{4}A_2{}^3\cos(3\omega_2 t) + \frac{1}{4}A_3{}^3\cos(3\omega_3 t) + \frac{3}{4}\sum_{i,j} A_{ij}\cos(2\omega_i - \omega_j)t + \frac{3}{2}A_1A_2A_3\cos(\omega_1 \pm \omega_2 \pm \omega_3)t\right]$$

Usually in multicarrier systems, $A_1 = A_2 = A_3 = \ldots A_n$, and the composite triple beat amplitude is higher than the regular third-order Intermodulation by 6 dB.

$$\alpha_2 V_{in}{}^2(t) \rightarrow \alpha_2 \left[\sum_{i,j} A_{ij}\cos(\omega_i \pm \omega_j)t + \sum_i A_i{}^2\cos(2\omega_i t)\right]$$

Index

Advanced Integrated Communication Microsystems, edited by Joy Laskar, Sudipto Chakraborty, Manos Tentzeris, Franklin Bien, and Anh-Vu Pham

ANALYSIS OF MULTICONDUCTOR TRANSMISSION LINES • *Clayton R. Paul*

INTRODUCTION TO ELECTROMAGNETIC COMPATIBILITY, Second Edition • *Clayton R. Paul*

ADAPTIVE OPTICS FOR VISION SCIENCE: PRINCIPLES, PRACTICES, DESIGN AND APPLICATIONS • *Jason Porter, Hope Queener, Julianna Lin, Karen Thorn, and Abdul Awwal (eds.)*

ELECTROMAGNETIC OPTIMIZATION BY GENETIC ALGORITHMS • *Yahya Rahmat-Samii and Eric Michielssen (eds.)*

INTRODUCTION TO HIGH-SPEED ELECTRONICS AND OPTOELECTRONICS • *Leonard M. Riaziat*

NEW FRONTIERS IN MEDICAL DEVICE TECHNOLOGY • *Arye Rosen and Harel Rosen (eds.)*

ELECTROMAGNETIC PROPAGATION IN MULTI-MODE RANDOM MEDIA • *Harrison E. Rowe*

ELECTROMAGNETIC PROPAGATION IN ONE-DIMENSIONAL RANDOM MEDIA • *Harrison E. Rowe*

HISTORY OF WIRELESS • *Tapan K. Sarkar, Robert J. Mailloux, Arthur A. Oliner, Magdalena Salazar-Palma, and Dipak L. Sengupta*

PHYSICS OF MULTIANTENNA SYSTEMS AND BROADBAND PROCESSING • *Tapan K. Sarkar, Magdalena Salazar-Palma, and Eric L. Mokole*

SMART ANTENNAS • *Tapan K. Sarkar, Michael C. Wicks, Magdalena Salazar-Palma, and Robert J. Bonneau*

NONLINEAR OPTICS • *E. G. Sauter*

APPLIED ELECTROMAGNETICS AND ELECTROMAGNETIC COMPATIBILITY • *Dipak L. Sengupta and Valdis V. Liepa*

COPLANAR WAVEGUIDE CIRCUITS, COMPONENTS, AND SYSTEMS • *Rainee N. Simons*

ELECTROMAGNETIC FIELDS IN UNCONVENTIONAL MATERIALS AND STRUCTURES • *Onkar N. Singh and Akhlesh Lakhtakia (eds.)*

ANALYSIS AND DESIGN OF AUTONOMOUS MICROWAVE CIRCUITS • *Almudena Suárez*

ELECTRON BEAMS AND MICROWAVE VACUUM ELECTRONICS • *Shulim E. Tsimring*

FUNDAMENTALS OF GLOBAL POSITIONING SYSTEM RECEIVERS: A SOFTWARE APPROACH, Second Edition • *James Bao-yen Tsui*

RF/MICROWAVE INTERACTION WITH BIOLOGICAL TISSUES • *André Vander Vorst, Arye Rosen, and Youji Kotsuka*

InP-BASED MATERIALS AND DEVICES: PHYSICS AND TECHNOLOGY • *Osamu Wada and Hideki Hasegawa (eds.)*

COMPACT AND BROADBAND MICROSTRIP ANTENNAS • *Kin-Lu Wong*

DESIGN OF NONPLANAR MICROSTRIP ANTENNAS AND TRANSMISSION LINES • *Kin-Lu Wong*

PLANAR ANTENNAS FOR WIRELESS COMMUNICATIONS • *Kin-Lu Wong*

FREQUENCY SELECTIVE SURFACE AND GRID ARRAY • *T. K. Wu (ed.)*

ACTIVE AND QUASI-OPTICAL ARRAYS FOR SOLID-STATE POWER COMBINING • *Robert A. York and Zoya B. Popović (eds.)*

OPTICAL SIGNAL PROCESSING, COMPUTING AND NEURAL NETWORKS • *Francis T. S. Yu and Suganda Jutamulia*

SiGe, GaAs, AND InP HETEROJUNCTION BIPOLAR TRANSISTORS • *Jiann Yuan*

ELECTRODYNAMICS OF SOLIDS AND MICROWAVE SUPERCONDUCTIVITY • *Shu-Ang Zhou*